18 0051744 7
WITHDRAWN FROM STOCK

AF598335

The Mammalian Preimplantation Embryo

Regulation of Growth and Differentiation *in Vitro*

The Mammalian Preimplantation Embryo

Regulation of Growth and Differentiation *in Vitro*

Edited by

Barry D. Bavister

Department of Veterinary Science
and Wisconsin Regional Primate Research Center
University of Wisconsin
Madison, Wisconsin

Plenum Press • New York and London

Library of Congress Cataloging in Publication Data

The Mammalian preimplantation embryo.

Includes bibliographies and index.
1. Embryology—Mammals. 2. Fertilization in vitro. 3. Cell culture. I. Bavister, Barry D.
QL959.M264 1987 599′.033 87-13986
ISBN 0-306-42595-5

A Division of Plenum Publishing Corporation
233 Spring Street, New York, N.Y. 10013

Printed in the United States of America

CONTRIBUTORS

BARRY D. BAVISTER Department of Veterinary Science, University of Wisconsin, 1655 Linden Drive, Madison, Wisconsin 53706, USA, and the Wisconsin Regional Primate Research Center, 1223 Capitol Court, Madison, Wisconsin 53715, USA

JOHN D. BIGGERS Department of Physiology and Biophysics, Laboratory of Human Reproduction and Reproductive Biology, Harvard Medical School, Boston, Massachusetts 02115, USA

DOROTHY E. BOATMAN Wisconsin Regional Primate Research Center, University of Wisconsin, 1223 Capitol Court, Madison, Wisconsin 53715, USA

FOLKMAR ELSAESSER Institut für Tierzucht und Tierverhalten (FAL), Mariensee, 3057 Neustadt 1, Federal Republic of Germany

CHARLES J. EPSTEIN Departments of Pediatrics and of Biochemistry and Biophysics, University of California, San Francisco, California 94143, USA

YVES HEYMAN I.N.R.A., Station de Physiologie Animale, 78350, Jouy-en-Josas, France

SUSAN HEYNER Department of Obstetrics and Gynecology, Albert Einstein Medical Center, Northern Division, York and Tabor Roads, Philadelphia, Pennsylvania 19141, USA

MICHAEL T. KANE Physiology Department, University College, Galway, Ireland

GERALD M. KIDDER Department of Zoology, University of Western Ontario, London, Ontario, N6A 5B7, Canada

TERRY MAGNUSON Department of Developmental Genetics and Anatomy, School of Medicine, Case Western Reserve University, Cleveland, Ohio 44106, USA

YVES MÉNÉZO I.N.S.A., Laboratoire de Biologie, 69621, Villeurbanne Cédex, France

HEINER NIEMANN Institut für Tierzucht und Tierverhalten (FAL), Mariensee, 3057 Neustadt 1, Federal Republic of Germany

JAMES V. O'FALLON Department of Animal Sciences, Washington State University, Pullman, Washington 99164-6332, USA

ERIC W. OVERSTRÖM Department of Anatomy and Cellular Biology, Tufts University, Schools of Medicine and Veterinary Medicine, 136 Harrison Avenue, Boston, Massachusetts 02111, USA

ANGIE RIZZINO Eppley Institute for Research in Cancer and Related Diseases, University of Nebraska Medical Center, 42nd and Dewey Avenue, Omaha, Nebraska 68105, USA

HORST SPIELMANN Max v. Pettenkofer-Institut, Bundesgesundheitsamt (BGA), P.O. Box 33 00 13, 1 Berlin 33, West Germany

LYNN M. WILEY Division of Reproductive Biology and Medicine, Department of Obstetrics and Gynecology, University of California, Davis, California 95616, USA

RAYMOND W. WRIGHT, JR. Department of Animal Sciences, Washington State University, Pullman, Washington 99164-6332, USA

CAROL A. ZIOMEK Worcester Foundation for Experimental Biology, 222 Maple Avenue, Shrewsbury, Massachusetts 01545, USA

PREFACE

With a few notable exceptions, mammalian preimplantation embryos grown *in vitro* are likely to exhibit sub-optimal or retarded development. This may be manifested in different ways, depending on the species and on the stage(s) of embryonic development that are being examined. For example, bovine embryos often experience difficulty in cleaving under *in vitro* conditions, and usually cease development at about the 8-cell stage (Wright and Bondioli, 1981). The block to development is stage-dependent; embryos cultured for 24 hr from the 1-cell stage are much more capable of developing into viable blastocysts after transfer to oviducts than embryos cultured for 24 hr from the 4-cell stage prior to transfer (Eyestone *et al.*, 1985). Similar problems with *in vitro* embryo development are encountered in other species. Pig embryos can be grown up to the 4-cell stage *in vitro* but usually no further (Davis and Day, 1978). In the golden hamster, in the rat and in many outbred strains of mice, development of zygotes *in vitro* is blocked at the 2-cell stage (Yanagimachi and Chang, 1964; Whittingham, 1975). Even with some inbred mouse strains, embryo development is reduced if very early cleavage stages are used as the starting point for *in vitro* culture (Spielmann *et al.*, 1980). A common finding is that embryos grown *in vitro* have reduced cell counts (Harlow and Quinn, 1982; Kane, 1985) and their viability is reduced (Bowman and McLaren, 1970; Papaioannou and Ebert, 1986) compared to equivalent developmental stages recovered from mated animals.

All of these difficulties show that we have a great deal to learn about culture conditions suitable for sustaining normal growth of preimplantation embryos. Resolution of these technical problems should be a high priority for investigators who are interested in analyzing mechanisms of embryo development in different species. Knowledge of culture requirements for embryos of various species will tell us much about their metabolism and the regulation of development, just as in the pioneering work of Biggers *et al.* (1971), Brinster (1971), Whitten (1957, 1971) and others using mouse embryos. Because of these pioneering efforts, mouse embryos have become firmly established as *the* models for the study of early (preimplantation) development. The notable advances in our understanding of the regulation of development that have derived from studies on mouse embryos, some of which are presented in this book, stand as elegant testimony to the validity of this approach. An enormous amount of information has been obtained on the cellular, molecular and genetic aspects of early embryonic development in the mouse. Yet the heavy emphasis placed on the study of the inbred mouse has fostered neglect of other important species. Cleavage stage embryos from some of these species, such as the domesticated animals, may be of considerable commercial interest, while others, such as non-human primate embryos, have great theoretical and practical interest because of their similarity to human embryos. Thus, there is an urgent need for comparative data on preimplantation embryogenesis using a wide variety of animal species.

It is rather ironic that human embryos, derived from *in vitro* fertilized (IVF) eggs, appear to be very easy to grow *in vitro*, even to the blastocyst stage. This would seem to undermine one major justification for studies with animal embryos, namely that information gained from such studies may help to increase understanding of human embryogenesis. However, the situation concerning experimental embryology in humans is quite complex. In many parts of the world, invasive experimentation on human preimplantation embryos, or even culture of embryos to the blastocyst stage, is proscribed for a variety of ethical and medico-legal reasons. Since the primary objective of human IVF is to transfer embryos, usually at the 4- to 8-cell stage, back to the infertile patient, "hard" information concerning (*e.g.*) optimal culture conditions for human embryos is difficult to obtain; factorially-designed experiments can hardly be done under these conditions. Moreover, about 90% of all IVF human embryos fail to develop to term following transfer, and we do not know what proportion of these failures is due to abnormalities of early embryo development *in vitro*. It is quite common for IVF primate embryos to undergo apparently normal cleavage *in vitro*, only to cease development before the morula stage (Bavister *et al.*, 1983a; Boatman *et al.*, 1987). Early cessation of embryo development also seems to be common *in vivo* (Enders *et al.*, 1982). In view of all these considerations, a strong case can be made for the necessity of studies using animal embryos in order to obtain information that is potentially useful for understanding human reproductive problems.

The present situation is that we know a great deal about regulation of embryo development in the mouse and very little in other species. This difference could soon be eliminated if techniques (*e.g.*, culture media) devised for mouse embryos were applicable to other mammalian species, and if information obtained on (*e.g.*) the nutritional requirements for mouse embryos were found to be representative of mammals in general. Unfortunately, neither of these situations appears to be true. From the limited amount of data available, it appears that preimplantation embryos of several mammals are much more demanding in their requirements for growth *in vitro* than are mouse embryos, and perhaps are also more sensitive to the trauma associated with collection and culture. For example, mouse embryos have been grown from the 2-cell stage to blastocysts in the absence of protein (Cholewa and Whitten, 1970) and do not require any amino acid source for development *in vitro* up to the late zonal blastocyst stage. In contrast, rhesus monkey embryos (at least from the 8-cell stage) need a complex culture medium with protein supplement (Morgan *et al.*, 1984), while growth of rabbit and hamster embryos is very dependent on exogenous amino acids (Kane and Foote, 1970; Bavister *et al.*, 1983b).

Information is urgently needed that will allow us to bridge the gulf of knowledge between embryos of the mouse and those of other mammals, not only in terms of our ability to culture these embryos successfully, but also in reaching a consensus about the similarities and differences between species. Then advances made using mouse embryos could rapidly be tested in other species, and embryos of the latter group could also be investigated for their own particular developmental characteristics. The production of this book represents one effort to build such a bridge. Each contributing author is in the forefront of his or her particular area of embryogenesis research. Subjects of the chapters were chosen to represent a range of topics, in terms of species and of analytical approach. It will be obvious, for reasons already mentioned, that research described using mouse embryos is, in general, at a more advanced stage than studies using other species. Authors were requested to provide technical details of their research as well as the embryological data obtained using these methods; this dual emphasis should help other investigators to confirm or extend the work and to facilitate start-up for those who are beginning to study embryo development *in vitro*.

Mechanisms involved in the fertilization process have not received attention in this book, partly because this would have excessively broadened the scope of the work, and partly because several books dealing specifically with fertilization have been published in the last few years. For similar reasons, I have largely avoided inclusion of IVF techniques, although there is presently considerable interest in IVF and the consequences for embryogenesis. Readers interested in the topics of fertilization and IVF are referred to works by Mastroianni and Biggers (1981), Beier and Lindner (1983), Hartmann (1983), Trounson and Wood (1984) and Seppälä and Edwards (1985). By focusing on developmental events following fertilization and up to peri-implantation stages, the contributing authors have brought a wide variety of interests and techniques to bear on a single topic: the mechanisms regulating growth of preimplantation embryos *in vitro*. The contributors have generally described their own most recent work in specialized areas of interest, while presenting this information against a background review of work from other laboratories. The bibliographic lists compiled by each contributor should by themselves be valuable aids to research, since the literature on preimplantation embryogenesis tends to be scattered widely among many different journals, books and symposium proceedings.

At the end of the book are two Appendices. In Appendix I, I have attempted to synergize some of the ideas of the contributing authors in order to point out some problem areas and lines of enquiry that should help to accelerate the pace of comparative research using preimplantation embryos. In addition, I have provided some technical notes on embryo culture, focusing on areas cited by the authors as being critically important, and also drawing from experiences in my own laboratory. Appendix II lists the names, and the addresses as far as possible, of suppliers of materials and equipment used in embryo culture research. The sole criterion for inclusion of names and items in this list was that the supplier or manufacturer was cited by one or more of the contributing authors. Not only is it convenient to group this information in one place, but it also avoids unnecessary duplication and (I hope) makes the text more readable.

A major reason for undertaking this work was to stimulate wider interest in the study of preimplantation embryogenesis. Time will tell if we have succeeded in this goal; however, the authors have all risen to the occasion splendidly in their attempts to meet it, and I thank each of them for their unstinting efforts.

I am indebted to Mary Born, Kirk Jensen, John Matzka and their staff at Plenum Press for their advice and practical help with the editorial process; to Amy Magulski for her tireless efforts in helping me to prepare the book for publication; and to Jean Lasecki for her heroic work on the index. I am grateful to the Department of Veterinary Science and to the Regional Primate Research Center, University of Wisconsin-Madison, for support during the preparation of this book.

Finally, I would like to dedicate this book to my Ph.D. supervisor, Dr. C. R. "Bunny" Austin, who is *emeritus* Charles Darwin Professor of Animal Embryology at Cambridge University. I am immensely indebted to him for his help in getting my research career started, and for his advice and generous encouragement of my work.

Barry D. Bavister
Madison, Wisconsin

References

Bavister, B.D., Boatman, D.E., Leibfried, M.L., Loose, M., and Vernon, M.W., 1983a, Fertilization and cleavage of rhesus monkey oocytes *in vitro*, *Biol. Reprod.* 28: 983-999.

Bavister, B.D., Leibfried, M.L., and Leiberman, G., 1983b, Development of preimplantation embryos of the golden hamster in a defined culture medium, *Biol. Reprod.* 28: 235-247.

Beier, H.M., and Lindner, H.R. (eds.), 1983, *Fertilization of the Human Egg In Vitro*, Springer-Verlag, Berlin.

Biggers, J.D., Whitten, W.K., and Whittingham, D.G., 1971, The culture of mouse embryos *in vitro*, in: *Methods in Mammalian Embryology* (J.C. Daniel, ed.), Freeman & Co., San Francisco, pp. 86-116.

Boatman, D.E., Morgan, P.M., and Bavister, B.D., 1987, Culture of *in vitro* fertilized rhesus monkey oocytes to peri-implantation stages of embryo development, *Biol. Reprod.* (submitted).

Bowman, P., and McLaren, A., 1970, Viability and growth of mouse embryos after *in vitro* culture and fusion, *J. Embryol. Exp. Morph.* 23: 693-704.

Brinster, R.L., 1971, *In vitro* culture of the embryo, in: *Pathways to Conception: the Role of the Cervix and the Oviduct in Reproduction* (A.I. Sherman, ed.), Charles C. Thomas, Springfield, pp. 245-277.

Cholewa, J.A., and Whitten, W.K., 1970, Development of 2-cell mouse embryos in the absence of a fixed nitrogen source, *J. Reprod. Fertil.* 22: 553-555.

Davis, D.L., and Day, B.N., 1978, Cleavage and blastocyst formation by pig eggs *in vitro*, *J. Anim. Sci.* 46: 1043-1053.

Enders, A.C., Hendrickx, A.G., and Binkerd, P.E., 1982, Abnormal development of blastocysts and blastomeres in the rhesus monkey, *Biol. Reprod.* 26: 353-366.

Eyestone, W.H., Northey, D.L., and Leibfried-Rutledge, M.L., 1985, Culture of 1-cell bovine embryos in the sheep oviduct, *Biol. Reprod.* 32 (Suppl. 1): 100a.

Harlow, G.M., and Quinn, P., 1982, Development of preimplantation mouse embryos *in vivo* and *in vitro*, *Aust. J. Biol. Sci.* 35: 187-193.

Hartmann, J.F. (ed.), 1983, *Mechanism and Control of Animal Fertilization*, Academic Press, New York.

Kane, M.T., 1985, A low molecular weight extract of bovine serum albumin stimulates rabbit blastocyst cell division and expansion *in vitro*, *J. Reprod. Fertil.* 73: 147-150.

Kane, M.T., and Foote, R.H., 1970, Culture of two- and four-cell rabbit embryos to the expanding blastocyst stage in synthetic media, *Proc. Soc. Exp. Biol. Med.* 133: 921-925.

Mastroianni, L., Jr., and Biggers, J.D. (eds.), 1981, *Fertilization and Embryonic Development In Vitro*, Plenum Press, New York.

Morgan, P.M., Boatman, D.E., Collins, K., and Bavister, B.D., 1984, Complete preimplantation development in culture of *in vitro* fertilized rhesus monkey oocytes, *Biol. Reprod.* (Suppl. 1): 96a.

Papaioannou, V.E., and Ebert, K.M., 1986, Development of fertilized embryos transferred to oviducts of immature mice, *J. Reprod. Fertil.* 76: 603-608.

Seppälä, M., and Edwards, R.G. (eds.), 1985, *In Vitro Fertilization and Embryo Transfer*, *Ann. N. Y. Acad. Sci.*, Volume 442, New York Academy of Sciences, New York.

Spielmann, H., Eibs, H.G., and Jacob-Müller, U., 1980, *In vitro* methods for the study of the effect of teratogens on preimplantation embryos, *Acta Morphologica Acad. Sci. Hung.* 28: 105-115.

Trounson, A., and Wood, C. (eds.), 1984, *In Vitro Fertilization and Embryo Transfer*, Churchill Livingstone, Inc., New York.

Whitten, W.K., 1957, Culture of tubal ova, *Nature (London)* 179: 1081-1082.

Whitten, W.K., 1971, Nutrient requirements for the culture of preimplantation embryos *in vitro*, in: *Schering Symposium on Intrinsic and Extrinsic Factors in Early Mammalian Development*, *Advances in the Biosciences*, Vol. 6 (G. Raspé, ed.), Pergamon Press, Oxford, pp. 129-141.

Whittingham, D.G., 1975, Fertilization, early development and storage of mammalian ova *in vitro*, in: *The Early Development of Mammals* (M. Balls, and A.E. Wild, eds.), Cambridge University Press, Cambridge, U.K., pp. 1-24.

Wright, R.J., Jr., and Bondioli, K.R., 1981, Aspects of *in vitro* fertilization and embryo culture in domestic animals, *J. Anim. Sci.* 53: 702-728.

Yanagimachi, R., and Chang, M.C., 1964, *In vitro* fertilization of golden hamster ova, *J. Exp. Zool.* 156: 361-376.

CONTENTS

Chapter 1

PIONEERING MAMMALIAN EMBRYO CULTURE

John D. Biggers

Chapter 2

CELL POLARITY IN THE PREIMPLANTATION MOUSE EMBRYO

Carol A. Ziomek

Chapter 3

INTERCELLULAR COMMUNICATION DURING MOUSE EMBRYOGENESIS

Gerald M. Kidder

Chapter 4

DEVELOPMENT OF THE BLASTOCYST: ROLE OF CELL POLARITY IN CAVITATION AND CELL DIFFERENTIATION

Lynn M. Wiley

Chapter 5

Eric W. Overström

IN VITRO ASSESSMENT OF BLASTOCYST DIFFERENTIATION

Chapter 6

STEROID HORMONES IN EARLY PIG EMBRYO DEVELOPMENT

Heiner Niemann and **Folkmar Elsaesser**

Chapter 7

GENETIC EXPRESSION DURING EARLY MOUSE DEVELOPMENT

Terry Magnuson and **Charles J. Epstein**

Chapter 8

DEFINING THE ROLES OF GROWTH FACTORS DURING EARLY MAMMALIAN DEVELOPMENT

Angie Rizzino

Chapter 9

INTERACTION OF TROPHOBLASTIC VESICLES WITH BOVINE EMBRYOS DEVELOPING IN VITRO

Yves Heyman and **Yves Ménézo**

Chapter 10

IN VITRO GROWTH OF PREIMPLANTATION RABBIT EMBRYOS

Michael T. Kane

Chapter 11

STUDIES ON THE DEVELOPMENTAL BLOCKS IN CULTURED HAMSTER EMBRYOS

Barry D. Bavister

Chapter 12

GROWTH OF DOMESTICATED ANIMAL EMBRYOS IN VITRO

Raymond W. Wright, Jr. and **James V. O'Fallon**

Chapter 13

IN VITRO GROWTH OF NON-HUMAN PRIMATE PRE- AND PERI- IMPLANTATION EMBRYOS

Dorothy E. Boatman

Chapter 14

ANALYSIS OF EMBRYOTOXIC EFFECTS IN PREIMPLANTATION EMBRYOS

Horst Spielmann

Chapter 15

APPLICATIONS OF ANIMAL EMBRYO CULTURE RESEARCH TO HUMAN IVF AND EMBRYO TRANSFER PROGRAMS

Susan Heyner

APPENDIX I

APPENDIX II

Chapter 1

PIONEERING MAMMALIAN EMBRYO CULTURE

JOHN D. BIGGERS

1. INTRODUCTION

In the fast pace and competitiveness of modern science, there is less and less time to teach students the background on how the currently used scientific methods and ideas came about. This fact is unfortunate, for it is salutary to realize that many so-called discoveries and inventions are not new and that the old literature is replete with ideas. In doing our work we often stand on the shoulders of those who have gone before. The study of early mammalian development *in vitro*, which has undergone explosive growth since 1960, provides many illustrations of these facts. In addition, to minimize the tendency of investigators to "rediscover the wheel", it is pertinent at this time to reflect on the roots of the field. The subject is closely intertwined with the history of embryo transfer (see Adams, 1982) and of *in vitro* fertilization [see Austin (1961) and Biggers (1984)].

The culture of early mammalian embryos has been studied predominantly in two species: the rabbit and the mouse. Up to 1949, the rabbit was used almost exclusively, with media composed of ill-defined biological fluids. After this time, techniques for the culture of preimplantation mouse embryos rapidly developed using simple defined media with relatively few components. The work in this species was fostered by the increasing availability of genetic strains of mice which could be exploited in embryological studies. In the 1960s, there was renewed interest in the culture of early rabbit embryos and a marked difference in the nutritional requirements of the two species was found. In contrast to the mouse, preimplantation rabbit embryos require more complex media. [see also Chapter 10 (Ed.).]

John D. Biggers Laboratory of Human Reproduction and Reproductive Biology, and Department of Physiology and Biophysics, Harvard Medical School, Boston, Massachusetts 02115, USA.

This historical survey will deal almost exclusively with the development of culture methods for embryos from these two species only, since they provided the basis for the culture of embryos from other species. By 1975, embryos of the rabbit, sheep, ferret and some genetic strains of the mouse could be cultured continuously from the one-cell stage to the blastocyst, while human, cow, pig, rat and hamster embryos could be cultured in part (see Whittingham [1975] for a review.)

2. EARLY PERIOD USING MEDIA PREPARED FROM BIOLOGICAL FLUIDS

The usefulness of the ability to culture mammalian embryos outside the female genital tract was recognized not long after the birth of tissue culture in the first decade of the twentieth century (see Harrison [1969] for an account of the origin of tissue culture methods). In 1912, Mark and Long, working on the development of rats and mice at the University of California, suggested that it would be instructive "*to study the course of early development in each species, both under normal conditions and also under artificial conditions which simulate natural ones as closely as possible*" (Mark and Long, 1912). These investigators built a special chamber with the facility to change the bathing fluids for the observation of mammalian ova and sperm. They concentrated on *in vitro* fertilization, which they failed to achieve, and therefore did not obtain development.

2.1. Rabbit

At the same time, a more successful study was undertaken by Brachet (1912, 1913) at the University of Brussels on the culture of the rabbit blastocyst. This investigator was concerned with the age-old problem of the relative importance of nature and nurture in the control of development, a topic which had earlier stimulated Walter Heape's embryo transfer experiments (Heape, 1891, 1897). Brachet's work was summarized by Maximow (1925) as follows: "*He explanted 5 to 7 day old blastodermic vesicles of the rabbit into glass dishes, filled with a large quantity of coagulated blood plasma... Brachet's technique was not perfect enough to allow the blastoderms to develop for longer than 40 hours* in vitro. *Still, even during this short time, he was able to observe the normal process of development. Apart from occasional necrosis of the wall of the vesicle, resulting in shrinkage and collapsing of the whole structure, there appeared in most cases, in due time, a quite normal looking primitive groove with typical mesodermic outgrowths and a head process. Around the caudal end of the embryonic shield there developed a normal horseshoe-shaped area with distinct trophoblastic formations, showing amitosis, pluripolar mitosis, a brush border, and finally a typical syncytium, the rudiment of the ectoplacenta. ... On the other areas of the trophoblastic surface, especially on the pole opposite to the embryo, papillary excrescences were found, which corresponded exactly to the usual outgrowths found in this place and normally penetrating into the uterine mucous membrane. Thus Brachet concluded that the presence of the uterine medium is not necessary for the production of placental structures. Phylogenetically they certainly are the result of the action of external influences, but ontogenetically they have to be looked upon merely as the result of hereditary transmission. Another very important observation of Brachet was the specificity of the germ layers, even in the very young stages. He observed that in those cases where the primitive groove was destroyed accidentally no formation of the mesoderm occurred."*

Soon after Brachet published his work, Maximow began studies in Leningrad on the culture of rabbit embryos but did not publish his results until 1925, after he moved to Chicago. In studies of explants of the germinal disc of $6\frac{1}{2}$-day old rabbit blastocysts grown on rabbit plasma clots, he observed histiotropic outgrowths (Fig. 1). These outgrowths are very similar to those described about 40 years later in the rabbit (Cole and Paul, 1965) and in the mouse (Gwatkin, 1966a; Gwatkin and Meckley, 1966).

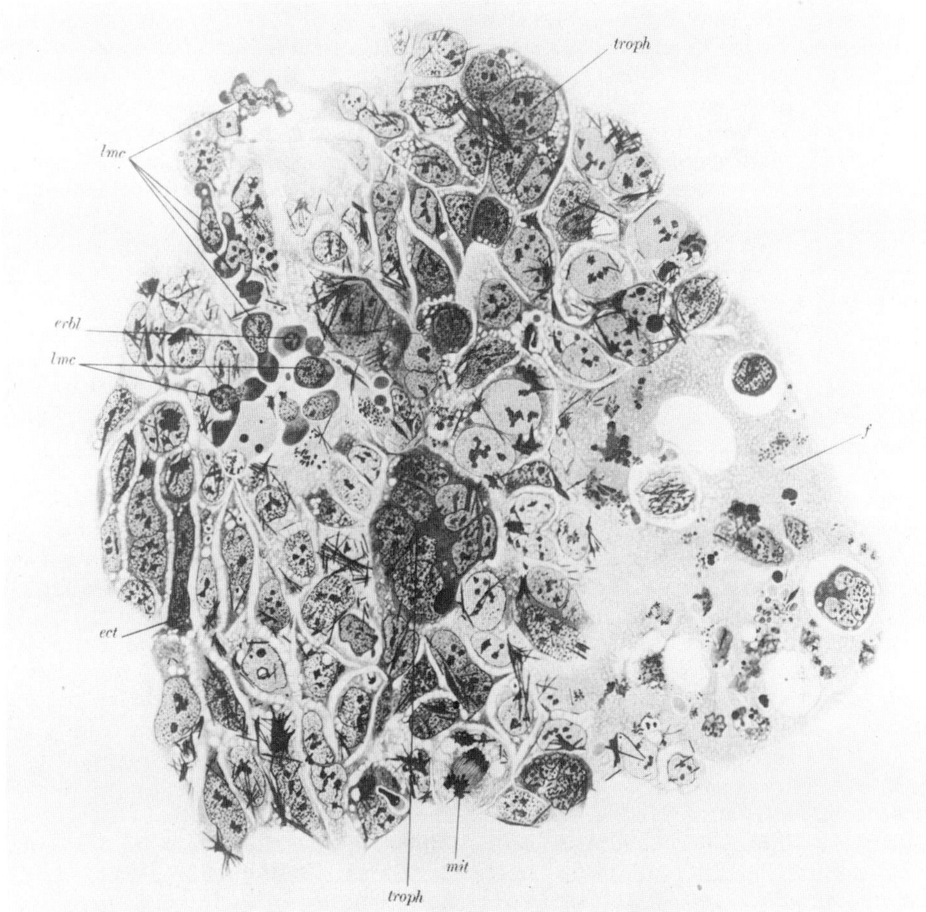

Figure 1. "Culture of a [rabbit] embryonic shield of 6.5 days; 4 days *in vitro*. At the periphery of the explant the extraembryonic ectoderm (*ect*) produces large uninucleated and multinucleated trophoblastic giant cells with crystals in the protoplasm (*troph*); *lmc* and *erbl*, large ameboid lymphocytes and primitive erythroblasts from the destroyed blood-vessels of the area vasculosa." From Maximow (1925), Plate 12.

The next important advances were made by P.W. Gregory and G. (Gregory) Pincus. Towards the end of the second decade of the century, these investigators, who both worked with the geneticist W.E. Castle in the Biological Laboratories at Harvard University, became involved in the culture of early rabbit preimplantation embryos.

Gregory had obtained training in the recovery and handling of early developmental stages of the rabbit from members of the staff of the Department of Embryology of the Carnegie Institute of Washington in Baltimore. He worked with Castle in trying to confirm the hypothesis that, at equivalent stages of development, embryos of large strain rabbits contained more cells than those of small strain rabbits. In their first paper, Castle and Gregory (1929) reported that 48 hr after mating, embryos of large strain rabbits were at the 32-cell stage while those of small strain rabbits were only at the 16-cell stage.

The first time-lapse cinematographic studies of the cleavage of mammalian embryos resulted from a collaborative study between Gregory and Warren Lewis in Baltimore using rabbit embryos in short-term cultures in blood plasma (Lewis and Gregory, 1929; Gregory, 1930). One cell embryos were observed developing to the 8-cell stage within 48 hr; older embryos near the morula stage developed for several days. In a subsequent paper, Gregory and Castle (1931) wrote: *"If it were possible to secure fertilization outside the body of the mother and to maintain the eggs under normal conditions while timing the development accurately from the moment of insemination, more definitive knowledge would be available about the comparative rate of development of A and B eggs up to the thirty-hour stage."*

Gregory Pincus worked under Castle's supervision for his doctoral dissertation on the karyotype of the rat (Pincus, 1927). During this time, Gregory showed him unfertilized rabbit eggs that had apparently fragmented, and Pincus examined the chromosomes in one of these eggs (Pincus, 1930). After completing his work at Harvard, Pincus took up a National Research Council Fellowship at the School of Agriculture, University of Cambridge under the supervision of John Hammond, Sr., where he studied the behavior of unfertilized and fertilized rabbit embryos both *in vivo* and *in vitro*. It was fortuitous that, at the same time, Honor Fell was pioneering the techniques of tissue culture at the Strangeways Research Laboratory in Cambridge, and its facilities were placed at Pincus' disposal. Pincus' interest was primarily in parthenogenesis of rabbit ova *in vitro* and the problem of distinguishing it from a process that had been described earlier in the mouse (Charlton, 1917) and rat (Long and Evans, 1922) called "fragmentation". During this work, he tried various culture techniques in use at the time: "..*the hanging drop with the ovum held in a plasma clot on a coverslip over a fluid-free cavity; a plasma clot occupying the total area under a raised coverslip; the Carrel flask; and the watch glass technique in which the sterile watch glass containing the culture medium is contained in a moist chamber*" (Pincus, 1936). Several media were used made up of mixtures of rabbit plasma, chick plasma, rabbit embryo extract and chick embryo extract. Cleavage was observed in several early embryos, and a few 2- and 4-cell embryos developed into morulae (Pincus, 1930).

Also at the Strangeways Research Laboratory, attempts were made to culture expanded rabbit blastocysts in order to try to repeat some of the classical embryological experiments done on amphibian and avian embryos (such as primordial isolation and transplantation). Waddington and Waterman (1933) cultured rabbit embryos from the pre-primitive streak stage to somite stages on a medium prepared from adult chick plasma and chick embryo extract. Pre-primitive streak stages failed to differentiate while later stages did so. Attempts to find other types of media were not particularly successful (Waterman, 1934). Later, Mather (1950) reported that primitive streak stage embryos would undergo considerable differentiation when cultivated on clots prepared from Tyrode's solution, egg albumin and agar.

Pincus continued his work on mammalian embryos after returning from Cambridge, England to Harvard and became interested in the factors controlling the growth of the rabbit blastocyst. In particular, he was intrigued by the possible parallel between the failure of rabbit blastocysts to expand *in vivo* following ovariectomy of the mother and the failure of blastocysts to expand *in vitro* in media that support cleavage. Could the blastocysts be deprived of some hormonally induced nutrient? Pincus and Werthessen (1938) examined the effect of progesterone on the development of rabbit blastocysts cultured in rabbit blood plasma or serum using three culture techniques. Two of these techniques were static and the third involved incubating the embryos in a perfusion chamber. Although the experiments did not demonstrate any direct effect of progesterone on blastocyst expansion, they did demonstrate the beneficial effects of the perfusion system. With this perfusion technique, early rabbit blastocysts expanded to a diameter of 4 mm, a size comparable to that occurring *in vivo*, though at a slower rate. The failure to demonstrate a direct stimulatory effect of progesterone on blastocyst expansion led to the hypothesis that the steroid had an indirect effect by stimulating the secretion of some other substance in the uterus. Glutathione was a fashionable candidate, for its growth-promoting properties described by Hammett (1930) had already been invoked by Castle and Gregory (1929) to explain the difference in the size of embryos from their large and small strains of rabbit. When glutathione was added to rabbit blastocysts cultured in serum, considerable expansion occurred (Pincus, 1937). Subsequently, Miller and Reimann (1940) showed that L-cysteine and DL-methionine would also stimulate expansion of rabbit blastocysts in short-term culture.

In subsequent studies, Pincus (1941) blocked expansion of the rabbit blastocyst by adding potassium cyanide to the medium and concluded that the energy for growth is derived from aerobic oxidative systems. He then studied the effect of various components of the Embden-Meyerhoff pathway and the Kreb's citric acid cycle on the medium to see if they affected blastocyst expansion. The results indicated that energy for expansion is derived from the glycolytic pathway. Nevertheless, he realised that the unequivocal determination of the nutritive requirements of preimplantation embryos must await the development of chemically defined media.

The first practical application of the culture of preimplantation embryos was as a test for viability, in some of the pioneering work on the low temperature storage of the cleavage stages (Chang, 1947, 1948; Smith, 1949, 1952, 1953).

2.2. Other Species

The work on the rabbit with Gregory at the Carnegie Institution's Department of Embryology in Baltimore stimulated Warren Lewis to attempt the culture of eggs of other species. He failed to culture mouse ova (Lewis, 1931) but together with Carl Hartman had limited success with ova of the rhesus monkey, filming cleavage of a 2-cell stage to an 8-cell stage maintained on a clot prepared from the monkey's plasma (Lewis and Hartman, 1933). Squier (1932), working with Lewis and Hartman, failed to obtain development of the guinea pig ovum in culture and concluded that the culture conditions required for rodents were very different from those required by the rabbit. In an independent study at Yale University, Defrise (1933) failed to obtain development of rat embryos of stages up to the blastocyst in a variety of physiological salt solutions, some of which were supplemented with biological fluids. Washburn (1951) also failed in his attempts to culture early rat embryos.

3. SUCCESSFUL CULTURE OF A RODENT PREIMPLANTATION EMBRYO

3.1. Initial Successes

The first major success in the culture of a rodent preimplantation embryo was made by John Hammond, Jr. (1949), working at the Strangeways Research Laboratory in Cambridge. This investigator was interested in using embryo culture in studies of embryo transfer, and decided to use a medium based on a physiological saline supplemented with hen egg white and yolk. Using this medium, Hammond made the important observation that development of mouse embryos *in vitro* depends on the age at which they are explanted. Eight-cell embryos developed into blastocysts, while 2-cell embryos, with one exception, did not cleave at all. Wesley Whitten (1956a), working at the Australian National University in Canberra, confirmed Hammond's observations. Because of difficulties in the control of pH, Whitten adopted Krebs-Ringer bicarbonate (Krebs and Henseleit, 1932) as the physiological saline, a solution common in the study of tissue metabolism by manometric techniques (Umbreit *et al.*, 1949). This saline was supplemented with glucose, penicillin, streptomycin and 1% fresh, thin egg white, and gassed with 5% carbon dioxide to give a pH of 7.4 (Table I). The medium provided good culture conditions for the development of 8-cell mouse embryos to blastocysts. Development did not occur in Krebs-Ringer bicarbonate alone. Whitten also showed that the essential factors in egg white were non-dialysable, but attempts to fractionate the egg white into active components failed. The search stopped, however, when it was shown that the egg white could be replaced with crystalline bovine albumin in concentrations ranging from 0.03 to 6%.

Whitten (1956b, 1957a) subsequently used the development of 8-cell mouse embryos *in vitro* to study the effect of progesterone on the development of the blastocyst, as Pincus and Werthessen (1938) had done earlier on the rabbit. Whitten found that the hormone had no effect at low concentrations and was toxic at high concentrations. Whitten (1956b) also studied the action of various agents (Triton W.R. 1339 and suramin) on the 8-cell embryo that might have contraceptive properties. The results he obtained with suramin led him to suggest with foresight the need to test the effect of new drugs on pregnant animals.

Table I
Media Based on Krebs-Ringer Bicarbonate for the Culture of Early Mouse Embryos[a]

Component	Whitten (1956a)	Whitten (1957b)	Brinster (1963)	Brinster (1965c)	Brinster (1971)[b]	Biggers *et al.* (1971) [BWW]
NaCl	118.46	118.46	109.23	119.23	94.89	94.59
KCl	4.74	4.74	4.78	4.78	4.78	4.78
KH_2PO_4	1.18	1.18	1.19	1.19	1.19	1.19
$CaCl_2$	2.54	-	1.71	1.71	1.70	-
$MgSO_4$	1.18	1.18	1.19	1.19	1.19	1.19
Ca lactate (L+)	-	2.54	-	-	-	1.71
Na lactate (DL)	-	-	10.15	25.00	25.00	21.58
Na pyruvate	-	-	-	0.25	0.51	0.25
$NaHCO_3$	24.88	24.88	25.07	25.07	25.07	25.07
Glucose	5.55	5.55	-	-	5.55	5.56
Crystalline BSA	1 mg/ml	1 mg/ml	1 mg/ml	1 mg/ml	5 mg/ml	1 mg/ml
Penicillin	10 μg/ml	10 μg/ml	100 IU/ml	100 IU/ml	100 IU/ml	100 IU/ml
Streptomycin	10 μg/ml	10 μg/ml	50 μg/ml	50 μg/ml	5 μg/ml	50 μg/ml

[a]Values given are mM.
[b]Medium available from Gibco Laboratories.

3.2. Proof that Cultured Preimplantation Embryos Could Develop into Normal Adults

Whitten (1956a) showed that mouse blastocysts produced from 8-cell embryos *in vitro* are capable of further development when placed under the kidney capsule. Soon after, Adams (1956), working at the Agricultural Research Council Unit of Reproductive Physiology and Biochemistry in Cambridge, England, found viable fetuses one week after embryo transfer in the uterus of a recipient female rabbit; this animal had been mated to a vasectomized male 3 days before the transfer. Prior to transfer, these 16-cell rabbit embryos had been cultured for 21.5 hr in Krebs-Ringer bicarbonate supplemented by plasma albumin (fraction V). Two years later, McLaren and Biggers (1958) and Biggers and McLaren (1958), working at the Royal Veterinary College in London, provided the essential evidence needed to show that mouse blastocysts produced by Whitten's technique could develop into normal adults. The work was facilitated by the technique of embryo transfer in the mouse that was perfected by McLaren and Michie (1956). Eight-cell embryos were recovered from superovulated females, which were homozygous for albinism, and cultured to blastocysts using Whitten's medium. The blastocysts were then transferred to the uteri of females, homozygous for full color, that had been mated to fertile males $2\frac{1}{2}$ days earlier. Of the 8 surrogates that became pregnant, 6 were killed on the 16th day after mating. They contained two types of fetus, those with pink eyes, which were derived from the transferred blastocysts produced *in vitro*, and those with pigmented eyes, which were native embryos. The two remaining pregnant animals were delivered by Caesarian section on the 19th day. One mother provided two albino young which were successfully adopted by a second lactating foster mother (Fig. 2). These grew into normal adults and gave rise to young of their own. Biggers and McLaren (1958) recognized the significance of this work as follows: "*It is inevitable that the thoughts of anyone who has worked on the*

Figure 2. The first adult mice produced by embryo transfer in which part of the preimplantation stage of development occurred *in vitro* (McLaren and Biggers, 1958). The two albino mice were cultured from the 8-cell embryo to the blastocyst stage in Whitten's medium.

subjects outlined in this article should turn to Aldous Huxley's fantasy 'Brave New World', where he describes completely artificial fertilization and development of human embryos. Fortunately we are far removed from this frightening prospect. The study of the cultivation and transfer of embryos is none the less of the greatest interest, both from the point of view of pure science, and because the techniques associated with it are potentially of immense value in the investigation of many biological problems in medicine and agriculture."

3.3. Further Development of Culture Media for Mouse Embryo Development

Whitten (1956a), like Hammond before him, failed to obtain development of embryos prior to the 8-cell stage in Krebs-Ringer bicarbonate supplemented with glucose and albumin. However, when calcium lactate, a non-deliquescent substance, was used to replace the very deliquescent calcium chloride in the medium, it was found that 2-cell embryos developed into blastocysts (Table I; Whitten, 1957b). Since then a whole family of related media has arisen from relatively minor changes made in different laboratories. Important contributions were made to the optimization of the medium for the development of 2-cell stages by Brinster, working as a graduate student in Biggers' laboratory at the Wistar Institute in Philadelphia and later at the University of Pennsylvania. The work was based on methods outlined by Biggers *et al.* (1957) for the optimization of culture media in general, using the study of concentration response curves and surfaces. A simple example of this type of approach is the variation of the concentration of sodium chloride to find the optimum osmolality to support development (Brinster, 1965a). Some of the biometrical aspects involved in these types of study were discussed by Biggers and Brinster (1965) and led to the medium described by Brinster in 1963 (Table I).

Other preliminary work of Whitten (1957b) had suggested that some metabolic intermediates could support the early development of preimplantation mouse embryos while others could not. Brinster (1965b) examined various members of the Embden-Meyerhoff pathway and the Krebs citric acid cycle one by one over a range of concentrations. He found that only 3 compounds (pyruvate, phosphoenolpyruvate and oxaloacetate) could substitute for lactate in Whitten's medium for the development of 2-cell embryos to blastocysts. In another study, Brinster (1965c) used a factorial experimental design to study the joint effect of lactate and pyruvate; pyruvate enhanced the response to lactate at low concentrations. This work led to a medium that was often called "Brinster's medium" (Table I). Two-cell embryos that developed to blastocysts *in vitro* in this medium were subsequently shown to develop into normal 14-day fetuses after transfer into uterine foster mothers (Biggers *et al.*, 1965). It was later found that, for the routine production of expanded blastocysts from 2-cell embryos, it was advantageous to include 3 carbon (energy) sources in the medium: pyruvate, lactate and glucose. Thus, two more modifications of Whitten's medium were proposed. One medium (Biggers *et al.*, 1971) is often called "BWW" (Table I), and another (Brinster, 1971) is the medium commercially available from Gibco Laboratories (Table I).

That mouse preimplantation embryos can also develop in media containing major amounts of undefined body fluids was shown by Beatrice Mintz (1964), working at the Institute for Cancer Research in Philadelphia. Her medium consisted of 50% fetal calf serum and 50% Earle's balanced salt solution (Earle, 1943) supplemented with lactate. This medium supports the development of 2-cell mouse embryos to the blastocyst stage, and was used by Mintz for the production of chimeras by the fusion of 8-cell stages.

Whitten (1956a, 1957b) cultured early mouse embryos in small test tubes. This method is not particularly convenient for the frequent observation of developing embryos. A more suitable method, the microdroplet method, was introduced by Ralph Gwatkin, also working in Biggers' laboratories in Philadelphia, who was studying whether the Mengo virus, and other viruses, could infect early mouse embryos (Gwatkin, 1963). The method was adapted from the work of Lwoff *et al.* (1955), who used it for studying the kinetics of release of the poliomyelitis virus from single cells. Using the microdroplet method, in which the embryos are cultivated in a drop of medium under mineral oil, it is possible to observe them directly, manipulate them, and microsample the medium for analysis. The method was then used routinely for the culture of early mouse embryos (Brinster, 1963).

3.4. Blocks to Development

In 1964, McLaren and Biggers proposed to the Ciba Foundation that a conference be held on the preimplantation stages of pregnancy. This conference, held in London the following year, set the stage for many of the advances that have now occurred in the study of early mammalian development (see Wolstenholme and O'Connor, 1965). At this conference, there was extensive discussion about the "2-cell block", *i.e.*, the observation that 1-cell mouse embryos would divide into 2 cells but would then degenerate (Cole and Paul, 1965), although late 2-cell embryos would readily develop into blastocysts, and fertilized ova would develop into blastocysts in organ cultures of the oviduct (Biggers *et al.*, 1962). These facts led Whittingham, working as a graduate student in Biggers' laboratory at the University of Pennsylvania, to test the viability of embryos that had undergone the first cleavage division using the technique of oviducal organ culture. Newly fertilized ova were cultured as far as arrested 2-cell stages in the medium of Brinster (1963) and then placed into organ cultures of the ampullary region of the oviduct, where they developed into blastocysts. Some of these blastocysts were transferred into the uteri of pseudopregnant recipients, where they had developed into normal 17-day fetuses when the experiments were terminated (Whittingham and Biggers, 1967). These experiments demonstrated that the first cleavage division of the zygote *in vitro* is normal. In some later experiments, Whittingham (1968) showed that 2-cell embryos would develop into blastocysts in organ cultures only of the ampullary region of the oviduct and not of the isthmic region. These experiments suggested that the ampullary region of the oviduct provides either special nutrients or special environmental conditions for the cleavage of late 2-cell mouse embryos into 4-cell stages. Other work showed that the first cleavage division is supported in Krebs-Ringer bicarbonate only with the addition of pyruvate or oxaloacetate but not lactate or phosphoenolpyruvate (Biggers *et al.*, 1967). The problem of the 2-cell block became even more complicated when Whitten and Biggers (1968) reported that some strains of mice could be cultured without interruption from the zygote to the blastocyst while others could not, thus suggesting that genetic factors play a role in the occurrence of the 2-cell block. For the first time, however, it was shown that under some circumstances complete preimplantation development in the mouse could be obtained *in vitro*.

Another feature of the culture of mouse preimplantation embryos that puzzled the early investigators was the fact that the media based on Krebs-Ringer bicarbonate did not support differentiation of the blastocyst. Such media containing pyruvate, lactate and glucose supported the development of 2-cell embryos only to blastocysts, after which they hatched and remained free-floating until they collapsed and degenerated (Gwatkin, 1966a). A preliminary report by Mintz (1964) showed that mouse blastocysts grew out on the surface of a culture dish in a medium consisting of equal parts of fetal calf serum and Earle's balanced salt solution supplemented with lactate. This observation prompted Gwatkin (1966a) to examine the phenomenon in more detail. He used Ham's F10 medium supplemented with fetal calf serum (Ham, 1963). Since preimplantation mouse embryos developed poorly in F10, blastocysts were first produced from 2-cell stages by cultivation in Krebs-Ringer bicarbonate containing pyruvate, lactate and glucose. These blastocysts were then transferred to Ham's F10 medium containing 10% fetal calf serum. The blastocysts attached to the bottom of the dish and produced extensive outgrowth over the next 3 to 4 days. In later experiments, it was shown that the simpler Eagle's medium (Eagle, 1959), containing only 30 components compared to the 46 in F10, would also support outgrowths from mouse blastocysts. The method was first used to study the karyotype of mouse blastocysts (Gwatkin and Meckley, 1966). Later, Gwatkin (1966b) showed that 10 amino acids (arginine, cystine, histidine, leucine, lysine, methionine, phenylalanine, threonine, tryptophane and tyrosine) were needed in the medium for maximal outgrowth from the mouse blastocyst.

4. CULTURE OF RABBIT EMBRYOS REVISITED

After 1960, research on the culture of the rabbit preimplantation embryo followed two paths. One route was to improve the efficiency of the technique using media prepared from biological fluids, and to assess the viability of the embryos produced *in vitro*. The other route was to develop chemically defined media that would support preimplantation development.

Following the comparison of several media, Purshottam and Pincus (1961), while at Clark University, reported that undiluted rabbit serum provided the best conditions for the development of the cleavage stages of the rabbit embryo to the early blastocyst stage. Onuma *et al.* (1968), working in Robert Foote's laboratory at Cornell University, confirmed that 2- and 4-cell rabbit embryos will readily develop into blastocysts and hatch in either rabbit or bovine serum supplemented with glucose. Other work by Maurer *et al.* (1969), also in Foote's laboratory, showed that rabbit zygotes would also develop into blastocysts in glucose-enriched bovine serum.

It had been suggested earlier by Adams (1965), however, that the viability of early rabbit blastocysts declined the longer they were maintained in culture, as assessed by the number of transferred embryos developing to term. Defects occurred in the embryonic disc while the trophoblast remained normal. In a detailed study by Maurer *et al.* (1970), 2- and 4-cell rabbit embryos cultured in glucose-enriched rabbit and bovine sera for 62 hr developed into viable blastocysts, as assessed by transfer into surrogate mothers; viability decreased, however, if the embryos were cultured for longer than 62 hr. Adams (1970) also transferred rabbit blastocysts produced from morulae *in vitro* and found that their viability was diminished after 48 hr in culture.

Pincus' interest of 20 years earlier in chemically defined media was rekindled by the successful development of preimplantation mouse embryos in such media. Purshottam and Pincus (1961), culturing rabbit embryos from the 2-cell stage in several chemically defined media using a shaker culture technique, found that no development occurred in Krebs-Ringer bicarbonate supplemented with glucose and albumin (fraction IV) and the protein-free Waymouth's medium (Waymouth, 1959). However, in Eagle's basal medium with no serum supplement, embryos cleaved to the morula stage. When Eagle's medium was supplemented with horse, human and particularly rabbit serum, the embryos developed into blastocysts. They concluded that, prior to the formation of the blastocyst, the nutritional requirements of the embryo are relatively simple, but once the blastocyst stage is reached, other serum components are required.

During the 1960s, in the course of trying to derive cell lines from preimplantation embryos, Cole *et al.* (1964), Edwards (1964) and Cole and Paul (1965), at the University of Glasgow, tried culturing rabbit cleavage stages and blastocysts after removing the zona pellucida with pronase (Mintz, 1962a). They used several chemically-defined media, including Fischer's medium (Fischer, 1947), Waymouth's medium, and Ham's F10, but these were all supplemented with up to 10% serum. One finding reported by Robert Edwards was that rabbit zygotes underwent several cleavage divisions, particularly in serum-supplemented Waymouth's medium (Edwards, 1964). The research also showed that, while the cells from cleavage stage embryos did not attach to the wall of the culture chamber, those from the blastocyst attached and gave trophoblast outgrowths. These cultures were similar to those obtained by Maximow (1925), a largely forgotten study (Fig. 1).

The use of undiluted plasma or the need for supplementing chemically-defined media with serum raised the possibility that plasma steroids might influence the development of preimplantation rabbit embryos in culture. Of particular concern was progesterone, which was known to be toxic to the mouse embryo (Whitten, 1957a). Daniel and Levy (1964), at the University of Colorado, exposed cleavage stage rabbit embryos and blastocysts to various concentrations of progesterone in Ham's F10 medium containing 10% rabbit serum. High concentrations of the steroid reversibly inhibited cleavage but did not interfere with development when added to cultures of blastocysts. Daniel (1964) found that, in high concentrations, testosterone also inhibited cleavage stages of the rabbit, while estrogens caused the embryos to fragment.

In the course of the work with progesterone, Daniel and Levy (1964) discovered that the inhibitory effect of progesterone could be overcome by the addition of amino acids to the medium. This caused Daniel (1965) to turn his attention to improving the growth and expansion of 5-day rabbit blastocysts *in vitro*, by adding various substances over a range of concentrations to Ham's F10 medium supplemented with 15% rabbit serum. On the basis of the results obtained, Daniel recommended a new medium called "modified F10" for supporting the growth of the rabbit blastocyst. The modified medium is regular F10 to which lactate and glycogen are added, the glucose concentration is reduced by 80% and the concentrations of glycine, alanine, glutamic acid, threonine, serine and pyruvate are increased (Table II). This medium is now used by several investigators, particularly since live young have been born after short term culture of blastocysts for up to 16 hr in modified F10 containing 10% rabbit serum and transfer to uterine foster mothers (Staples, 1967). In a further study, Daniel and Krishnan (1967) reported that 10 essential amino acids (arginine, histidine, leucine, lysine, methionine, phenylalanine, serine, threonine, tryptophane and valine) are necessary for blastocyst expansion. However, since blastocyst growth was reduced when the non-essential amino acids were omitted, both essential and non-essential amino acids were included in the medium (Table II). Although this work has attempted to optimize the conditions for the stage in which the rabbit blastocyst expands, the medium may not be optimum for other stages, since amino acid requirements may increase as the rabbit embryo develops from the zygote (Daniel and Olson, 1968).

A different approach to the design of a chemically-defined medium was taken by Michael Kane, who worked as a graduate student in Robert Foote's laboratory at Cornell University, with the objective of finding a medium that would support development of early rabbit embryos from the 2-cell stage to the expanding blastocyst stage (Kane and Foote, 1970). The earlier observation in this laboratory that 2- and 4-cell rabbit embryos would develop to expanded blastocysts in Ham's F10 medium supplemented with bovine serum albumin, instead of the usual blood sera, caused these investigators to examine the components of medium F10. Two control media were used: a simple defined medium which was a slight modification of the original medium of Brinster (1963) for the mouse, and Ham's F10 containing 1.5% crystallized bovine plasma albumin. The components of Ham's medium were split up into 4 blocks: amino acids, vitamins, trace elements and nucleic acid precursors. A complete synthetic medium was defined as one in which all 4 blocks of compounds were added to the simple Brinster-type medium. This complete medium, as well as Ham's F10 medium supplemented with bovine albumin, equally supported the development of early rabbit embryos into blastocysts. However, no blastocysts formed in the simple defined medium after 4 days in culture. With the omission of the amino acid block from the complete medium, no blastocysts formed. The omission of the vitamin block permitted only slight development and leaving out the nucleic acid precursors or the trace elements had little if any effect (see Chapter 10, Table I [Ed.]). In a parallel experiment, using the same medium described by Kane and Foote (1970), with the omission of the nucleic acid precursors, Naglee *et al.* (1969) demonstrated that the optimum osmolality for the production of hatched blastocysts from 2-cell stages is 270 mOsmols. Subsequently, by adding pyruvate to their medium, Kane and Foote (1971) showed that zygotes will also cleave and develop into expanding blastocyts. The composition of this final medium is also shown in Table II.

The development of 2- and 4-cell rabbit embryos up to morulae can be accomplished in simple defined media only, provided it contains bovine serum albumin (Kane and Foote, 1970). In agreement was an observation made at about the same time by Brinster (1970), who found that 2-cell rabbit embryos develop into morulae in a modified Krebs Ringer bicarbonate supplemented with an amino-nitrogen source such as bovine serum albumin, oxidized glutathionine, glutamine, proline or alanine. By 1970, it had been clearly

Table II
Composition of Ham's F10 Medium and Comparison of Several Modifications for Rabbit Embryo Culture[a]

Component	F10[b]	MF10a[c]	MF10b[d]	KF[e]
Basic physiological saline (mM):				
NaCl	120.00	120.00	120.00	**102.98**
KCl	3.8	3.8	3.8	**4.77**
$CaCl_2.2H_2O$	0.3	0.3	0.3	**1.71**
KH_2PO_4	0.48	0.48	0.48	**1.19**
$Na_2HPO_4.7H_2O$	1.1	1.1	1.1	-
$MgSO_4.7H_2O$	0.62	0.62	0.62	**1.19**
$NaHCO_3$	14.3	14.3	14.3	**25.07**
Glucose	6.1	**1.11**	**1.11**	**10.0**
Sodium pyruvate	1.0	**9.09**	**9.09**	-
Sodium lactate	-	**7.14**	**7.14**	-
Amino acids (mM):				
L-alanine	0.1	**1.12**	0.1	0.1
L-arginine.HCl	1.0	1.0	**10.0**	1.0
L-asparagine.H_2O	0.1	0.1	**1.0**	0.1
L-cysteine	0.2	0.2	**2.0**	0.2
L-glutamic acid	0.1	**1.36**	0.1	0.1
Glycine	0.1	**2.66**	**10.0**	0.1
L-isoleucine	0.02	0.02	**0.2**	0.02
L-lysine.HCl	0.1	0.1	**1.0**	0.1
L-serine	0.1	**0.95**	0.1	0.1
L-threonine	0.03	**0.84**	**0.3**	0.03
L-valine	0.03	0.03	**0.3**	0.03
L-glutamine	1.0	1.0	1.0	1.0
L-aspartic acid	0.1	0.1	0.1	0.1
L-histidine.HCl	0.1	0.1	0.1	0.1
L-leucine	0.1	0.1	0.1	0.1
L-methionine	0.03	0.03	0.03	0.03
L-phenylalanine	0.03	0.03	0.03	0.03
L-proline	0.1	0.1	0.1	0.1
L-tryptophane	0.003	0.003	0.003	0.003
L-tyrosine	0.01	0.01	0.01	0.01
Other components (mM):				
Hypoxanthine	0.03	0.03	0.03	0.03
Thymidine	0.003	0.003	0.003	0.003
Phenol red	0.003	0.003	0.003	-
Glycogen	-	***0.1 mg/L***	***0.1 mg/L***	-
Trace elements (μM):				
$CuSO_4.5H_2O$	0.01	0.01	0.01	0.01
$FeSO_4.7H_2O$	3.00	3.00	3.00	3.00
$ZnSO_4.7H_2O$	0.10	0.10	0.10	0.10

(cont.)

Table II (cont.)

Component	F10[b]	MF10a[c]	MF10b[d]	KF[e]
Vitamins (μM):				
Biotin	0.1	0.1	0.1	0.1
DL-calcium pantothenate	3.0	3.0	3.0	3.0
Choline chloride	5.0	5.0	5.0	5.0
i-Inositol	3.0	3.0	3.0	3.0
Niacinamide	5.0	5.0	5.0	5.0
Pyridoxine HCl	1.0	1.0	1.0	1.0
Riboflavin	1.0	1.0	1.0	1.0
Thiamine HCl	3.0	3.0	3.0	3.0
Vitamin B_{12}	1.0	1.0	1.0	1.0
Lipoic acid	1.0	1.0	1.0	1.0

[a]Values in bold type show modifications of original F10 formulation.
[b]Ham (1963).
[c]Daniel (1965).
[d]Daniel and Krishnan (1967).
[e]Kane and Foote (1970); Naglee *et al.* (1969).

established that a marked change in the nutritive requirements of the rabbit blastocyst occurs at the time of blastocyst formation.

5. CONCLUSION

As we look at the history of the development of methods for the culture of early mammalian embryos, it becomes clear that a continuous thread often connects the work of one laboratory with another. Investigators have indeed stood on each others' shoulders.

The coupling of the technique for the culture of preimplantation embryos with the technique of embryo transfer has paved the way for the exploration of many new areas of pure and applied research in experimental mammalian embryology, as predicted by Biggers and McLaren (1958). The production and study of artificial chimeras exemplifies one such advance. This new field was heralded by the independent demonstrations of Andrzej Tarkowski, at the University of Warsaw, and Beatrice Mintz that whole 8-cell mouse embryos could be fused in culture to form artificial chimeras (Tarkowski, 1961, 1963; Mintz, 1962b, 1964). Later, alternative procedures were introduced, such as the production of chimeras by injecting cells into the blastocoel (Gardner, 1968), by injecting inner cell masses or parts of these structures into the blastocoel (Gardner and Johnson, 1973), and by reconstituting blastocysts from isolated inner cell masses and trophoblastic vesicles (Gardner *et al.*, 1973). The subsequent development of these artificial embryos in culture and their transfer into surrogate mothers for further development has played a major role in studies of cell lineages in the mouse (see Gardner, 1985a,b).

An alternative technique for studies of cell movement and fate is the use of cell lineage markers, first introduced by Wilson *et al.* (1972). Here, a label is introduced into a cell of the embryo, which is then cultured. After a period of time, the position of the marker is examined. Many other phenomena could not have been studied without the availability of successful culturing techniques. For example, the process of polarization, the initial phase of compaction, involves the study of blastomere interaction *in vitro* (Ziomek and Johnson, 1980). [See Chapter 2 (Ed.).] Likewise, factors that control compaction and decompaction have been investigated using culture methods (Ducibella and Anderson, 1975). An example of embryo bisection is the separation of the blastomeres from a 2-cell mouse embryo; they are then cultured to give two half blastocysts. One half can then be karyotyped to diagnose the sex of the embryo, the other used for biochemical analysis (Epstein *et al.*, 1978). Such techniques are useful in the analysis of sex-linked processes. Blastocyst biopsies or other embryo fragments can be used for diagnosing sex of the embryo, first demonstrated by Gardner and Edwards (1968) in the rabbit. The use of techniques for freezing embryos, developed in the mouse by Whittingham (1972) and Wilmut (1972), depends strongly on the use of culture methods. [See Chapter 13 (Ed.).] The technique is widely used in human *in vitro* fertilization and embryo transfer, in the livestock industry, and in research laboratories for the preservation of genetically valuable animals, thus eliminating the need for expensive breeding colonies.

Despite these successes, there is still a need for further research on methods for the culture of preimplantation mammalian embryos. We have not yet solved the same basic problems that faced the pioneers. The so-called block to development, in which development becomes arrested at species-specific stages, was discovered in the mouse and is well known in the hamster, cow and pig. [See Chapters 6, 9, 11 and 13 (Ed.).] The causes of these blocks are still unknown. In culture, the rate of development falls behind the normal rate of development *in vivo*.

Though this developmental delay may or may not be serious, according to the particular application, if solved it would open up new avenues of research. For instance, if it were possible to understand the factors that regulate the rate of development *in vitro*, an embryo could be bisected, one half held to develop at a reduced rate and the other half stimulated to grow rapidly, so that it could be used for diagnostic purposes, such as sexing. Such procedures could be exploited for world-wide transportation of genetically valuable species. A third problem, and one of major basic biological interest, is the loss in culture of the normal spatial organization of the embryo after the blastocyst stage. Spreading of trophoblast outgrowths on the culture dish is an example of this phenomenon. Another well known example is the type of growth described by Hsu (1971), in which specific tissues develop, often in a disorganized fashion. While observations on these types of preparation have been invaluable in the study of cell differentiation in early mammalian embryos, they have been of little use in the study of factors that maintain the normal three-dimensional pattern of development. This field may advance in the next few years as our knowledge of growth factors and the nature of extracellular matrices is applied to the study of early mammalian development. Techniques are becoming so sensitive that the study of critical biochemical processes in single embryos is now within the realm of possibility.

As a final *caveat*, it is important to remember that embryo culture methods are a tool in biological research, and the results obtained with them should always be interpreted in the light of other knowledge on the reproductive process. This practice was rigorously followed by the pioneers of the field and should be strictly adhered to today.

ACKNOWLEDGMENTS

I am indebted to Dr. Betsey S. Williams for invaluable criticism of the manuscript and to Carol Kountz in its preparation.

6. REFERENCES

Adams, C.E., 1956, Egg transfer and fertility in the rabbit, *Proc. Int. Congr. Anim. Reprod. (Cambridge)*, Section III, pp. 5-6.

Adams, C.E., 1965, The influence of maternal environment on preimplantation stages of pregnancy in the rabbit, in: *Preimplantation Stages of Pregnancy* (G.E.W. Wolstenholme, and M. O'Connor, eds.), Churchill, London, pp. 345-377.

Adams, C.E., 1970, The development of rabbit eggs after culture *in vitro* for 1-4 days, *J. Embryol. Exp. Morphol.* 23: 21-34.

Adams, C.E., 1982, Egg transfer: historical aspects, in: *Mammalian Egg Transfer* (C.E. Adams, ed.), CRC Press, Boca Raton, FL, pp. 1-18.

Austin, C.R., 1961, *The Mammalian Egg*, Charles C. Thomas, Springfield, IL.

Biggers, J.D., 1984, *In vitro* fertilization and embryo transfer in historical perspective, in: *In Vitro Fertilization and Embryo Transfer* (A. Trounson, and C. Wood, eds.), Churchill Livingstone, London, pp. 3-15.

Biggers, J.D., and Brinster, R.L., 1965, Biometrical problems in the study of early mammalian embryos *in vitro*, *J. Exp. Zool.* 158: 39-47.

Biggers, J.D., and McLaren, A., 1958, "Test tube" animals. The culture and transfer of early mammalian embryos, *Discovery (London)* 19: 423-426.

Biggers, J.D., Rinaldini, L.M., and Webb, M., 1957, The study of growth factors in tissue culture, *Symp. Soc. Exp. Biol.* 11: 264-297.

Biggers, J.D., Gwatkin, R.B.L., and Brinster, R.L., 1962, Development of mouse embryos in organ cultures of fallopian tubes on a chemically defined medium, *Nature (London)* 194: 747-749.

Biggers, J.D., Moore, B.D., and Whittingham, D.G., 1965, Development of mouse embryos *in vivo* after cultivation from two-cell ova to blastocysts *in vitro*, *Nature (London)* 206: 734-735.

Biggers, J.D., Whittingham, D.G., and Donahue, R.P., 1967, The pattern of energy metabolism in the mouse oocyte and zygote, *Proc. Natl. Acad. Sci. USA* 58: 560-567.

Biggers, J.D., Whitten, W.K., and Whittingham, D.G., 1971, The culture of mouse embryos *in vitro*, in: *Methods in Mammalian Embryology* (J.C. Daniel, Jr., ed.), Freeman, San Francisco, pp. 86-116.

Brachet, A., 1912, Développement *in vitro* de blastodermes et de jeunes embryons de mammifères, *C.R. Hebd. Seances Acad. Sci.* 155: 1191-1193.

Brachet, A., 1913, Recherches sur le déterminisme héréditaire de l'oeuf des mammifères. Développement *in vitro* de jeunes vésicules blastodermiques du lapin, *Arch. Biol. (Paris)* 28: 447-503.

Brinster, R.L., 1963, A method for *in vitro* cultivation of mouse ova from two-cell to blastocyst, *Exp. Cell Res.* 32: 205-207.

Brinster, R.L., 1965a, Studies on the development of mouse embryos *in vitro*. I. The effect of osmolality and hydrogen ion concentration, *J. Exp. Zool.* 158: 49-57.

Brinster, R.L., 1965b, Studies on the development of mouse embryos *in vitro*. II. The effect of energy sources, *J. Exp. Zool.* 158: 59-68.

Brinster, R.L., 1965c, Studies on the development of mouse embryos *in vitro*. IV. Interaction of energy sources, *J. Reprod. Fertil.* 10: 227-240.

Brinster, R.L., 1970, Culture of two-cell rabbit embryos to morulae, *J. Reprod. Fertil.* 21:17-22.

Brinster, R.L., 1971, *In vitro* culture of the embryo, in: *Pathways to Conception* (A.I. Sherman, ed.), Charles C. Thomas, Springfield, IL, pp. 245-277.

Castle, W.E., and Gregory, P.W., 1929, The embryological basis of size inheritance in the rabbit, *J. Morphol.* 48: 81-103.

Chang, M.C., 1947, Normal development of fertilized rabbit ova stored at low temperature for several days, *Nature (London)* 159: 602-603.

Chang, M.C., 1948, Transplantation of fertilized rabbit ova: the effect on viability of age, *in vitro* storage period, and storage temperature, *Nature (London)* 161: 978-979.

Charlton, H.H., 1917, The fate of the unfertilized egg in the white mouse, *Biol. Bull.* 33: 321-338.

Cole, R.J., Edwards, R.G., and Paul, J., 1964, Cytodifferentiation in cell colonies and cell strains derived from cleaving ova and blastocysts of the rabbit, *Exp. Cell Res.* 37: 501-504.

Cole, R.J., and Paul, J., 1965, Properties of cultured preimplantation mouse and rabbit embryos, and cell strains derived from them, in: *Preimplantation Stages of Pregnancy* (G.E.W. Wolstenholme, and M. O'Connor, eds.), Churchill, London, pp. 82-122.

Daniel, J.C., Jr., 1964, Some effects of steroids on cleavage of rabbit eggs *in vitro*, *Endocrinology* 75: 706-710.

Daniel, J.C., Jr., 1965, Studies on the growth of 5-day old rabbit blastocysts *in vitro*, *J. Embryol. Exp. Morphol.* 13: 83-95.

Daniel, J.C., Jr., and Krishnan, R.S., 1967, Amino acid requirements for growth of the rabbit blastocyst *in vitro*, *J. Cell. Comp. Physiol.* 70:155-160.

Daniel, J.C., Jr., and Levy, J.D., 1964, Action of progesterone as a cleavage inhibitor of rabbit ova *in vitro*, *J. Reprod. Fertil.* 7:323-329.

Daniel, J.C., Jr., and Olson, J.D., 1968, Amino acid requirements for cleavage of the rabbit ovum, *J. Reprod. Fertil.* 15: 453-455.

Defrise, A., 1933, Some observations on living eggs and blastulae of the albino rat, *Anat. Rec.* 57: 239-250.

Ducibella, T., and Anderson, E., 1975, Cell shape and membrane changes in the eight-cell mouse embryo: prerequisites for morphogenesis of the blastocyst, *Dev. Biol.* 45: 231-250.

Eagle, H., 1959, Amino acid metabolism in mammalian cell cultures, *Science* 130: 432-437.

Earle, W.R., 1943, Production of malignancy *in vitro*. IV. The mouse fibroblast cultures and changes seen in the living cells, *J. Natl. Cancer Inst.* 4: 165-212.

Edwards, R.G., 1964, Cleavage of one- and two-celled rabbit eggs *in vitro* after removal of the zona pellucida, *J. Reprod. Fertil.* 7: 413-415.

Epstein, C.J., Smith, S., Travis, B., and Tucker, G., 1978, Both X-chromosomes function before visible X-chromosome inactivation in female mouse embryos, *Nature (London)* 274: 500-503.

Fischer, A., 1947, *Biology of Tissue Cells,* Cambridge University Press, Cambridge, England.

Gardner, R.L., 1968, Mouse chimeras obtained by the injection of cells into the blastocyst, *Nature (London)* 220: 596-597.

Gardner, R.L., 1985a, Origin and development of the trophectoderm and inner cell mass, in: *Implantation of the Human Embryo* (R.G. Edwards, J.M. Purdy, and P.C. Steptoe, eds.), Academic Press, London, pp. 155-178.

Gardner, R.L., 1985b, Clonal analysis of early mammalian development, *Phil. Trans. Roy. Soc. Lond. B* 312: 163-178.

Gardner, R.L., and Edwards, R.G., 1968, Control of the sex ratio at full term in the rabbit by transferring sexed blastocysts, *Nature (London)* 218: 346-348.

Gardner, R.L., and Johnson, M.H., 1973, Investigation of early mammalian development using interspecific chimaeras between rat and mouse, *Nature (London) New Biol.* 246: 86-89.

Gardner, R.L., Papaioannou, V.E., and Barton, S.C., 1973, Origin of the ectoplacental cone and secondary giant cells in mouse blastocysts reconstituted from isolated trophoblast and inner cell mass, *J. Embryol. Exp. Morphol.* 30: 561-572.

Gregory, P.W., 1930, The early embryology of the rabbit, *Carnegie Inst. Wash. Publ.* 21: 141-168.

Gregory, P.W., and Castle, W.E., 1931, Further studies on the embryological basis of size inheritance in the rabbit, *J. Exp. Zool.* 59: 199-211.

Gwatkin, R.B.L., 1963, Effect of viruses on early mammalian development, I. Action of Mengo encephalitis virus on mouse ova cultivated *in vitro, Proc. Natl. Acad. Sci. USA* 50: 576-581.

Gwatkin, R.B.L., 1966a, Defined media and development of mammalian eggs *in vitro, Ann. N.Y. Acad. Sci.* 139: 79-90.

Gwatkin, R.B.L., 1966b, Amino acid requirements for attachment and outgrowth of the mouse blastocyst *in vitro, J. Cell Physiol.* 68: 335-344.

Gwatkin, R.B.L., and Meckley, P.E., 1966, Chromosomes of the mouse blastocyst following its attachment and outgrowth *in vitro, Ann. Med. Exp. Biol. Fenn.* 44: 125-127.

Ham, R.G., 1963, An improved nutrient solution for diploid Chinese hamster and human cell lines, *Exp. Cell Res.* 29: 515-526.

Hammett, F.S., 1930, The natural chemical regulation of growth by increase in cell number, *Proc. Amer. Phil. Soc.* 69: 217-223.

Hammond, J., Jr., 1949, Recovery and culture of tubal mouse ova, *Nature (London)* 163: 28-29.

Harrison, R.G., 1969, *Organization and Development of the Embryo,* Yale University Press, New Haven, CT, pp. 67-116.

Heape, W., 1891, Preliminary note on the transplantation and growth of mammalian ova within a uterine foster-mother, *Proc. Roy. Soc. Lond. B Biol. Sci.* 48: 457-458.

Heape, W., 1897, Further note on the transplantation and growth of mammalian ova within a uterine foster-mother, *Proc. Roy. Soc. Lond. B Biol. Sci.* 62: 178-183.

Hsu, Y., 1971, Post-blastocyst differentiation *in vitro*, *Nature (London)* 231: 100-102.

Kane, M.T., and Foote, R.H., 1970, Culture of two- and four-cell rabbit embryos to the expanding blastocyst stage in synthetic media, *Proc. Soc. Exp. Biol. Med.* 133: 921-925.

Kane, M.T., and Foote, R.H., 1971, Factors affecting blastocyst expansion of rabbit zygotes and young embryos in defined media, *Biol. Reprod.* 4: 41-47.

Krebs, H.A., and Henseleit, K., 1932, Untersuchungen über die Harnstoffbildung im Tierkorper, *Z. Phys. Chem.* 210: 33-66.

Lewis, W.H., 1931, Living mouse eggs, *Anat. Rec.* 48: 52 (abs).

Lewis, W.H., and Gregory, P.W., 1929, Cinematographs of living developing rabbit-eggs, *Science* 69: 226-229.

Lewis, W.H., and Hartman, C.G., 1933, Early cleavage stages of the egg of the monkey (*Macacus rhesus*), *Carnegie Inst. Wash. Publ.* 24: 187-203.

Long, J.A., and Evans, H.M., 1922, The oestrous cycle in the rat and its associated phenomena, *Mem. Univ. Calif.* 6: 1-148.

Lwoff, A., Dulbecco, M., Vogt, M., and Lwoff, M., 1955, Kinetics of the release of poliomyelitis virus from single cells, *Virology* 1: 128-139.

Mark, E.L, and Long, J.A., 1912, The living eggs of rats and mice with a description of apparatus for obtaining and observing them, *Univ. Calif. Publ. Zool.* 9: 105-136.

Mather, W.B., 1950, The technique of rabbit blastoderm culture, *Univ. Queensland Pap. Dept. Biol.* 2 (No. 15): 1-8.

Maurer, R.R., Whitener, R.H., and Foote, R.H., 1969, Relationship of *in vivo* gamete aging and exogenous hormones to early embryo development in rabbits, *Proc. Soc. Exp. Biol. Med.* 131: 882-885.

Maurer, R.R., Onuma, H., and Foote, R.H., 1970, Viability of cultured and transferred rabbit embryos, *J. Reprod. Fertil.* 21: 417-422.

Maximow, A., 1925, Tissue-cultures of young mammalian embryos, *Carnegie Inst. Wash. Publ.* 16: 47-113.

McLaren, A., and Biggers, J.D., 1958, Successful development and birth of mice cultivated *in vitro* as early embryos, *Nature (London)* 182: 877-878.

McLaren, A., and Michie, D., 1956, Studies in the transfer of fertilized mouse eggs to uterine foster mothers. I. Factors affecting the implantation and survival of native and transferred eggs, *J. Exp. Biol.* 33: 394-416.

Miller, B.J., and Reimann, S.P., 1940, Effect of DL-methionine and L-cysteine on the cleavage rate of mammalian eggs, *Arch. Pathol.* 29: 181-188.

Mintz, B., 1962a, Experimental study of the developing mammalian egg; removal of the zona pellucida, *Science* 138: 594.

Mintz, B., 1962b, Experimental recombination of cells in the developing mouse egg. Normal and lethal mutant genotypes, *Amer. Zool.* 2: 145 (abs).

Mintz, B., 1964, Formation of genetically mosaic mouse embryos, and early development of "lethal (t^{12}/t^{12}) - normal" mosaics, *J. Exp. Zool.* 157: 273-285.

Naglee, D.L., Maurer, R.R., and Foote, R.H., 1969, Effect of osmolarity on *in vitro* development of rabbit embryos in a chemically defined medium, *Exp. Cell Res.* 58: 331-333.

Onuma, H., Maurer, R.R., and Foote, R.H., 1968, *In vitro* culture of rabbit ova from early cleavage stages to the blastocyst stage, *J. Reprod. Fertil.* 16: 491-494.

Pincus, G., 1927, Comparative study of the chromosomes of the Norway rat (*Rattus norvegius ersel.*) and the black rat (*Rattus rattus L.*), *J. Morphol.* 44: 515-540.

Pincus, G., 1930, Observations on the living eggs of rabbits, *Proc. Roy. Soc. Lond. B Biol. Sci.* 107: 132-167.

Pincus, G., 1936, *The Eggs of Mammals*, Macmillan, New York, pp. 1-160.

Pincus, G., 1937, The metabolism of ovarian hormones, especially in relation to the growth of the fertilized ovum, *Cold Spring Harbor Symp.* 5: 44-56.

Pincus, G., 1941, Factors controlling the growth of rabbit blastocysts, *Amer. J. Physiol.* 133: P412-P413.

Pincus, G., and Werthessen, N.T., 1938, The comparative behaviour of mammalian eggs *in vivo* and *in vitro*, *J. Exp. Zool.* 78: 1-18.

Purshottam, N., and Pincus, G., 1961, *In vitro* cultivation of mammalian eggs, *Anat. Rec.* 140: 51-55.

Smith, A.U., 1949, Cultivation of rabbit eggs and cumuli for phase-contrast microscopy, *Nature (London)* 164: 1136-1137.

Smith, A.U., 1952, Behaviour of fertilized rabbit eggs exposed to glycerol and to low temperatures, *Nature (London)* 170: 374-375.

Smith, A.U., 1953, *In vitro* experiment with rabbit eggs, in: *Mammalian Germ Cells* (G.E.W. Wolstenholme, M.P. Cameron, and S.J. Freeman, eds.), Churchill, London, p. 217.

Squier, R.R., 1932, The living egg and early stages of its development in the guinea-pig, *Carnegie Inst. Wash. Publ.* 21: 225-250.

Staples, R.E., 1967, Development of 5-day rabbit blastocysts after culture at 37°C, *J. Reprod. Fertil.* 13: 369-372.

Tarkowski, A.K., 1961, Mouse chimeras developed from fused eggs, *Nature (London)* 190: 857-860.

Tarkowski, A.K., 1963, Studies on mouse chimeras developed from eggs fused *in vitro*, in: *Symposium on Organ Culture, Natl. Cancer Inst. Monograph No. 11* (C.J. Dawe, ed.), National Cancer Institute, Bethesda, MD, pp. 51-70.

Umbreit, W.W., Burris, R.H., and Stauffer, J.F., 1949, *Manometric Techniques and Tissue Metabolism*, Burgess, Minneapolis, MN, pp. 118-119.

Waddington, C.H., and Waterman, A.J., 1933, The development *in vitro* of young rabbit embryos, *J. Anat.* 67: 356-370.

Washburn, W.W., 1951, A study of the modifications in rat eggs observed *in vitro* and following tubal retention, *Arch. Biol.* 62: 439-458.

Waterman, A.J., 1934, Survival of young rabbit embryos on artificial media, *Proc. Natl. Acad. Sci. USA* 20: 145-146.

Waymouth, C., 1959, Rapid proliferation of sublines of NCTC clone 929 (Strain L) mouse cells in a simple chemically defined medium (MB752/1), *J. Natl. Cancer Inst.* 22: 1003-1016.

Whitten, W.K., 1956a, Culture of tubal ova, *Nature (London)* 177: 96.

Whitten, W.K., 1956b, Physiological control of population growth, *Nature (London)* 178: 992.

Whitten, W.K., 1957a, Culture of tubal ova, *Nature (London)* 179: 1081-1082.

Whitten, W.K., 1957b, The effect of progesterone on the development of mouse eggs *in vitro*, *J. Endocrinol.* 16: 80-85.

Whitten, W.K., and Biggers, J.D., 1968, Complete development *in vitro* of the pre-implantation stages of the mouse in a simple chemically defined medium, *J. Reprod. Fertil.* 17: 399-401.

Whittingham, D.G., 1968, Development of zygotes in cultured mouse oviducts. I. The effect of varying oviductal conditions, *J. Exp. Zool.* 169: 391-398.

Whittingham, D.G., 1975, Fertilization, early development and storage of mammalian ova *in vitro*, in: *The Early Development of Mammals* (M. Balls, and A.E. Wild, eds.), Cambridge University Press, Cambridge, England, pp. 1-24.

Whittingham, D.G., and Biggers, J.D., 1967, Fallopian tube and early cleavage in the mouse, *Nature (London)* 213: 942-943.

Whittingham, D.G., Leibo, S.P., and Mazur, P., 1972, Survival of mouse embryos frozen to -196°C and -296°C, *Science* 113: 247.

Wilmut, I., 1972, Effect of cooling rate, warming rate, cryoprotective agent and stage of development on survival of mouse embryos during cooling and thawing, *Life Sci.* 11 (pt. 2): 1071-1079.

Wilson, I.B., Bolston, E., and Cuttler, R.H., 1972, Preimplantation differentiation in the mouse egg as revealed by microinjection of vital markers, *J. Embryol. Exp. Morphol.* 27: 467-479.

Wolstenholme, G.E.W., and O'Connor, M., 1965, *Preimplantation Stages of Pregnancy*, Churchill, London.

Ziomek, C.A., and Johnson, M.H., 1980, Cell surface interaction induces polarization of mouse 8-cell blastomeres at compaction, *Cell* 21: 935-942.

Chapter 2

CELL POLARITY IN THE PREIMPLANTATION MOUSE EMBRYO

CAROL A. ZIOMEK

1. INTRODUCTION

The union of two highly polarized cells, the sperm and the egg, initiates a series of dramatic cellular transformations that culminate, during the first 4 days of mouse preimplantation development, in the production of a multicellular blastocyst (reviewed by Wiley, Chapter 4) having two distinct and committed tissues. The outer layer of transporting epithelial cells, which surround and generate the blastocoelic cavity, are the trophectodermal cells that will give rise to extraembryonic structures. Attached to the interior of the trophectodermal layer at one end of the blastocoelic cavity is a cluster of relatively undifferentiated cells, the inner cell mass, that will subsequently give rise to the embryo proper. Attempts to elucidate the mechanisms by which these two distinct cell types diverge from a common pathway during the first 3 days of preimplantation development have yielded not only clues as to possible differentiative signals operating in development, but also detailed structural information on the morphology and properties of cells from the early embryo. One such piece of structural information gained from these studies is that many embryonic cells are architecturally polarized both at their cell surface and in their cytoplasmic domain; these cell asymmetries may play important roles in embryonic development.

2. THE UNFERTILIZED EGG

2.1. Methodology

Female mice (CF-1) were superovulated with 5 IU pregnant mare's

Carol A. Ziomek Worcester Foundation for Experimental Biology, 222 Maple Avenue, Shrewsbury, Massachusetts 01545, USA.

serum gonadotropin (PMSG) followed 48 hr later with 5 IU human chorionic gonadotropin (hCG). Egg masses were excised from the oviducts at 16-20 hr post hCG. The egg masses were dissociated in 0.1% hyaluronidase and washed through 3 changes of Hanks balanced salt solution containing 4 mg/ml bovine serum albumin (HBSS + BSA). Although the zona pellucida of the unfertilized egg can be removed mechanically (Wolf *et al.*, 1976), thermally (Cholewa-Stewart and Massaro, 1977) or by incubations in pronase (Mintz, 1962), trypsin, ficin (Smithberg, 1953), chymotrypsin (Smithberg, 1953; Boldt and Wolf, 1982) or 2-mercaptoethanol (Inoue and Wolf, 1974), the technique most commonly employed in our laboratory is a brief exposure to acidic (pH 2.7) Tyrode's solution (Nicolson *et al.*, 1975).

The denuded egg plasma membrane can be labeled with a variety of fluorescent probes including fluorescein and rhodamine-labeled concanavalin A (Con A), succinyl Con-A, trinitrobenzene sulfonic acid followed by a rhodamine Fab fragment of sheep anti-TNP (Wolf and Ziomek, 1983), and a series of fluorescent lipid probes of the indocarbocyanine family ($C_N$dil: Wolf *et al.*, 1981, 1982; Wolf, 1983). Such labeled eggs were then examined for their fluorescent staining patterns and for the mobility parameters of the membrane components by the technique of fluorescence photobleaching and recovery (FPR; reviewed in Wolf and Edidin, 1981). In this technique, an attenuated (non-bleaching) laser beam is focused onto a <1 μm region of the fluorescent labeled cell membrane and the intensity of the fluorescence emanating from this region is quantitated using a photomultiplier. The laser power is then briefly raised 1000-fold, resulting in an irreversible bleaching of some of the fluorescent molecules in that spot. The attenuated laser and photomultiplier measure the rate and extent of the return of fluorescence to the bleached spot (by diffusion of unbleached molecules into the spot and bleached molecules out of the spot). Three mobility parameters can be obtained by this approach: 1) the fluorescence intensity (degree of labeling) in the 1 μm spot, 2) the diffusion coefficient of the fluorescent labeled molecules, and 3) the degree of freedom (% recovery) of the molecules. If no fluorescence returns to the spot, then all the fluorescent labeled molecules are immobile. If fluorescence recovers to only 20% of the original value, 20% of molecules are free to diffuse, whereas 80% are immobilized in some way.

The cytoskeletal arrangement of the egg cytocortex was investigated using antibodies or agents specific for various cytoskeletal elements, on appropriately fixed eggs attached to poly-D-lysine (5 mg/ml in dH_2O, mol. wt. > 540 kD) coated coverslips (Lepire and Ziomek, unpublished). The cytoskeletal elements that we have studied so far include: f-actin, total actin, myosin, cytokeratin and tubulin. We are currently studying the microtubule associated proteins (MAPs).

The egg surface was examined by both SEM analysis of zona-free eggs and TEM analysis of zona-enclosed eggs. Eggs were fixed with 0.3% glutaraldehyde and 0.5% paraformaldehyde in HBSS. For SEM analysis, the zona was removed from fixed embryos by 0.5% pronase, the eggs attached to poly-D-lysine coated coverslips, processed by a modified osmium-thiocarbohydrazide technique (Malick and Wilson, 1975; Adler and Ziomek, 1986a), critical point dried from liquid CO_2, and sputter coated with 10 nm of gold/palladium. For TEM analysis, eggs were processed in suspension and embedded in Spurr's embedding medium.

2.2. Microvillus Distribution

As alluded to in the Introduction, the female gamete is indeed a highly polarized cell. Transmission and scanning electron microscopic studies have detailed the architectural polarity both at the cell surface and in the cytocortex of the mature tubal egg (Thompson *et al.*, 1974; Eager *et al.*, 1976; Wolf *et al.*, 1976; Nicosia *et al.*, 1977; Wabik-Sliz and Kujat, 1979; Longo and Chen, 1985; Allworth, Lepire and Ziomek, unpublished). It has been observed that although 80-90% of the egg surface (Fig. 1a) is covered with a uniform, dense layer of microvilli (3.7 ± 0.3 microvilli/μm^2 of egg surface; 0.9 ± 0.2 μm in length by 0.1 ± 0.03 μm in diameter; Longo and Chen, 1985), one pole of the egg, overlying the second meiotic spindle, is virtually devoid of microvilli (Thompson *et al.*, 1974; Eager *et al.*, 1976; Longo and Chen, 1985; Ziomek and Lepire, unpublished). This polarity in microvillus distribution arises during meiotic maturation of the oocyte and it has been suggested that the microvillus-free region is induced by the meiotic chromosomes or by elements of the meiotic apparatus (Maro *et al.*, 1984; Longo and Chen, 1985). Displacement of the meiotic apparatus from its position beneath the plasma membrane at one end of the egg also displaces the microvillus-free region (Longo and Chen, 1985).

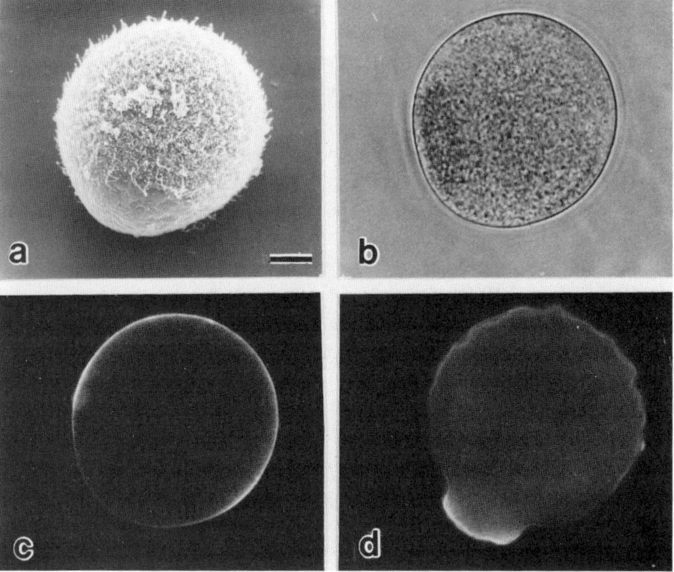

Figure 1. Zona-free unfertilized eggs. (a) In SEM the egg appears uniformly covered with microvilli except for the protruding microvillus-free region (nipple) overlying the second meiotic apparatus (in the lower region of the micrograph); (b and c) phase-contrast and fluorescence micrographs of an egg labeled with FITC-Con A showing the diminished staining of the nipple region; (d) fluorescence micrograph of a fixed egg stained with NBD-phallicidin to determine f-actin localization. Although a cortical ring of actin is seen throughout the egg, the nipple protrusion exhibits the most intense staining for f-actin. Bar = 10 μm (all micrographs at same magnification).

2.3. Membrane Polarity

Polarity of the egg surface has also been demonstrated at the light microscope level utilizing fluorescent-lectins and antibodies. Fluorescent-Con A, succinyl-Con A (S-Con A) and TNBS-antibody stain the microvillus membrane (body) of the egg more intensely than the microvillus-free region or 'nipple' (Fig. 1c). These fluorescent intensities (FI) have been quantitated in the FPR instrument and the ratio of intensities in the two regions determined for S-Con A and TNBS. For S-Con A, we found that FI (body)/FI (nipple) was $\geq$ 4.1 $\pm$ 1.1, while for TNBS, FI (body)/FI (nipple) = 2.0 $\pm$ 0.3 (Wolf and Ziomek, 1983). If all membrane components are homogeneously distributed in the egg plasma membrane, then the ratio of FI (body)/FI (nipple) should be equal to a constant, the surface area (body)/surface area (nipple) within the beam. Based on the calculations of Longo and Chen (1985), the microvilli effectively double the surface area of the egg and the ratio should be approximately 2.0. Therefore, although the observed TNBS labeling ratio of approximately 2.0 may reflect membrane amplification due to microvilli on the egg body, Con A receptors (ratio $\geq$ 4.1) show a true concentration difference between the two regions, in addition to surface amplification due to microvilli. It is also interesting to note that the unfertilized egg exhibits a polarity in sperm binding (Johnson *et al.*, 1975; Nicosia *et al.*, 1977) and incorporation (Nicosia *et al.*, 1977), both of which occur exclusively on the microvillus main body. It is, as yet, unknown whether this polarity in sperm binding and/or incorporation is due to a lack of sperm receptors or to the absence of microvilli (sperm interact with microvilli prior to sperm-egg fusion) from the nipple region.

In addition to the existence of a true concentration gradient in the distribution of some membrane components, there is also a difference in at least one of the mobility parameters (% recovery) measured by FPR for protein components in the two regions. The diffusion coefficients for the S-Con A and TNBS labels were determined to be $1.5 \times 10^{-10} \pm 0.8$ cm^2/sec and $0.7 \times 10^{-10} \pm 0.1$ cm^2/sec, respectively, with no difference between the body and nipple regions. The fraction of membrane label which was free to diffuse on the nipple was much greater than that measured on the body (S-Con A, 57% $\pm$ 22 *vs.* 22% $\pm$ 3; TNBS, 71% $\pm$ 4 *vs.* 39% $\pm$ 3), suggesting some type of protein anchorage in the microvillus membrane. Although a complete understanding of the factors that govern membrane protein diffusibility has yet to be attained, cytoplasmic factors, in particular the cytoskeleton, have been proposed to play a role in the deviation of protein mobility parameters from ideality (Tank *et al.*, 1982; Wu *et al.*, 1982).

2.4. Cytoplasmic Polarity

Underlying this surface polarity is a cytoplasmic polarity in the localization of organelles, granules and cytoskeletal elements. Cortical granules and Golgi complexes are localized to the cell cortex beneath the microvillus membrane and are absent from the cortex of the microvillus-free, meiotic spindle region (Nicosia *et al.*, 1977). In addition, a 100 to 400 μm thick layer of fine filamentous material is restricted to the cortical cytoplasm of the nipple region (Thompson *et al.*, 1974; Nicosia *et al.*, 1977; Longo and Chen, 1985). Labeling of microfilaments of mouse eggs by immunofluorescence or NBD-phallicidin (Maro *et al.*, 1984; Longo and Chen, 1985; Allworth and Ziomek, unpublished) revealed a cortical ring of f-actin with the most intense staining in the microvillus-free area (Fig. 1d) coincident with this thick filamentous layer. It therefore appears that, contrary to the suggestions of Tank *et al.* (1982) and Wu *et al.* (1982), there is no simple correspondence in the number of microfilaments underlying a given region of plasma membrane and the fractional diffusion of components in that region. In our case, the region having the highest protein mobile fraction also has the greatest concentration of microfilaments. Upon exposure to either cytochalasin B or D (two microfilament disruptors), the protruding nipple region collapses with a loss of the f-actin staining polarity and a reduction in surface polarity as measured by the binding of fluorescent Con A. We are currently measuring the mobility parameters of the S-Con A receptors in cytochalasin treated eggs by FPR. Colchicine, a microtubule disruptor, has no effect on nipple protrusion, microvillus distribution (Longo and Chen, 1985; Allworth and Ziomek, unpublished) or surface polarity of Con A binding (Allworth and Ziomek, unpublished). The cytoplasmic localization of non-spindle tubulin, myosin and cytokeratin do not follow the same pattern as that described for actin. These components have a rather uniform cortical distribution in the egg and, although they do stain the nipple region, it is with an intensity similar to that seen for the body (Allworth and Ziomek, unpublished). The absence of intermediate filaments from early preimplantation embryos has been reported in the literature (Jackson *et al.*, 1980). Our positive staining with an affinity purified antibody to cytokeratin supports the more recent observations of other workers (Lehtonen *et al.*, 1983; Oshima *et al.*, 1983) who detected both the synthesis of the cytokeratins and the presence of intermediate-sized filaments in early embryos.

3. THE FERTILIZED EGG

3.1. Methodology

The methods used to probe the fertilized egg are identical to those detailed under the unfertilized egg, except that fertilized eggs were obtained from females mated overnight with males. Successful fertilization was judged by the presence in the eggs of 2 pronuclei and/or 2 or 3 polar bodies.

3.2. Microvillus and Cytoplasmic Polarity

Scanning and transmission electron microscopy (Thompson *et al.*, 1974: Eager *et al.*, 1976; Nicosia *et al.*, 1977; Wolf *et al.*, 1979) of the fertilized egg reveals a uniform distribution of microvilli over the egg surface, after the extrusion of the microvillus-free nipple region of the unfertilized egg as the second polar body. There appears to be no detectable cytoplasmic polarity, except that which might be dictated by the position of the sperm mitochondrial complex in the egg cytoplasm. Most of the cortical granules have been exocytosed prior to and subsequent to fertilization and the Golgi complexes and microfilaments appear to be uniformly distributed (Nicosia *et al.*, 1977; Wolf *et al.*, 1979; Maro *et al.*, 1984).

3.3. Membrane Parameters

There is a loss of cellular polarity and a change in a variety of membrane parameters associated with fertilization. The membrane of the fertilized egg exhibits an increased permeability to glycerol (Jackowski *et al.*, 1980) and an increased ability of Con A receptors to be patched by ligand (Johnson and Calarco, 1980). In addition, fertilized eggs are more agglutinable by tetravalent Con A than unfertilized eggs (Siracusa *et al.*, 1978). There is, however, no polarity of binding of Con A, TNBS or a variety of lectins and antibodies to the fertilized egg: labeling is uniform in all cases.

As a result of the insertion of new membrane into the egg surface by the cortical reaction or by other membrane changes triggered by fertilization (Nicosia *et al.*, 1977; Wolf, 1978; Johnson and Edidin, 1978; Johnson and Calarco, 1980), it might be expected that there are changes in the diffusion of egg membrane components. FPR measurements on fertilized eggs using the protein probes, succinyl-Con A and TNBS (Wolf and Ziomek, 1983) and a variety of lipid probes (Wolf *et al.*, 1981, 1982), has revealed that there is no generalizable effect on membrane protein or lipid diffusion associated with fertilization. This is in direct contrast to the report of Johnson and Edidin (1978) who found reduced protein and lipid diffusions after fertilization of mouse eggs. In our studies, TNBS showed an increase in diffusion coefficient (from 0.73 ± 0.1 to 2.2 ± 0.5) and a decrease in the mobile fraction (from 39% $\pm$ 3 to 21% $\pm$ 2) after fertilization, whereas for S-Con A, there was no significant change in either parameter. This latter finding suggests that the ability of membrane proteins to patch and their freedom to diffuse in the plane of the membrane do not necessarily correlate. Other mechanisms not requiring diffusion may play a role in ligand induced protein patching, particularly in view of the observation that sodium azide can block the patching of Con A receptors in mouse embryos (A.H. Handyside, personal communication; Ziomek and Wolf, unpublished). Diffusionally governed patching is insensitive to azide inhibition. The lipid probes have yielded results which suggest that bulk membrane viscosity does not change with fertilization but that an alteration in the ensemble of lipid domains does occur. The magnitude and direction of the change is highly dependent upon the alkyl chain length of the lipid probe (reviewed in Wolf, 1983). The developmental significance of these changes has not been determined.

4. 2-CELL AND 4-CELL STAGES

4.1. Methodology

Two-cell embryos were flushed with HBSS + BSA from the oviducts of successfully mated CF-1 females at 38 to 54 hr and 4-cell embryos at 54 to 62 hr post hCG. We have determined that the CF-1 mouse which we use (Harlan-Sprague Dawley) has an unusually long 2-cell cycle. Although division to the 2-cell stage begins around 36 hr post hCG, division to the 4-cell stage does not commence until 54 hr post hCG. Embryos isolated before 42 hr post hCG exhibit a block to further development *in vitro* similar to that seen in other mouse strains but not in certain F1 hybrids (Whittingham and Biggers, 1967; Whitten and Biggers, 1968; Abramczuk *et al.*, 1977; Muggleton-Harris *et al.*, 1982; Goddard and Pratt, 1983).

4.2. Morphological and Cytoplasmic Asymmetries

The free surfaces of the blastomeres of 2-cell (Fig. 2a) and 4-cell embryos (Reeve and Ziomek, 1981) are covered with a uniform layer of microvilli. In the region of cell-cell contact, the spherical shape of the blastomeres is decidedly flattened and exhibits a higher microvillus density (Thompson *et al.*, 1974). Ultrastructural studies have also shown that sometime during the 2-cell stage, blastomeres acquire the ability to phagocytose supernumerary sperm (Thompson and Zamboni, 1974), suggesting some type of alteration in membrane and/or cytoplasmic properties at this stage.

Cytoplasmic polarity in cytoskeletal organization has been reported for α-spectrin (Sobel and Alliegro, 1985) and myosin (Sobel, 1983a,b, 1984). In the early 2-cell embryo, an antibody to α-spectrin of avian erythrocytes labeled only the cytoplasmic region underlying areas of cell-cell contact. This polarized labeling pattern disappeared during the late 2-cell stage and a continuous cortical ring of labeling was observed in the 4-cell embryo. Myosin, on the other hand, failed to label the contact region between cells but intensely labeled the cytoplasm beneath the free surface of the blastomeres. Unlike the

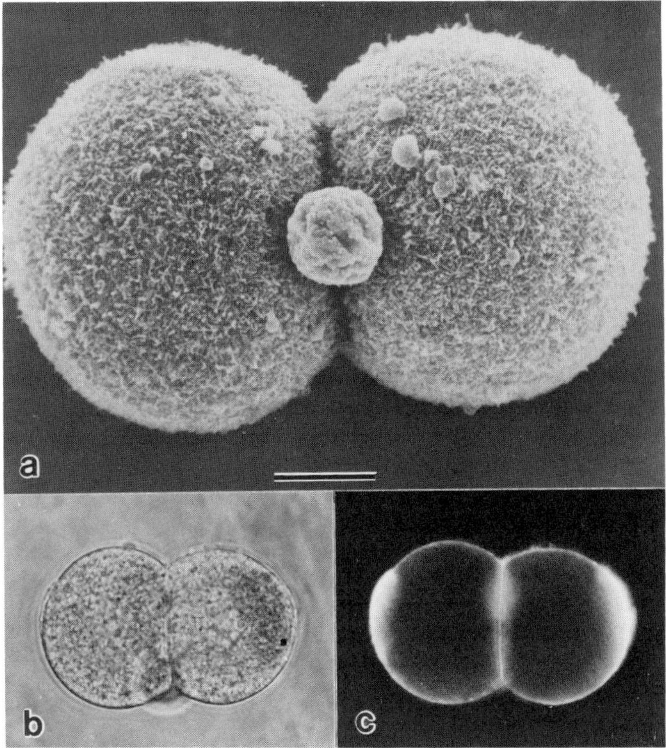

Figure 2. Zona-free 2-cell mouse embryos. (a) SEM micrograph showing uniform distribution of microvilli over the surface of 2-cell blastomeres. A polar body is seen in the region of cell-cell contact. Bar = 10 μm. (b and c) Phase and fluorescence photomicrographs of a late 2-cell mouse embryo labeled with FITC-Con A showing the polarized Con A binding that occurs during this period (optical magnification for b and c x 475).

case for mechanical zona removal, the use of acidic Tyrode's solution or pronase for zona removal abolished the polarized staining pattern with the antibody to myosin. This finding suggests that these treatments have a deleterious effect on some cellular processes and that results obtained using these methods should always be compared to those obtained after mechanical zona removal.

4.3. Membrane Polarity

During the early 2-cell stage, the embryo binds ligands such as FITC-Con A uniformly over the surface of the blastomeres. During the course of a timing experiment in which pregnant females were killed at 2 hr intervals over a period of 3 days, their embryos were examined for the pattern of FITC-Con A binding and for their ability to develop further *in vitro*. We found that a transient polarity in FITC-Con A binding (Fig. 2c) developed late in the 2-cell stage (49 to 53 hr post hCG). In some cases, there appeared to be some type of membrane blebbing in the region of the pole. We are currently attempting to determine the nature of this 2-cell polarity, its developmental significance, whether it is ligand induced or simply reflects a morphological alteration of the free surface of the blastomeres, and whether it might represent an area for the insertion/deletion of new/old membrane or membrane components prior to cell division. Should it be discovered that this polarity is ligand induced, it would still be significant since its restriction to such a short period of the 2-cell cycle would suggest that it reflects some type of membrane or cytoskeletal alteration during that period. The FITC-Con A binding to 4-cell embryos was uniform over the surface of the blastomeres throughout the entire 4-cell stage.

Expression of the embryonic genome is first detected at the 2-cell stage (Flach *et al.*, 1982; reviewed in Johnson *et al.*, 1984; see also Chapter 7). During the late 2-cell stage, several proteins make their first appearance in the blastomere plasma membrane. Two such proteins are the ectoenzymes, alkaline phosphatase and 5'-nucleotidase (Vorbrodt *et al.*, 1977; Mulnard and Huygens, 1978; Nizeyimana-Rugina and Mulnard, 1979; Izquierdo *et al.*, 1980; Izquierdo and Ebensperger, 1982). It has been reported that both of these enzymes have a polarized distribution in the membrane of the early embryo. The localization of alkaline phosphatase (AP) activity has been the subject of much controversy. Mulnard and Huygens (1978) and Izquierdo *et al.* (1980) determined that AP was restricted to the membrane in regions of cell-cell contact and was absent from the free surface (facing the external milieu) of the blastomeres. On the other hand, Vorbrodt *et al.* (1977) reported that the enzyme occurred over the entire surface of each cell of the embryo.

Due to our interest in membrane polarity in the embryo, we decided to try to resolve this conflict in the reported appearance and localization of AP. Using a fluorescent histochemical stain for AP (Ziomek and Lepire, 1984; Ziomek and Lepire, 1986) and an antibody directed against human placental alkaline phosphatase (Ziomek and Lepire, 1986), we have confirmed that alkaline phosphatase does, indeed, appear on the cell surface at the late 2-cell stage, but that its distribution is uniform over the surface of both 2- and 4-cell stage embryos, rather than being polarized as was previously suggested. [Later stage embryos also exhibit AP activity on both free and apposed blastomere surfaces.] The 5'-nucleotidase has also been the subject of some controversy. Vorbrodt *et al.* (1977) and Izquierdo and Ebensperger (1982) found the activity was restricted to the apposed contact surfaces of blastomeres and was absent from the free surface of the cells, while Nizeyimana-Rugina and Mulnard (1979) reported that 5'-nucleotidase was distributed on both the free and apposed surfaces. The resolution of this conflict about localization requires further experimentation, but it would appear that, except for the transient polarity in FITC-Con A binding and a polarity in monoclonal antibody binding (Hahnel and Eddy, 1982) seen at the late 2-cell stage, there are few established polarities in the distribution of membrane components in either the 2- or 4-cell embryo.

5. THE 8-CELL STAGE EMBRYO

5.1. Methodology

Late 4-cell embryos were flushed from the oviducts of pregnant females at 61 hr post hCG. Zonae were removed with acidic Tyrode's and the embryos were dissociated to single cells (1/4-cells) after 15 min exposure to Ca^{2+}-free Medium 16 containing 6 mg/ml BSA (M16 + BSA: Pratt *et al.*, 1982). The single cells were cultured in microdrops of M16 + BSA, the microdrops were examined hourly for cell division, and the newly divided cells were harvested and designated as natural 2/8 pairs. These pairs were either cultured further as natural pairs or disaggregated to single cells (1/8), as above. The single 1/8 cells could be cultured alone, reaggregated to a companion 1/8 cell (1/8 + 1/8) or to other embryonic cells (Ziomek and Johnson, 1980; Johnson and Ziomek, 1981a; Adler and Ziomek, 1986a,b), tumor cells or to various beads or dishes. After 10 hr in culture, the 8-cell blastomeres were labeled with FITC-Con A and/or subjected to analysis of microvillus distribution by SEM.

5.2. Change in Microvillus Distribution

Both TEM and SEM analysis of various age 8-cell embryos revealed that, although the early 8-cell embryo is composed of cells uniformly covered with microvilli, the blastomeres of the late 8-cell embryo are highly polarized (Ducibella *et al.*, 1977; reviewed in Reeve and Ziomek, 1981). Permanent structural microvilli become restricted to a single outward facing pole on the free surface (facing the external milieu) of each blastomere. Although there are microvilli in the cell-cell contact region, they appear to be labile and disappear upon embryo dissociation. Coincident with the polarization of microvillus distribution, the embryo undergoes a dramatic transformation in overall morphology termed "compaction" (Lewis and Wright, 1935; Ducibella and Anderson, 1975). The 8 normally spherical blastomeres flatten upon each other and assume a blunted wedge-shaped appearance.

5.3. Cytoplasmic Asymmetries

In addition to the polarization of microvillus distribution, the transition from early to late 8-cell embryo is hallmarked by the development of a polarity in the cytoplasmic domain of each blastomere. Reeve (1981) has demonstrated that cells of the early 8-cell embryo internalized exogenously-added horseradish peroxidase (HRP) and exhibited a random display of HRP-containing vesicles in their cytoplasm. However, late 8-cell embryos, under the same conditions, developed a cytoplasmic polarity in the distribution of the HRP-containing vesicles, which became restricted to a column extending from the basal nucleus to the apical microvillus pole.

It was observed that nuclei were randomly located in the blastomeres of the early 8-cell embryo but migrated to a basal position in each cell during the course of the 8-cell stage (Reeve and Kelly, 1983). It has also been reported that mitochondria and microtubules become aligned in the cytocortex parallel to the plasma membrane in the cell-cell contact region between blastomeres, although the microtubules were more randomly arranged near the apical microvillus pole of the cell (Ducibella and Anderson, 1975).

The cytoskeletal elements that have been most studied in the 8-cell embryo are actin (Johnson and Maro, 1984), myosin (Sobel, 1983a,b, 1984) and spectrin (Sobel and Alliegro, 1985). Cytoplasmic actin, which is homogeneously distributed in the early 8-cell embryo, becomes excluded from contact regions and restricted to the apical portion of each blastomere during the 8-cell stage. Likewise, myosin was excluded from beneath the cell-cell contact region and was restricted to the apical portion of the cells. Dissociation of the cell contacts caused a reversion of the myosin localization to a uniform pattern. Spectrin distribution was observed to be non-polarized with the spectrin organized in a continuous peripheral layer in each blastomere.

5.4. Membrane Polarity

A variety of fluorescent lectins and antibodies have been used to demonstrate the dramatic transformation of the plasma membrane of the 8-cell blastomere termed "polarization". While the cells of the early 8-cell embryo bind these agents uniformly over their surfaces, blastomeres of the late 8-cell embryo display a highly polarized staining pattern with the most intense fluorescence restricted to a single outward facing pole on each cell (Handyside, 1980; Reeve and Ziomek, 1981). The transition from the non-polar to the polar staining pattern appears to be a progressive process since intermediate levels of polarity (1/2 or 3/4 cells bright) are observed during the mid-8-cell stage with a shift in the population toward fully polar with advanced age (Reeve and Ziomek, 1981).

Although the pole of fluorescent ligand binding and the pole of microvilli appear to coincide, it is not yet clear whether the fluorescent pole results merely from membrane amplification at the cell apex due to microvilli or whether a true concentration gradient in the distribution of components in the membrane also exists. In an attempt to distinguish between these two possibilities, we have used the FPR technique to determine the mobility parameters in the intact embryo of the lipid probe, C_{16}diI, the general protein label, TNBS (Ziomek *et al.*, 1984), and a more specific protein probe, a monovalent antibody directed against alkaline phosphatase. The fluorescence intensity ratio was determined for the apex (free surface) *vs.* the base (cell-cell contact region) of the blastomeres. In the early 8-cell embryo, the FI (apex)/FI (base) was 1.11 for C_{16}diI, 1.22 for TNBS and 1.3 for AP. The late 8-

cell embryo had ratios of 1.45 for C_{16}dil, 3.16 for TNBS, and 3.1 for alkaline phosphatase, suggesting a true protein concentration difference in the two regions. The diffusion coefficients for all 3 probes showed little difference between the early *vs.* late or apical *vs.* basal regions. The percentage recoveries were identical between apical and basal surfaces on early 8-cell embryos, but on late 8-cell embryos the mobile fraction free to diffuse in the basal membrane was significantly greater than the mobile fraction of the apical microvillus membrane (Ziomek, Moynihan and Wolf, in preparation).

5.5. Induction of 8-Cell Polarity

Due to the asynchrony of cell division from the 4-cell to 8-cell stages (Ziomek and Johnson, 1980), it is impossible to probe the processes responsible for the induction of 8-cell polarization by the use of intact embryos. The realization that synchronized pairs of 8-cell blastomeres could be obtained by the 1/4 cell culture technique described in the methodology, and the finding that polarity was induced in both cells of a pair during subsequent culture (Ziomek and Johnson, 1980) paved the way for *in vitro* aggregation experiments designed to probe the mechanisms involved in the induction of 8-cell polarization. Single 8-cell blastomeres (1/8) cultured in isolation failed to polarize. Pairs of 8-cell blastomeres, either natural (2/8) or reaggregated (1/8 + 1/8), that were cultured for 8 to 10 hr displayed high levels of polarity (85-95%). In addition, the position of the pole of intense FITC-Con A binding was always directly opposite the region of cell-cell contact (Fig. 3a,b). When 3 or more cells were aggregated together in various geometrical arrays, all contacts were respected in determining the position of the poles on each cell, and the pole occurred in that area which was most distant from all contacts. If a cell was given symmetrical contacts, such as surrounding a single cell with 20 companion cells, polarity was suppressed in the central cell (Johnson and Ziomek, 1981a). These results strongly suggest that polarity is not a cell autonomous event but that it depends upon asymmetric cell-cell contacts (Ziomek and Johnson, 1981).

The specificity of the cell-cell interaction was explored utilizing aggregates made from non-polar 1/8 cells and single cells from earlier embryonic stages. Unfertilized and fertilized eggs could not induce the polarization of companion 8-cell blastomeres, whereas 30% of 2-cell, 70% of 4-cell and 90-100% of 8- and 16-cell blastomeres were able to induce a companion 8-cell blastomere to polarize (Johnson and Ziomek, 1981a). More recently, we have tested both the inner and the outer cells from later embryonic stages for their inducing ability (Adler and Ziomek, 1986a). [From the 16-cell stage onward, there are at least two morphologically and behaviorally distinct cell populations in the embryo, the inner, non-polar cells and the outer, polar cells (Johnson and Ziomek, 1981b, 1982, 1983; Ziomek and Johnson, 1981, 1982).] Using an immunosurgical isolation procedure (Solter and Knowles, 1975), inside cell clusters from late morula and inner cell masses from blastocysts expressed high levels (78% and 82%, respectively) of 8-cell polarity inducing ability, comparable to control 1/8 + 1/8 pairs (85%). When 1/8 cells were aggregated to the outer cell layer of the intact morula, 73% of the 8-cell blastomeres polarized in response to this contact. The outer cells of the blastocyst can be divided into two subpopulations based on their positions and

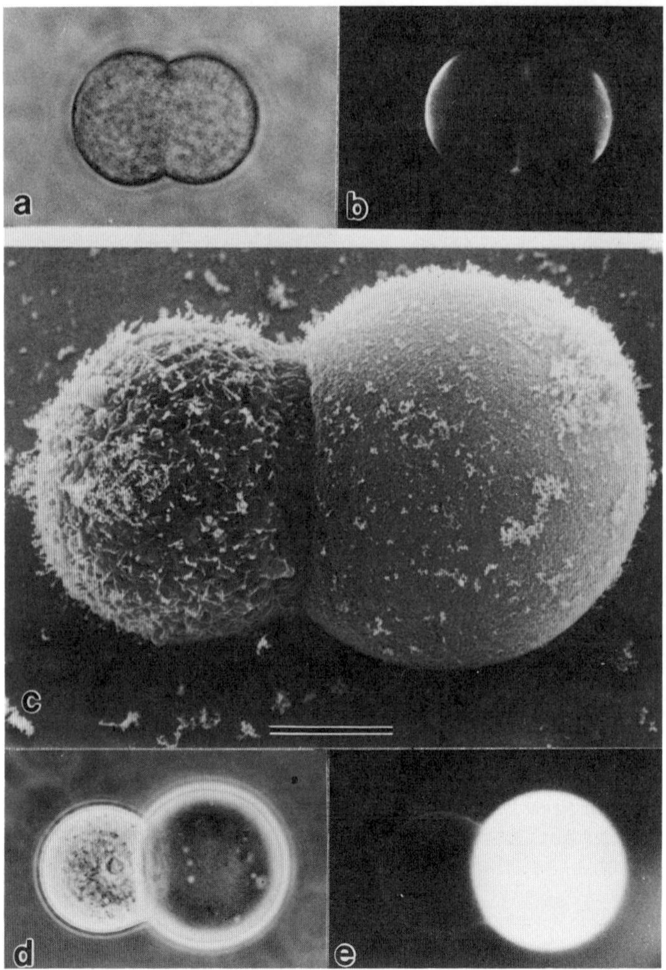

Figure 3. Eight-cell stage blastomeres. (a and b) Phase and fluorescence photomicrographs of a natural 2/8 pair cultured for 10 hr prior to decompaction in Ca^{2+}-free medium and staining with FITC-Con A. Note that polarity develops in both cells opposite to the region of contact (optical magnification x 535). (c, d and e) Scanning, phase and fluorescence micrographs of a 1/8 cell attached to a Con A-bead for 10 hr prior to fixation or staining with FITC-Con A. Note the extensive area of cell-bead apposition and the absence of a polarized microvillus distribution or polarized Con A binding on the 1/8 cell (optical magnification x 515); bar = 10 μm (3c only).

properties. The proliferative cells that overlie the inner cell mass are the polar trophectoderm. The cells that surround the blastocoelic cavity, which are the most differentiated cells of the blastocyst, are the mural trophectoderm. We aggregated 1/8 cells to both the mural and polar trophectoderm layer of intact early, late and hatched blastocysts. The polar trophectoderm expressed moderate but decreasing levels of inducing ability in early, late and hatched blastocysts (65%, 56% and 41%, respectively). The mural trophectoderm had little inducing ability at any stage (8%, 7%, and 17%, respectively).

This result suggests that inducing ability decreases with differentiation. The hierarchy of differentiation seems to parallel the hierarchy of inducing capacity, with inside cell clusters and outer morula cells ≥ ICM > polar trophectoderm > mural trophectoderm (Adler and Ziomek, 1986a). This hypothesis has gained additional support from the our recent observations that the teratocarcinoma cell line F-9 has 8-cell polarity inducing ability and that this ability is reduced when the F-9 cells are induced to differentiate in the presence of retinoic acid (Adler and Ziomek, 1986b).

During the course of these experiments, it was noted that in some cases, single 1/8 blastomeres adhered to the bottom of the culture dish. When these cells were scored for FITC-Con A binding, they displayed a polarized binding pattern. A striking feature of this polarization was that, in all cases, the pole formed directly opposite a midbody remnant left behind when the cytoplasmic bridge between sister blastomeres was broken. It appears that the midbody remnant is very sticky, adheres to the culture dish and might be perceived by the 8-cell blastomere as an inducing companion (Ziomek and Johnson, 1980; Ziomek and Lepire, in preparation). To explore the possibility that some non-specific interactions might be inductive for polarization, we tested a variety of culture dishes for their adhesiveness for single 8-cell blastomeres. In addition, we aggregated non-polar 1/8 cells to a variety of beads (lectin or protein conjugated, charged and glass) and fixed cells that were induction competent in the living state. The least adhesive dish that we tested was the Costar® tissue culture dish on which only 5 to 13% of the cells were adherent and/or polarized after 10 hr of culture. None of the beads (Con A, lotus tetragonolobus, wheat germ agglutinin, peanut agglutinin, lentil lectin, alkaline phosphatase, charged polystyrene, glass) or fixed cells (1/4 or 1/8) was able to induce a companion 8-cell blastomere to polarize, even though in some cases, adhesion of the cell to the bead was so strong (Fig. 3c-e) that attempts to remove the bead resulted in cell lysis (Adler *et al.*, 1985; Ziomek and Lepire, in preparation). Therefore, it appears that the contact required for successful polarity induction is highly specific.

What else do we know about the inductive interaction? It is clear that gap junctional communication (see Chapter 3) is not involved in polarity induction since junctions do not form between 4-cell and 8-cell blastomeres (Goodall and Johnson, 1982, 1984) and yet this interaction induces polarity in the 8-cell blastomere. Small but measurable transcellular ion currents have been detected in polarized 8-cell blastomeres (Nuccitelli and Wiley, 1985). It has been suggested that these currents may be involved in the establishment of blastomere polarity analogous to their role in effecting cellular asymmetries in other types of embryos (Nuccitelli, 1983). Although the intriguing report of Shirayoshi *et al.* (1983) suggested that the calcium-dependent cell adhesion system played a role in the polarity induction process, their results are open to question since polarity was measured solely by Con A binding at a single time point with no SEM analysis of microvillus distribution.

Previously, it was shown that the extent of cell-cell contact required for successful polarity induction need not be extensive since polarity could develop in intact zona-enclosed embryos cultured in Ca^{2+}-free medium (Pratt *et al.*, 1982), which effectively blocks compaction. Recent experiments in our laboratory have addressed the question of the length of contact time required for successful induction. Timed, natural 2/8 pairs were cultured for 0, 2, 4, 6, 8 and 10 hr as natural pairs before disaggregation and further cultured for 10,

8, 6, 4, 2, and 0 hr, respectively, as single 1/8 cells. At each time point, some natural 2/8 pairs were stained with FITC-Con A to determine the baseline level of polarity for that contact period. We found that continuous cell contact was required for successful induction until such time that the poles were fully established. Cells that were only 1/2 or 3/4 polarized reverted back to a non-polar morphology upon subsequent culture without contacts. Fully established poles remained so, even upon subsequent single cell culture. Although these experiments suggest that continuous cell contact is required during the period of pole establishment, they do not give us any information on the actual time required to establish a pole. To explore this question, we disaggregated newly-formed pairs of 2/8 cells to single 1/8 cells and cultured the non-polar 1/8 blastomeres for 0, 2, 4, 6, 8, and 10 hr before aggregating them to companion 8-cell blastomeres of the equivalent age for an additional 10, 8, 6, 4, 2, and 0 hr of culture. Surprisingly, we found that 2 hr of contact, such as in the "8 hr single plus 2 hr as pair" case, was sufficient to induce and establish polarity in 79% of the cells. [A total of 58 cells were scored, with 31% fully polar, 26% 1/2 polar, 22% 3/4 polar and 17% non-polar.] In addition, we found that compaction occurred almost immediately after aggregation of 1/8 cells greater than 4 hr of age, indicating that the ability to compact, as well as the ability to polarize, develops in these cells in a contact-independent fashion (Ziomek and Lepire, in preparation).

5.6. Establishment and Maintenance of Polarity

The mechanisms involved in the actual physical reorganization of the cell at polarization are still for the most part unknown. Disruptors of microfilaments and microtubules (Pratt *et al.*, 1982; Sutherland and Calarco-Gillam, 1983; Maro and Pickering, 1984; Johnson and Maro, 1984) alter the polarity in some respects but do not totally block its occurrence. An inhibitor of protein glycosylation, tunicamycin, delays but does not inhibit polarization. The only agent that we have found that inhibits polarization when applied at the 8-cell stage is sodium azide, which interferes with the energy system of the cell. Sodium vanadate, an inhibitor of ATP-dependent reactions, inhibits 8-cell polarization if it is applied at the 4-cell stage (it does not inhibit cell division), but not if applied at the 8-cell stage (Lepire and Ziomek, unpublished). It would, therefore, appear that polarization is an energy-dependent process that involves microfilaments and microtubules in some aspects of the cellular reorganization.

Once established, polarity appears to be stable even if the cell is cultured further in isolation. None of the agents discussed above disrupt the pole once it is established.

6. DEVELOPMENTAL SIGNIFICANCE OF EMBRYONIC POLARITIES

Although cellular polarity is found in the unfertilized egg and transiently in the 2-cell embryo, it does not become permanently established during preimplantation development until the 8-cell stage. The polarity of the egg may function to safeguard the egg from loss of the genetic contribution of the sperm, which might occur if the sperm bound to and became incorporated at the site of polar body extrusion, the microvillus-free nipple region. Similarly, cortical granule contents and Golgi complexes would be similarly lost if they were located in the nipple region. Based on our limited knowledge of the 2-cell stage, the transient polarity at the late 2-cell stage would seem to serve no useful purpose but may prove important as we gain a more thorough understanding of the changes occurring in the membrane at this time.

The permanent cellular polarity established at the 8-cell stage may be crucial for successful embryonic development since it provides the foundation for the formation of the two distinct cell lineages leading to the trophectoderm and inner cell mass of the blastocyst. Polar 8-cell stage blastomeres divide to give rise to either two polar progeny or one polar and one non-polar offspring. These polar and non-polar 16-cell blastomeres have different behavioral and morphological properties and come to occupy outer and inner positions, respectively, in the 16-cell embryo (Johnson and Ziomek, 1981b, 1982, 1983; Ziomek and Johnson, 1981, 1982). The properties of these polar and non-polar 16-cell blastomeres are reinforced by their positions in the compacted 16-cell embryo, such that polar, outer cells give rise to descendants that preferentially, but not exclusively, occupy the trophectoderm layer of the blastocyst, whereas inner, non-polar cell progeny preferentially, but not exclusively, occupy the inner cell mass.

ACKNOWLEDGEMENTS

The author would like to thank Merry Lepire, Ann Allworth, Margaret Moynihan, Dr. Richard Adler, and Dr. David Wolf, for permission to quote their unpublished data. This research was supported in the author's laboratory by NICHHD Grant HD 17674, Cancer Center Grant P30 12708 and institutional grants to the Worcester Foundation from the Andrew Mellon Foundation and the Edward John Noble Foundation.

7. REFERENCES

Abramczuk, J., Solter, D., and Koprowski, H., 1977, The beneficial effect of EDTA on development of mouse one-cell embryos in chemically defined medium, *Dev. Biol.* 61: 378-383.

Adler, R.A., and Ziomek, C.A., 1986a, Cell specific loss of polarity-inducing ability by later stage mouse preimplantation embryos, *Dev. Biol.* 114, in press.

Adler, R., and Ziomek, C.A., 1986b, Blastomere polarization by embryonal carcinoma cells, *Biol. Bull.*, in press.

Adler, R., Lepire, M., and Ziomek, C.A., 1985, 8-cell blastomere polarity-inducing ability of a variety of embryonic cells and non-cellular materials, *J. Cell Biol.* 101: 343a.

Boldt, J., and Wolf, D.P., 1982, A rapid high yield method for preparing zona-free mouse eggs, *J. Cell Biol.* 97: 184a.

Cholewa-Stewart, J., and Massaro, E.J., 1972, Thermally induced dissolution of the murine zona pellucida, *Biol. Reprod.* 7: 166-169.

Ducibella, T., and Anderson, E., 1975, Cell shape and membrane changes in eight cell mouse embryo: prerequisite for morphogenesis of the blastocyst, *Dev. Biol.* 47: 45-58.

Ducibella, T., Ukena, T., Karnovsky, M.J., and Anderson, E., 1977, Changes in cell surface and cortical cytoplasmic organization during early embryogenesis in the preimplantation mouse embryo, *J. Cell Biol.* 74: 153-167.

Eager, D.D., Johnson, M.H., and Thurley, K.W., 1976, Ultrastructural studies on the surface membrane of the mouse egg, *J. Cell Sci.* 22: 345-353.

Flach, G., Johnson, M.H., Braude, P.R., Taylor, R.A.S., and Bolton, V.N., 1982, The transition from maternal to embryonic control in the 2-cell mouse embryo, *The EMBO Journal* 1: 681-686.

Goddard, M.J., and Pratt, H.P.M., 1983, Control of events during early cleavage of the mouse embryo: an analysis of the "2-cell block", *J. Embryol. Exp. Morphol.* 73: 111-123.

Goodall, H., and Johnson, M.H., 1982, Use of carboxyfluorescein diacetate to study formation of permeable channels between mouse blastomeres, *Nature (London)* 295: 524-526.

Goodall, H., and Johnson, M.H., 1984, The nature of intercellular coupling within the preimplantation mouse embryo, *J. Embryol. Exp. Morphol.* 79: 53-76.

Hahnel, A.C., and Eddy, E.M., 1982, Three monoclonal antibodies against cell surface components on early mouse embryos, *J. Cell Biol.* 95: 156a.

Handyside, A.H., 1980, Distribution of antibody- and lectin-binding sites on dissociated blastomeres from mouse morulae: evidence for polarization at compaction, *J. Embryol. Exp. Morphol.* 69: 99-116.

Inoue, M., and Wolf, D.P., 1974, Comparative solubility properties of the zonae pellucidae of unfertilized and fertilized mouse ova, *Biol. Reprod.* 4: 558-565.

Izquierdo, L., and Ebensperger, C., 1982, Cell membrane regionalization in early mouse embryos as demonstrated by 5'-nucleotidase activity, *J. Embryol. Exp. Morphol.* 69:115-126.

Izquierdo, L., Lopez, T., and Marticorena, P., 1980, Cell membrane regions in preimplantation mouse embryos, *J. Embryol. Exp. Morphol.* 59: 89-102.

Jackowski, S., Leibo, S.P., and Mazur, P., 1980, Glycerol permeabilities of fertilized and unfertilized mouse ova, *J. Exp. Zool.* 212: 329-341.

Jackson, B.W., Grund, C., Schmid, E., Burki, K., Franke, W.W., and Illmensee, K., 1980, Formation of cytoskeletal elements during mouse embryogenesis, *Differentiation* 17: 161-179.

Johnson, L.V., and Calarco, P.G., 1980, Mammalian preimplantation development: the cell surface, *Anat. Rec.* 196: 201-219.

Johnson, M.H., and Edidin, M., 1978, Lateral diffusion in plasma membrane of mouse egg is restricted after fertilization, *Nature (London)* 272: 448-450.

Johnson, M.H., and Maro, B., 1984, The distribution of cytoplasmic actin in mouse 8-cell blastomeres, *J. Embryol. Exp. Morphol.* 82: 97-117.

Johnson, M.H., and Ziomek, C.A., 1981a, Induction of polarity in mouse 8-cell blastomeres: specificity, geometry and stability, *J. Cell Biol.*, 91: 303-308.

Johnson, M.H., and Ziomek, C.A., 1981b, The foundation of two distinct cell lineages within the mouse morula, *Cell* 24: 71-80.

Johnson, M.H., and Ziomek, C.A., 1982, Cell subpopulations in the late morula and early blastocyst of the mouse, *Dev. Biol.* 91: 431-439.

Johnson, M.H., and Ziomek, C.A., 1983, Cell interactions influence the fate of mouse blastomeres undergoing the transition from the 16- to 32-cell stage, *Dev. Biol.* 211-218.

Johnson, M.H., Eager, D.D., Muggleton-Harris, A.L., and Grave, M.H., 1975, Mosaicism in the organization of concanavalin A receptors on surface membrane of mouse egg, *Nature (London)* 257: 321-322.

Johnson, M.H., McConnell, J., and Van Blerkom, J., 1984, Programmed development in the mouse embryo, *J. Embryol. Exp. Morphol.* 83 Suppl.: 197-231.

Lehtonen, E., Lehto, V.P., Vartio, T., Badley, R.A., and Virtanen, I., 1983, Expression of cytokeratin polypeptides in mouse oocytes and preimplantation embryos, *Dev. Biol.* 100: 158-165.

Lewis, W.H., and Wright, E.S., 1935, On the early development of the mouse egg, *Contrib. Embryol. Carnegie Instn.* 148: 115-143.

Longo, F.J., and Chen, D-Y, 1985, Development of cortical polarity in mouse eggs: Involvement of the meiotic apparatus, *Dev. Biol.* 107: 382-394.

Malick, I.E., and Wilson, R.B., 1975, Evaluation of a modified technique for SEM examination of vertebrate specimens without evaporated metal layers, in: *Scanning Electron Microscopy* (O. Johari, ed.), IITRI, Chicago, IL, pp. 259-266.

Maro, B., Johnson, M.H., Pickering, S.J., and Flach, G., 1984, Changes in actin distribution during fertilization of the mouse egg, *J. Embryol. Exp. Morphol.* 81: 211-237.

Maro, B., and Pickering, S.J., 1984, Microtubules influence compaction in preimplantation mouse embryos, *J. Embryol. Exp. Morphol.* 84: 217-232.

Mintz, B., 1962, Experimental study of the developing mammalian egg; removal of the zona pellucida, *Science* 138: 594-595.

Muggleton-Harris, A., Whittingham, D.G., and Wilson, L., 1982, Cytoplasmic control of preimplantation development *in vitro* in the mouse, *Nature (London)* 299: 460-462.

Mulnard, J., and Huygens, R., 1978, Ultrastructural localization of nonspecific alkaline phosphatase during cleavage and blastocyst formation in the mouse, *J. Embryol. Exp. Morphol.* 44: 121-131.

Nicolson, G.L., Yanagimachi, R., and Yanagimachi, H., 1975, Ultrastructural localization of actin-binding sites on the zonae pellucidae and plasma membranes of mammalian eggs, *J. Cell Biol.* 66: 263-274.

Nicosia, S.V., Wolf, D.P., and Inoue, M., 1977, Cortical granule distribution and cell surface characteristics in mouse eggs, *Dev. Biol.* 57: 56-74.

Nizeyimana-Rugina, E., and Mulnard, J., 1979, Ultrastructural localization of 5'-nucleotidase in preimplantation mouse embryos, *Arch. Biol.* 90: 131-140.

Nuccitelli, R., 1983, Transcellular ion currents: Signals and effectors of cell polarity, in: *Modern Cell Biology*, Vol. 2 (J.R. McIntosh, ed.), Alan R. Liss, New York, pp. 451-481.

Nuccitelli, R., and Wiley, L.M., 1985, Polarity of isolated blastomeres from mouse morulae: Detection of transcellular ion currents, *Dev. Biol.*, 109: 452-463.

Oshima, R.G., Howe, W.E., Klier F.G., Adamson, E.D., and Shevinsky, L.H., 1983, Intermediate filament protein synthesis in preimplantation murine embryos, *Dev. Biol.* 99: 447-455.

Pratt, H.P.M., Ziomek, C.A., Reeve, W.J.D., and Johnson, M.H., 1982, Compaction of the mouse embryo: an analysis of its components, *J. Embryol. Exp. Morphol.* 70: 113-132.

Reeve, W.J.D., 1981, Cytoplasmic polarity develops at compaction in rat and mouse embryos, *J. Embryol. Exp. Morphol.* 62: 351-367.

Reeve, W.J.D., and Kelly, F.P., 1983, Nuclear position in the cells of the mouse early embryo, *J. Embryol. Exp. Morphol.* 75: 117-139.

Reeve, W.J.D., and Ziomek, C.A., 1981, Distribution of microvilli on dissociated blastomeres from mouse embryos: evidence for surface polarization at compaction, *J. Embryol. Exp. Morphol.* 62: 339-350.

Shirayoshi, Y., Okada, P.S., and Takeichi, M., 1983, The calcium-dependent cell-cell adhesion system regulates inner cell mass formation and cell surface polarization in early mouse development, *Cell* 35: 631-638.

Siracusa, G., DeFelici, M., Coletta, M., and Vivarelli, E., 1978, Inhibitors of protein synthesis increase the agglutinability mediated by concanavalin A of the unfertilized mouse oocyte, *Dev. Biol.* 62: 530-533.

Smithberg, M., 1953, The effect of different proteolytic enzymes on the zona pellucida of mouse ova, *Anat. Rec.* 117: 554.

Sobel, J.S., 1983a, Localization of myosin in the preimplantation mouse embryo, *Dev. Biol.* 95: 227-231.

Sobel, J.S., 1983b, Cell-cell contact modulation of myosin organization in the early mouse embryo, *Dev. Biol.* 100: 207-213.

Sobel, J.S., 1984, Myosin rings and spreading in mouse blastomeres, *J. Cell Biol.* 99: 1145-1150.

Sobel, J.S., and Alliegro, M.A., 1985, Changes in the distribution of a spectrin-like protein during development of the preimplantation mouse embryo, *J. Cell Biol.* 100: 333-336.

Solter, D., and Knowles, B.B., 1975, Immunosurgery of mouse blastocyst, *Proc. Natl. Acad. Sci. USA* 72: 5009-5102.

Sutherland, A.E., and Calarco-Gillam, P.G., 1983, Analysis of compaction in the preimplantation mouse embryo, *Dev. Biol.* 100: 328-338.

Tank, D.W., Wu, E-S, and Webb, W.W., 1982, Enhanced molecular diffusibility in muscle membrane blebs: release of lateral constraints, *J. Cell Biol.* 92: 207-212.

Thompson, R.S., and Zamboni, L., 1974, Phagocytosis of supernumerary spermatozoa of the mouse by two-cell mouse embryos, *Anat. Rec.* 178: 3.

Thompson, R,S., Smith, D.M., and Zamboni, L., 1974, Fertilization of mouse ova *in vitro*: an electron microscopic study, *Fertil. Steril.* 25: 222-249.

Vorbrodt, A., Konwinski, M., Solter, D., and Koprowski, H., 1977, Ultrastructural cytochemistry of membrane-bound phosphatases in preimplantation mouse embryos, *Dev. Biol.* 55: 117-134.

Wabik-Sliz, B., and Kujat, R., 1979, The surface of mouse oocytes from two inbred strains differing in efficiency of fertilization, as revealed by scanning electron microscopy, *Biol. Reprod.* 20: 405-408.

Whitten, W.K., and Biggers, J.D., 1968, Complete development *in vitro* of the preimplantation stages of the mouse in a simple chemically defined medium, *J. Reprod. Fertil.* 17: 399-401.

Whittingham, D.G., and Biggers, J.D., 1967, Fallopian tube and early cleavage in the mouse, *Nature (London)* 213: 942.

Wolf, D.E., 1983, The plasma membrane in early embryogenesis, in: *Development in Mammals*, Vol. 5 (M.H. Johnson, ed), Elsevier Science Publishers, pp. 187-208.

Wolf, D.E., and Edidin, M., 1981, Diffusion and mobility of molecules in surface membranes, in: *Techniques in Cellular Physiology*, P105, Part 1 (P.F. Baker, ed.), Elsevier Biomedical Press, Amsterdam, pp. 1-14.

Wolf, D.E., and Ziomek, C.A., 1983, Regionalization and lateral diffusion of membrane proteins in unfertilized and fertilized mouse eggs, *J. Cell Biol.* 96: 1786-1790.

Wolf, D.E., Edidin, M., and Handyside, A.H., 1981, Changes in the organization of the mouse egg plasma membrane upon fertilization and first cleavage: indications from the lateral diffusion rates of fluorescent lipid analogs, *Dev. Biol.* 85: 195-198.

Wolf, D.E., Handyside, A.H., and Edidin, M., 1982, Effect of microvilli on lateral diffusion measurements made by the fluorescence photobleaching recovery technique, *Biophys. J.* 38: 295-297.

Wolf, D.P., 1978, The block to polyspermy in zona-free mouse eggs, *Dev. Biol.* 64: 1-10.

Wolf, D.P., Nicosia, S.V., and Mastroianni, L., Jr., 1976, Surface topography of mouse eggs before and after insemination, *Biol. Bull.* 151: 435a.

Wolf, D.P., Nicosia, S.F., and Hamada, M., 1979, Premature cortical granule loss does not prevent sperm penetration of mouse eggs, *Dev. Biol.* 71: 22-32.

Wu, E-S, Tank, D.W., and Webb, W.W., 1982, Unconstrained lateral diffusion of concanavalin A receptors of bulbous lymphocytes, *Proc. Natl. Acad. Sci. USA* 79: 4962-4966.

Ziomek, C.A., and Johnson, M.H., 1980, Cell surface interaction induces polarization of mouse 8-cell blastomeres at compaction, *Cell* 21: 935-942.

Ziomek, C.A., and Johnson, M.H., 1981, Properties of polar and apolar cells from the 16-cell mouse morula, *Wilhelm Roux's Archiv.* 190: 287-296.

Ziomek, C.A., and Johnson, M.H., 1982, The roles of phenotype and position in guiding the fate of 16-cell mouse blastomeres, *Dev. Biol.* 91: 440-447.

Ziomek, C.A., and Lepire, M., 1984, The fluorescent histochemical demonstration of alkaline phosphatase in the preimplantation mouse embryo, *J. Cell Biol.* 99: 267a.

Ziomek, C.A, and Lepire, M.L., 1986, Fluorescent histochemical and immunofluorescent localization of cell surface alkaline phosphatase on mouse preimplantation embryos, *J. Embryol. Exp. Morphol.* (submitted).

Ziomek, C.A., Lepire, M., Moynihan, M., and Wolf, D.E., 1984, Membrane protein and lipid diffusion and regionalization in 8-cell mouse embryos, *J. Cell Biol.* 99: 279a.

CHAPTER 3

INTERCELLULAR COMMUNICATION DURING MOUSE EMBRYOGENESIS

GERALD M. KIDDER

1. INTRODUCTION

The preimplantation mouse embryo has been the subject of years of intense scrutiny, such that today we have a more complete understanding of its molecular, genetic, cellular, and intercellular control mechanisms than we have for any other mammal. Investigators have been able to manipulate the arrangements of cells in a variety of ways in order to explore the relationships that govern cell commitment and differentiation. Although it is a relatively simple system, we are still far from a complete understanding of the regulatory factors that predispose cells in the late blastocyst to become primitive ectoderm, primitive endoderm and polar or mural trophectoderm. Yet it is already clear that cell interactions involving intercellular communication play an important role. In one way or another, intercellular communication is involved in processes such as the induction of cell polarization (reviewed by Ziomek, Chapter 2), maintenance of the developmental lability of the ICM (Wiley, 1984), regulation of molecular differentiation within the ICM and its derivatives (Monk and Petzoldt, 1976; Hogan and Tilly, 1981), and the control of trophoblast proliferation and secondary giant cell transformation (reviewed by Kaufman, 1983). In addition, there is evidence for restriction of intercellular communication between groups of cells in the early postimplantation embryo which may be involved in the establishment of functional tissue domains (reviewed by Lo, 1980; Schultz, 1985).

Intercellular communication in the early mouse embryo could be mediated by at least 3 different mechanisms: by intercellular cytoplasmic bridges, by specialized permeable membrane channels (gap junctions) or by cell surface interactions not involving either of these. Although there is ample

Gerald M. Kidder Department of Zoology, University of Western Ontario, London, Ontario, N6A 5B7, Canada.

evidence for the existence of all of these modes of communication, the nature of the signals which they transmit and the mode of action of those signals in influencing cell behavior are unknown. Likewise, one can only speculate on the role of specific modes of intercellular communication in particular developmental events. The purpose of this chapter is to describe the methodology used for investigation of intercellular communication in mouse embryos, and to review the evidence for its various modes. Particular attention will be directed to the gap junction and the control of its assembly, since we have used this relatively simple organelle as a focal point for investigating the control of morphogenesis. The chapter will conclude with a consideration of the possible roles of intercellular communication in mouse embryogenesis. Given the broad similarities between preimplantation embryos of eutherian mammals, it is reasonable to hope that information gleaned from mouse embryos will shed light on developmental control mechanisms in other embryos of prime interest, including human embryos.

2. COMMUNICATION *VIA* CYTOPLASMIC BRIDGES

2.1. Methodology

Intercellular cytoplasmic bridges, maintained by midbodies persisting from the cleavage divisions, can be observed in living embryos (Lehtonen, 1980; Ziomek and Johnson, 1980) and in embryos embedded and sectioned for electron microscopy (Ducibella and Anderson, 1975; Sołtyńska, 1982). A third method involves microinjection of a relatively large tracer molecule into a blastomere, noting its subsequent distribution among other blastomeres. If the tracer is large enough, *i.e.*, exceeding the upper limit (roughly 1000 Daltons) for passage through intercellular membrane channels (see section 3.1), then its passage into a neighboring cell is evidence that a cytoplasmic bridge exists between the 2 cells. This method provides a straightforward means of mapping the distribution of cytoplasmic bridges present in a particular stage of cleavage (Goodall and Johnson, 1984) and can serve as a means of marking the descendents of a particular cell for short-term lineage studies (Bałakier and Pedersen, 1982).

The tracer most commonly used in work with early mouse embryos is horseradish peroxidase (HRP). This enzyme (40 kD) is too large to pass through membrane channels. Its distribution subsequent to iontophoretic injection can be visualized by a histochemical staining reaction (Graham and Karnovsky, 1966), as illustrated in Figure 1. Further details of procedure can be found in Lo and Gilula (1979a), Bałakier and Pedersen (1982), and Goodall and Johnson (1984). Recently, rhodamine-conjugated dextran has been used in conjunction with HRP; it affords the advantage of instantaneous monitoring of its distribution in living embryos by means of fluorescence microscopy (Pedersen *et al.*, 1986).

2.2. Evidence for Persistent Midbodies in Early Embryos

Lo and Gilula (1979a) were the first to call attention to the fact that sister blastomeres in the early mouse embryo remain connected *via* a cytoplasmic bridge, containing a midbody, after each cleavage division. These investigators used microinjected HRP to visualize cytoplasmic bridges;

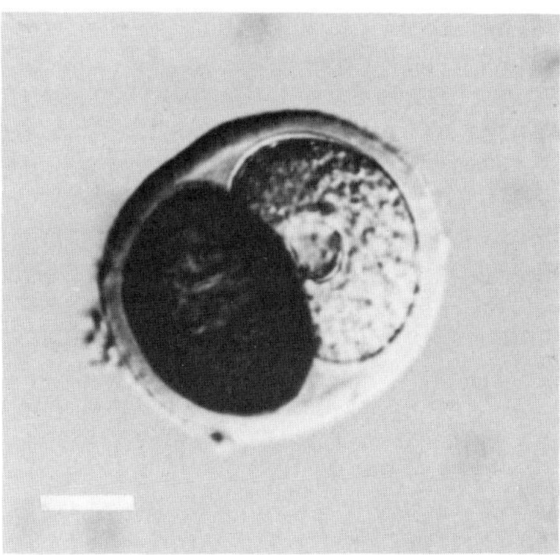

Figure 1. Mouse 2-cell embryo after microinjection of horseradish peroxidase (HRP) into the left blastomere. Confinement of the enzyme to one blastomere demonstrates the absence of a cytoplasmic bridge (persistent midbody) between the two cells. Scale bar indicates 20 μm. Micrograph provided by Dr. Roger A. Pedersen.

although the number of embryos tested was not reported, they concluded that sister blastomeres remain connected for a significant length of time before completion of cytokinesis. Thus, in early 2-cell embryos, HRP passed readily from the injected blastomere into its sister, whereas no transfer of the enzyme could be detected in late 2-cell embryos. Early 4-cell embryos were found to consist of two pairs of sister blastomeres, each pair connected by a bridge able to transfer HRP, but the enzyme was not observed to pass between non-sisters. These findings indicate that the first cleavage midbody has closed before initiation of the second cleavage, and that the second cleavage midbodies persist. They also found pairs of bridge-connected blastomeres in late 4-cell embryos collected early on day 3, suggesting that the second cleavage midbodies persist well into the third cell cycle. These midbodies appeared to have closed by the time of completion of the third cleavage, because once again only pairs of bridge-connected blastomeres were found in early (uncompacted or partially-compacted) 8-cell embryos. Overall, the results of Lo and Gilula (1979a) indicate that the completion of cytokinesis after each of the first 3 cleavage divisions is delayed until sometime in the succeeding cell cycle, providing a transient cytoplasmic bridge between sister pairs. These findings have been confirmed by using anti-tubulin immunofluorescence to detect midbody microtubule bundles: midbodies were found to connect sister blastomeres through the 8-cell stage, disappearing before each succeeding cell division cycle (Kidder and Olmsted, in preparation).

On the other hand, a rather different picture has emerged from the recent work of Goodall and Johnson (1984). They found that in approximately one-third of early 8-cell embryos injected with HRP, the enzyme passed to more than one neighboring cell; in some cases, to 4 of the 8 cells as if the second cleavage midbody had persisted beyond the third cleavage in those embryos. This interpretation was substantiated by experiments with embryos in transition between the 4- and 8-cell stages: HRP injected into a 1/4 blastomere (one of the blastomeres from the 4-cell stage) could be passed into both of the daughters of another 1/4 blastomere that had already divided to the 8-cell stage. The crucial difference between this study and that of Lo and Gilula (1979a), according to Goodall and Johnson (1984), is that an interval of 2 hr or more was allowed between HRP injection and fixation, providing extra time for the enzyme to diffuse through the (presumably) more tightly occluded second cleavage midbody. Indeed, when the time between injection and fixation was reduced to one hour or less, fewer embryos exhibited passage of the enzyme to more than one other blastomere. Goodall and Johnson (1984) concluded that the coupling of early 8-cell stage blastomeres *via* persistent midbodies is more extensive than previously suspected, involving the third cleavage midbodies in all embryos (coupling blastomeres in pairs), the second cleavage midbodies in roughly a third of embryos (coupling in quartets), and the first cleavage midbody in very few embryos (potentially coupling the entire embryo; only a very small proportion of embryos contained more than 4 HRP-positive cells).

These findings have interesting implications for understanding the timing of events in different blastomeres of the same embryo; they also complicate the interpretation of cell lineage studies using large tracer molecules (Bałakier and Pedersen, 1982). Because of this, Pedersen *et al.* (1986) carried out experiments to try to confirm the findings of Goodall and Johnson (1984). Surprisingly, they were unable to demonstrate the existence of persistent second cleavage midbodies in any but a few 8-cell embryos: the vast majority (33/37 or 89%) of HRP-injected blastomeres either retained the enzyme or passed it on to only one neighboring sister cell, even when fixation was delayed for 2 hr after injection. Since the investigators attempted to reproduce the iontophoresis conditions used by Goodall and Johnson (1984), the discrepancy between these two sets of results is difficult to explain. It may simply reflect a difference in timing of midbody occlusion between different strains of mice. Pedersen *et al.* (1986) also investigated the existence of cytoplasmic bridges in embryos undergoing the fourth cleavage (8- to 16-cell stage). When fixed for HRP visualization immediately after injection, the proportion of embryos in which the enzyme passed into more than one additional blastomere was 8/77 (10%) for embryos in transition to the 16-cell stage and 9/110 (8%) for those which had completed that division cycle. Most early 16-cell stage blastomeres (68%) were connected to a sister cell *via* a cytoplasmic bridge, the proportion declining slightly (to 57%) in the late 16-cell stage. Thus, persistent midbodies appear to provide a potential avenue of intercellular communication through at least the 16-cell stage, but one that is incomplete at best, being limited in most embryos to sister blastomeres remaining connected through one cleavage cycle.

3. COMMUNICATION *VIA* INTERCELLULAR MEMBRANE CHANNELS

3.1. Structure of Gap Junctions

Gap junctions are specialized cell-cell membrane channels that provide for direct passage of small molecules between adjacent cells (reviewed by Lo, 1980; Loewenstein, 1981). In mammals, the upper limit on the size of molecules that can pass through is on the order of 1000 Daltons; below this limit, permeability to a particular molecule is also affected by its electrical charge. The channel diameter has been estimated to be 1.6 to 2.0 nm. The basic unit of the gap junction is the connexon, a particle consisting of 6 subunits arranged hexagonally, surrounding an aqueous channel. Connexons within the plasma membranes of adjoining cells align, generating channels connecting the cytoplasm of the two cells (the intercellular spacing is tight: only 2 to 3 nm separate the plasma membranes). The gap junction itself consists of an array of connexon pairs forming a plaque; this structure can be seen to best advantage in freeze fracture replicas, as illustrated in Figure 2. Recently, several investigators have succeeded in isolating gap junction plaques from plasma membrane fractions, and have purified a protein (27 kD) which is believed to be the connexon subunit (Hertzberg and Skibbens, 1984; Dermietzel *et al.*, 1984).

3.2. Methodology

A variety of methods has been used to investigate intercellular membrane channels in early mouse embryos. Gap junctions have been identified with the electron microscope, both in thin sections in which they appear as regions of closely-apposed plasma membranes separated by a narrow gap

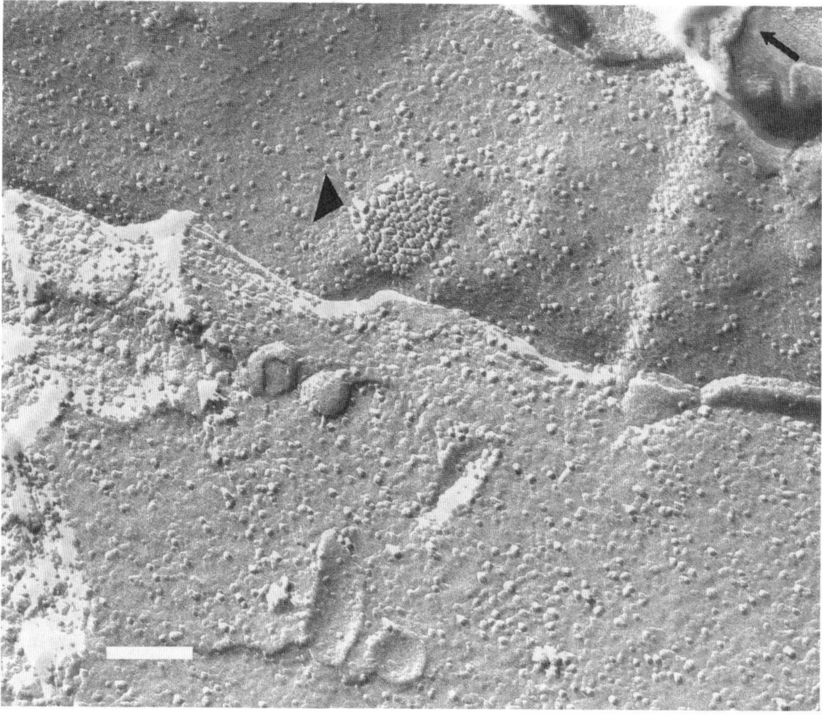

Figure 2. Gap junctional plaque (arrowhead) from a mouse blastocyst as seen in freeze-fracture. The plaque is an aggregation of intramembranous particles; other, similar, particles can be seen scattered throughout the exposed plasma membrane surface. Scale bar indicates 100 nm; arrow in the top right corner indicates the direction of platinum shadowing. Micrograph provided by Dr. Jeanne R. McLachlin.

(Ducibella *et al.*, 1975; Ducibella and Anderson, 1975), and in freeze-fracture replicas (Fig. 2) in which aggregations or plaques of connexon-like particles (but not individual widely-spaced particles) can reasonably be assumed to be gap junctions (Magnuson *et al.*, 1977). These morphological approaches have two major weaknesses, however: because of the haphazard nature of the sectioning or fracturing procedure, it is difficult to sample large regions of membrane. Even when identified by morphological criteria, gap junctions cannot unequivocally be assumed (without functional evidence) to be providing open channels for the intercellular transfer of molecules. Indeed, Loewenstein (1981) has suggested that the terms "gap junction" and "connexon" be restricted to designating the structural entity, with terms like "permeable junction" or "cell-cell channel" being used only when functional data are available.

As with cytoplasmic bridges, the presence of permeable intercellular membrane channels can be demonstrated by microinjection of tracers; in this case, using ions or small molecules. Investigators working with early mouse embryos have used two general approaches. In one, ionic (electrical) coupling, an ion current is pulsed into a blastomere *via* a microelectrode filled with a KCl solution, and the change in membrane potential in another blastomere due to intercellular current passage is recorded *via* a second microelectrode (Fig. 3). The second approach is dye coupling, which usually involves iontophoretic injection of a fluorescent dye such as fluorescein (mol. wt. 332), carboxy-fluorescein (mol. wt. 376), or Lucifer yellow CH (mol. wt. 443), and following its spread throughout the embryo (Fig. 4). Technical details of these procedures can be found in Lo and Gilula (1979a), McLachlin *et al.* (1983), and Goodall and Johnson (1984). The passage of either current or dye from one blastomere to another provides functional evidence that intercellular membrane channels are present, provided that transfer of the tracer *via* cytoplasmic bridges (persistent midbodies) can be ruled out. With dye coupling, this is fairly straightforward: since midbodies connect sister blastomeres in pairs (or, at most, quartets), the spread of dye throughout an embryo of more than 4 cells (see Fig. 4) leaves no doubt that channels and not bridges are involved. When testing for electrical coupling, the simplest way to avoid impaling sister blastomeres with the microelectrodes is to position them to enter the embryo on opposite sides, thus avoiding contiguous pairs. A further precaution, one that can be applied equally well to embryos having fewer than 8 blastomeres, is to record the transmitted current from two or more blastomeres in succession (Fig. 3). In practice, it is not always possible to record from all blastomeres of an embryo, especially beyond the 4-cell stage, hence electrical coupling does not easily provide the embryo-wide picture of cell coupling that can be obtained from dye injection experiments. Thus, although both electrical coupling and dye coupling observations can be used to substantiate the presence of permeable membrane channels, the two techniques are not equivalent.

Another consideration is that, of the two methods, dye coupling is generally regarded as less sensitive, since it relies on visual detection of fluorescence in a recipient cell. This is especially a problem with early embryos, in which a small amount of dye entering from a neighboring cell is diluted into a large volume of cytoplasm. Hence, investigators have noted instances of electrical coupling in the apparent absence of dye coupling in mouse embryos, the simplest (but not the only) interpretation being that the

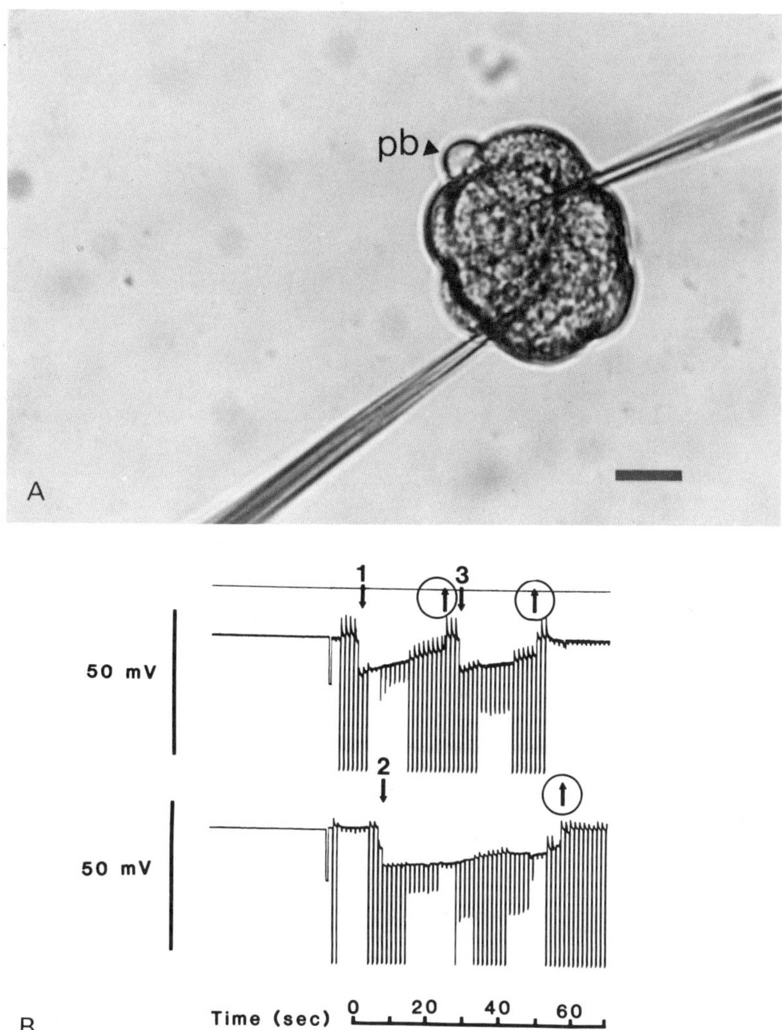

Figure 3. Detection of ionic coupling between nonsister blastomeres of a compacted 8-cell mouse embryo (slight decompaction has occurred after prolonged pulsing of the ion current). A. Phase-contrast micrograph showing the final positions of the electrodes; the polar body (pb) is also indicated. Scale bar indicates 20 μm. B. Chart recording showing that current pulsed into the right electrode (lower trace) is detected in the left electrode (upper trace), and *vice-versa*. The long voltage deflections correspond to the pulsing electrode and the short deflections to the recording electrode in each case. At points 1 and 2 the left and right electrodes, respectively, were inserted into non-adjacent blastomeres, the downward deflection in the baselines indicating the membrane potentials. At point 3 the left electrode was reinserted into a third blastomere to demonstrate that ionic coupling involved more than 2 blastomeres. Circled arrows indicate removal of the electrodes. From McLachlin (1984) with permission.

number of effective channels is too small to permit the passage of a detectable amount of dye, but sufficient to conduct a current (Lo and Gilula, 1979b; McLachlin *et al.*, 1983). Goodall and Johnson (1984) directly compared the

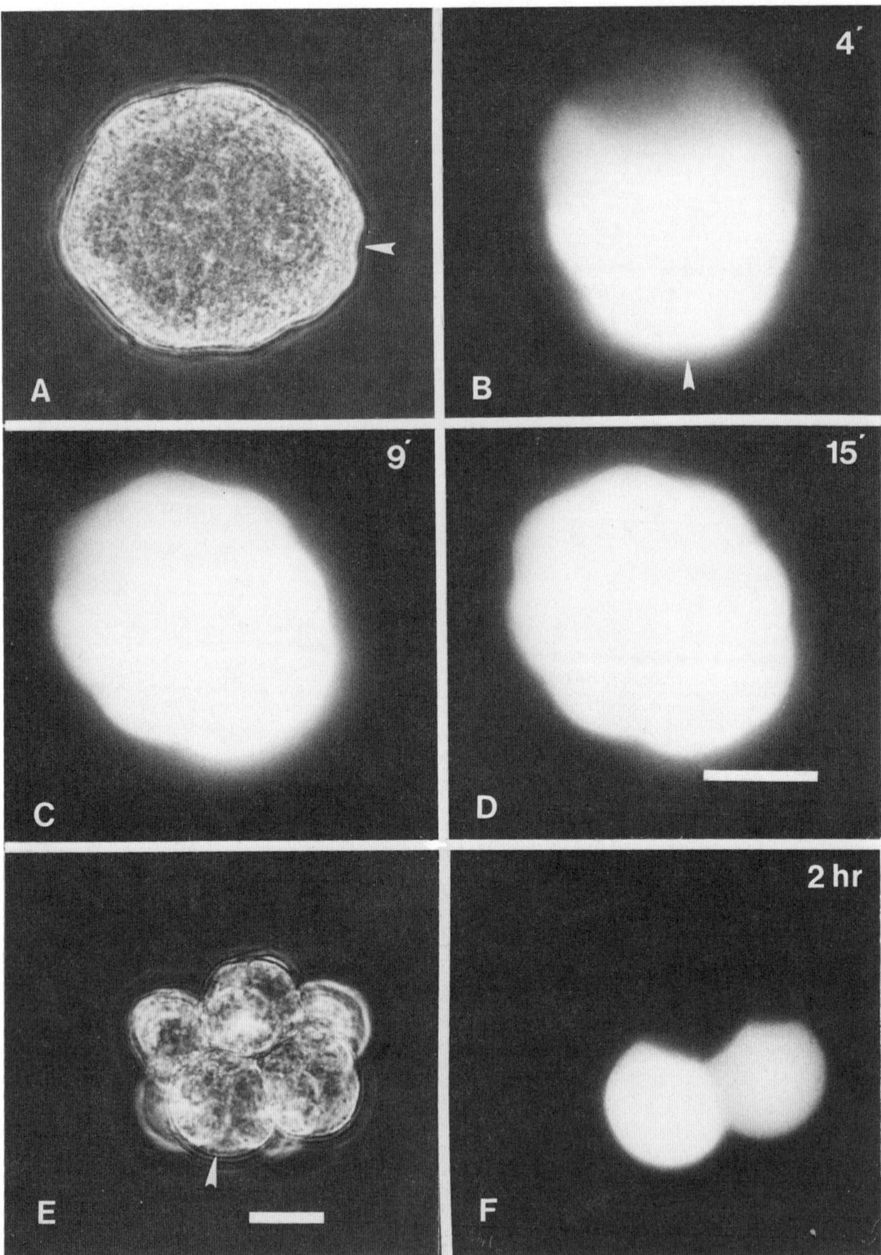

Figure 4. Demonstration of embryo-wide dye coupling by microinjection of Lucifer yellow into a mouse morula (8- to 16-cells; A-D), and its absence in an uncompacted 8-cell embryo (E,F). Arrowheads indicate the point of dye injection, and scale bars indicate 20 μm. Time after injection is indicated in the upper right corner of each fluorescence micrograph. Whereas the dye equilibrated rapidly among all the blastomeres of the morula, in the uncompacted embryo it was passed to only a single blastomere, presumably *via* a midbody. From McLachlin (1984) with permission.

sensitivities of ionic coupling and dye coupling (carboxyfluorescein injection) for detecting the formation of intercellular membrane channels, and found the first method to be superior. It is to the investigator's advantage, then, to employ both methods in order to allow the strength of one to complement the weakness of the other.

An additional dye coupling method was introduced recently for studying the capacity of mouse embryo blastomeres to form permeable membrane channels between them (Goodall and Johnson, 1982). It involves loading an embryo or an isolated blastomere with carboxyfluorescein diacetate (CFDA), which simply requires incubation in a solution of the dye in culture medium or buffered saline. Being membrane permeable, the dye readily enters the cells where it is cleaved by esterases to form the hydrophilic dye, carboxyfluorescein, which is trapped within the cells. Treatment with CFDA at moderate concentrations (100 μg/ml or less) has no apparent deleterious effects on viability. The labeled embryo or blastomere is then aggregated with an unlabeled counterpart, and the aggregate is observed over a period of several hours for transfer of dye from donor to recipient (Fig. 5). The principal advantage of this method is its simplicity, since it does not require microinjection. According to Goodall and Johnson (1984), it is as sensitive as electrical coupling and more sensitive than dye injection for detecting the formation of permeable junctions between aggregated blastomeres. It can also be used to test for the formation of junctions between blastomeres or embryos of different developmental stages (the significance of this will be discussed in section 3.4). Lastly, since the channels being formed are between reaggregated blastomeres or between embryos, the aggregation protocol itself has an advantage in that it removes any uncertainty that the dye is being transferred *via* membrane channels and not *via* intercellular bridges.

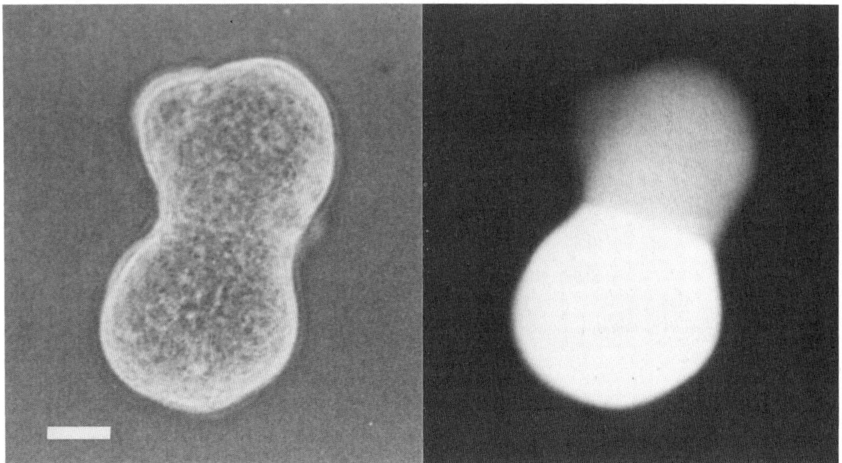

Figure 5. Demonstration of dye coupling between aggregated, compacted 8-cell mouse embryos using carboxyfluorescein diacetate. The phase-contrast and fluorescence photographs were made at the same time, 7 hr after aggregation. The lower embryo is the one originally labelled with CFDA, and it has passed carboxyfluorescein *via* new interembryonic membrane channels to the upper embryo. Scale bar indicates 20 μm.

3.3. Timing of Gap Junction Appearance in Early Development

Membrane specializations having the attributes of gap junctions were first detected in 8- to 16-cell mouse embryos prepared for thin-section electron microscopy (Ducibella *et al.*, 1975). Prior to the 8-cell stage, blastomere plasma membranes may be closely apposed in places, but no specialized junctions can be seen. As blastomeres flatten against one another during compaction, both focal tight and gap junctions appear (Ducibella and Anderson, 1975; Magnuson *et al.*, 1977). The appearance of gap junctions at the 8-cell stage has been confirmed by freeze-fracture. Although connexon-like intramembranous particles have been seen in fracture replicas of membranes from 2- and 4-cell embryos (Magnuson *et al.*, 1977; Shivers and McLachlin, 1984), it is not until the 8-cell stage that they appear in the closely-packed arrays characteristic of gap junctions (Magnuson *et al.*, 1977). As development continues through the morula stage, gap junction plaques grow and become more numerous, many of them in association with an expanding network of tight junctions. In the blastocyst, they appear to connect trophoblast cells with each other, ICM cells with each other, and ICM with trophoblast (Ducibella *et al.*, 1975; Ducibella and Anderson, 1975; Magnuson *et al.*, 1977, 1978).

Although the appearance of gap junctions in the 8-cell stage had been demonstrated morphologically, data of a functional nature were needed to substantiate the notion that these could serve as intercellular pathways of communication. Lo and Gilula (1979a) provided those data by showing that embryo-wide dye coupling occurs for the first time in embryos undergoing compaction; prior to this, neither ionic nor dye coupling could be detected except between sister blastomeres connected by a midbody capable of transferring HRP. As embryos passed from the uncompacted to the compacted state, microinjected HRP remained confined to one or two blastomeres, while fluorescein diffused freely throughout all 8 blastomeres. Embryo-wide intercellular ionic and dye coupling were found to be maintained thereafter throughout preimplantation development and into the earliest stages of implantation (Lo and Gilula, 1979a,b). The data from electrophysiological experiments, therefore, are in excellent agreement with the morphological evidence and demonstrate that the morphological entities identified as gap junctions could, in fact, function as intercellular channels for the passage of small molecules.

Because of the apparent abruptness of gap junction appearance, and the possibility that such membrane channels play a role in cellular regulation (see Section 5), this event has attracted considerable interest. Investigators have sought to determine more accurately the timing of assembly of membrane channels in relation to cleavage and compaction, and to explore the genetic and cellular control mechanisms involved. One approach to define the time of acquisition of the capacity to assemble permeable membrane channels has been the CFDA labeling-blastomere aggregation method. By aggregating labeled and unlabeled 2/8 couplets of known age after division of 1/4 blastomeres, Goodall and Johnson (1982) demonstrated that blastomeres acquire the ability to form membrane channels beginning 2 to 3 hr after the third cleavage. This finding was later confirmed by dye injection along with ionic coupling, using the same experimental design (Goodall and Johnson, 1984). Thus embryos become *capable* of assembling permeable membrane channels early in the fourth cell cycle, just at the time that compaction is beginning. However, it is possible that the cell flattening brought about by the lectin-mediated aggregation step in these experiments might cause junctional channels to form somewhat earlier than they otherwise would in the intact embryo. In light of this, we have attempted to define more precisely the timing of assembly of permeable membrane channels in intact embryos undergoing compaction.

Lo and Gilula (1979a) had reported embryo-wide dye coupling in embryos that had just begun to compact, whereas we had failed to confirm this in a small group of such embryos by ionic coupling (McLachlin *et al.*, 1983). It now appears that embryos undergoing compaction comprise a heterogeneous population with respect to embryo-wide cell coupling: whereas virtually all uncompacted 8-cell embryos lack permeable membrane channels and all fully-compacted 8-cell embryos have them, partially-compacted embryos fall in between, with about two-thirds possessing membrane channels that allow injected Lucifer yellow to spread to all 8 blastomeres (Table I; McLachlin and Kidder, 1986). Furthermore, the apparent rate of dye spread among the blastomeres of those partially-compacted embryos that were coupled was noticeably slower than in fully-compacted embryos (Table II). This finding corroborates the conclusion from electron microscopical studies that the size and frequency of gap junctional plaques increases as embryos proceed through compaction (Magnuson *et al.*, 1977).

To summarize, mouse embryo blastomeres are able to assemble permeable cell-cell membrane channels within a few hours of cleaving to the 8-cell stage. Both morphological and functional evidence indicate that this capability is utilized during the process of compaction, as blastomeres flatten against one another, although the precise timing may vary from one embryo to another. The finding that not all embryos undergoing compaction possess functional membrane channels raises the possibility that compaction and junction

Table I

Analysis of Intercellular Junctional Coupling in 8-Cell Mouse Embryos[a]

	8-Cell embryos		
	Uncompacted	Partially compacted	Fully compacted
No. embryos ionically-coupled/ no. tested	1/20	19/28	95/95
No. embryos dye-coupled/ no. tested	0/16	9/15	15/15

[a]Data from McLachlin (1984) with permission.

Table II
Relative Rates of Spreading of Injected Lucifer Yellow Among Mouse Embryo Blastomeres[a]

	Time to fill all cells[b]		Time to equilibrate[b]	
Stage	≤ 5 min	10-30 min	≤ 30 min	> 30 min
8-Cell, partially compacted	0	7	0	7
8-Cell, fully compacted	8	2	7	2
Morula	2	0	2	0

[a]Data from McLachlin and Kidder (1986).
[b]The time-course of dye spread was photographed at intervals for later analysis. "Time to fill" refers to the time needed for the dye to enter all blastomeres; "Time to equilibrate" refers to the time needed for the fluorescence intensity to equalize throughout the embryo. These were assessed from the photographs.

assembly may not be obligatorily linked. This and other issues concerning the control of junction assembly are considered in the next section.

3.4. Control of Gap Junction Assembly

The abrupt appearance of gap junctions early in the 8-cell stage is one of a series of morphogenetic events (others include cell flattening, cell polarization and cavitation) which presumably prepare the embryo for implantation and the formation of divergent cell lineages. An important goal is to understand how such events are programmed. We have focussed our attention on the possible control mechanisms governing the appearance of gap junctions, as a model for cellular morphogenesis. We have sought to ascertain the relationship between junction assembly and such potential control factors as transcription, protein synthesis, cell flattening (compaction) and the cleavage cycle.

The simplest approach to examine the role of macromolecular synthesis in programming morphogenetic events is to use metabolic inhibitors. We have used two such agents, cycloheximide, an inhibitor of protein synthesis, and α-amanitin, an inhibitor of RNA synthesis whose primary target is RNA polymerase II, the enzyme that transcribes structural genes (reviewed by Warner, 1977). In mouse embryos, however, α-amanitin also interrupts the production of rRNA, hence the interpretation of its effects is not as straightforward as one might hope (Warner and Hearn, 1977; Levey and Brinster, 1978). Using either of these 2 inhibitors at appropriate concentrations (10-50 μg/ml for cycloheximide; 100 μg/ml for α-amanitin), we can achieve interruption of macromolecular synthesis within 2 hr of the start of treatment (McLachlin *et al.*, 1983; Kidder *et al.*, 1985). Surprisingly, mouse embryos can withstand such a metabolic blockade by either inhibitor for many hours, such that treatment begun in the 4-cell stage allows many embryos in the population to continue through the third cleavage and compaction in apparently normal fashion (Kidder and McLachlin, 1985). Embryos treated in this way were even found to maintain membrane potentials that were not different from controls (McLachlin and Kidder, 1986). When tested for ionic coupling between non-adjacent (nonsister) blastomeres, almost all such embryos were found to have assembled permeable membrane channels (Table III). Dye coupling experiments, however, indicated that the number or effective area of such channels was severely reduced by either inhibitor, since dye transfer was weak or undetectable (McLachlin and Kidder, 1986). Our interpretation is that precursors of gap junctions are present in 4-cell embryos, many hours in advance of the time of junction assembly. However, this store of precursors is insufficient to allow the full extent of cell coupling to be established in the absence of further synthesis during the 4-cell and early 8-cell stages.

This hypothesis predicts that the connexon protein, the principal precursor of gap junctions, is present in 4-cell embryos (or even earlier). Recently, antibodies against the 27 kD protein of mouse liver gap junctions have become available (Hertzberg and Skibbens, 1984; Dermietzel *et al.*, 1984). These antibodies provide a direct means of testing our hypothesis. Until such tests have been completed, the conclusions drawn from experiments with inhibitors should be regarded as tentative.

A second strategy for exploring the control of assembly of membrane channels has been to test their inducibility in various stages of cleavage by embryo or blastomere aggregation. It has been known for some time that co-cultured or aggregated somatic cells, even those from different organs or species, can form communicating junctions between them (reviewed by Loewenstein, 1981). Similarly, when zona-free mouse embryos that have completed compaction (and thus are communication-competent) are aggregated together, permeable membrane channels form between them (McLachlin and Kidder, 1981; McLachlin *et al.*, 1983). When tested for ionic coupling, such channels could be detected within 3 to 5 hr after aggregation. Since this involves the assembly of membrane channels in a portion of the plasma mem-

Table III

Effects of Metabolic Inhibitors on Intercellular Junctional Coupling in Compacted 8-Cell Mouse Embryos[a]

Inhibitor	Treatment duration (hr)	Ionic coupling (no. coupled/ no. tested)	Dye coupling[b] (no. coupled/ no. tested)
Cycloheximide	17	8/8	1/17
	14	12/13	1/9
α-Amanitin	21	7/9	5/11
	18	11/12	1/17
	17	ND[c]	10/11
	7	ND[c]	8/8

[a]Data from McLachlin and Kidder (1986).
[b]Dye coupling was assessed using iontophoretically-injected Lucifer yellow.
[c]ND: not determined.

brane (the external-facing membrane) that did not previously contain them, it must necessitate either the mobilization of connexons from elsewhere, or the synthesis and insertion of new junctional components. The latter possibility is the more likely, since the formation of permeable channels between aggregated embryos was blocked by cycloheximide. When embryo aggregation was used to attempt induction of membrane channels in 2-cell or 4-cell embryos, however, the results were negative. Even after 10 hr of aggregation with compacted 8-cell embryos, no interembryonic membrane channels could be detected (McLachlin *et al.*, 1983). The same result was obtained when oocytes, zygotes, or isolated blastomeres from 2-cell or 4-cell embryos were aggregated with blastomeres from compacted 8-cell embryos, despite the fact that in this situation, many of the pairings would involve that portion of the 8-cell blastomere surface already bearing connexons (Goodall and Johnson, 1982, 1984). It seems that, even if channel precursors are present well in advance of their time of assembly into junctions, it is not simply the absence of close membrane apposition which prevents them from assembling. We favor the notion that junction assembly is a precisely timed event, and have begun to explore the factors which might influence its timing.

Our first step was to re-examine the relationship between compaction (cell flattening) and channel assembly. We had noticed previously (McLachlin *et al.*, 1983) that decompaction caused by prolonged pulsing of current into an embryo does not disrupt ionic coupling. Furthermore, Ducibella and Anderson (1979) had reported that gap junctions could still be detected by electron microscopy in embryos prevented from completing compaction in low calcium medium, although it was not clear whether those embryos had already initiated junction assembly before being transferred into the experimental medium. We have used the CFDA/embryo aggregation method to test for the capacity to assemble permeable membrane channels in embryos prevented from initiating compaction by continuous culture from the early 8-cell stage in cytochalasin B (Granholm *et al.*, 1979; Surani *et al.*, 1980). Such embryos do not undergo cell flattening, yet they remain viable and can compact and resume morphogenesis when later removed from the drug. When left in cytochalasin until age-matched controls had compacted, and then aggregated in the continued presence of the drug with untreated communication-competent, CFDA-labeled embryos (which soon underwent decompaction), the treated embryos were able to assemble permeable membrane channels and receive the dye (Fig. 6A; Kidder *et al.*, 1986). Not only did the carboxyfluorescein pass from the labeled embryo into the unlabeled one, but it spread throughout the recipient embryo, demonstrating that intercellular, as well as interembryonic, channels had developed. Thus, cell flattening is not a prerequisite for assembly of intercellular channels; compaction and gap junction assembly must therefore be regarded as coincident but independent events. Furthermore, the capacity to assemble membrane channels is acquired independently of cytokinesis, which normally occurs just a few hours before. Figure 6B illustrates the result of the same type of experiment carried out with 4-cell embryos prevented by treatment with cytochalasin B from undergoing the third cleavage and cell flattening. The treated embryos were allowed to reach the age of fully-compacted 8-cell embryos before the aggregation step. Such embryos are capable of receiving dye passed from previously-compacted, untreated 8-cell embryos. Subsequent experiments have demonstrated that zygotes prevented from undergoing cytokinesis by continuous cytochalasin

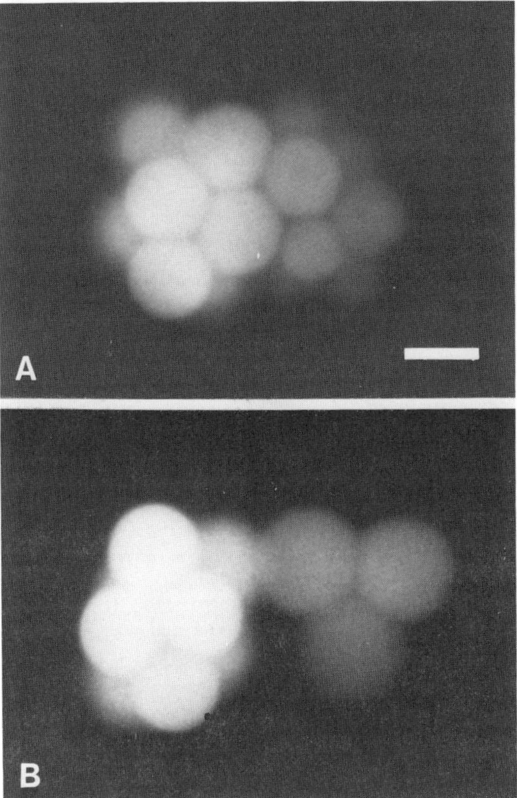

Figure 6. Demonstration of interembryonic dye coupling (CFDA method) involving embryos blocked in the 4-cell or uncompacted 8-cell stage by treatment with cytochalasin B (10 μg/ml). (A) Early 8-cell embryo (on the right) was treated with cytochalasin for 1 hr, then aggregated with a CFDA-labelled, compacted 8-cell embryo (on the left) in the continued presence of cytochalasin. The fluorescence micrograph was made 7 hr after aggregation; it shows dye transfer to the recipient embryo, which had been prevented from initiating cell flattening. (B) A 4-cell embryo (on the right) was treated with cytochalasin for 11 hr to block cytokinesis and cell flattening, during which time controls from the same population became compacted 8-cell embryos. The treated embryo was then aggregated in the presence of cytochalasin with a CFDA-labelled, compacted embryo (on the left), resulting in dye transfer. The fluorescence micrograph was made 10 hr after aggregation. Scale bar indicates 20 μm.

treatment from the 1- or 2-cell stage can likewise assemble interembryonic junctions when they reach the appropriate age (Kidder *et al.*, 1986). This must mean that chronological age, and not the number of cells or the progress of compaction, is the principal factor governing the timing of assembly of intercellular membrane channels.

In summary, the evidence available to date indicates that the appearance of gap junctions in the mid-8-cell stage of the mouse is programmed by previously-existing transcripts, present at least as early as the 4-cell stage. Likewise, sufficient gap junction protein is presumed to exist in the 4-cell stage to permit the assembly of at least a few membrane channels in the absence of protein synthesis. Despite this, embryos in the 4-cell stage cannot be induced to form junctions by aggregation with communication-competent embryos or blastomeres. In our search for the limiting factor that triggers channel assembly, we have eliminated cell flattening and cytokinesis. Experiments in progress are designed to investigate the possible role of mitotic cycles and DNA replication.

4. COMMUNICATION *VIA* CELL SURFACE INTERACTION

4.1. Methodology

Because of its very nature, communication *via* cell contact not involving intercellular channels is difficult to demonstrate and to study. Cell contact is known to be required in many instances for the transmission of inductive signals during development (reviewed by Saxén *et al.*, 1980), but in many such cases, gap junctions have been implicated as the transmission pathway (Saxén and Lehtonen, 1978; Model *et al.*, 1981; see also review by Caveney, 1985). Although models of cell communication *via* non-junctional cell surface interaction have been proposed (Edelman, 1983), it is extremely difficult to rule out the participation of communicating channels. The early mouse embryo provides one of the clearest examples of cell communication mediated by cell contact but not requiring the formation of intercellular channels.

The example in question is polarization of blastomeres in the 8-cell stage (reviewed by Ziomek, Chapter 2). Cell polarity can be demonstrated by a variety of approaches, summarized by Ziomek and Johnson (1980); the simplest means is to label embryos with the lectin concanavalin A, conjugated with fluorescein isothiocyanate (FITC-Con A; Handyside, 1980). Most blastomeres disaggregated from compacted 8-cell embryos show a polarized distribution of lectin binding sites, with the fluorescence concentrated in that portion of the blastomere surface that had been exposed at the surface of the compacted embryo. The formation of this microvillous, lectin-binding pole is dependent on contact with neighboring blastomeres, the orientation of the pole being dictated by the geometry of cell contacts (Ziomek and Johnson, 1980). The nature of these cell contacts has been explored in blastomere aggregation experiments in which the polarization-inducing ability of blastomeres from different cleavage stages has been tested (Johnson and Ziomek, 1981a).

4.2. Evidence that Induction of Cell Polarization Does Not Require Intercellular Channels

When isolated 1/4 blastomeres are allowed to cleave in culture to form natural 2/8 couplets, each member of such a couplet develops a microvillous pole in a large proportion of cases (Ziomek and Johnson, 1980). Likewise, reaggregated early 1/8 blastomeres (artificial 2/8 couplets) become similarly polarized. In the vast majority of cases, the axis of polarity is normal to the plane of cell contact, regardless (in artificial couplets) of the original orientation of the two blastomeres. In contrast, the proportion of single 1/8 blastomeres that polarize is much lower, unless they are allowed to remain in contact with a highly adhesive surface for several hours. Johnson and Ziomek (1981a,b) have presented evidence that the polarized phenotype is stable, and that it generates, through subsequent asymmetric cleavages, two blastomere populations representing the foundations of the inner (ICM) and outer (trophectoderm) cell lineages. Thus, the cell interactions leading to polarization may have far-reaching effects on subsequent cell differentiation. How is this influence exerted?

An obvious answer to this question would be that gap junctions developing in the 8-cell embryo transmit morphogenetic signals which somehow predispose blastomeres to develop an axis of polarity. However, this cannot be the case. In a series of blastomere reaggregation experiments, Johnson and Ziomek (1981a) demonstrated that polarization of early 8-cell blastomeres can be induced by contact with blastomeres from both 2-cell and 4-cell embryos, which, it will be recalled, cannot participate in the formation of permeable intercellular membrane channels (Goodall and Johnson, 1982; McLachlin *et al.*, 1983). Here, then, is an example of intercellular communication, one that has important effects on cell morphology and behavior, which is mediated by cell contact but without the participation of cell-cell channels (either gap junctions or midbodies). Shirayoshi *et al.* (1983) have identified a cell surface component (124 kD), part of a calcium-dependent cell-cell adhesion system, that plays an important role in this process: treatment of embryos with monoclonal antibody against this protein interfered with both compaction and polarization, resulting in the formation of blastocyst-like vesicles lacking ICM. Such a clear-cut example of cell surface-mediated intercellular communication demands further close scrutiny.

5. ON THE ROLE OF INTERCELLULAR COMMUNICATION PATHWAYS IN EMBRYOGENESIS

In the foregoing sections, we have established that at least three modes of intercellular communication are available to the blastomeres of the early mouse embryo. What (if any) use is made of them? Do persistent midbodies serve to provide metabolic coupling between the early blastomeres prior to the assembly of gap junctions? Do gap junctions, arising abruptly soon after the third cleavage, transmit small signal molecules that, if they do not participate in the establishment of cell polarity, serve to establish or to perpetuate the differences between the resulting inside and outside cells? Do cell surface interactions, such as that which induces polarization, govern any other aspects of embryogenesis? We have only vague answers to such questions at present, but there is reason to expect that clarification will not be long in coming.

As mentioned in the Introduction, there is ample evidence for the involvement of *some* type of intercellular communication in numerous aspects of mouse embryogenesis. In a few cases, a potentially testable hypothesis can be formulated, invoking a particular mode of communication. For example, polarization of 16-cell blastomeres, like that of 8-cell blastomeres, can be induced by cell contact (Johnson and Ziomek, 1983), indicating that communication *via* cell-cell adhesion molecules plays a continuing role in governing cell behavior in the morula. In addition, differences in cell surface adhesion between inside and outside cells appear to be important for maintaining the respective positions and shapes of these two cell populations, outside cells being less adhesive and spreading over inside cells (Randle, 1982; Kimber *et al.*, 1982; Surani and Handyside, 1983; Surani and Barton, 1984). One hypothesis would be that the induction of 8-cell blastomere polarization is but the beginning of a continuum of cell interactions, dependent on cell-cell adhesion molecules, which maintain cell position and cell shape, thus contributing to the process that commits blastomeres to an ICM or trophectoderm fate. Wiley (1984; and see Chapter 4) has proposed that a blastomere's fate in the morula might be influenced by the proportion of its cell surface that is apposed (adherent to other cells) *vs.* free. The failure of ICM to develop in embryos treated during the 16-cell stage with antibody to the 124kD cell-cell adhesion molecule is consistent with this hypothesis (Shirayoshi *et al.*, 1983).

Do gap junctions play a role in this process? Gap junction-mediated transmission of ions or small metabolites between closely-adhering blastomeres could be part of the mechanism whereby cell position (inside *vs.* outside) is recognized and translated into divergent pathways of cell differentiation. Gap junctions have been invoked many times as participating in the generation of positional information by transmitting morphogenetic signals, but hard evidence for such a function is lacking (Caveney, 1985). Recent success in preparing antibodies against the 27kD gap junction protein provides a means of experimentally blocking the participation of these channels in blastomere interactions (Warner *et al.*, 1984; Hertzberg *et al.*, 1985). By comparing the effects of blocking the formation or function of membrane channels with the effects of interrupting cell-cell adhesion, it should be possible to clarify the relative contributions of these two modes of intercellular communication.

As pointed out by Caveney (1985), it is unlikely that gap junctions transmit regulatory molecules exclusively; most of the flux through these channels (or through cytoplasmic bridges) probably consists of essential metabolites being shared among a population of cells. Indeed, this is a well-documented function of the gap junctions that couple the growing mammalian oocyte with its surrounding follicle cells (reviewed by Moor and Cran, 1980; Schultz, 1985). Probably the most convincing evidence that gap junctions do not simply allow for the sharing of metabolites during embryogenesis is the fact that as cells become organized into tissue domains and undergo differentiation, the junctional connections between domains become attenuated (reviewed by Caveney, 1985; Schultz, 1985). In mouse embryos "implanting" *in vitro*, for example, dye coupling is gradually lost (but ionic coupling is maintained) between the ICM and the spreading trophoblast; furthermore, a restriction was noted in dye coupling between embryonic ectoderm (the core of the ICM) and peripheral cells, probably extraembryonic endoderm (Lo and Gilula, 1979b). We have recently confirmed that dye coupling is restricted between embryonic ectoderm and visceral extraembryonic endoderm in such *in vitro* "implanted" embryos (Caveney and Kidder, unpublished). Perhaps attenuation of gap junctions, reflected in loss of dye coupling, serves to stabilize differentiating tissue domains by restricting the influx of developmental signals emanating from neighboring domains.

Obviously, much experimental work remains to be done. Given the number of events influenced by cell interactions, there is ample opportunity for exploring the involvement of specific modes of intercellular communication in mouse embryogenesis. The next significant advances in this field will undoubtedly come from the use of antibodies, which can selectively interfere with pathways of intercellular communication. In addition, it is to be hoped that embryos of mammalian species other than the mouse will receive greater attention, since important insights concerning structure-function relationships can be gained by examining the similarities and differences between events in various species. To cite just one example, compaction (and presumably blastomere polarization) is delayed in primate embryos relative to the mouse, occurring in the 16-cell stage in the human and even later, beyond the 24-cell stage, in baboon and monkey embryos (reviewed by Lopata *et al.*, 1983). Is gap junction assembly likewise delayed? Recent evidence suggests that it is (Enders and Schlafke, 1984). If so, then a more convincing case can be made for gap junction involvement in processes ensuing from compaction. Such questions await the wider application of methods described in this chapter.

ACKNOWLEDGMENTS

I would like to thank Drs. Jeanne R. McLachlin and Roger A. Pedersen for making available unpublished data and micrographs, and Dr. Pedersen and Dr. Stan Caveney for reviewing the manuscript. The work from my laboratory was supported by NSERC Canada, and was carried out with the expert technical assistance of Cynthia Pape. The manuscript was typed by Callie Cesarini and Becky Bannerman.

6. REFERENCES

Balakier, H., and Pedersen, R.A., 1982, Allocation of cells to inner cell mass and trophectoderm lineages in preimplantation mouse embryos, *Dev. Biol.* 90: 352-362.

Caveney, S., 1985, The role of gap junctions in development, *Ann. Rev. Physiol.* 47: 319-335.

Dermietzel, R., Leibstein, A., Frixen, U., Janssen-Timmen, U., Traub, O., and Willecke, K., 1984, Gap junctions in several tissues share antigenic determinants with liver gap junctions, *EMBO J.* 3: 2261-2270.

Ducibella, T., and Anderson, E., 1975, Cell shape and membrane changes in the eight-cell mouse embryo: Prerequisites for morphogenesis of the blastocyst, *Dev. Biol.* 47: 45-58.

Ducibella, T., and Anderson, E., 1979, The effects of calcium deficiency on the formation of the zonula occludens and blastocoel in the mouse embryo, *Dev. Biol.* 73: 46-58.

Ducibella, T., Albertini, D.F., Anderson, E., and Biggers, J.D., 1975, The preimplantation mammalian embryo: Characterization of intercellular junctions and their appearance during development, *Dev. Biol.* 45: 231-250.

Edelman, G.M., 1983, Cell adhesion molecules, *Science* 219: 450-457.

Enders, A.C., and Schlafke, S., 1984, Morphology of development in the primate: Blastocyst to villous placental stage, *J. Biosci.* 6, Suppl. 2: 53-61.

Goodall, H., and Johnson, M.H., 1982, Use of carboxyfluorescein diacetate to study formation of permeable channels between mouse blastomeres, *Nature (London)* 295: 524-526.

Goodall, H., and Johnson, M.H., 1984, The nature of intercellular coupling within the preimplantation mouse embryo, *J. Embryol. Exp. Morphol.* 79: 53-76.

Graham, R.C., Jr., and Karnovsky, M.J., 1966, The early stages of absorption of injected horseradish peroxidase in the proximal tubules of mouse kidney: Ultrastructural cytochemistry by a new technique, *J. Histochem. Cytochem.* 14: 291-302.

Granholm, N.H., Brenner, G.M., and Rector, J.T., 1979, Latent effects on *in vitro* development following cytochalasin B treatment of 8-cell mouse embryos, *J. Embryol. Exp. Morphol.* 51: 97-108.

Handyside, A.H., 1980, Distribution of antibody- and lectin-binding sites on dissociated blastomeres from mouse morulae: Evidence for polarization at compaction, *J. Embryol. Exp. Morphol.* 60: 99-116.

Hertzberg, E.L., and Skibbens, R.V., 1984, A protein homologous to the 27,000 Dalton liver gap junction protein is present in a wide variety of species and tissues, *Cell* 39: 61-69.

Hertzberg, E.L., Spray, D.C., and Bennett, M.V.L., 1985, Reduction of gap junctional conductance by microinjection of antibodies against the 27-kDa liver gap junction polypeptide, *Proc. Natl. Acad. Sci. USA* 82: 2412-2416.

Hogan, B.L.M., and Tilly, R., 1981, Cell interactions and endoderm differentiation in cultured mouse embryos, *J. Embryol. Exp. Morphol.* 62: 379-394.

Johnson, M.H., and Ziomek, C.A., 1981a, Induction of polarity in mouse 8-cell blastomeres: Specificity, geometry, and stability, *J. Cell Biol.* 91: 303-308.

Johnson, M.H., and Ziomek, C.A., 1981b, The foundation of two distinct cell lineages within the mouse morula, *Cell* 24: 71-80.

Johnson, M.H., and Ziomek, C.A., 1983, Cell interactions influence the fate of mouse blastomeres undergoing the transition from the 16- to the 32-cell stage, *Dev. Biol.* 95: 211-218.

Kaufman, M.H., 1983, The origin, properties and fate of trophoblast in the mouse, in: *Biology of Trophoblast* (Y.W. Loke, and A. Whyte, eds.), Elsevier Science Publishers B.V., Amsterdam, pp. 23-68.

Kidder, G.M., and McLachlin, J.R., 1985, Timing of transcription and protein synthesis underlying morphogenesis in preimplantation mouse embryos, *Dev. Biol.*, 112: 265-275.

Kidder, G.M., Green, A.F., and McLachlin, J.R., 1985, On the use of α-amanitin as a transcriptional blocking agent in mouse embryos: A cautionary note, *J. Exp. Zool.* 233: 155-159.

Kidder, G.M., Rains, J., and McKeon, J., 1986, Gap junction assembly in early mouse embryos is independent of microtubules, microfilaments, cell flattening and cytokinesis, *Proc. Natl. Acad. Sci. USA* (submitted).

Kimber, S.J., Surani, M.A.H., and Barton, S.C., 1982, Interactions of blastomeres suggest changes in cell surface adhesiveness during the formation of inner cell mass and trophectoderm in the preimplantation mouse embryo, *J. Embryol. Exp. Morphol.* 70: 133-152.

Lehtonen, E., 1980, Changes in cell dimensions and intercellular contacts during cleavage-stage cell cycles in mouse embryonic cells, *J. Embryol. Exp. Morphol.* 58: 231-249.

Levey, I.L., and Brinster, R.L., 1978, Effects of α-amanitin on RNA synthesis by mouse embryos in culture, *J. Exp. Zool.* 203: 351-359.

Lo, C.W., 1980, Gap junctions and development, in: *Development in Mammals*, Volume 4 (M.H. Johnson, ed.), Elsevier/North-Holland Biomedical Press, New York, pp. 39-80.

Lo, C.W., and Gilula, N.B., 1979a, Gap junctional communication in the preimplantation mouse embryo, *Cell* 18: 399-409.

Lo, C.W., and Gilula, N.B., 1979b, Gap junctional communication in the post-implantation mouse embryo, *Cell* 18: 411-422.

Loewenstein, W.R., 1981, Junctional intercellular communication: The cell-to-cell membrane channel, *Physiol. Rev.* 61: 829-913.

Lopata, A., Kohlman, D., and Johnston, I., 1983, The fine structure of normal and abnormal human embryos developed in culture, in: *Fertilization of the Human Egg In Vitro* (H.M. Beier, and H.R. Lindner, eds.), Springer-Verlag, Berlin, pp. 189-210.

Magnuson, T., Demsey, A., and Stackpole, C.W., 1977, Characterization of intercellular junctions in the preimplantation mouse embryo by freeze-fracture and thin-section electron microscopy, *Dev. Biol.* 61: 252-261.

Magnuson, T., Jacobson, J.B., and Stackpole, C.W., 1978, Relation between intercellular permeability and junction organization in the preimplantation mouse embryo, *Dev. Biol.* 67: 214-224.

McLachlin, J.R., 1984, Control of compaction and junctional communication in preimplantation mouse embryos, *Doctoral Dissertation*, The University of Western Ontario, London, Canada.

McLachlin, J.R., and Kidder, G.M., 1981, Genetic control of intercellular communication in the early mouse embryo, *J. Cell Biol.* 91: 169a.

McLachlin, J.R., and Kidder, G.M., 1986, Intercellular junctional coupling in preimplantation mouse embryos: Effect of blocking transcription or translation, *Dev. Biol.*, in press.

McLachlin, J.R., Caveney, S., and Kidder, G.M., 1983, Control of gap junction formation in early mouse embryos, *Dev. Biol.* 98: 155-164.

Model, P.G., Jarrett, L.S., and Bonazzoli, R., 1981, Cellular contacts between hindbrain and prospective ear during inductive interaction in the axolotl embryo, *J. Embryol. Exp. Morphol.* 66: 27-41.

Monk, M., and Petzoldt, U., 1976, Control of inner cell mass development in cultured mouse blastocysts, *Nature (London)* 265: 338-339.

Moor, R.M., and Cran, D.G., 1980, Intercellular coupling in mammalian oocytes, in: *Development in Mammals*, Volume 4 (M.H. Johnson, ed.), Elsevier/North-Holland Biomedical Press, New York, pp. 3-37.

Pedersen, R.A., Wu, K., and Bałakier, H., 1986, Origin of the inner cell mass in mouse embryos: cell lineage analysis by microinjection, *Dev. Biol.*, in press.

Randle, B.J., 1982, Cosegregation of monoclonal antibody reactivity and cell behaviour in the mouse preimplantation embryo, *J. Embryol. Exp. Morphol.* 70: 261-278.

Saxén, L., and Lehtonen, E., 1978, Transfilter induction of kidney tubules as a function of the extent and duration of intercellular contacts, *J. Embryol. Exp. Morphol.* 47: 97-109.

Saxén, L., Ekblom, P., and Thesleff, I., 1980, Mechanisms of morphogenetic cell interactions, in: *Development in Mammals*, Volume 4 (M.H. Johnson, ed.), Elsevier/North-Holland Biomedical Press, New York, pp. 161-202.

Schultz, R.M., 1985, Roles of cell-to-cell communication in development, *Biol. Reprod.* 32: 27-42.

Shirayoshi, Y., Okada, T.S., and Takeichi, M., 1983, The calcium-dependent cell-cell adhesion system regulates inner cell mass formation and cell surface polarization in early mouse development, *Cell* 35: 631-638.

Shivers, R.R., and McLachlin, J.R., 1984, Freeze-fracture of 2-cell mouse embryos. A new method for fracture of very small and scarce biological samples. *J. Submicrosc. Cytol.* 16: 423-430.

Sołtyńska, M.S., 1982, The possible mechanism of cell positioning in mouse morulae: An ultrastructural study, *J. Embryol. Exp. Morphol.* 68: 137-147.

Surani, M.A.H., and Barton, S.C., 1984, Spatial distribution of blastomeres is dependent on cell division order and interactions in mouse morulae, *Dev. Biol.* 102: 335-343.

Surani, M.A.H., and Handyside, A.H., 1983, Reassortment of cells according to position in mouse morulae, *J. Exp. Zool.* 225: 505-511.

Surani, M.A.H., Barton, S.C., and Burling, A., 1980, Differentiation of 2-cell and 8-cell mouse embryos arrested by cytoskeletal inhibitors, *Exp. Cell Res.* 125: 275-286.

Warner, A.E., Guthrie, S.C., and Gilula, N.B., 1984, Antibodies to gap-junctional protein selectively disrupt junctional communication in the early amphibian embryo, *Nature (London)* 311: 127-131.

Warner, C.M., 1977, RNA polymerase activity in preimplantation mammalian embryos, in: *Development in Mammals*, Vol. 1 (M.H. Johnson, ed.), Elsevier/North-Holland Biomedical Press, New York, pp. 99-136.

Warner, C.M., and Hearn, T.F., 1977, The effect of α-amanitin on nucleic acid synthesis in preimplantation mouse embryos, *Differentiation* 7: 89-97.

Wiley, L.M., 1984, The cell surface of the mammalian embryo during early development, in: *Ultrastructure of Reproduction* (J. Van Blerkom, and P.M. Motta, eds.), Martinus Nijhoff Publishers, Boston, pp. 190-204.

Ziomek, C.A., and Johnson, M.H., 1980, Cell surface interaction induces polarization of mouse 8-cell blastomeres at compaction, *Cell* 21: 935-942.

Chapter 4

DEVELOPMENT OF THE BLASTOCYST: ROLE OF CELL POLARITY IN CAVITATION AND CELL DIFFERENTIATION

LYNN M. WILEY

1. INTRODUCTION

1.1. Blastocyst Formation in the Mouse Embryo: Cavitation

Blastocyst formation entails two interrelated events: 1) cavitation, whereby the embryo develops a blastocoele, and 2) differentiation of the first two cell types to appear in the mammalian embryo, *i.e.*, trophectoderm and inner cell mass (ICM). The trophectoderm, which forms the wall of the blastocoele, is the only tissue that can implant the embryo in the uterus and protect the fetus against maternal immunological rejection. The ICM, which is a cluster of 3 to 10 cells that adheres to the luminal surface of the trophectoderm, is the sole cellular source of the entire embryo (Gardner, 1975; Gardner and Rossant, 1976). The blastocoele is thought to be essential for the normal development and function of the trophectoderm and ICM, since there has never been a recorded case of both these two cell types developing in the absence of cavitation.

Cavitation begins at the morula stage when the embryo consists of 16-32 closely apposed blastomeres (Fig. 1). The morula exhibits two topographical features that are important to cavitation and to the differentiation of trophectoderm and ICM. First, some blastomeres are completely enclosed so that there are "outer" and "inner" blastomeres. Second, outer blastomeres will have two types of cell surfaces: free (facing the zona pellucida) and apposed (flattened against adjacent blastomeres) making them "polar". On the other hand, inner blastomeres will have only one type of cell surface (apposed) making them "apolar". As cavitation proceeds, inner apolar cells become ICM while outer polar blastomeres become trophectoderm.

Lynn M. Wiley Division of Reproductive Biology and Medicine, Department of Obstetrics and Gynecology, University of California, Davis, California 95616, USA.

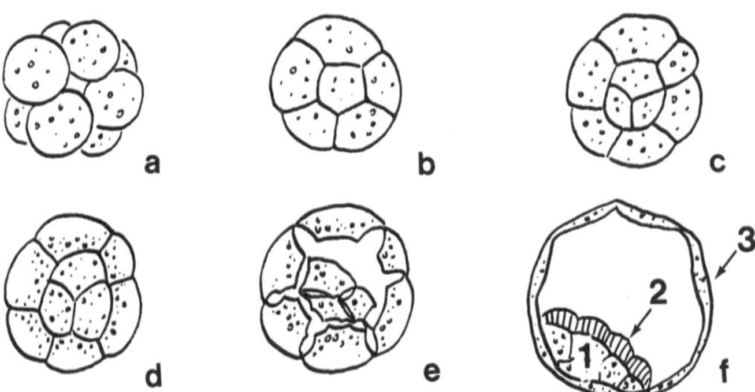

Figure 1. Schematic of cavitation in the mouse morula. At the 8-cell stage, initially spherical blastomeres (a) flatten against one another to maximize cell-cell apposition (= compaction, b). After another round of mitosis to the 16-cell stage (c), randomly distributed refractile lipid droplets and mitochondria in outer blastomeres become localized to basolateral cell borders (d), which then begin to separate due to accumulation of nascent blastocoele fluid (e). The fluid coalesces into a definitive blastocoele to transform the morula into a blastocyst (f), which has an inner cell mass (1), whose luminal cell layer becomes primitive endoderm (2), and a trophectodermal outer layer (3) enclosing the blastocoele.

1.2. Why Does the Mammalian Embryo Cavitate?

A survey of metazoan embryology reveals that fertilized ova from several phyla first cleave to form a characteristic number of blastomeres and then cavitate to form a cystic structure comprised of a fluid-filled cavity enclosed by an epithelial layer. Consequently, intraembryonic cavities and epithelial layers are the first structures to develop in such embryos and the subsequent modification of these structures establishes the three-dimensional architecture of the embryo. In the case of the mouse preimplantation embryo, I have proposed the existence of two specific purposes for cavitation: 1) to facilitate the development of the outer layer of blastomeres into the first epithelial layer of the mammalian embryo, *i.e.*, trophectoderm, and 2) to produce a new population of "outer" cells with a free surface on the luminal face of the ICM within the blastocoele to cue the differentiation of the second epithelial layer appearing in the mouse embryo, primitive endoderm (Fig. 1).

1.3. The Cell Apposition Hypothesis

The Cell Apposition Hypothesis functionally links cavitation to the formation of epithelial layers, in particular trophectoderm and primitive endoderm (Wiley, 1984b). The primary tenet of this hypothesis is that the genetic activity of an outer, polar blastomere is sensitive to the ratio of free to apposed cell surface. This hypothesis predicts that within an aggregate of pluripotent embryonic stem cells, differentiated cells will always appear at the periphery of the aggregate as a direct result of possessing a portion of their cell surface that is not in contact with adjacent cells. This is indeed what is observed during cell differentiation in morulae, in isolated ICMs

(Wiley *et al.*, 1978) and in embryoid bodies comprised of embryonal carcinoma stem cells (Martin, 1980).

2. THREE MODELS FOR CAVITATION: CELL POLARITY AND THE PRODUCTION OF NASCENT BLASTOCOELE FLUID

The above discussion provided the rationale for assigning a functional relationship between cavitation and the formation of epithelial layers. I will now discuss cavitation in terms of three experimental cavitation models to provide: 1) an historical perspective on our current understanding of cavitation, and 2) the rationale for assigning a functional relationship between cell polarity and the ability of the embryo to cavitate and form epithelial layers.

2.1. Secretion Cavitation Model

This model is based on the observations of Melissinos (1907) that the initial appearance of blastocoele fluid at the 16-cell stage is preceded by the localization of refractile cytoplasmic droplets to the basolateral borders of the outer blastomeres (Fig. 2). As fluid accumulates between the blastomeres and coalesces to form the blastocoele, the droplets decrease in number. This observation led Melissinos to suggest that the droplets contain nascent blastocoele fluid that is secreted into the intercellular spaces. Droplet-mediated secretion was subsequently widely accepted as the mechanism for the production of nascent blastocoele fluid (see Calarco and Brown, 1969; Wiley and Eglitis, 1980). Another observation favoring a droplet-mediated secretion mechanism is that cavitation is reversibly inhibited by drugs that depolymerize microtubules (Wiley and Eglitis, 1980). In other systems, these drugs inhibit the translocation of vesicles from their sites of synthesis to their sites of secretion (Hoffstein *et al.*, 1977; Malaisse *et al.*, 1975; Poisner and Cooke, 1975).

However, three additional observations do not support the Secretion Cavitation Model. First, electron microscopic studies have failed to produce morphological evidence of droplet exocytosis (*e.g.*, Calarco and Brown, 1969). Second, the droplets resemble lipid droplets morphologically (Fig. 3; Calarco and Brown, 1969), while autoradiographic and biochemical studies suggest that the droplets consist of neutral lipids (Flynn and Hillman, 1978; Nadajcka and Hillman, 1975). Third, morphological and physiological evidence indicates that the blastocoele fluid is aqueous, not lipid (Calarco and Brown, 1969; Wiley and Eglitis, 1980; Borland *et al.*, 1977). It becomes necessary, then, to determine whether (and how) droplets, which are most likely lipid, are converted into an extracellular fluid that is aqueous.

2.2. Transport Cavitation Model

The extensive cell-cell apposition of the morula results from an earlier event called "compaction", which occurs at the 8-cell stage when initially spherical blastomeres flatten against one another to obliterate intercellular spaces. Compaction is accompanied by the formation of specialized cell junctions (tight junctions) that unite the apical ends of the outer blastomeres

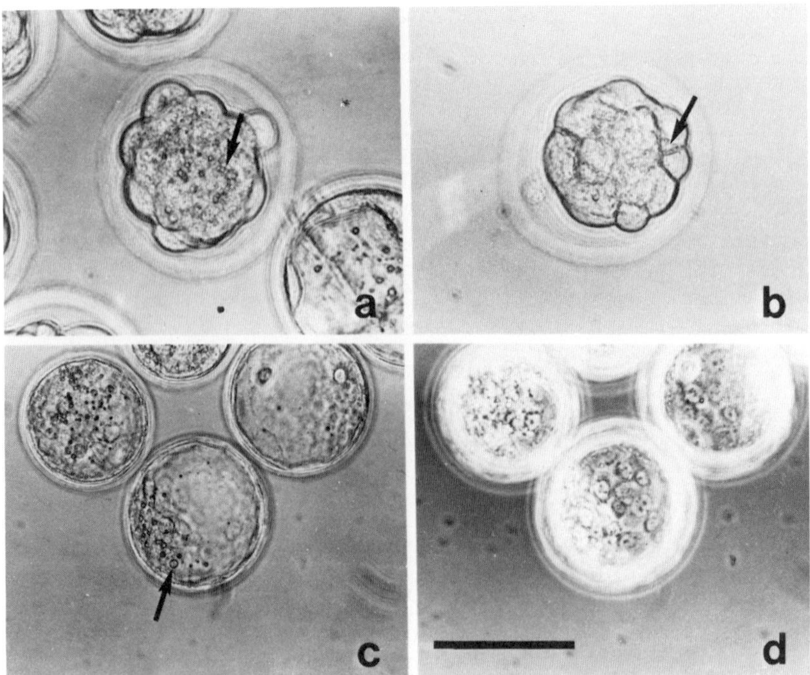

Figure 2. Phase contrast micrographs of mouse morulae cavitating into blastocysts. (a) Late morula approximately 80 hr after fertilization (egg activation) with refractile cytoplasmic vesicles (lipid; arrow) adjacent to basolateral plasma membrane; (b) morula approximately 86 hr after fertilization with nascent blastocoele fluid (arrow) separating adjacent blastomeres; (c) expanded blastocysts approximately 125 hr after fertilization showing retention of refractile vesicles within the ICM (arrow), suggesting that inner cells destined to become ICM do not contribute to the formation of blastocoele fluid by catabolism of refractile lipid droplets; (d) same field as in "c" but with the microscope focused on the trophectoderm: note absence of refractile vesicles within trophectoderm cells, suggesting that the lipid droplets in outer blastomeres have been catabolized, possibly to provide nascent blastocoele fluid according to the Metabolic Cavitation Model. Bar in "d" represents 100 μm (all micrographs made at same optical magnification).

(Ducibella *et al.*, 1975). Earlier ultrastructural studies with permeability tracers suggested that these tight junctions become zonular and form a permeability seal which isolates the inner blastomeres (and potential intercellular spaces) from the extraembryonic milieu before cavitation begins (Mintz, 1965; Ducibella and Anderson, 1975). These observations were incorporated into the Transport Cavitation Model, the primary tenet of which is that zonular tight junction formation and establishment of a permeability seal are prerequisites of cavitation, and that they, together with transepithelial movement of ions and water across the outer blastomeres, are responsible for the origin of nascent blastocoele fluid (Ducibella and Anderson, 1975). Indeed, the existence of ion and fluid transport has been shown to be mediated by ouabain-sensitive ion transport in mature mouse blastocysts (Borland *et al.*, 1977; DiZio and Tasca, 1977).

However, in the case of the 16-cell stage morula, recent studies have shown that zonular tight junctions of mouse blastocysts are still permeable to

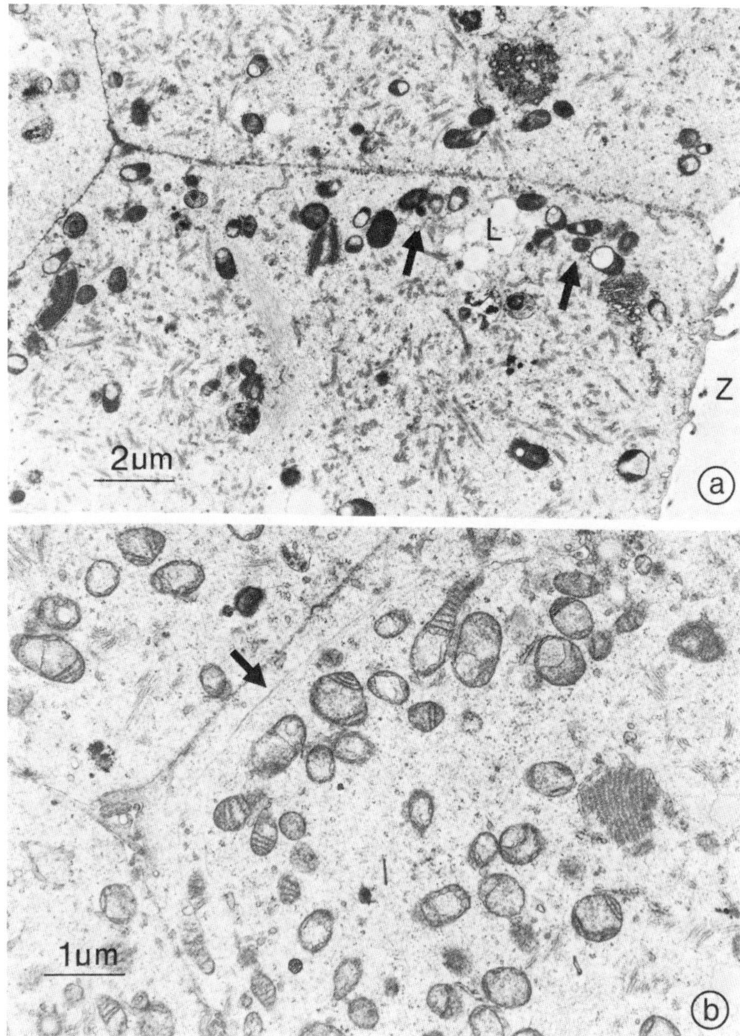

Figure 3. Electron micrographs of localization of lipid droplets and mitochondria to apposed cell borders of outer blastomeres of mouse morulae approximately 80 hr after fertilization (egg activation). (a) Lipid (L) and mitochondria (arrows) adjacent to an apposed cell border of an outer blastomere: the moderate electron opacity and lack of a limiting unit membrane identify the lipid; (b) a population of mitochondria along the basolateral border of an outer blastomere: arrow points to microtubules running parallel with the plasma membrane.

proteins, as shown by the ability to lyse ICM cells with immune serum and complement placed in the culture medium 14 hr after cavitation has begun, when the blastocoele has already formed (McLaren and Smith, 1977). In addition, embryo volume does not begin to increase until the diameter of the developing blastocoele is one third the diameter of the embryo, making it difficult to envision how the embryo could take up extraembryonic fluid without increasing in volume during nascent blastocoele fluid production (Wiley and Eglitis, 1981). Finally, this model does not account for the decrease in the number of refractile cytoplasmic droplets that accompanies the formation of nascent blastocoele fluid.

2.3. Metabolic Cavitation Model

To overcome the shortcomings of the Secretion and Transport Cavitation Models, a new model has been proposed called the Metabolic Cavitation Model, so named because it invokes beta-oxidation of lipid in the production of nascent blastocoele fluid (Wiley, 1984a).

As mentioned above, by the time the morula begins to cavitate, the apical and basolateral surfaces of outer blastomeres display several morphological, biochemical and functional differences (reviewed in Wiley, 1984b). Of special significance to the Metabolic Cavitation Model is the distribution of the ion transport enzyme Na^{+}/K^{+}-ATPase, which is cytochemically detectable on the outer surface of the plasma membrane of basolateral aspects but not on the apical aspects of outer blastomeres (Vorbrodt *et al.*, 1977). Why this observation is significant will become clear momentarily.

For the present, however, I will return to the problem of how a reduction in the number of refractile (lipid) droplets can be related to the appearance of aqueous nascent blastocoele fluid between the blastomeres. Lipid catabolism (Flynn and Hillman, 1980) and ATP utilization (Ginsberg and Hillman, 1973) both increase significantly in morulae at the beginning of cavitation and available data make it likely that the droplets in mouse morulae consist predominantly of neutral lipid, including palmitic acid (Nadajcka and Hillman, 1975). Catabolism of neutral lipids occurs by the intramitochondrial pathway of beta-oxidation so that one mole of palmitic acid, for example, yields 131 moles of ATP and 146 moles of water. Mitochondria in the outer blastomeres co-localize with the lipid droplets to the basolateral aspects just prior to the appearance of nascent blastocoele fluid (Wiley, 1984a).

The Metabolic Cavitation Model proposes that it is the juxtaposition of mitochondria, lipid droplets and Na^{+}/K^{+}-ATPase located on the basolateral plasma membrane, that is responsible for the production of nascent blastocoele fluid (Fig. 4). It is proposed that the specific purpose of this juxtaposition is to generate water and ATP in close proximity to the Na^{+}/K^{+}-ATPase that is located on the basolateral plasma membrane. The ATP would then be utilized by the Na^{+}/K^{+}-ATPase to pump Na^{+} out of the cytoplasm into the intercellular spaces. The water would then passively follow the Na^{+} to become nascent blastocoele fluid. This mechanism explains how lipid could give rise to aqueous intercellular fluid without invoking the participation of droplet exocytosis. In this way, we can account for the observed decrease in refractile lipid droplets and the maintenance of embryo diameter during nascent blastocoele fluid formation.

2.4. The Effects of Ouabain and Extracellular Potassium on Cavitation

Two important predictions can be made based on the proposed role of Na^{+}/K^{+}-ATPase activity in moving Na^{+} out of the cytoplasm into the inter-

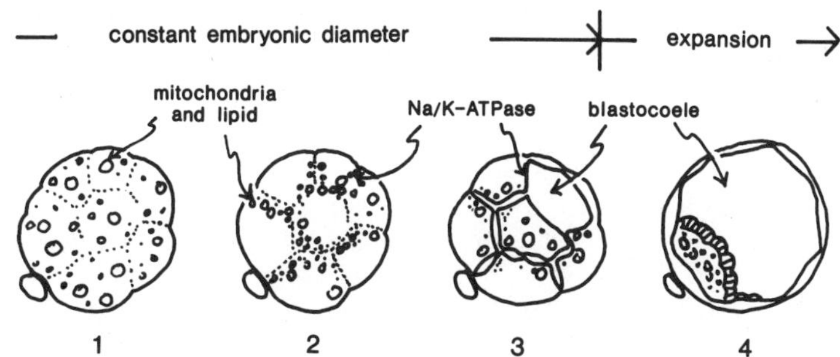

Figure 4. Schematic of the Metabolic Cavitation Model. (1) Organelles are dispersed in early morulae; (2) refractile lipid droplets and mitochondria become localized to the basolateral cell borders of outer blastomeres where the Na^+/K^+-ATPase is located on the outer face of the plasma membrane; (3) beta-oxidation of the lipid by mitochondria generates water and ATP; the ATP is utilized by Na^+/K^+-ATPase to pump cytoplasmic sodium into the intercellular spaces and water follows passively to become nascent blastocoele fluid; (4) continued blastocoele expansion involving Na^+/K^+-ATPase-dependent transport of extraembryonic ions and water across the maturing trophectoderm into the blastocoele (see Benos, 1981a,b).

cellular spaces. The first prediction is that cavitation should be affected by treatments that modify Na^+/K^+-ATPase activity. The second prediction is based on the fact that if Na^+ is indeed leaving the basolateral cytoplasm then Na^+ must be leaking into the apical cytoplasm. The movement of Na^+ across the blastomere would be accompanied by an electro-osmotic flow of water that could sweep the lipid droplets and mitochondria towards the basolateral cytoplasm. Consequently, if Na^+/K^+-ATPase activity were modified, then the second prediction is that the basolateral localization of these organelles should also be modified along with cavitation.

Two ways of modifying Na^+/K^+-ATPase activity are: 1) to vary the concentration of extracellular potassium (K^+), which affects Na^+/K^+-ATPase activity indirectly by altering plasma membrane potentials and passive ion fluxes (Cohen *et al.*, 1976), and 2) to treat with the cardiac glycoside ouabain, an inhibitor of Na^+/K^+-ATPase activity (Glynn, 1957, 1964).

When mouse morulae are incubated in ouabain for 40 hr, low concentrations of ouabain (10^{-5}M) accelerate the rate at which cavitation occurs, while at higher concentrations (10^{-4}M) ouabain delays the rate at which cavitation occurs (Fig. 5). This biphasic response is normally observed with ouabain and is an indication of the specificity of the drug's action on Na^+/K^+-ATPase activity (Glynn, 1964; Cohen *et al.*, 1976). In addition, sensitivity of cavitation rate to ouabain is decreased as K^+ increases (Fig. 5). Because K^+ competes with ouabain for the same binding site on Na^+/K^+-ATPase (Glynn, 1964; Fossel and Solomon, 1978; Stekhoven and Bonting, 1981), this observation also indicates that the effects of ouabain on cavitation are related to modified Na^+/K^+-ATPase activity. Finally, the rate of cavitation is inversely related to the concentration of K^+, implying that hyperpolarization accelerates the rate of cavitation while hypopolarization slows the rate of cavitation (Fig. 6). These observations imply that cavitation rate can be modified indirectly by membrane potential changes affecting Na^+/K^+-ATPase activity. Collectively,

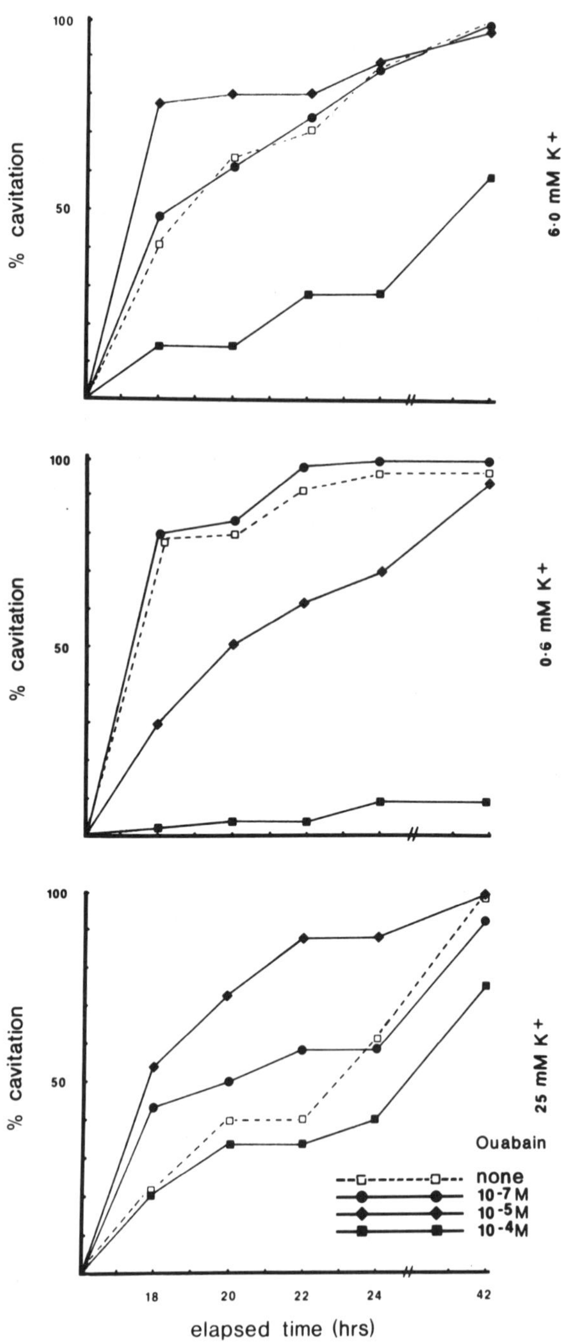
100
50
% cavitation
6·0 mM K+
100
50
% cavitation
0·6 mM K+
100
50
% cavitation
25 mM K+
Ouabain
none
10-7 M
10-5 M
10-4 M
18
20
22
24
42
elapsed time (hrs)

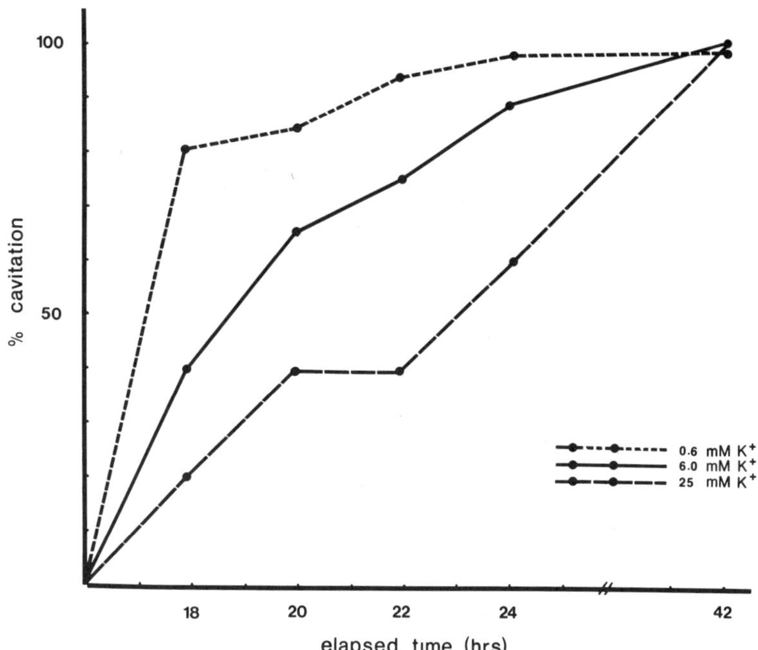

Figure 6. Effects of varying the concentration of extracellular potassium (0.6, 6.0 or 25 mM) on cavitation. Each point represents the response of 40 to 50 embryos pooled from 3 to 5 experiments. Horizontal axis represents elapsed hours of the experimental culture period. Vertical axis represents the percentage of embryos with an intraembryonic extracellular cavity discernible using an inverted phase contrast microscope at 200 x magnification. The first observations were performed at 18 hr; all curves have been extrapolated to the origin. (From Wiley, 1984a).

these three observations indicate that the production of nascent blastocoele fluid is dependent on Na^+/K^+-ATPase activity, thus confirming the first prediction that cavitation would be affected by treatments which modify Na^+/K^+-ATPase activity.

When morulae treated with ouabain and with varying concentrations of K^+ are prepared for morphometric analyses at the microscopic level (Fig. 7), one finds that treatments accelerating cavitation increase the population density of mitochondria along basolateral plasma membranes, while those treatments that slow cavitation decrease the population density of mitochondria along basolateral plasma membranes of outer blastomeres (Tables I and II;

Figure 5. Effects of ouabain (10^{-7}, 10^{-5} or 10^{-4}M or none) on cavitation when extracellular potassium is 0.6, 6.0 and 25 mM. Each point represents the response of 40 to 55 embryos pooled from 2 experiments. Horizontal axes represent elapsed hours in the presence of ouabain. Vertical axes represent the percentage of embryos with an intraembryonic extracellular cavity discernible using an inverted phase contrast microscope at 200 x magnification. The first observations were performed at 18 hr: all curves have been extrapolated to the origin. (From Wiley, 1984a).

Table I
Number of Mitochondria per Millimeter of Apposed or Free Cell Border in Outside Cells[a]

	Treatment (mean ± SD)[b,c]			
	Standard K -ouabain	Standard K +ouabain	Low K -ouabain	Low K +ouabain
Cell border				
apposed	0.1119 ± 0.0693	0.0974 ± 0.0423	0.1590 ± 0.0889	0.1121 ± 0.0630
free	0.0503 ± 0.0518	0.0516 ± 0.0407	0.0865 ± 0.0562	0.0836 ± 0.0514
N (number of cells counted)				
apposed cell border	15	29	30	25
free cell border	15	26	29	24

[a]Data from Wiley (1984a).
[b]Values represent number of mitochondria per millimeter of plasma membrane in a montage like that shown in Figure 7.
[c]Standard K = medium with 6.0 mM K; low K = 0.6 mM. Ouabain (10^{-5}M) is present (+) or absent (-). Values for apposed and free cell borders for treatment group [low K -ouabain] are statistically comparable (Students' t-test, $p < 0.05$); this treatment significantly delayed cavitation. For all other treatment groups, values for apposed and free cell borders are significantly different ($p < 0.01$). Treatments with standard K +ouabain and low K -ouabain both accelerated cavitation.

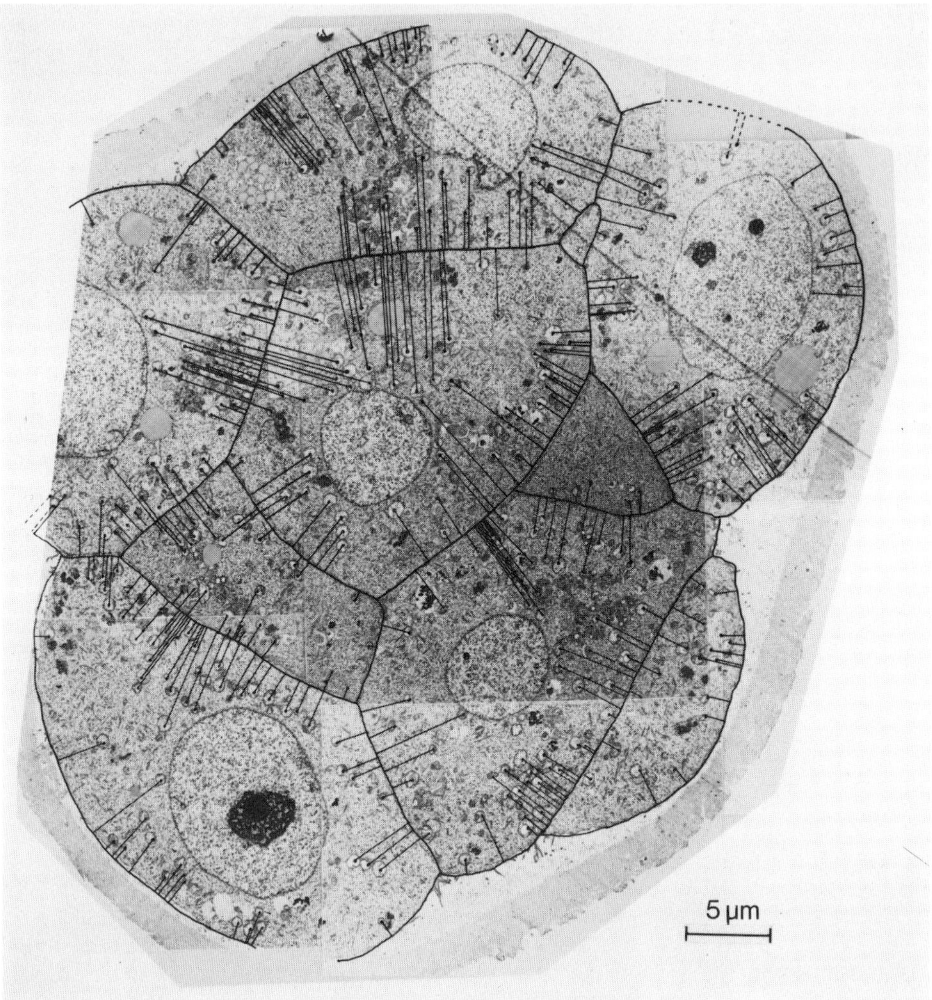

Figure 7. Morphometric analyses: Montage of a morula (approximately 80 hr after fertilization, *i.e.*, after egg activation) with corresponding tissue paper overlay. The blastomeres are outlined and each mitochondrion is represented by a dot. Each dot is connected by a perpendicular straight line to the nearest cell border. (From Wiley, 1984a).

Wiley, 1984a). This observation confirms the prediction that treatments modifying cavitation rate would also modify the basolateral localization of organelles in outer blastomeres.

2.5. A Mechanism for the Basolateral Localization of Organelles in Outer Blastomeres

By modifying the Metabolic Cavitation Model to accommodate these

Table II
Descriptive Statistics for Distances (in Microns) of Mitochondria from Apposed Cell Borders in Outside Cells[a,b]

Treatment	Percentile[c] 25	50	75	IQ distance[d]
Standard K -ouabain	0.93	1.61	3.13	2.20
Standard K +ouabain	0.70	1.32*	2.45	1.75
Low K -ouabain	0.82	1.61	3.30	2.48
Low K +ouabain	1.15	2.28**	4.03	2.88

[a]Data from Wiley (1984a).
[b]Number of cells from which measurements were obtained per treatment: standard K -ouabain: 16; standard K +ouabain: 33; low K -ouabain: 25; low K +ouabain: 30. For each treatment, these cells were selected from at least 5 different embryos. Number of mitochondria that were measured per treatment: standard K -ouabain: 380; standard K +ouabain: 1062; low K -ouabain: 1104; low K +ouabain: 1063. Concentrations of K and ouabain given in Table I.
[c]Levels of significance: *$p < 0.025$; **$p < 0.01$.
[d]IQ = interquartile distance: in microns, the distance between the 25th and 75th percentile medians that are given in the first and third columns of values in the table.

results obtained with ouabain, we gain a mechanism for the localization of lipid droplets and mitochondria to the basolateral borders of outer blastomeres. This mechanism consists of the transcellular flow of extraembryonic ions and osmotic water through the outer blastomeres that enters their apical surfaces and leaves their basolateral surfaces. Paracellular shunting (Berridge and Oschman, 1972) is proposed to explain how the extraembryonic ions and water could leave the embryo so that during the beginning of cavitation embryo diameter would remain constant.

Two predictions can be made, based on the hypothesis that there is a transcellular flow of ions and water across the outer blastomeres that is involved in the basolateral localization of lipid droplets and mitochondria. First, outer blastomeres will have apical-basal transcellular ionic currents. Second, since these currents must leak back between the blastomeres to complete the current loop (*i.e.*, paracellular shunting), they will generate an electric field along their lateral surfaces. Such a field could move glycoproteins in the plane of the membrane by lateral electrophoresis (Jaffe, 1977), and the distribution of such glycoproteins could have intracellular effects mediated through cell surface-cytoskeletal interactions and/or through additional mechanisms. Consequently, an *imposed* electric field might impart a morphological polarity on isolated, unpolarized blastomeres.

Transcellular ionic currents and attendant electric fields have been observed around the Xenopus oocyte (Robinson, 1979), the fucoid embryo (Quatrano *et al.*, 1978) and for embryos of many other species (Stern, 1982; see also Jaffe, 1982). In the fucoid embryo, it is hypothesized that ion currents induce the localization of specific organelles to specific cell surfaces and are thought to produce a self-electrophoresis-like action and/or an electro-osmotic flow of water across the cells to cause certain organelles to migrate (Quatrano *et al.*, 1978). In fact, the notion that cellular electrical polarity serves as an effector of developmental events by inducing cell polarity has acquired a substantial body of supportive experimental evidence (Jaffe, 1982; Nuccitelli, 1983). The first direct confirmation that electrical polarity can indeed redistribute organelles within an animal cell has just recently been obtained with isolated mouse blastomeres in an experiment that will be described below (Wiley and Nuccitelli, 1985).

3. ELECTRICAL POLARITY OF OUTER BLASTOMERES FROM MOUSE MORULAE

3.1. Isolated Outer Blastomeres Have Apical-Basal Transcellular Ionic Currents

To test the prediction that outer blastomeres have transcellular ionic currents, isolated outer blastomeres were examined with an extracellular vibrating probe (Nuccitelli and Wiley, 1985). Because the magnitude of an electric field around a spherical cell diminishes with the inverse cube of the distance from the center of the cell, current detectability increases with an increase in the ratio of the cell radius divided by the distance from the cell

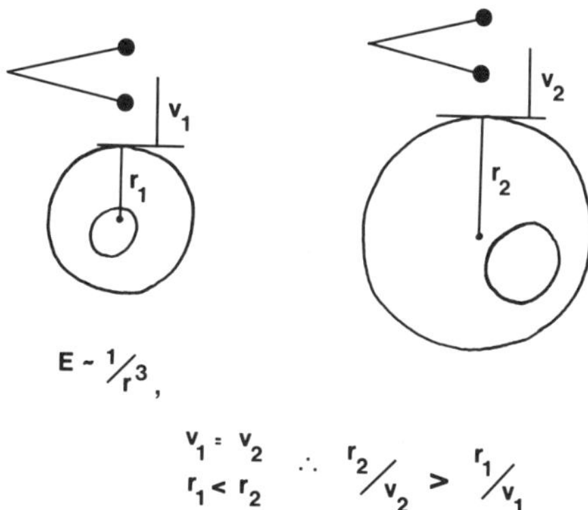

Figure 8. Relationship between the detectability of an electric field (E) and the size of a spherical cell. The magnitude of E diminishes with the inverse cube of the distance from the cell surface to the cell center (r). Thus, detectability of E of given magnitude increases in proportion to the ratio of that distance between the midpoint of the distance traveled by the probe tip as it vibrates and the tangent of the cell surface (v) divided by the distance from the cell surface to the cell center (r). Therefore, if v is held constant, then as r increases, r/v increases and so does detectability of E.

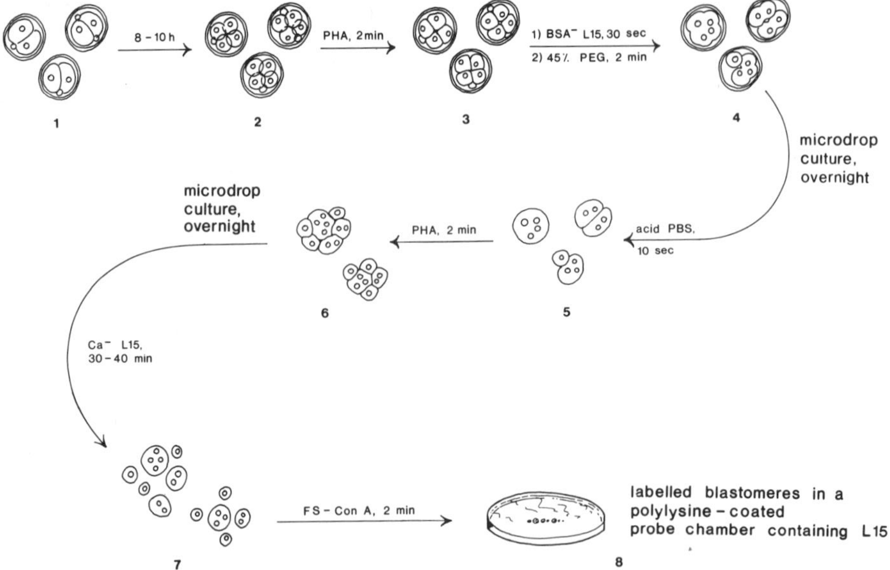

Figure 9. Preparation of enlarged blastomeres. Two-cell stage embryos (1) are cultured to the 4-cell stage (2) and treated with phytohemagglutinin (PHA) to maximize cell-cell apposition (3). After a quick rinse in BSA-free medium L15, embryos are treated with a 45% solution of polyethylene glycol (PEG) to initiate blastomere fusion (4). Treated embryos are subsequently cultured overnight and briefly rinsed with acidified phosphate-buffered saline (PBS) to remove their zonae pellucidae (5) and aggregated together with another PHA treatment (6). The aggregates are cultured overnight and then dissociated the following morning with calcium-free medium to recover enlarged blastomeres (7), which are then labelled with fluoresceinated succinylated concanavalin A (FS-Con A) and transferred to a probe chamber, the glass bottom of which has been rinsed with a 1% aqueous solution of polylysine and filled with the desired medium for probe analyses (8). (From Nuccitelli and Wiley, 1985).

surface to the midpoint of the distance traveled by the probe tip as it vibrates (Fig. 8). The diameter of natural outer 16-cell stage blastomeres is 20 to 30 μm, which proves to be too small to provide reproducible current detectability. To increase current detectability using present technology, it was necessary to enlarge the blastomeres, which was accomplished here by polyethylene-glycol induced fusion of blastomeres in intact 4-cell embryos (Fig. 9; Eglitis and Wiley, 1981).

The resulting enlarged blastomeres were then aggregated with carrier blastomeres so they could participate as 'outer' blastomeres in the process of compaction and so become polarized. Enlarged blastomeres were recovered from the aggregates and then treated with fluoresceinated succinylated Concanavalin A (FS-Con A, 150 μg/ml: Ziomek and Johnson, 1981). This treatment marks the apical poles of the enlarged blastomeres with fluorescent apical 'caps', which result from membrane amplification due to the apical microvilli (Fig. 10).

The extracellular vibrating probe operates on the principle that a net flux of ions moving into or out of a cell will generate a local voltage gradient in the extracellular medium. The actual current density generating this voltage can be calculated given the resistivity of the medium. The resolution of

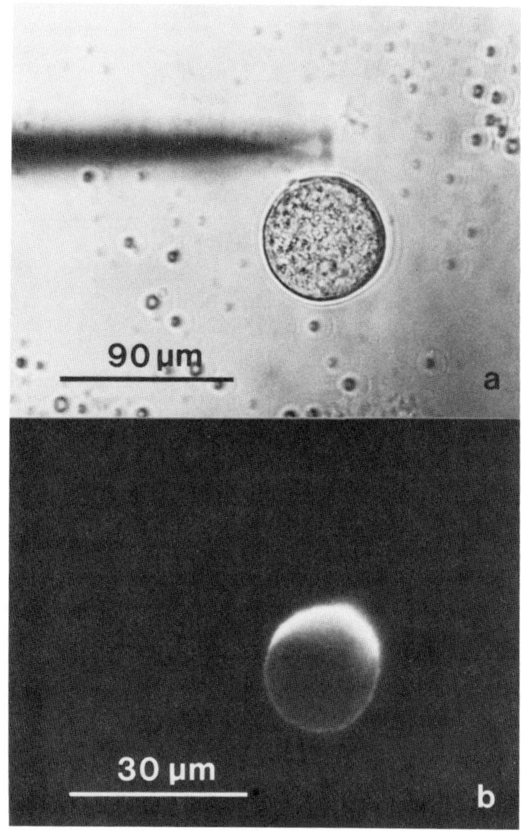

Figure 10. Measuring current density crossing the apical cap of an FS-Con A-labelled enlarged blastomere. (a) Bright-field image of an enlarged blastomere with tip of the vibrating probe positioned over the apical cap; (b) fluorescence micrograph of a different enlarged blastomere to illustrate fluorescence of the apical cap used to determine the apical-basal axis. (From Nuccitelli and Wiley, 1985).

the probe depends on electrode noise, which can be reduced by increasing probe tip size (losing spatial resolution) or by averaging individual measurements over longer time intervals (losing temporal resolution). For this experiment, the compromise that was struck was the use of a probe tip diameter of 8-10 µm, vibrating with an amplitude of 15 µm at a frequency of 350 Hz in medium with 72 ohm-cm resistivity using a 3-sec time constant. This approach yielded a resolution of 0.2 $\mu A/cm^2$ (for additional theoretical and experimental information, see Jaffe and Nuccitelli, 1974; Nuccitelli and Wiley, 1985).

As expected, the average current density (sum of absolute values of apical and basal current densities divided by 2) around natural blastomeres (20-30 µm in diameter) approached the resolution of the probe system (*i.e.*, 0.2 $\mu A/cm^2$) and was undetectable in 57% of the cases (Table III). However, with enlarged blastomeres (30-65 µm in diameter), the average current density was 0.8 $\mu A/cm^2$ and currents were detectable in 70% of the cases.

Table III
Ion Currents Through Isolated, 8- to 16- Cell Stage Mouse Blastomeres[a]

Condition	Position (avg. current density)[b] $\mu A/cm^2 \pm$ SEM (n)	Cases of inward current	Cases of outward current	Current undetected
Normal	Apical [0.3 ± 0.03 (12)]	3/14 (21%)	3/14 (21%)	8/14 (57%)
	Basal [0.2 ± 0.07 (8)]	0/14 (0%)	3/14 (21%)	11/14 (79%)
Enlarged	Apical [0.8 ± 0.1 (38)]	22/53 (42%)	16/53 (30%)	15/53 (28%)
	Basal [0.9 ± 0.2 (35)]	6/50 (12%)	29/50 (58%)	15/50 (30%)
	Lateral [0.6 ± 0.1 (34)]	13/72 (18%)	21/72 (29%)	38/72 (53%)

[a]Data from Nuccitelli and Wiley (1985).
[b]Mean of absolute values of detected current densities.

To determine whether the currents were truly transcellular, the component of current crossing perpendicular to the plasma membrane of each blastomere was measured in the horizontal plane at 4 positions: apically (marked by the FS-Con A-labelled 'cap'), basally, and at the two lateral points midway between the apical and basal ends (Fig. 11). Of those blastomeres 40 µm in diameter or larger, the pattern was inward apical (11/16 or 69%) and outward basal (15/16 or 94%) current. Lateral currents were detected in only 39% of cases, of which 18% were outward and 21% were inward (Table IV). Of major importance here is the *ratio* of the average apical-basal current to average lateral current. This ratio was 3.0 ± 0.6 (SEM, n = 13), which means that, on average, the transcellular current through the apical-basal axis was 3 times larger than the current perpendicular to this axis. Collectively, these results confirm the prediction that the outer blastomeres from mouse morulae do have a transcellular ionic current with an apical-basal axis.

3.2. Outer Blastomeres Become Morphologically Polarized by an Imposed Electric Field

As mentioned earlier, that portion of a current loop passing over the lateral surface of an outer blastomere will generate an extracellular electric field along the lateral surface of that blastomere. The magnitude of this field can be on the order of a few millivolts for the same amount of current that produces an intracellular field of but a few microvolts. This is because of the greatly increased resistance imposed on the current extracellularly from being squeezed between adjacent blastomeres (recall Ohm's Law, V=IR, where V is voltage, I is current and R is resistance). If it is indeed possible for extracel-

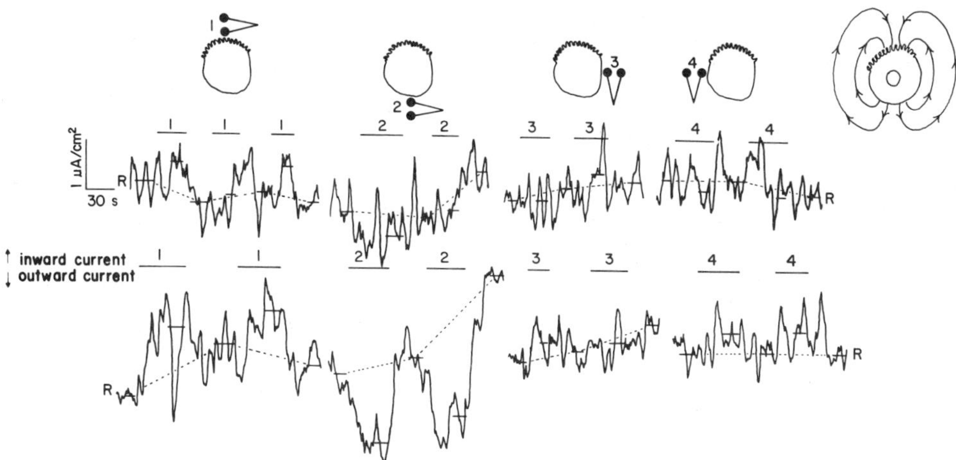

Figure 11. Two examples of probe measurements in 4 positions around isolated, enlarged 16-cell stage blastomeres. Top row: probe positions are illustrated and numbered. Middle row: representative recording of probe output in the 4 positions illustrated from studies of a blastomere 52 μm in diameter between 24 and 36 min after dissociation. The breaks between measurements are less than 1 min in duration. These records indicate current densities (in $\mu A/cm^2$ ± SEM) of 0.4 ± 0.3, -0.4 ± 0.1, 0.1 ± 0.3 and 0.2 ± 0.1 in positions 1, 2, 3 and 4, respectively. Bottom row: similar recordings from a blastomere 55 μm in diameter that exhibited a larger transcellular current. The breaks between measurements are 10 min between positions 1 and 2; 7 min between 2 and 3; 30 sec between 3 and 4. These records indicate current densities (in $\mu A/cm^2$ ± SEM) of 1 ± 0.1, -2.1 ± 0.5, 0.1 ± 0.2 and 0.5 ± 0.1 in positions 1, 2, 3 and 4, respectively. Time constant: 3 sec; probe diameter: 10 μm; vibration amplitude: 15 μm. The transcellular current pattern indicated by these measurements is illustrated in the upper right of this figure. (From Nuccitelli and Wiley, 1985).

Table IV

Ion Currents Through Enlarged 8- to 16-Cell Stage Mouse Blastomeres with Diameters Greater than 40 Microns[a]

Position (avg. current density)[b] $\mu A/cm^2$ ± SEM (n)	Cases of inward current	Cases of outward current	Current undetected
Apical [1.1 ± 0.3 (12)]	11/16 (69%)	2/16 (13%)	3/16 (19%)
Basal [1.0 ± 0.2 (14)]	1/16 (6%)	15/16 (94%)	0/16 (0%)
Lateral [0.7 ± 0.1 (12)]	6/28 (21%)	5/28 (8%)	17/28 (61%)

ratio of avg. (apical + basal)/avg. lateral current = 3.0 ± 0.6 (SEM, n = 13)

[a]Data from Nuccitelli and Wiley (1985).
[b]Mean of absolute values of detected current densities.

lular electric fields to segregate plasma membrane components into functional domains that can subsequently affect cytoplasmic polarity, then the prediction can be made that an *imposed* electric field of a few millivolts will cause non-polar isolated blastomeres to become morphologically polarized. To test this prediction, the following experiment was performed.

Isolated, non-polar 8-cell stage blastomeres were immobilized onto plastic coverslips coated with a sandwich of fibronectin and PHA (Wiley *et al.*, 1985). The coverslips had already been placed in specially designed chambers built for subjecting cells to electric fields under physiological conditions (see Erickson and Nuccitelli, 1984). When the field chambers were sealed and placed onto the heated (35°C) stage of an inverted phase contrast microscope, agar bridges were brought into place and current was passed across the chambers to create a field strength of 5-10 mV per cell diameter (30 μm). After 5 hr, the blastomeres were fixed and further processed for scanning or transmission electron microscopy.

Three major changes were seen in the topographical morphology of blastomeres after 5 hr in the electric field. First, there was an elongation of the blastomeres parallel to the field axis (Fig. 12). Second, the blastomeres developed a structure resembling a lamellopodium at their ends facing the positive pole (anode) of the field (Fig. 13). Finally, attenuated processes extended from the sides of the blastomeres down to the coverslip, perpendicular to the field axis (Fig. 13).

Similarly, there were three major changes in the cytoplasmic morphology of blastomeres after 5 hr in the electric field (Figs. 14 and 15). First, the nucleus was displaced towards the negative pole (cathode). Second, when mitosis occurred during the 5 hr treatment period, the plane of division was perpendicular to the field axis. Finally, morphometric analyses indicated that

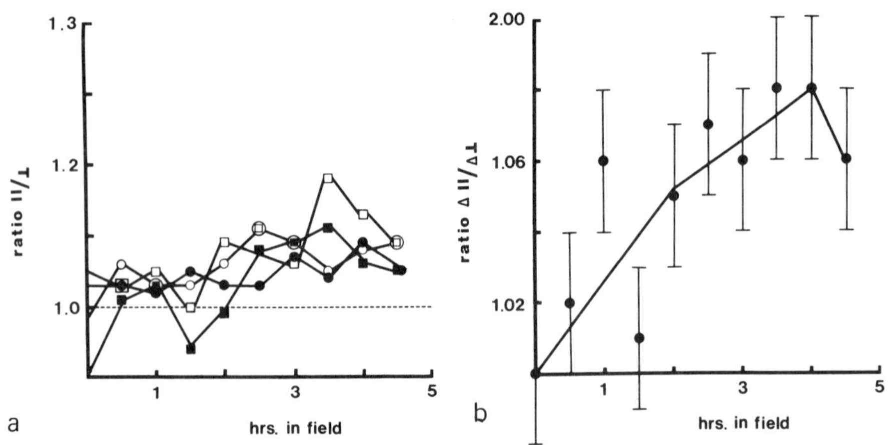

Figure 12. Elongation of isolated blastomeres along the axis of an imposed electric field. Data expressed as the ratio of blastomere diameter along the field axis divided by blastomere diameter perpendicular to the field axis at designated time intervals during the 5 hr period of an experiment. Measurements were taken from continuous videotapings of the experiments. Vertical axes represent the ratio of the two diameters: (diameter parallel to field axis II; diameter perpendicular to field axis ⊥). Horizontal axes represent hours in the electric field. (a) Ratios of 4 different blastomeres plotted separately; (b) the ratios in "a" summed and plotted as the average % standard deviation for each time point.

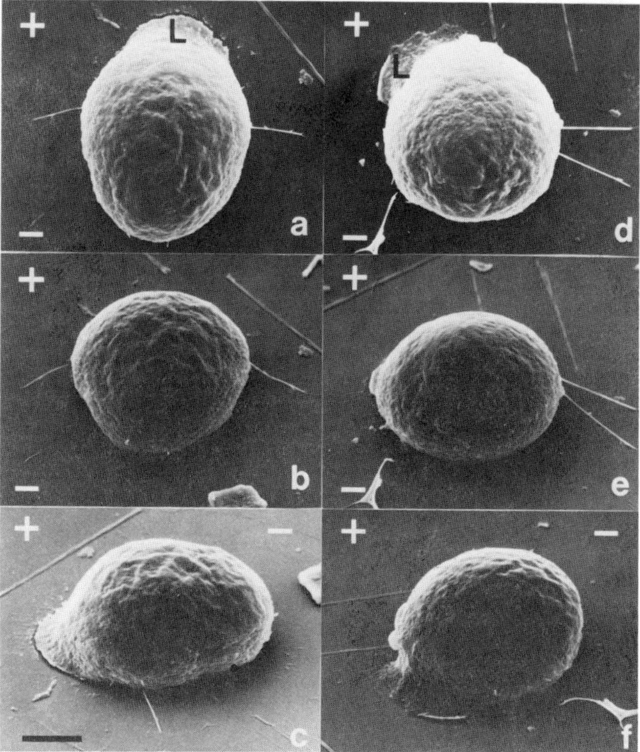

Figure 13. Scanning electron micrographs of isolated blastomeres after 5 hr in an imposed electric field. One blastomere shown in a-c and another in d-f. (a and d) Views of each blastomere parallel to the coverslip; a lamellopodium-like structure (L) is present at the positive end of the blastomeres while attenuated processes extend outward at right angles to the field axis (denoted by the minus and plus signs); (b and e) frontal views of the surface of each blastomere facing the negative pole of the field, 60° tilt from the plane of the coverslip; (c and f) views at right angles to the field axis, 60° tilt from the plane of the coverslip. Bar in lower left of "c" represents 2.2 μm.

mitochondria tended to segregate towards the positive pole (anode; Table V). It is interesting to note here that in control blastomeres (Table VI), the mitochondria and the nucleus segregate to the same hemisphere, rather than to opposite hemispheres of an isolated blastomere as was observed for those blastomeres subjected to the electric field. Collectively, these observations confirm the prediction that an imposed electric field will morphologically polarize an isolated blastomere consistent with the field axis. In addition, these observations provide the first direct experimental evidence that electrical polarity of an animal cell can regulate its morphological polarity including the distribution of specific organelles.

3.3. Relating the Detected Electrical Polarity of the Blastomere to the Intact Morula

The vibrating probe experiment reveals that the outer blastomeres of

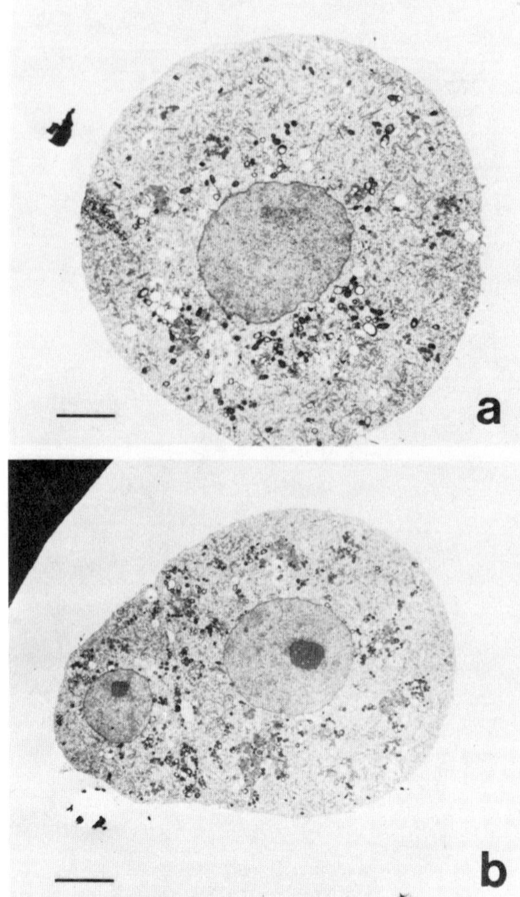

Figure 14. Transmission electron micrographs: control blastomeres. These blastomeres received identical treatment to blastomeres shown in Figure 15 except they were not exposed to an imposed electric field during the 5 hr treatment period. (a) Single 8-cell stage blastomere. Bar represents 4 μm; (b) blastomere pair. Bar represents 6 μm. Note near-central location of nuclei in all 3 blastomeres.

mouse morulae are electrically polarized. The electric field experiment shows that electrical polarity can directly influence morphological polarity. The implication that emerges is that morphological and electrical polarity are functionally related, perhaps in a causal way.

The vibrating probe experiment shows that the outer blastomere drives an apical-basal transcellular ionic current through itself that returns along a basal-apical extracellular path between adjacent blastomeres. This current can affect many cellular components, but two specific targets are proposed here. The first is the cytoplasm. The apical flow of ions (Na^+ in particular, although other ion species may be involved) and the accompanying water will generate both an ion concentration gradient and a fluid flow through the blastomere. Either of these could influence organelle distribution indirectly by

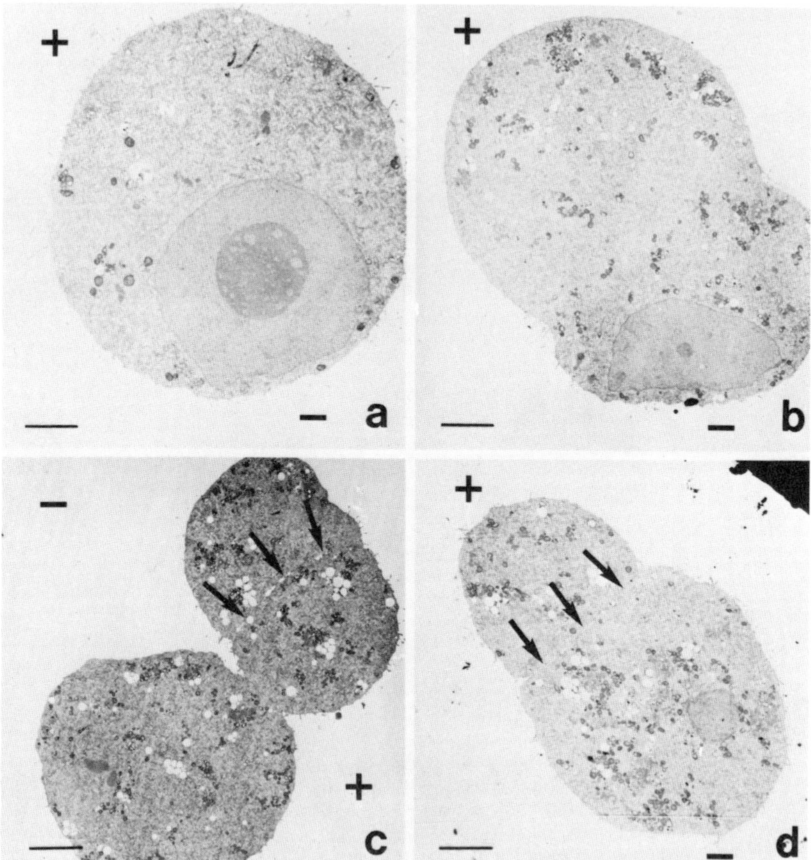

Figure 15. Transmission electron micrographs of 8-cell stage blastomeres after 5 hr in an imposed electric field. For each micrograph, the field axis is indicated by a plus (anode) and minus (cathode) sign. (a) Single blastomere. Bar represents 3 μm; (b) another single blastomere that has elongated parallel to the field axis. Bar represents 5 μm; (c) blastomere pair, of which one divided in the electric field. Note that the plane of division (arrows) is perpendicular to the field axis. Bar represents 6 μm; (d) single blastomere that divided while in the field. Again, the plane of division (arrows) is perpendicular to the field axis. Bar represents 3.7 μm.

orienting cytoskeletal components or by an electro-osmotic sweeping action of water to translocate organelles directly. The second of these targets is the plasma membrane. Glycoproteins in the membrane that have charged groups (*i.e.*, Na^+/K^+-ATPase, molecules participating in cell junction formation and cell-cell adhesion) could be laterally electrophoresed along the membrane by the external field generated by the basal-apical extracellular component of the current loop (see Jaffe, 1977). One millivolt per cell diameter is sufficient to segregate proteins in the plane of the plasma membrane (Jaffe, 1977).

The electric field experiment shows that an externally imposed electric field of a few millivolts per cell diameter can indeed influence the distribution of organelles within the cytoplasm, specifically the nucleus and mito-

Table V
Number of Mitochondria Within the Positive and Negative Hemispheres of 8-Cell Stage Blastomeres that Were in an Electric Field (5 mV/Cell Diameter) for 5 Hr

		Number (% total)		
		+	−	Total
1		18 (62%)	11 (38%)	29
2		135 (56.2%)	105 (43.8%)	240
3	A:	104 (55.9%)	82 (44.1%)	186
	B:	29 (29.5%)	69 (70.5%)	98
	C:	18 (26.4%)	50 (73.6%)	68
	A:	27 (47.3%)	30 (52.7%)	57
	B:	106 (53.5%)	92 (46.5%)	198
1 + 2 + 3		257 (56.4%)	198 (43.5%)	455

Data were collected from 7 blastomeres whose outlines and orientation in the field are indicated by drawings on the left. A line was drawn down the center of each blastomere connecting the 2 poles of the field and then crossed by another line at right angles that bisected the cells into positive (+) and negative (-) halves. The numbers of mitochondria lying either in the (+) or (-) half of each blastomere were then tabulated separately. The nuclear position is also indicated in those blastomeres whose nucleus was included in sections used to obtain these data.

chondria. As mentioned previously, a likely target of the field will be the plasma membrane, particularly those plasma membrane regions nearest the poles of the field and those where the field is tangential to the cell. In the case of the blastomere *in situ*, these tangential regions can be substantial due to flattening against adjacent blastomeres. In the polar regions, the field will modify the voltage across the plasma membrane, depolarizing it slightly at the site facing the negative pole and hyperpolarizing it slightly at the site facing the positive pole. This slight localized perturbation of membrane potential may influence local ion permeabilities and thus generate ion concentration gradients. The plasma membrane regions tangential to the field will exhibit the greatest lateral electrophoresis, and the imposed field may mimic the natural extracellular electric field most closely in this respect. If the field

Table VI
Distribution of Mitochondria in "Control" Isolated 8-Cell Stage Blastomeres

Blastomere		A	B	C	D	Total
1	Number	55	20	25	26	126
	(% total)	(43.7%)	(15.9%)	(19.8%)	(20.6%)	
		A+B 75 (59.5%)		C+D 51 (40.5%)		
2	Number	57	43	24	14	138
	(% total)	(41.3%)	(31.2%)	(17.4%)	(10.5%)	
		A+B 100 (72.5%)		C+D 38 (27.5%)		
3	Number	45	61	12	39	157
	(% total)	(28.7%)	(38.9%)	(7.6%)	(24.8%)	
		A+B 106 (67.5%)		C+D 51 (32.5%)		

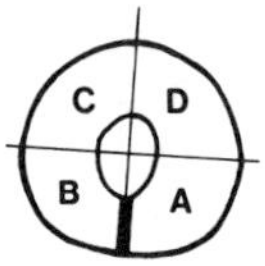

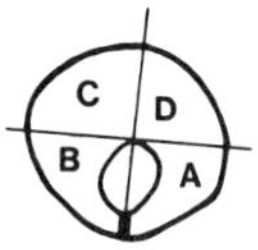

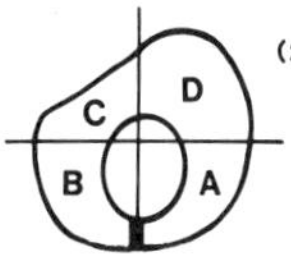

Data were collected from 3 blastomeres whose outlines and nuclear positions are indicated on the left. Each blastomere was divided into quadrants A, B, C and D by taking as a reference point the shortest distance from the nucleus to the nearest cell border (thick line). Mitochondria were then counted and tabulated for each quadrant. Note that the majority of the nucleus resides within the 2 quadrants (*i.e.*, cell half) that also contain the majority of the mitochondria.

is indeed segregating membrane components and creating functionally discrete membrane domains, then such domains may interact with cytoskeletal components in concert with localized modified ion permeabilities to translocate mitochondria to the positive pole and the nucleus to the negative pole of the imposed field, as was observed in this electric field experiment.

If we now construct an outer blastomere that incorporates the data from these two experiments and insert it into a morula, the following paradigm will emerge (Fig. 16). This blastomere will have its nucleus residing in the apical hemisphere, which faces the negative pole of the extracellular electric field, while the majority of its mitochondria will reside in its basolateral hemisphere, which faces the positive pole of the electric field. The results of the ouabain experiments (Wiley, 1984a) verify that the majority of mitochondria do indeed reside in the basolateral hemisphere of the outer blastomere (see

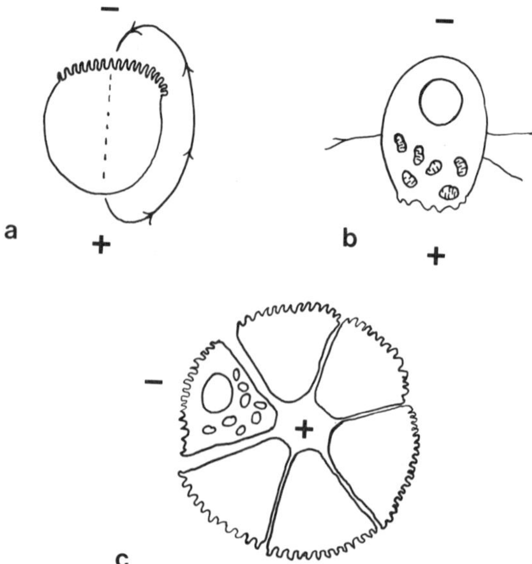

Figure 16. Relating the electrical polarity of the isolated blastomere to the intact morula. (a) The vibrating probe experiment reveals that isolated outer blastomeres exhibit a transcellular ionic apical-basal current that is accompanied by an extracellular basal-apical ionic current and associated electric field. (b) The imposed electric field experiment indicates that isolated blastomeres elongate along the field axis, display a lamellopodium-like structure at their positive ends and attenuated cell processes perpendicular to the field axis. In addition, in such isolated blastomeres, the nucleus tends to occupy the negative-facing half of the cell while the mitochondria tend to occupy the positive-facing end of the cell. (c) Combining observations from the vibrating probe and imposed electric field experiments and relating them to the outer blastomere *in situ*, it can be predicted that the nucleus will be located apically, the mitochondria basally. These predictions are borne out by morphometric analyses.

Table VII
Position of Nuclei Along the Apical-Basal Axis of Outer Blastomeres in Intact Mouse Morulae

Total number of morulae	11
Total number of outer blastomeres (n)	26
$\bar{x}$ of A/B (n = 26)	0.9129
Range of A/B	0.04 to 3.65
Number of cells for which A/B <1	18
Number of cells for which A/B >1	8

For each blastomere, the shortest distance from the nuclear margin to the apical cell membrane (A) and to the basal cell membrane (B) was measured.

section 2.4). The placement of the nucleus in the apical hemisphere is confirmed by another morphometric analysis of morulae, in which the ratio was taken of the distance between the apical plasma membrane and the nucleus divided by the distance between the basal plasma membrane and the nucleus (Table VII). In more than two thirds of the cases, this ratio is less than 1, indicating that in outer blastomeres the nucleus will tend to reside in the apical hemisphere.

The placement of the positive pole of the electric field at the basal end of the blastomere makes yet another prediction, namely, that the blastocoele will be positive with respect to the outside of the blastocyst. Trans-trophectodermal voltage potentials have been obtained for the mouse blastocyst, but they make the blastocoele negative with respect to the outside of the blastocyst (Cross *et al.*, 1973). However, this information may be inaccurate because of leakage problems, thus requiring that additional measurements be made of the voltage drop across the mouse trophectoderm.

3.4. The Ion Current Polarization Hypothesis

How does the outer blastomere acquire its electrical polarity in the first place? The Ion Current Hypothesis deals with the establishment of electrical polarity and proposes that it produces morphological polarity and that both types of polarity become mutually enhancing (Fig. 17; Nuccitelli and Wiley, 1985). The main assumption of this hypothesis is that asymmetric cell-cell contacts will produce a reduced accessibility of extracellular ions to the plasma membrane facing adjacent cells. This in turn will produce asymmetric ion fluxes so that less Na^+, for example, will enter the blastomere across the basal plasma membrane than across the apical plasma membrane. As a result, the direction of net current will be apical to basal across the cytoplasm of the blastomere.

To complete the current loop, ions must leak back along the sides of the blastomere and will produce an electric field similar to what was imposed on isolated blastomeres (as described in section 3.2). The electric field can then electrophorese ion channels and ion pumps along the plane of the plasma membrane to segregate them according to charge. By moving more channels to the inward current region, positive feedback occurs and this segregation further enhances the transcellular ionic currents and attendant extracellular electric fields, which can then produce the cytoplasmic effects described previously (see section 3.3). In this way, the Ion Current Polarization Hypothesis takes into account the observation that morphological polarity is established in response to asymmetric cell contact (see Chapter 2). The hypothesis provides a mechanism whereby asymmetric cell contact can result in cell polarity.

4. SUMMARY

We began with the observation that the first two epithelial layers to develop in the mouse embryo do so after the embryo cavitates to form the blastocoele. We then recalled that the earliest morphogenetic event for many

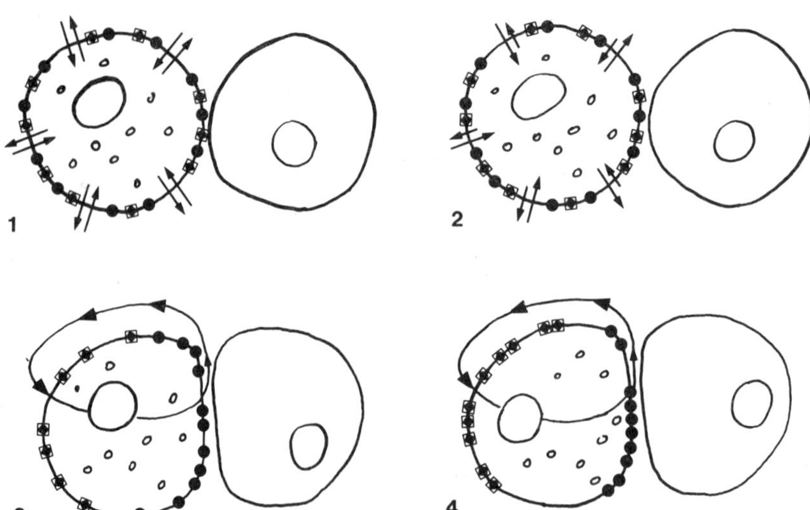

Figure 17. Ion Current Polarization Hypothesis. Asymmetric cell-cell contacts reduce accessibility of extracellular ions to the plasma membrane facing adjacent blastomeres (1). This, in turn, produces asymmetric ion fluxes so than less of a given ion will enter the blastomere across the basolateral plasma membrane than across the apical plasma membrane (2). As a result, the direction of net ion current will be apical to basal across the blastomere. To complete the current loop, ions must leak back along the sides of the blastomere and the resulting extracellular electric field can electrophorese charged components such as ion pumps and ion channels in the plane of the plasma membrane (3). Positive feedback enhances the continued segregation of plasma membrane components and hence the magnitude of the associated ionic currents and electric fields, which can then act on the cytoplasm to segregate cytoplasmic components (4).

types of metazoan embryos is also cavitation, forming a cystic structure comprised of a fluid-filled cavity enclosed by an epithelial layer. The Cell Apposition Hypothesis was put forth to link cavitation functionally to the formation of a specific epithelial layer, primitive endoderm, by proposing that cavitation creates new asymmetric cell contacts on the luminal cells of the inner cell mass. This would initiate a differentiation process that is sensitive to the ratio of free-to-apposed cell surface, *i.e.*, differentiation is cued by the cell perceiving that it is topographically polarized. This ratio served to initiate a discussion on how cavitation occurs in terms of three cavitation models. The most recent of these models, the Metabolic Cavitation Model, was proposed to overcome the deficiencies of the two previous models in accounting for all of the physiological, biochemical and morphological data presently available on cavitation. In addition, the Metabolic Cavitation Model highlights the potential importance of cell polarity in cavitation, blastocoele formation and the formation of epithelial layers.

We proceeded to consider what causes polarity of the outer blastomeres of the mouse morula and reviewed the experimental evidence leading to the observation that the outer blastomeres are electrically polarized. This polarity entails currents loops that consist of intracellular apical-basal transcellular currents and extracellular basal-apical currents that generate electric fields of several millivolts. These fields have the capability to induce the formation of discrete plasma membrane functional domains that could, in turn, induce cytoplasmic polarity. To test this possibility, isolated blastomeres were placed for 5 hr in an imposed electric field of 5 millivolts per cell diameter. This treatment caused the blastomeres to become morphologically polarized such that the nucleus migrated towards the end of the blastomere facing the negative pole of the field, while the majority of the mitochondria were found in the end of the blastomere facing the positive pole.

Placing our hypothetical electrically and morphologically polarized blastomere back into an outer position in a morula, we found that experimental data places the nucleus in the apical hemisphere and the majority of the mitochondria in the basal hemisphere of the blastomere. The negative pole of the extracellular electric field surrounding our blastomere is at its apical end, the positive pole at its basal end, which in turn predicts that the blastocoele will be positive with respect to the outside of the blastocyst. Existing data confirms all of the above except for the blastocoele being positive, a disputed point that is currently being investigated. Collectively, the data support the notion that electric fields generated by individual blastomeres play a role in cell differentiation and in morphogenesis in the developing mouse embryo.

ACKNOWLEDGMENTS

The author wishes to thank Drs. Susan Heyner, Richard Nuccitelli and Glen Winkel for their many helpful discussions regarding the data and concepts described in this chapter.

5. REFERENCES

Benos, D.J., 1981a, Developmental changes in epithelial transport characteristics of preimplantation rabbit blastocysts, *J. Physiol. (London)* 316: 191-202.

Benos, D.J., 1981b, Ouabain binding to preimplantation rabbit blastocysts, *Dev. Biol.* 83: 69-78.

Berridge, M.J., and Oschman, J.L., 1972, *Transporting Epithelia*, Academic Press, New York.

Borland, R.M., Biggers, J.D., and Lechene, C.P., 1977, Studies on the composition and formation of mouse blastocoele fluid using electron probe microanalysis, *Dev. Biol.* 55: 1-8.

Calarco, P.G., and Brown, E.H., 1969, An ultrastructural and cytological study of preimplantation development of the mouse, *J. Exp. Zool.* 171: 253-284.

Cohen, I., Daut, J., and Noble, D., 1976, An analysis of the actions of low concentrations of ouabain on membrane currents in Purkinje fibers, *J. Physiol. (London)* 260: 75-103.

Cross, M.H., Cross, P.C., and Brinster, R.L., 1973, Changes in membrane potential during mouse egg development, *Dev. Biol.* 33: 412-416.

DiZio, S.M., and Tasca, R.J., 1977, Sodium-dependent amino acid transport in preimplantation mouse embryos. III. Na^+-K^+-ATPase-linked mechanism in blastocysts, *Dev. Biol.* 59: 198-205.

Ducibella, T., and Anderson, E., 1975, Cell shape and membrane changes in the eight-cell mouse embryo: prerequisites for morphogenesis of the blastocyst, *Dev. Biol.* 47: 45-58.

Ducibella, T., Albertini, D.F., Anderson, E., and Biggers, J.D., 1975, The preimplantation mammalian embryo: Characterization of intercellular junctions and their appearance during development, *Dev. Biol.* 45: 231-250.

Eglitis, M.A., and Wiley, L.M., 1981, Tetraploidy and early development: Effects on developmental timing and embryonic metabolism, *J. Embryol. Exp. Morphol.* 66: 91-108.

Erickson, C.A., and Nuccitelli, R., 1984, Embryonic fibroblast motility and orientation can be influenced by physiological electric fields, *J. Cell Biol.* 98: 296-307.

Flynn, T.J., and Hillman, N., 1978, Lipid synthesis from [U-^{14}C] glucose in preimplantation mouse embryos in culture, *Biol. Reprod.* 19: 922-926.

Flynn, T.J., and Hillman, N., 1980, The metabolism of exogenous fatty acids by preimplantation mouse embryos developing *in vitro*, *J. Embryol. Exp. Morphol.* 56: 157-168.

Fossel, E.T., and Solomon, A.K., 1978, Ouabain-sensitive interaction between human red cell membrane and glycolytic enzyme complex in cytosol, *Biochim. Biophys. Acta* 570: 99-111.

Gardner, R.L., 1975, Analyses of determination and differentiation in the early mammalian embryo using intra- and inter-specific chimaeras, in: *The Developmental Biology of Reproduction* (C.L. Markert, and J. Papaconstantinou, eds.), Academic Press, New York, pp. 207-236.

Gardner, R.L., and Rossant, J., 1976, Determination during embryogenesis, in: *Embryogenesis in Mammals*, CIBA Foundation Symposium 40, Elsevier, North-Holland, Amsterdam, pp. 5-26.

Ginsberg, L., and Hillman, N., 1973, ATP metabolism in cleavage-stage mouse embryos, *J. Embryol. Exp. Morphol.* 30: 267-282.

Glynn, I.M., 1957, The action of cardiac glycosides on sodium and potassium movements in human red cells, *J. Physiol. (London)* 136: 148-173.

Glynn, I.M., 1964, The action of cardiac glycosides on ion movements, *Pharmacol. Rev.* 16: 381-407.

Hoffstein, S., Goldstein, I.M., and Weissmann, G., 1977, Role of microtubule assembly in lysosomal enzyme secretion from human polymorphonuclear leukocytes. A reevaluation, *J. Cell Biol.* 73: 242-256.

Jaffe, L., 1977, Electrophoresis along cell membranes, *Nature (London)* 265: 600-602.

Jaffe, L.F., 1982, Developmental currents, voltages, and gradients, in: *Developmental Order: Its Origin and Regulation* (S. Subtelny, and P.B. Green, eds.), Alan R. Liss, New York, pp. 183-215.

Jaffe, L.F., and Nuccitelli, R., 1974, An ultrasensitive vibrating probe for measuring steady extracellular currents, *J. Cell Biol.* 63: 614-628.

Malaisse, W. J., Malaisse-Lange, F., Van Obberghen, E., Somers, G., DeVis, G., Ravazzola, M., and Orci, L., 1975, Role of microtubules in the phasic pattern of insulin release, *Ann. N.Y. Acad. Sci.* 253: 630-652.

Martin, G.R., 1980, Teratocarcinomas and mammalian embryogenesis, *Science* 209: 768-776.

McLaren, A., and Smith, R., 1977, Functional test of tight junctions in the mouse blastocyst, *Nature (London)* 267: 351-353.

Melissinos, K., 1907, Die entwicklung des eles der mause, *Arch. Mikr. Anat.* 70: 577-628.

Mintz, B., 1965, Experimental genetic mosaicism in the mouse, in: *Preimplantation Stages of Pregnancy* (G.E.W. Wolstenholme, and M. O'Connor, eds.), Churchill Ltd., London, pp. 194-207.

Nadajcka, M., and Hillman, N., 1975, Autoradiographic studies of t^n/t^n mouse embryos, *J. Embryol. Exp. Morphol.* 33: 725-730.

Nuccitelli, R., 1983, Transcellular ion currents: Signals and effectors of cell polarity, in: *Spatial Organization of Eukaryotic Cells* (J.R. McIntosh, ed.), Alan R. Liss, New York, pp. 451-482.

Nuccitelli, R., and Wiley, L.M., 1985, Polarity of isolated blastomeres from mouse morulae: Detection of transcellular ion currents, *Dev. Biol.* 109: 452-463.

Poisner, A.M., and Cooke, P., 1975, Microtubules and the adrenal medulla, *Ann. N.Y. Acad. Sci.* 253: 653-669.

Quatrano, R.S., Brawley, S.H., and Hogsett, W.E., 1978, The control of the polar deposition of a sulfated polysaccharide in *Fucus* zygotes, in: *Determinants of Spatial Organization* (S. Subtelny, and I.R. Konigsberg, eds.), Academic Press, New York, pp. 77-96.

Robinson, K.R., 1979, Electrical currents through full-grown and maturing *Xenopus* oocytes, *Proc. Natl. Acad. Sci. USA* 76: 837-841.

Stekhoven, F.S., and Bonting, S.L., 1981, Transport adenosine triphosphatases: properties and functions, *Physiol. Rev.* 61: 1-76.

Stern, C.D., 1982, Experimental reversal of polarity in chick embryo epiblast sheets *in vitro*, *Exp. Cell Res.* 140: 468-471.

Vorbrodt, A., Konwinski, M., Solter, D., and Koprowski, H., 1977, Ultrastructural cytochemistry of membrane-bound phosphatases in preimplantation mouse embryos, *Dev. Biol.* 55: 117-134.

Wiley, L.M., 1984a, Cavitation in the mouse preimplantation embryo: Na/K-ATPase and the origin of nascent blastocoele fluid, *Dev. Biol.* 105: 330-342.

Wiley, L.M., 1984b, The cell surface of the mammalian embryo during early development, in: *Ultrastructure of Reproduction* (J. Van Blerkom, and P. Motta, eds.), Martinus Nijhoff Publishers, Boston, pp. 190-204.

Wiley, L.M., and Eglitis, M.A., 1980, Effects of colcemid on cavitation during mouse blastocoele formation, *Exp. Cell Res.* 127: 89-101.

Wiley, L.M., and Eglitis, M.A., 1981, Cell surface and cytoskeletal elements: Cavitation in the mouse preimplantation embryo, *Dev. Biol.* 86: 493-501.

Wiley, L.M., and Nuccitelli, R., 1985, Polarity of isolated blastomeres from mouse 8-cell stage embryos: Effect of an applied electric field, *J. Cell Biol.* 101: 472a.

Wiley, L.M., Spindle, A.I., and Pedersen, R.A., 1978, Morphology of isolated mouse inner cell masses *in vitro*, *Dev. Biol.* 63: 1-10.

Wiley, L.M., Takaki, K.K., and Yamagata, M., 1985, Electron microscopy of mouse preimplantation embryos and isolated blastomeres immobilized on coverslips, *Gamete Res.* 11: 51-58.

Ziomek, C.A., and Johnson, M.H., 1981, Properties of polar and apolar cells from the 16-cell mouse morula, *Wilhelm Roux's Arch. Dev. Biol.* 190: 287-296.

Chapter 5

IN VITRO ASSESSMENT OF BLASTOCYST DIFFERENTIATION

ERIC W. OVERSTRÖM

1. INTRODUCTION - OVERVIEW OF BLASTOGENESIS

It is generally agreed that one of the most critical periods of embryonic development in mammals occurs during the first weeks of gestation. During this early window of development, which is characterized by embryo cleavage, blastulation and gastrulation events, up to 60 percent of pregnancies fail, either totally (uniparous animals) or in part (multiparous animals), depending on species. Hammond (1914) first reported that a significant proportion of pig and rabbit embryos die during early intrauterine development. It was suggested that embryonic death reflected nature's evolutionary selection process by favoring those embryos which possessed a more competent genetic character. More recently Biggers (1983), in re-evaluating earlier data of Hertig *et al.* (1959) concerning embryonic mortality in humans, concluded that the intrauterine mortality rate is highest (42-55%) during the period of preimplantation development. Similarly, on average, greater than 33% of fertilized pig embryos do not survive the first 3 weeks of gestation (Day *et al.*, 1959; Perry and Rowlands, 1962; Hafez, 1974; Hafez *et al.*, 1965; Anderson, 1978). Since the incidence of fertilization is high in most mammals (90 to 100%), it is generally agreed that a significant percentage of pregnancy loss is due to early embryonic death. Thus, while our overall understanding of the mechanisms that regulate embryogenesis has increased markedly over the years, the direct relationship between specific cellular processes and embryo mortality/ survivability rates *in vivo* have not, as yet, been clearly delineated.

Intrauterine embryonic mortality is the net effect of both intrinsic and extrinsic factors. Extrinsic (maternal) factors include endocrine, hereditary, nutritional and environmental components. While studies of environmental and

Eric W. Overström Department of Anatomy and Cellular Biology, Tufts University, Schools of Medicine and Veterinary Medicine, 136 Harrison Avenue, Boston, Massachusetts 02111, USA.

nutritional influences on reproductive function in mammals have been undertaken, comprehensive studies of early embryonic development, *i.e.*, intrinsic factors, have lagged behind for lack of applicable methodologies. Only recently have techniques and procedures been developed whereby physiological, morphological and biochemical analyses of individual *live* embryos are now possible. The application of these procedures has been undertaken primarily in laboratory animals and several comprehensive reviews have been published (Biggers and Stern, 1973; Wales, 1975; Pike, 1981; Rieger, 1984). The important aspects of mammalian blastocyst development are summarized below.

During mammalian embryogenesis, a multitude of distinct morphological and physiological changes occur. In the embryo, regionalization of cells first occurs following the formation of tight junctions between adjacent blastomeres at the 8- to 16-cell stage (Ducibella *et al.*, 1975; Hastings and Enders, 1975) [see also Chapter 3 (Ed.)]. This process, called "compaction" (Lewis and Gregory, 1929), while establishing a permeability seal on the outside of the embryo, spatially segregates embryonic cells into two subsets, the inner cell mass and the trophectoderm. The inner cell mass subsequently differentiates to form the embryo proper in an environment isolated from direct maternal influence. The outer embryonic cells that comprise the trophectoderm differentiate to form a functional, unilaminar epithelium. As such, the trophectoderm acts as a selective barrier to solute exchange, thereby regulating the composition of extracellular fluid which bathes the developing inner cell mass (Biggers *et al.*, 1977, 1978). Cavitation occurs soon after a permeability seal is established between adjacent blastomeres and is characterized by the accumulation of variable amounts of blastocoelic fluid within the embryo prior to attachment to the uterine epithelium (Burgoyne and Ducibella, 1977) [see also Chapter 4 (Ed.)]. Although all mammalian blastocysts expand during preimplantation development (Borland *et al.*, 1977), the degree of fluid formation appears to be species dependent. Rodent and primate embryos accumulate small volumes of fluid and are considered minimally expanding blastocysts. In contrast, blastocysts of the rabbit, pig, horse and the domestic ungulates increase their volume by several orders of magnitude prior to implantation/attachment and are appropriately termed "maximal expanders" (Benos and Biggers, 1981). Since the chronology of maximal embryonic mortality parallels the period of astonishingly rapid blastocyst growth and differentiation *in vivo*, our increasing understanding of the cellular processes that regulate "normal" blastocyst development may also lend insight into the causative mechanisms of aberrant embryogenesis and subsequent embryonic death. Such information will have a significant impact on the development of increasingly successful human infertility and IVF-ET programs as well as programs intended to maximize the reproductive performance of domestic food animals.

This chapter focusses on four *in vitro* experimental techniques for assessing mammalian blastocyst differentiation. Each described procedure is designed to characterize a specific parameter of blastocyst metabolism by evaluating one or more living blastocysts per observation. For example, the technique for assessing oxygen consumption rates (QO_2) of individual blastocysts is particularly valuable as a rapid, non-perturbing assay of oxidative metabolism. Thus, additional experiments can be performed on the same blastocyst, thereby allowing for comparison and correlation of QO_2 with one or more additional physiological parameters of blastocyst differentiation such as trans-trophectodermal Na^+ transport and protein synthesis activity. Previous studies are briefly reviewed and the details of each technique are described. For each parameter, our results are discussed and compared with those reported in other species. Concluding remarks seek to convey to the reader both the present utility and the future potential of these techniques to basic and applied studies of cellular differentiation of the mammalian blastocyst.

2. OXIDATIVE METABOLISM

Like all adult mammalian cells, blastomeres of the developing blastocyst are capable of deriving metabolic energy, in the form of ATP, by way of the mitochondrial Krebs cycle. Glucose and its metabolites serve as minor energy substrates while aerobic respiration provides 85 to 93 percent of total metabolic energy during blastocyst differentiation (Benos and Balaban, 1983). Concomitant with this process is the uptake and utilization of oxygen (QO_2) by the blastocyst and the subsequent formation and release of CO_2 (Fridhandler, 1961; Mills and Brinster, 1967; Brinster, 1968; O'Fallon and Wright, 1986) [see also Chapter 12 (Ed.)]. The ability to measure blastocyst QO_2 has several inherent advantages in accurately assessing embryonic development. First, the procedure is non-invasive, so the functional integrity of the embryo is not violated; second, accurate measurements can now be made on individual blastocysts; and third, the protocol is a relatively simple and rapid procedure that allows for subsequent experimentation with the same embryo, including transfer to a foster mother.

Early techniques to measure QO_2 (Cartesian diver, Warburg apparatus, *etc.*) lacked sufficient sensitivity to evaluate single embryos (Fridhandler *et al.*, 1957; Daniel, 1967; Mills and Brinster, 1967). More recently, Benos and Balaban (1980) utilized a Clark-type O_2 probe to determine QO_2 of pre-implantation rabbit blastocysts. Still, 4 to 8 embryos were required to obtain a reproducible tracing. The problem limiting sensitivity is that the probe itself consumes O_2. Recently, a micro-cathode electrode has become available (Microelectrodes, Inc.; Instech, Inc.) which consumes 1000-fold less O_2 than a standard Clark probe (Fig. 1). Using this probe, we are now able to quantitate the QO_2 of a single mouse blastocyst within 30 min (Fig. 2) and a single pig blastocyst in 15 to 20 min.

2.1. Techniques

On the fourth to the tenth day after breeding, we routinely recover blastocysts from uterine flushings which are collected surgically (pig, rabbit), non-surgically (cow) or immediately after cervical dislocation (mouse). The standard flushing medium consists of CO_2-saturated Krebs-Ringer bicarbonate solution supplemented with 5.5 mM glucose (KRBG, pH 7.2, 0.2 μm filter sterilized, 37-39°C). To maximize embryo survivability and to decrease the possibility of inducing metabolic artifacts *in vitro*, careful attention should be paid to ensure an adequate and constant temperature (37°C, except 39°C for the pig) and gas environment (5% CO_2 in humidified air) during all subsequent blastocyst manipulations *in vitro*. If, for technical reasons, blastocysts cannot be constantly gassed with CO_2 during experimentation, a non-bicarbonate

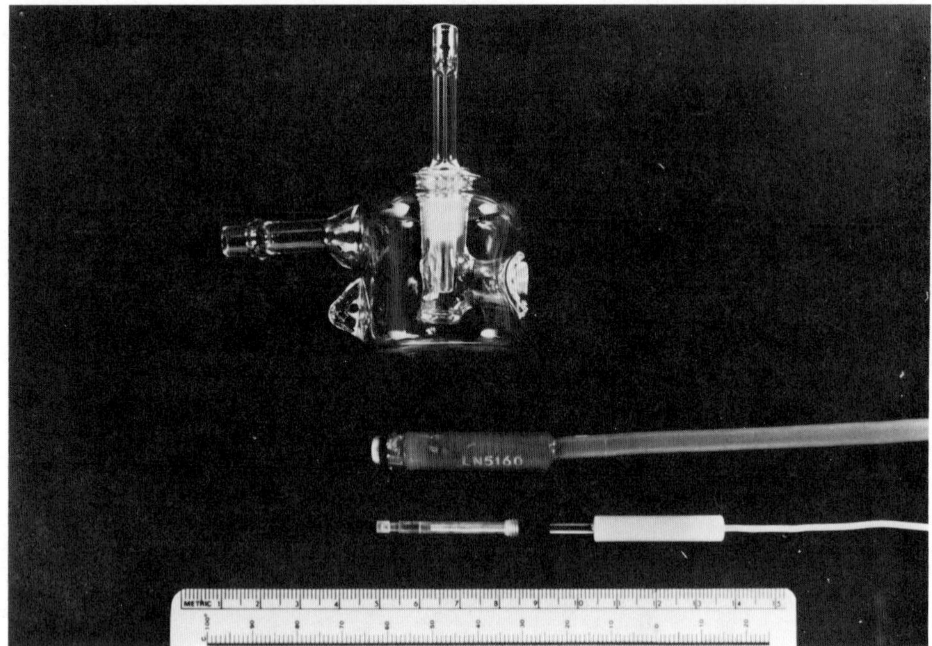

Figure 1. Apparatus to measure blastocyst oxygen consumption (QO_2) including: water-jacketed glass chamber with O_2 electrode port (top), standard Clark O_2 probe (center) and micro-cathode oxygen probe (bottom) which consumes 1000 times less O_2 than the standard Clark probe. This apparatus is interfaced with an oxygen monitor and strip-chart recorder.

buffer system, *i.e.*, HEPES or phosphate, can often be substituted for short periods (1 to 6 hr). Preliminary experiments should always be conducted, however, to test the suitability of a particular buffer with the stage and species of blastocyst to be investigated. In addition, if blastocysts are to be maintained (incubation or culture) for periods beyond 2 to 3 hr, we routinely transfer them to an enriched medium, *e.g.*, Ham's F10, F12 or NCTC-109, which often contains either bovine serum albumin (BSA, 1 to 10 mg/ml) or fetal bovine serum (FBS, 1 to 20%). Depending on species and age of blastocysts, such media can support continued growth and differentiation, albeit at *sub-in vivo* rates, for periods of 24 hr to 6 days *in vitro*. Individual embryos are measured (long and short axes) using a Zeiss SR stereoscope fitted with a calibrated eyepiece graticule and dual fiber-optic illumination. The blastocyst is assumed to be a prolate spheroid for calculations of area and volume.

Oxygen consumption is determined polarographically (Benos and Balaban, 1980; Overström and Ebert, 1985) utilizing either a standard Clarke-type O_2 probe (YSI, Inc., for a group of blastocysts) or a micro-cathode oxygen probe (Microelectrodes, Inc., for single blastocysts) mounted within a sealable, water-jacketed (37 to 39°C) glass chamber (0.2 to 1.5 ml, Wilber Scientific) containing CO_2-equilibrated KRBG. Each calibrated probe interfaces with a system-compatible oxygen monitor (YSI, Inc., model 53; Microelectrodes,

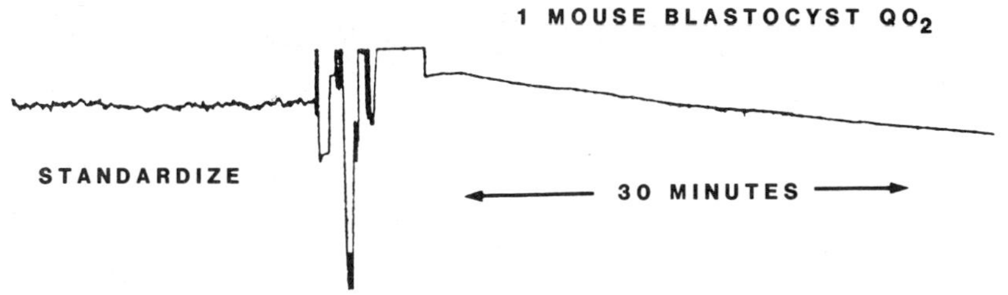

Figure 2. QO_2 measurement of a single 4-day mouse blastocyst. After recording a stable 20 to 30 min tracing, blastocyst QO_2 is calculated from the decrease in chamber medium O_2 tension (slope).

Inc., model OM-1). An individual blastocyst is placed in the chamber for periods of 10 to 40 min, depending on its developmental age and species. As the metabolic activity of the blastocyst consumes O_2, the relative decrease in oxygen tension of the bathing solution with time is detected by the O_2 probe and is electronically displayed on the O_2 monitor in digital or analog format. Simultaneously, a permanent, real-time tracing of blastocyst QO_2 is recorded with an interfaced multichannel strip-chart recorder (Fig. 2). By knowing the chamber volume, the O_2 tension of the bathing solution (μl/ml at 37°) and blastocyst area and volume, the QO_2 (μl/cm^2/hr) can be calculated. Each water-jacketed QO_2 chamber is fitted with a thermistor (±0.01°C, Omega, Inc.) to accurately record (on the strip-chart) chamber temperature at the same time as the QO_2 measurement. This is necessary as the O_2 probe is sensitive to minor temperature fluctuations (1 to 2°C).

The inherent flexibility of the QO_2 system is extremely useful in that agents which specifically alter metabolism, *e.g.*, reversible and irreversible drugs and hormones, can be added and their acute effects on blastocyst QO_2 determined quickly and with relative ease. Moreover, the system is relatively inexpensive and infinitely expandable. We are presently increasing our capabilities to be able to simultaneously record QO_2 of up to 8 individual blastocysts. As one generally has 8 to 50 blastocysts available on a given day, it is extremely important that QO_2 measurements be completed within a reasonable period (0.5 to 2 hr) if blastocysts are to be used in subsequent experiments.

2.2. Studies in the Mouse, Rabbit and Pig

The QO_2 characteristics of preimplantation mouse and rabbit blastocysts have been reported by several investigators and these data are summarized in Table I. Mouse and rabbit blastocysts consume O_2 at rates of 0.2 to 4.4 μl/cm^2/hr. These rates are comparable to the high aerobic respiration rates of kidney and brain. In general, rabbit blastocysts have higher QO_2 values than mouse blastocysts of comparable age. This is reasonable, since mouse blastocysts expand minimally prior to implantation and thus would expend proportionally less metabolic energy to support active mechanisms of blastocoel fluid formation, *i.e.*, Na^+/K^+-ATPase activity, as compared to the rabbit blastocyst. While the QO_2 of rabbit

Table I
Oxygen Consumption Rates of Mouse and Rabbit Blastocysts

Species	Day *post coitum*	QO_2 ($\mu l/cm^2/hr$)	References
Mouse	4	1.67 ± 0.07	Benos and Balaban (1983)
	3.5. - 4.5	0.48 - 0.56[a] (μl/blast/hr)	Mills and Brinster (1967)
Rabbit	4	0.61[b]	Fridhandler *et al.* (1959); Fridhandler (1961)
	5	0.65	
	6	0.23	
	4	4.0 ± 0.4	Benos and Balaban (1980)
	5	4.4 ± 0.7	
	6	2.5 ± 0.2	
	7	2.7 - 3.0 ± 0.2 - 0.3	
	6	1.73 ± 0.13	Benos and Balaban (1983)

[a]QO_2 calculations based on $\mu l/cm^2/hr$ assumed surface area of 2.44×10^{-4} cm^2 for 3.5- to 4.5-day mouse blastocysts (from Benos and Balaban, 1983).
[b]For calculation of QO_2 ($\mu l/cm^2/hr$), area values were from Benos and Biggers (1981).

blastocysts decreases 30 to 40 percent after day 5, blastocoelic fluid continues to accumulate at ever-increasing rates (Borland *et al.*, 1977). More recently, Benos and Balaban (1983) have reported that the 6-day rabbit blastocyst expends about 6 to 11 percent of its metabolic energy for active trans-cellular Na^+ transport. Thus, ion-driven blastocyst expansion can occur, even as blastocyst QO_2 decreases, since the net metabolic cost is low in relative terms.

Table II shows QO_2 values for expanding pig blastocysts between day 7 and 10 of pregnancy. These data (compiled from Overström *et al.*, 1984; Overström and Ebert, 1986) demonstrate that the QO_2 of pig blastocysts is higher than that of rabbit blastocysts but shows similar decreasing values during blastogenesis on a cm^2/hr basis. Of great interest is the observation that QO_2 decreases sharply on day 10, when blastocyst mortality begins to increase, while the between embryo variation increases dramatically at that stage of development. A plausible interpretation is that these changes reflect an increase in embryonic mortality *in vivo*. From these observations, we have developed our working hypothesis: that increased embryo mortality can be determined quantitatively based on the QO_2 of a single embryo as it compares with the established mean for its developmental age and species. Embryo transfer studies are underway to evaluate this possibility.

3. TRANS-TROPHECTODERMAL SODIUM TRANSPORT

As the process of blastogenesis progresses, the trophectoderm of the mammalian blastocyst differentiates as a polarized, unilaminar epithelium and

Table II
Oxygen Consumption Rates of Pig Blastocysts[a]

Day *post coitum*	area (cm^2)	QO_2 (μl/blast/hr)[b]	QO_2 ($\mu l/cm^2$/hr)[b]
7	0.0068	0.0788 ± 0.0067	11.59 ± 0.98
8	0.0113	0.0700 ± 0.0032	6.19 ± 0.28
9	0.0179	0.1618 ± 0.0015	8.96 ± 0.60
10	0.2350	0.2138 ± 0.1307	0.91 ± 0.56

[a]Data from Overström *et al.* (1984) and Overström and Ebert (1986).
[b]All values represent the mean ± SEM of 11 to 17 embryos per data point.

displays solute transport characteristics similar to those of other actively transporting epithelia. In classic ion-transporting epithelia, such as frog skin (Zerahn, 1956), toad bladder (Nellans and Finn, 1974) and kidney (Cohen and Kamm, 1976), a linear relationship exists between the rate of oxygen consumption and active sodium transport. Early studies, alluded to in Section 2, have similarly demonstrated that mammalian embryos possess a high aerobic metabolic rate during preimplantation development (Fridhandler *et al.*, 1957, 1961; Daniel, 1967; Brinster, 1967, 1968, 1969). More recently, several investigators have made progress in: 1) characterizing the dynamic trans-trophectodermal ion transport properties of the differentiating blastocyst, and 2) correlating such solute transport activities with blastocyst QO_2 during the preimplantation period (Cross and Brinster, 1970; Gamow and Daniel, 1970; Smith, 1970; Cross, 1973; Borland *et al.*, 1977; DiZio and Tasca, 1977; Benos and Balaban, 1980, 1983; Benos, 1981a; Overström *et al.*, 1984; Overström and Ebert, 1986; for reviews see Benos and Biggers, 1981; Benos *et al.*, 1985). These reports support the hypothesis that oxidative metabolism is coupled with mechanisms of active trans-trophectodermal solute transport, and as such, both activities are requisite mediators of normal blastocyst growth and differentiation.

3.1. Techniques

Since the mammalian blastocyst exists as a hollow ball of cells, the wall of which consists of a single layer epithelium, the trophectoderm, it is an ideal model for studies of epithelial transport during early development. By either incubating blastocysts in external medium containing radiolabeled solutes or introducing such agents into the blastocoelic cavity by microinjection procedures, one can directly assess trans-trophectodermal transport activity (both *ad*-coelic and *ab*-coelic) as a function of solute(s), time and stage (age) of preimplantation development. This is accomplished by sampling the appropriate fluid compartment (blastocoel or external medium) after

predetermined incubation periods. By combining blastocyst microinjection/ micromanipulation techniques with basic procedures for ion flux determination one can employ specific inhibitors to evaluate the trans-epithelial Na^+ transport activity of the trophectoderm. For example, drugs such as amiloride (which blocks Na^+ access to the intracellular compartment at the apical plasma membrane) and ouabain (which inhibits Na^+/K^+-ATPase function at the basolateral plasma membrane) are used to investigate the functional expression of transient and chronic Na^+-dependent and Na^+-independent transport mechanisms during blastocyst differentiation. In Figure 3, the apical and basolateral regions of the plasma membrane are indicated in the expanded drawing of adjacent trophoblast cells. Several known transport mechanisms associated with either apical or basolateral solute movement are illustrated.

In general, procedures to characterize blastocyst ion transport activities require the following equipment (see Appendix II): research quality stereomicroscope (*e.g.*, Wild, Zeiss) with magnification to 50 x and transmitted/ reflected light source; micromanipulators (2) (*e.g.*, Brinkman, Leitz, Narishige); micropipette puller (Kopf Instruments) microforge (Curtin Matheson); injection/suction micrometers (Gilson, Stoelting Co.)

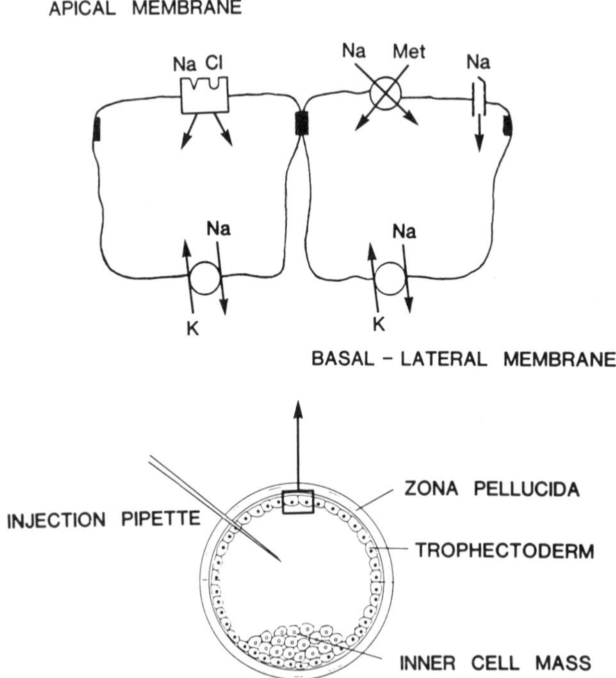

Figure 3. A diagrammatic representation of the mammalian blastocyst is illustrated (bottom). The route of microinjection/ micropuncture of the blastocoelic cavity is depicted. The apical and basolateral regions are indicated in the expanded drawing of adjacent trophoblast cells (top). Transport properties associated with either apical or basolateral solute movement are illustrated. (Reprinted from Benos *et al.*, 1985, with permission).

Blastocysts are recovered from uterine flushings, their dimensions recorded as described earlier and randomly assigned to control and treatment groups as required. To minimize between animal variability at each stage of development, we routinely treat all blastocysts collected from an individual dam as a separate and complete experiment, having both control and treatment group embryos. Ouabain-treatment blastocysts are placed in KRBG (37 to 39°C) and anchored by gentle suction with a holding pipet. The blastocoel cavity is gently pierced with a microbore glass needle and immediately injected with ouabain (Fig. 3; 5×10^{-4}M, 0.1 to 1.0 µl depending on blastocoel volume). To ensure saturation of all ouabain binding sites (Na^+/K^+-ATPase), blastocysts are preincubated for 45 min prior to subsequent experimentation. Control and ouabain treated embryos are placed in KRBG tracer medium containing 8 µCi/ml [^{22}Na]Cl (New England Nuclear) and the flux performed for 30 min at 37 to 39°C in a humidified atmosphere of 5% CO_2 in air. The effect of amiloride on Na^+ influx across the apical plasma membrane is determined by using identical tracer medium supplemented with 10^{-4}M amiloride. At the end of the flux, individual blastocysts are placed in ice-cold, tracer-free KRBG and a sample of blastocoelic fluid (1 to 20 µl) is obtained by micropuncture to quantitate [^{22}Na] radioactivity. By knowing blastocyst area, volume and blastocoel specific activity, the [^{22}Na] influx (µmole/cm^2/hr) can be calculated.

3.2. Studies in the Rabbit and Pig

The huge accumulation of blastocoelic fluid during preimplantation development of the rabbit blastocyst has been shown to be mediated, at least in part, by mechanisms of active transepithelial solute transport (Benos and Biggers, 1981). Table III summarizes the Na^+ transport characteristics of the differentiating rabbit trophoblast. Unidirectional [^{22}Na] influxes increase from 0.098 to between 1.0 and 3.9 µmole/cm^2/hr during preimplantation development. In the presence of ouabain, a specific inhibitor of membrane-bound

Table III
Trans-trophectodermal Na^+ Influxes in Rabbit Blastocysts

Day *post coitum*	Treatments			References
	Control	Ouabain (µmole/cm^2/hr)	Amiloride	
4	0.098			Borland *et al.* (1977)
5	0.342			Borland *et al.* (1976)
5		0.20		Benos (1981b)
6	0.452			Borland *et al.* (1977)
6	0.50	0.33	0.51	Benos (1981b)
7	0.820			Borland *et al.* (1977)
7	3.9	1.3	1.1	Benos (1981b)

Na^+/K^+-ATPase (EC 3.6.1.3), blastocyst fluid accumulation is partially inhibited (Smith, 1970; Biggers *et al.*, 1978). In measuring unidirectional [^{22}Na] influxes (Na^+ influx) across the trophectoderm of 5-, 6- and 7-day *p.c.* rabbit blastocysts in the presence or absence of 0.5 μM ouabain within the blastocoele, Benos (1981a) has reported that the magnitude of both total and ouabain-sensitive Na^+ influx increased between days 5 and 7 of development. In addition, the amount of specific [^{3}H] ouabain binding, a direct measure of the number of Na^+ pump sites, has been determined for 4-, 5-, 6- and 7-day *post-coitum (p.c.)* rabbit blastocysts (Benos, 1981b; Benos and Biggers, 1981). Ouabain binding was biphasic with regard to developmental age, with significant increases observed between days 4 and 5 and again after day 6 *p.c.* In similar experiments, we have recently reported that Na^+ is likewise actively transported across the trophectoderm of 7- to 10-day pig blastocysts (Overström *et al.*, 1984; Overström and Ebert, 1986). Between days 7 and 9 of pregnancy, the area (0.68 to 1.79 cm^2) and volume (0.005 to 0.23 ml) of pig blastocysts increase slightly, while Na^+ influx rates (approx. 0.26 μmole/cm^2/hr) show little change (Fig. 4a). However, on day 10, dramatic increases in blastocyst area (13-fold) and volume (46-fold) reflect a concomitant rise in Na^+ influx (1.20 μmole/cm^2/hr). Amiloride treatment reduces Na^+ transport only in day 9 blastocysts (approx. 50%, Fig. 4b), while ouabain partially inhibits total Na^+ influx (17 to 24%) in 8- to 10-day pig blastocysts. These results suggest: 1) that fluid accumulation during blastogenesis in the pig is mediated by a process of trans-trophectodermal Na^+ transport, and 2) that transtrophectodermal Na^+ transport is developmentally regulated at the level of the apical and basolateral trophoblast plasma membranes, as revealed by

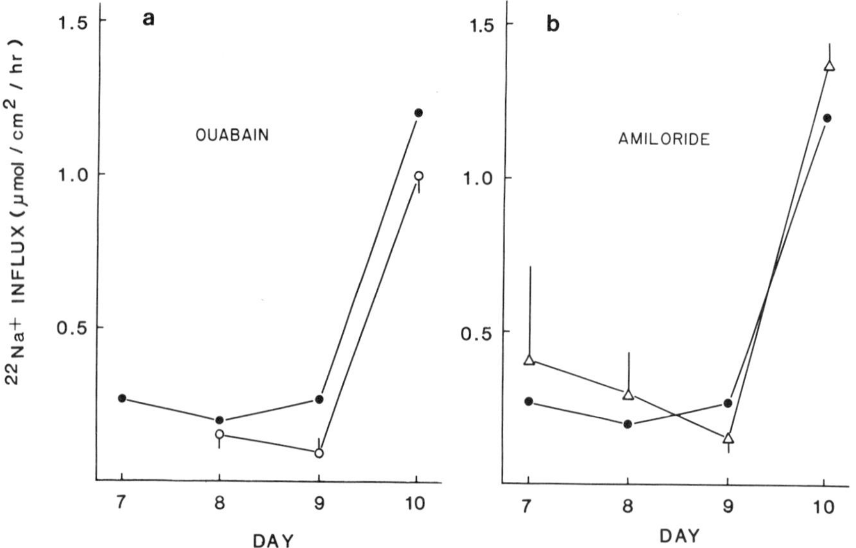

Figure 4. The trans-trophectodermal [$^{22}Na^+$] influx in 7, 8, 9 and 10 day pig blastocysts is shown (●). In addition, the effects of ouabain (a, ○) and amiloride (b, △) treatment on trans-trophectodermal [$^{22}Na^+$] influx of 7- to 10-day blastocysts are illustrated. Fluxes were performed as described in the text. Values represent the mean ± SEM of 11 to 19 blastocysts.

treatment with amiloride and ouabain, respectively. Further, since maximally expanding rabbit and pig blastocysts demonstrate similar Na^+ transport characteristics, both in magnitude and chronological expression, active trans-trophectodermal ion transport may be a universal physiological determinant of mammalian blastogenesis *in vivo*.

4. PROTEIN SYNTHESIS DURING BLASTOGENESIS

Several laboratories have characterized the changing patterns of protein synthesis in preimplantation mouse (Van Blerkom and Brockway, 1975; Handyside and Barton, 1977; Levinson *et al.*, 1978) and rabbit embryos (Manes and Daniel, 1969; Petzoldt, 1972, 1974; Van Blerkom and Manes, 1974) using techniques of isotopic labeling, 1- and 2-dimensional gel electrophoresis (O'Farrell, 1975) combined with autoradiography/fluorography (Bonner and Laskey, 1974). These studies have generally described qualitative aspects of protein synthesis during embryogenesis in an effort to identify temporal changes that might correlate with known morphological or physiological events of differentiation. To a lesser extent, quantitative measurements of protein synthesis have been reported for individual embryonic proteins of known (Abrew and Brinster, 1978; Schultz *et al.*, 1979) and unknown identity (Brinster *et al.*, 1979; Chen *et al.*, 1980). Using similar procedures, we have conducted experiments investigating quantitative changes in the rate of synthesis of membrane-bound Na^+/K^+-ATPase and several other actively synthesized blastocyst proteins during preimplantation development in the rabbit (Overström *et al.*, 1982, 1983, 1986; Benos *et al.*, 1985).

4.1. Techniques

Clearly, intrinsic anomalies of protein synthesis during blastocyst development may be manifested *in vivo* as an increase in blastocyst mortality with a resulting decrease in fecundity. Using the rabbit blastocyst as a model, we are investigating this possibility by characterizing the changing synthesis patterns of blastocyst proteins as a function of developmental age (day 4 to 7 *post coitum*). Analytical techniques to assess protein synthesis activity of mammalian embryos have been described in detail by Van Blerkom (1978). Briefly, blastocysts are collected as previously described and incubated (5% CO_2 in humidified air, 37 to 39°C) in a labeling solution consisting of Ham's F10 medium (0.1 μM methionine) supplemented with 200 μCi/ml [^{35}S] methionine (>1200 Ci/mmole). After labeling for 3 hr, blastocyst tissues (6 to 15 blastocysts per experiment) are washed extensively in ice-cold, tracer-free F10 medium containing 1mM phenylmethylsulfonyl fluoride (PMSF) and solubilized in 20 to 50 μl lysis buffer (O'Farrell, 1975). Labeled blastocyst proteins are resolved by 1- and 2-dimensional polyacrylamide gel electrophoresis (1D-, 2D-PAGE) as described by Van Blerkom (1978) with minor modification. Acrylamide used for isoelectric focussing gels is ultra-pure grade (Boehringer, Mannheim) and dithiothreitol (DTT, 40 mM) is substituted for 2-mercaptoethanol in all reagent solutions.

For qualitative analysis, separated proteins are visualized by staining with either Coomassie blue R-250 or silver nitrate (Morrissey, 1981). Gels are subsequently photographed (total protein pattern), dried and autoradiographs are prepared by exposing the radioactive gel to x-ray film (Kodak XAR-2) at -70°C for 4 days to 3 weeks to differentiate actively synthesized proteins. The amount of [^{35}S] methionine incorporated into individual proteins can be quantitated by cutting out protein spots of interest from dried gels. Labeled gel pieces are solubilized (Protosol®, H_2O_2) and the incorporated [^{35}S] methionine activity determined by liquid scintillation spectroscopy. A programmed quench curve allows for automatic quench correction of [^{35}S] in Coomassie and silver stained polyacrylamide gel. Prior to PAGE analysis, total acid precipitable [^{35}S] radioactivity of blastocyst tissue is determined from aliquots of radiolabeled tissue lysate using bovine serum albumin (0.1%) as a carrier.

4.2. Studies in the Rabbit

Figure 5 shows an autoradiograph composite developed from [^{35}S]-labeled 2D-PAGE gels of 4-, 5-, 6- and 7-day rabbit blastocyst proteins. Each gel contained 8.0 x 10^4 dpm (day 4) to 1.2 x 10^6 dpm (day 7) of [^{35}S] labeled blastocyst proteins. Up to 300 polypeptides are routinely resolved, of which 15 to 20 predominate. Thus, blastocyst growth and differentiation are reflected in an increasingly complex qualitative pattern of synthesized polypeptides. Using coelectrophoresis and immunoblotting techniques (O'Brien and Millette, 1984), we have identified Na^+/K^+-ATPase (A), γ actin (B) and β actin (C) [Fig. 5, panel D6] at each stage of development. The incorporation rates of [^{35}S] methionine into γ and β actin at 4, 5, 6 and 7 days *p.c.* are shown in Figure 6. Of all blastocyst polypeptides, β actin demonstrates the highest rate of incorporation at each day of development. Similarly, γ actin displays a parallel but reduced rate (approx. 1000-fold) of incorporation between days 4 and 7 *p.c.* Interestingly, the decrease in [^{35}S] incorporation into β and γ actin reflects an overall decrease in [^{35}S] incorporation into total blastocyst protein on day 7. The developmental significance of this observation is unclear and is currently being investigated.

4.3. Trophectodermal Na^+/K^+-ATPase

The rate of incorporation of [^{35}S] methionine into membrane-bound Na^+/K^+-ATPase is shown in Figure 7. These data reveal a 90-fold increase in rate of incorporation between the fourth and sixth day of development. Since this increase in Na^+/K^+-ATPase synthesis and the period of maximal blastocyst expansion coincide chronologically, it is reasonable to postulate that a direct relationship exists between these two processes of blastocyst differentiation. In addressing this hypothesis, we have recently identified blastocyst Na^+/K^+-ATPase using sensitive and specific immunoprecipitation procedures (Overström *et al.*, 1986). Rabbit blastocysts are radiolabeled ([^{35}S] methionine), washed and their tissues isolated as described previously. Labeled tissues (8 to 10 blastocysts) are placed in 1.5 ml microfuge tubes containing 0.75 ml of hypotonic Tris-buffered saline (TBS) -1% deoxycholate -1% NP40, 1 mM PMSF for 30 min at 4°C. Tissues are sonicated (3 x 10 sec) and lysates prepared by centrifugation at 12,000 x g. Twenty microliters of antiserum is added to 1.5 ml microfuge tubes containing 0.5 to 0.7 ml of blastocyst lysate. After incubation for 2 hr at 4°C, 50 μl of a 10% suspension of fixed *Staphylo-*

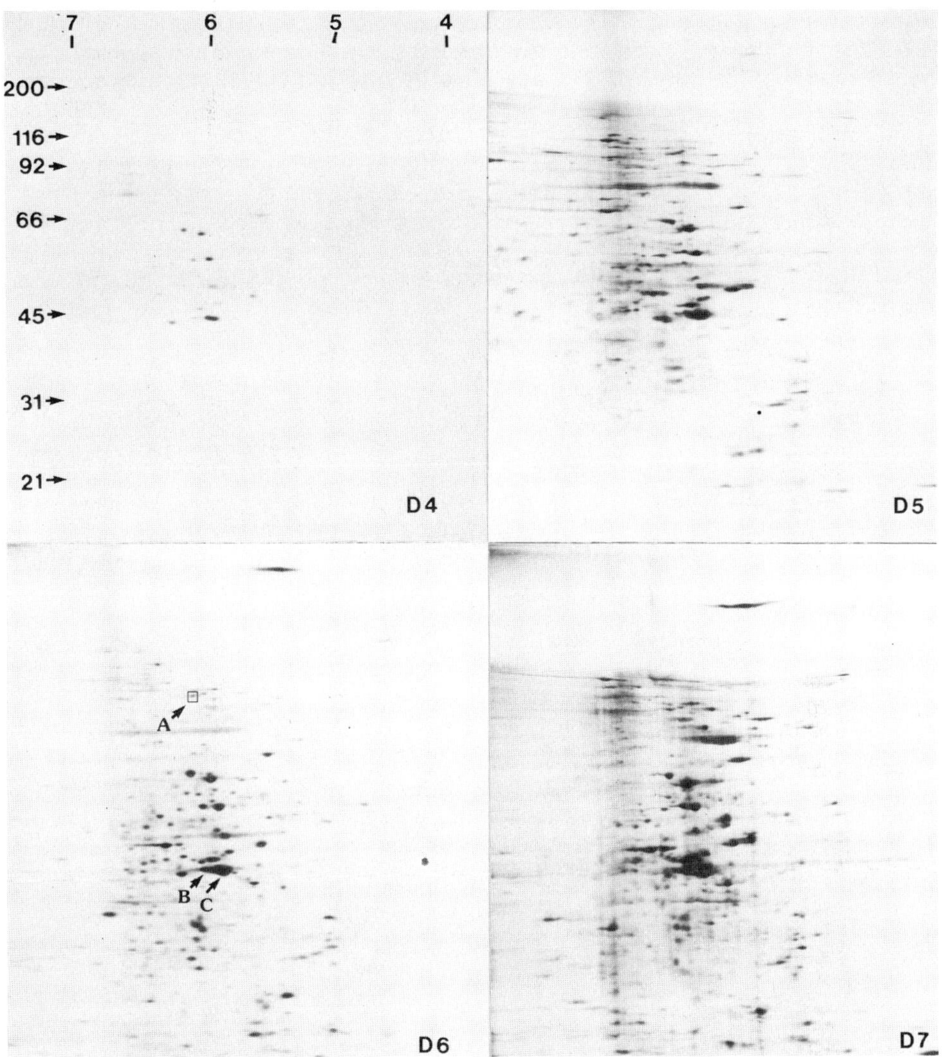

Figure 5. A composite two-dimensional gel autoradiograph of 4- to 7-day *p.c.* rabbit blastocysts labeled with [^{35}S] methionine. Na^+/K^+-ATPase (α subunit), γ and β actin are polypeptides A, B and C, respectively (plate D6). Molecular weights (daltons x 10^{-3}) and pI values are shown in the left and top margins, respectively.

coccus aureus cells (IgGsorb®) is added, mixed and incubated for 30 min (4°C). Immunoprecipitates are pelleted (10,000 x g) and washed 3 times with isotonic TBS -1% deoxycholate -1% NP40, 1 mM PMSF, and subjected to 1D-PAGE/ fluorographic analysis. Treatment of labeled blastocyst lysate with preimmune serum serves as a negative control.

The 1D-PAGE profiles of [^{35}S] labeled, immunoprecipitated Na^+/K^+-ATPase and total day 6 blastocyst protein are shown in Figure 8. The two

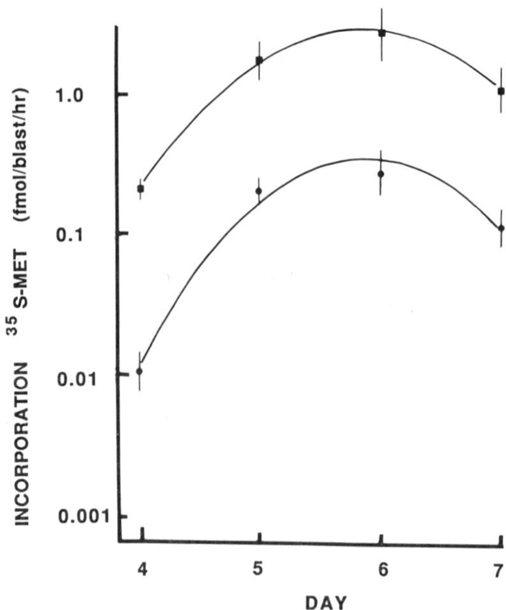

Figure 6. Rates of [^{35}S] methionine incorporation into blastocyst γ (●) and β (■) actin between days 4 and 7 of development in the rabbit.

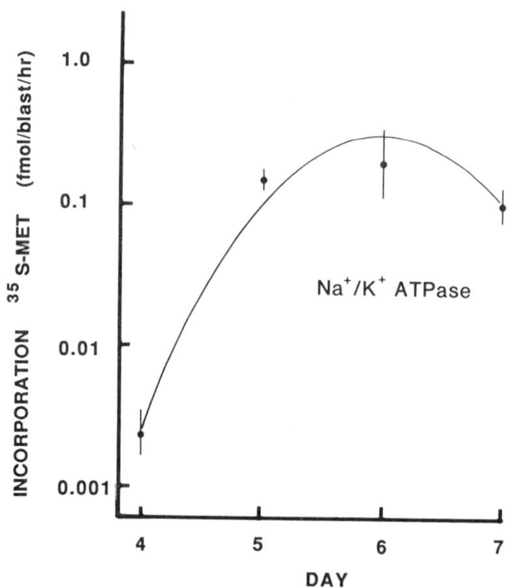

Figure 7. The rate of [^{35}S] methionine incorporation into blastocyst Na^+/K^+-ATPase *vs.* developmental age in the rabbit.

distinct bands seen in lane 1 represent the holoenzyme (A, approx. 340 kD) and the active α-subunit (B, approx. 100 kD) of the Na^+ pump. In contrast, a non-precipitated duplicate aliquot of total tissue lysate shows a very complex pattern of polypeptides (lane 2). At present, studies are underway with 4-, 5- and 7-day rabbit blastocysts in an effort to define more completely the quantitative expression of Na^+/K^+-ATPase during blastocyst development.

5. CHARACTERIZATION OF BLASTOCYST PLASMA MEMBRANES

Since there is considerable evidence indicating that dramatic changes in trans-trophectodermal solute transport mechanisms occur during mammalian blastocyst development (Benos, 1981a; Benos *et al.*, 1985), it is reasonable to postulate that these changes are a consequence of changing developmental expression of trophectodermal membrane constituents. To this end, we have initiated studies to isolate and characterize plasma membranes of the preimplantation rabbit blastocyst.

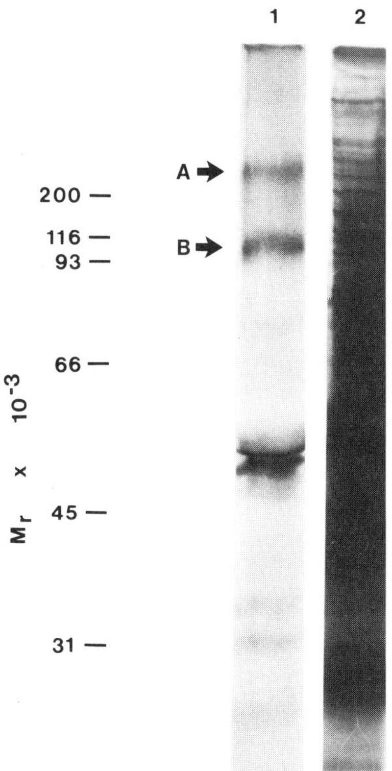

Figure 8. Autoradiograph of immunoprecipitated blastocyst Na^+/K^+-ATPase (lane 1) and total [^{35}S] radiolabeled polypeptides (lane 2) of day 6 rabbit blastocysts. Band A is the holoenzyme (approx. 340 kD) and band B is the active α-subunit (approx. 100 kD) of Na^+/K^+-ATPase.

5.1. Techniques

Due to its spherical geometry, the mammalian blastocyst avails itself to vectorial studies of apical *vs.* basolateral plasma membrane structure/ function. We have taken advantage of this to differentially radioiodinate apical or basolateral membrane proteins under physiologic conditions using the glucose~glucose-oxidase~lactoperoxidase reagent system. Blastocysts are either radiolabeled on their apical surface by incubation in a solution containing 100 μCi Na[^{125}I], 20 IU lactoperoxidase, 2 IU lactoperoxidase, 2 IU glucose oxidase per ml of phosphate buffered saline plus 5 mM glucose, or microinjected with the above reagents, thereby labeling protein constituents of the basolateral plasma membranes. A diagrammatic representation of the blastocyst illustrating the route of microinjection for [^{125}I]-labeling of basolateral membranes is shown in Figure 3. After extensive washing with ice-cold saline containing 0.1 mM NaI and 1 mM PMSF, blastocyst tissues are homogenized and their proteins solubilized and subjected to 1D- and 2D-PAGE and subsequent autoradiographic analysis. In an effort to isolate purified blastocyst plasma membranes, we have conducted experiments in which apically iodinated day-6 rabbit blastocysts are subsequently homogenized and their subcellular components fractionated by discontinuous sucrose density-gradient ultracentrifugation as described by Millette *et al.* (1980). Sucrose gradients containing separated blastocyst cell components are fractionated (ISCO Model 640) and the [^{125}I] radioactivity of each fraction is quantitated on a Beckman 4000 gamma counter.

5.2. Studies in the Rabbit

Figures 9a and 9b show autoradiographs of 1D- and 2D- gels of [^{125}I]-labeled apical and basolateral membrane proteins of 6-day rabbit blastocysts, respectively. Labeled apical membrane components are characterized by 3 major bands with apparent molecular weights between 45 kD and 85 kD. In contrast, iodinated basolateral membranes display a more complex pattern of labeled proteins with several major components resolved between 25 kD and 130 kD. Two-dimensional separation of iodinated blastocyst membrane proteins revealed 10 to 20 polypeptides which are characteristic of their apical or basolateral domain.

The profile of [^{125}I]-labeled day 6 blastocysts, fractionated on sucrose gradients, is shown in Figure 10. Two major peaks (2a, 2b) and two minor peaks (1, 3) of [^{125}I] radioactivity are resolved. Greater than 81% of the label is contained in peaks 2a and 2b. These 2 peaks, recovered at or near the 30 to 40% sucrose interface, represent labeled plasma membranes. Less than 9% of the label is associated with the nuclear pellet. These results suggest that the labeling procedure is indeed vectorial and specific for surface membrane proteins. These studies are continuing in an effort to characterize qualitative and quantitative changes in trophectodermal membrane constituents as a function of their apical or basolateral domain.

6. CONCLUSIONS: FUTURE PERSPECTIVES

Four methods have been described whereby a specific physiological, bio-

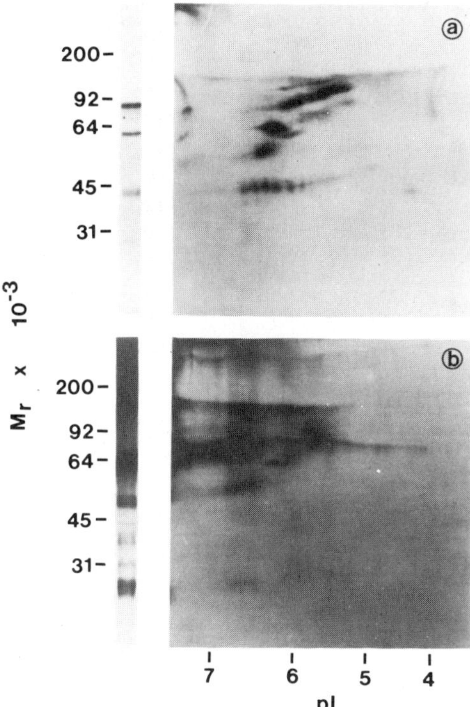

Figure 9. One- and two-dimensional gel composite of iodinated apical (a) and basolateral (b) plasma membranes of 6-day rabbit blastocysts. Molecular weights (daltons x 10^{-3}) and pI values are indicated.

chemical or morphological parameter of early mammalian development can be characterized in one or more living blastocysts. Of these techniques, the ability to assess the oxygen consumption rate of an individual blastocyst is particularly useful in that the procedure is rapid, non-invasive and, as such, allows for subsequent experimentation, including transfer to a synchronized recipient host. Since blastocyst QO_2 assessment is a direct quantitative measurement, it is amenable to development as a clinical diagnostic tool to screen for embryo viability prior to embryo transfer. From an applied perspective, QO_2 assessment has clear potential advantages over current subjective embryo screening procedures, based on the "normalcy" of embryo morphology, which are employed in human IVF-ET clinics and domestic animal embryo transfer units.

As described herein, analytical techniques, including characterization of trans-trophectodermal Na^+ transport properties, determination of Na^+/K^+-ATPase and total protein synthesis rates, and isolation and characterization of blastocyst plasma membranes, can independently provide much useful information concerning the magnitude and chronology of expression of specific cellular processes of blastocyst development. However, by integrating these biotechniques, *i.e.*, conducting 2 or more of these experimental procedures in sequence on an individual blastocyst, one can now begin to assess and

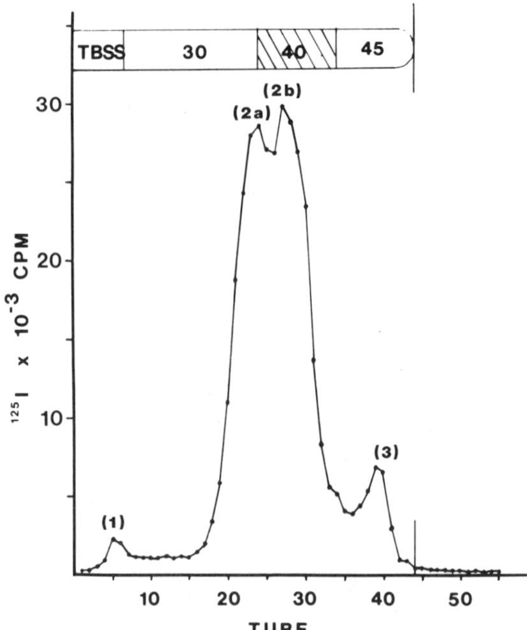

Figure 10. Profile of [^{125}I]-labeled apical plasma membranes fractionated by sucrose density-gradient centrifugation. Labeled plasma membranes migrate as 2 major peaks (2a, 2b) at the 30 to 40% sucrose interface.

correlate several cellular determinants of "normal" blastocyst development. Moreover, such information will hopefully provide clues to the identity of specific causative mechanisms that underlie aberrant blastocyst differentiation and subsequent embryonic death.

ACKNOWLEDGEMENTS

The excellent technical assistance of Ms. M. Tuson and Ms. A. Schmidt is gratefully acknowledged. The collaborative efforts of Dr. K.M. Ebert are gratefully acknowledged. I thank Ms. D. Lomber for typing the manuscript. This work was supported, in part, by funds from the Monsanto Corporation.

7. REFERENCES

Abrew, S.L., and Brinster, R.L., 1978, Synthesis of tubulin and actin during preimplantation development of the mouse, *Exp. Cell Res.* 114: 135-141.

Anderson, L.L., 1978, Growth, protein content and distribution of early pig embryos, *Anat. Rec.* 190: 143-154.

Benos, D.J., 1981a, Developmental changes in epithelial transport characteristics of preimplantation rabbit blastocysts, *J. Physiol. (London)* 316: 191-202.

Benos, D.J., 1981b, Ouabain binding to preimplantation rabbit blastocysts, *Dev. Biol.* 83: 69-78.

Benos, D.J., and Balaban, R.S., 1980, The energy requirements of the developing mammalian blastocyst for active ion transport, *Biol. Reprod.* 23: 941-947.

Benos, D.J., and Balaban, R.S., 1983, Energy metabolism of preimplantation mammalian blastocysts, *Amer. J. Physiol.* 245: C40-C45.

Benos, D.J., and Biggers, J.D., 1981, Blastocyst fluid formation, in: *Fertilization and Embryonic Development In Vitro* (L. Mastroianni, Jr., and J.D. Biggers, eds.), Plenum Press, New York, pp. 283-297.

Benos, D.J., Balaban, R.S., Biggers, J.D., Mills, J.W., and Overström, E.W., 1985, Developmental aspects of sodium dependent transport processes in preimplantation rabbit embryos, in: *Regulation and Development of Membrane Transport Processes* (J.S. Graves, ed.), John Wiley and Sons, New York, pp. 211-235.

Biggers, J.D., 1983, Risks of *in vitro* fertilization and embryo transfer in humans, in: *In Vitro Fertilization and Embryo Transfer* (P.G. Crosignani, and B.L. Rubin, eds.), Serono Clinical Colloquia on Reproduction, No. 4, Academic Press, London, pp. 393-410.

Biggers, J.D., and Stern, S., 1973, Metabolism of the pre-implantation mammalian embryo, *Adv. Reprod. Physiol.* 6: 1-59.

Biggers, J.D., Borland, R.M., and Powers, R.D., 1977, Transport mechanisms in the preimplantation mammalian embryo, in: *The Freezing of Mammalian Embryos*, Ciba Foundation Symposium 52 (K. Elliott, and J. Whelan, eds.), Elsevier, Amsterdam, pp. 129-153.

Biggers, J.D., Borland, R.M., and Lechene, C.P., 1978, Ouabain-sensitive fluid accumulation and ion transport by rabbit blastocysts, *J. Physiol. (London)* 280: 319-330.

Bonner, W.M., and Laskey, R.A., 1974, A film detection method for tritium-labeled protein and nucleic acids in polyacrylamide gels, *Eur. J. Biochem.* 46: 83-88.

Borland, R.M., Biggers, J.D., and Lechene, C.P., 1977, Fluid transport by rabbit preimplantation blastocysts, *J. Reprod. Fertil.* 52: 131-135.

Brinster, R.L., 1967, Carbon dioxide production from glucose by the preimplantation mouse embryo, *Exp. Cell Res.* 47: 271-277.

Brinster, R.L., 1968, Carbon dioxide production from glucose by the preimplantation rabbit embryo, *Exp. Cell Res.* 51: 330-334.

Brinster, R.L., 1969, Radioactive carbon dioxide production from pyruvate and lactate by the preimplantation rabbit embryo, *Exp. Cell Res.* 54: 205-209.

Brinster, R.L., Brunner, S., Joseph, X., and Levey, I.L., 1979, Protein degradation in the mouse blastocysts, *J. Biol. Chem.* 254: 1927-1931.

Burgoyne, P.S., and Ducibella, T., 1977, Changes in the properties of the developing mouse trophoblast as revealed by aggregation studies, *J. Embryol. Exp. Morphol.* 40: 143-157.

Chen, H.Y., Brinster, R.L., and Merz, E.A., 1980, Changes in protein synthesis following fertilization of the mouse ovum, *J. Exp. Zool.* 212: 355-360.

Cohen, J.J., and Kamm, D.F., 1976, Renal metabolism: Relation to renal function, in: *The Kidney* (B.M. Brenner, and F.C. Rector, eds.), Saunders, Philadelphia, pp. 126-214.

Cross, M.H., 1973, Active sodium and chloride transport across the rabbit blastocoel wall, *Biol. Reprod.* 8: 566-575.

Cross, M.H., and Brinster, R.L., 1970, Influence of ions, inhibitors, and anoxia on transtrophoblast potential of rabbit blastocyst, *Exp. Cell Res.* 62: 303-309.

Daniel, J.C., 1967, The pattern of utilization of respiratory metabolic intermediates by preimplantation rabbit embryos *in vitro*, *Exp. Cell Res.* 47: 619-623.

Day, B.N., Anderson, L.L., Emmerson, M.A., Hazel, L.N., and Melampy, R.M., 1959, Effect of estrogen and progesterone on early embryonic mortality in ovariectomized gilts, *J. Anim. Sci.* 18: 607-613.

DiZio, S., and Tasca, R.J., 1977, Sodium-dependent amino acid transport in preimplantation mouse embryos. III. Na-K-ATPase-linked mechanisms in blastocysts, *Dev. Biol.* 59: 198-205.

Ducibella, T., Albertini, D.F., Anderson, E., and Biggers, J.D., 1975, The preimplantation mammalian embryo: Characterization of intercellular junctions and their appearance during development, *Dev. Biol.* 445: 231-250.

Fridhandler, L., 1961, Pathways of glucose metabolism in fertilized rabbit ova at various preimplantation stages, *Exp. Cell Res.* 13: 132-139.

Fridhandler, L., Hafez, E.S.E., and Pincus, G., 1957, Developmental changes in the respiratory activity of rabbit ova, *Exp. Cell Res.* 13: 132-139.

Gamow, E., and Daniel, J.C., 1970, Fluid transport in the rabbit blastocyst, *Wilhelm Roux Archiv.* 164: 261-278.

Hafez, E.S.E., 1974, in: *Reproduction in Farm Animals* (E.S.E. Hafez, ed.), Lea and Febiger, Philadelphia, pp. 275-287.

Hafez, E.S.E., Jainudeen, M.R., and Lindsay, D.R., 1965, Gonadotropin-induced twinning and related phenomena in beef cattle, *Acta Endocrinol. Suppl.* 102: 1-43.

Hammond, J., 1914, On some factors controlling fertility in domestic animals, *J. Agricult. Sci. (Cambridge)* 6: 263.

Handyside, J.B., and Barton, S.C., 1977, Evaluation of the technique of immunosurgery for the isolation of inner cell masses from mouse blastocysts, *J. Embryol. Exp. Morphol.* 37: 217-226.

Hastings, R.A., and Enders, A.C., 1975, Junctional complexes in the preimplantation rabbit embryo, *Anat. Rec.* 181: 17-34.

Hertig, A.T., Rock, J., Adams, E.C., and Menkin, M.C., 1959, Thirty-four fertilized human ova, good, bad and indifferent, recovered from 210 women of known fertility, *Pediatrics* 23: 202-211.

Levinson, J., Goodfellow, P., Vadebonceom, M., and McDevitt, H., 1978, Identification of stage-specific polypeptide synthesized during murine preimplantation development, *Proc. Nat. Acad. Sci. USA* 75: 3332-3336.

Lewis, W.H., and Gregory, P.W., 1929, Cinematography of living developing rabbit eggs, *Science* 69: 226-229.

Manes, C., and Daniel, J.C., 1969, Quantitative and qualitative aspects of protein synthesis in the preimplantation rabbit embryo, *Exp. Cell Res.* 55: 261-268.

Millette, C.F., O'Brien, D.A., and Moulding, C.T., 1980, Isolation of plasma membranes from purified mouse spermatogenic cells, *J. Cell Sci.* 43: 279-299.

Mills, R.M., and Brinster, R.L., 1967, Oxygen consumption of preimplantation mouse embryos, *Exp. Cell Res.* 47: 337-344.

Morrissey, J.H., 1981, Silver stain for proteins in polyacrylamide gels:

Modified procedure with enhanced uniform sensitivity, *Anal. Biochem.* 117: 307-310.

Nellans, H.P., and Finn, A.F., 1974, Oxygen consumption and sodium transport in the toad urinary bladder, *Amer. J. Physiol.* 227: 670-675.

O'Brien, D.A., and Millette, C.F., 1984, Identification and immunochemical characterization of spermatogenic cell surface antigens that appear during early meiotic prophase, *Dev. Biol.* 101: 307-317.

O'Fallon, J.V., and Wright, W.W., 1986, Quantitative determination of the pentose phosphate pathway in preimplantation mouse embryos, *Biol. Reprod.* 34: 58-64.

O'Farrell, P.H., 1975, High resolution two-dimensional electrophoresis of proteins, *J. Biol. Chem.* 250: 4007-4021.

Overström, E.W., and Ebert, K.M., 1986, in preparation.

Overström, E.W., Benos, D.J., and Biggers, J.D., 1982, Quantitative changes in Na^+/K^+-ATPase in preimplantation rabbit blastocysts, *J. Cell Biol.* 95: 163.

Overström, E.W., Benos, D.J., and Biggers, J.D., 1983, Radio-iodination of apical and basolateral plasma membranes of preimplantation rabbit blastocysts, *Biol. Reprod.* 28 (Suppl. 1): 46.

Overström, E.W., Benos, D.J., Biggers, J.D., and Godkin, J.D., 1984, Trans-trophectodermal sodium transport during porcine blastogenesis, *Biol. Reprod.* 30 (Suppl. 1): 46.

Overström, E.W., Benos, D.J., and Biggers, J.D., 1986, Differentiation of the expanding preimplantation rabbit blastocyst: Identification and synthesis of Na^+/K^+-ATPase, *Dev. Biol.* (submitted).

Perry, J.S., and Rowlands, I.W., 1962, Early pregnancy in the pig, *J. Reprod. Fertil.* 2: 175-188.

Petzoldt, U., 1972, Protein patterns of the rabbit blastocyst tissues, *Cytobiologie* 6: 473-475.

Petzoldt, U., 1974, Micro-disc electrophoresis of soluble proteins in rabbit blastocysts, *J. Embryol. Exp. Morphol.* 31: 479-487.

Pike, I.L., 1981, Comparative studies of embryo metabolism in early pregnancy, *J. Reprod. Fertil.* Suppl. 29: 203-213.

Rieger, D., 1984, The measurement of metabolic activity as an approach to evaluating viability and diagnosing sex in early embryos, *Theriogenology* 21: 138-149.

Schultz, R.M., Letourneau, G.E., and Wasserman, P.M., 1979, Program of early development in the mammal: Changes in patterns and absolute rates of tubulin and total protein synthesis during oogenesis and early embryogenesis in the mouse, *Dev. Biol.* 68: 341-359.

Smith, M.W., 1970, Active transport in the rabbit blastocyst, *Experientia* 26: 736-738.

Van Blerkom, J., 1978, Methods for the high-resolution analysis of protein synthesis: Applications to studies of early mammalian development, in: *Methods in Mammalian Reproduction* (J.C. Daniel, Jr., ed.), Academic Press, New York, pp. 67-109.

Van Blerkom, J., and Brockway, G.O., 1975, Qualitative patterns of protein synthesis in the preimplantation mouse embryo. II. During release from facultative delayed implantation, *Dev. Biol.* 46: 446-451.

Van Blerkom, J., and Manes, C., 1974, Development of preimplantation rabbit embryos. II. A comparison of qualitative aspects of protein synthesis, *Dev. Biol.* 40: 40-51.

Wales, R.G., 1975, Maturation of the mammalian embryo: biochemical aspects, *Biol. Reprod.* 12: 66-81.

Zerahn, K., 1956, Oxygen consumption and active sodium transport in the isolated and short-circuited frog skin, *Acta Physiol. Scand.* 36: 300-312.

CHAPTER 6

STEROID HORMONES IN EARLY PIG EMBRYO DEVELOPMENT

HEINER NIEMANN and FOLKMAR ELSAESSER

1. INTRODUCTION

Several studies with rodent embryos have provided evidence that pre-morula stages were able to produce their own estrogens (see Dickmann *et al.*, 1976). These estrogens seem to be necessary for the transformation of the compacted morula into the cavitated blastocyst stage (Sengupta *et al.*, 1977, 1982, 1983; Paria *et al.*, 1984). In contrast, the role of sex steroids in the development of zona pellucida intact embryos from farm animals is not well defined. In the pig, it was suggested that estrogens are involved in the migration and spacing of porcine embryos on day 7 (Pope *et al.*, 1982). On days 11 to 12 of pregnancy, estrogens are considered to be the embryonic signal for the maternal recognition of pregnancy (Perry *et al.*, 1973; Heap *et al.*, 1979).

Recently, we investigated the specificity of progesterone and estradiol-17β uptake by early porcine blastocysts *in vitro*. The role of estradiol-17β in the development of zona pellucida intact pig embryos was studied by different experimental designs. Specifically, we looked at the effects of estradiol withdrawal on the development of porcine morula stages (Niemann and Elsaesser, 1984, 1985, 1986a,b).

2. METHODOLOGY

2.1. Superovulatory Regimen and Embryo Recovery in the Pig

Porcine embryos were recovered from normal cycling (n=26) or superovulated (n=134) German Landrace gilts. Superovulation was induced in

Heiner Niemann and **Folkmar Elsaesser** Institut fur Tierzucht und Tierverhalten (FAL), Mariensee, 3057 Neustadt 1, Federal Republic of Germany.

prepubertal gilts with intramuscular injections of 1500 IU pregnant mare's serum gonadotropin (PMSG: Seragon®), followed 72 hr later by 500 IU human chorionic gonadotropin (hCG: Ekluton®). The animals were either mated or inseminated with freshly diluted semen 24 and 48 hr after hCG injections and were slaughtered 4 days after the second breeding (day of second breeding = day 0) in our institute's slaughterhouse. The genital tract was removed immediately after bleeding and transported at 37°C to the laboratory. The number of corpora lutea (CL) was recorded. Each uterine horn was flushed with 80 ml phosphate buffered saline (PBS) containing 1% heat-inactivated newborn calf serum (NBCS). Embryos were also recovered surgically 4 days after mating according to the method described by Dziuk (1971) using PBS supplemented with NBCS as the flushing medium.

The fluid was examined under a stereomicroscope at 12- and 50-fold magnifications. Recovered embryos were placed in fresh PBS and NBCS and were evaluated at 50- and 100-fold magnifications for fertilization, developmental stage and morphology of the blastomeres and zona pellucida. Morphologically intact morulae (Fig. 1) with more than 10 to 12 compacted blastomeres and early blastocysts were used for the experiments. Unfertilized ova or retarded and/or degenerate embryos were discarded.

After superovulation, a mean number of about 30 ± 15 CL per gilt (mean ± SD) was counted and about 12 ± 2 CL after spontaneous estrus. The fertilization rate tended to be higher after spontaneous estrus (97.5%) compared to

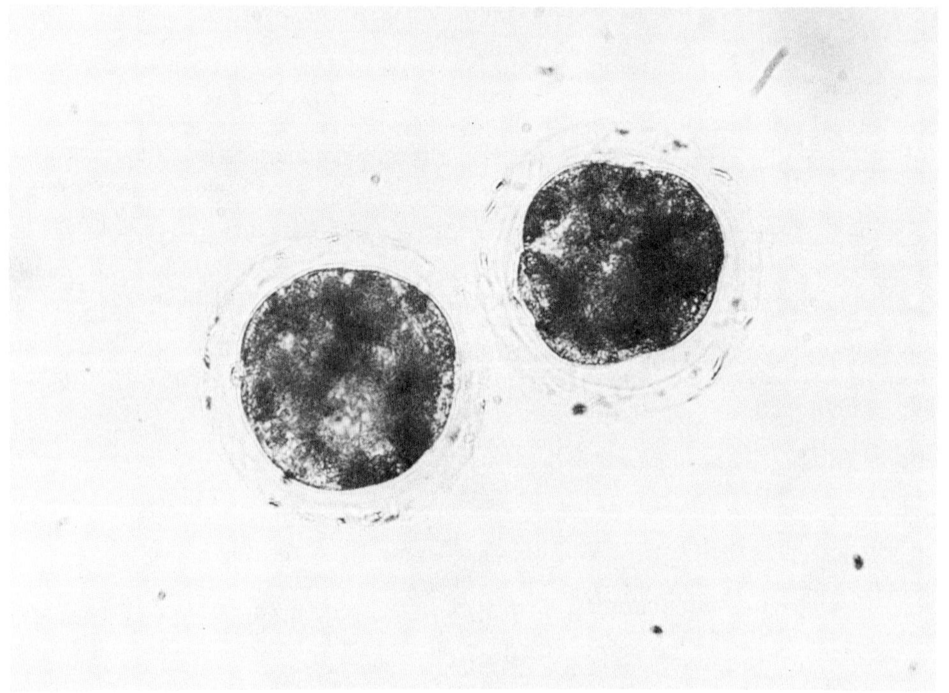

Figure 1. Two late pig morulae grown *in vivo* and collected on day 4 (day 0 = day of last breeding or insemination). Diameter of embryos approx. 157 μm (see Niemann *et al.*, 1983).

estrus after superovulatory treatment (90-94%). The number of degenerate and/or retarded embryos was increased after superovulatory treatment (8-10%) compared to normal cycling gilts (1-2%). Embryo recovery rates (= embryos/CL) were similar (85 to 90%) after spontaneous or superovulatory estrus as well as surgical recoveries or recoveries from isolated uterine horns. On average, the number of embryos could be increased by a factor of 2 to 3 by superovulatory treatment, without affecting embryonic developmental capacity.

2.2. *In Vitro* Culture of Preimplantation Embryos

Porcine embryos were cultured *in vitro* in petri dishes containing 2 ml Krebs-Ringer-bicarbonate (KRB) (Davis and Day, 1978) with or without bovine serum albumin (BSA) and supplemented with 10% heat-inactivated (30 min at 56°C) lamb serum. Cultures were maintained at 37°C under 5% CO_2 in air and in a humidified atmosphere. Embryonic development was recorded every 24 hr at 100- and 200-fold magnifications. Using these culture conditions, a high percentage of embryos recovered from the 4-cell to the early blastocyst stage developed to hatching and hatched blastocysts (Niemann *et al.*, 1983). By culturing pig embryos for at least 120 hr it was possible to distinguish 4 different phases during the early embryonic development: the 4-cell stage, 8-cell to morula stage, blastocyst stage (Figs. 2 and 3) and the hatched blastocyst (Fig. 4) after escaping from the zona pellucida.

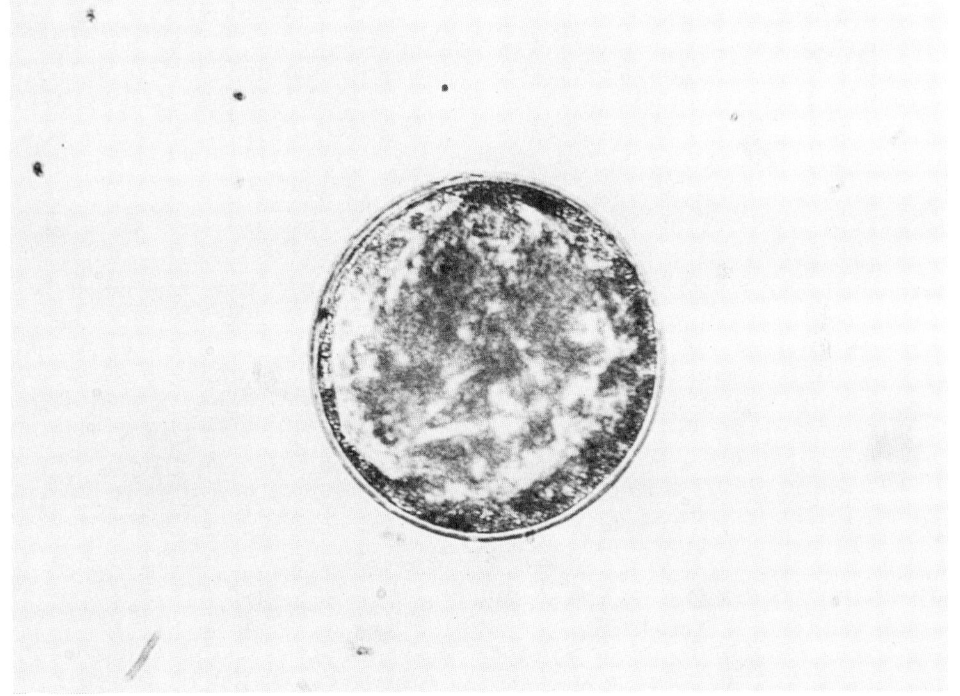

Figure 2. Expanded pig blastocyst, collected on day 5 to 5.5; diameter approx. 198 μm.

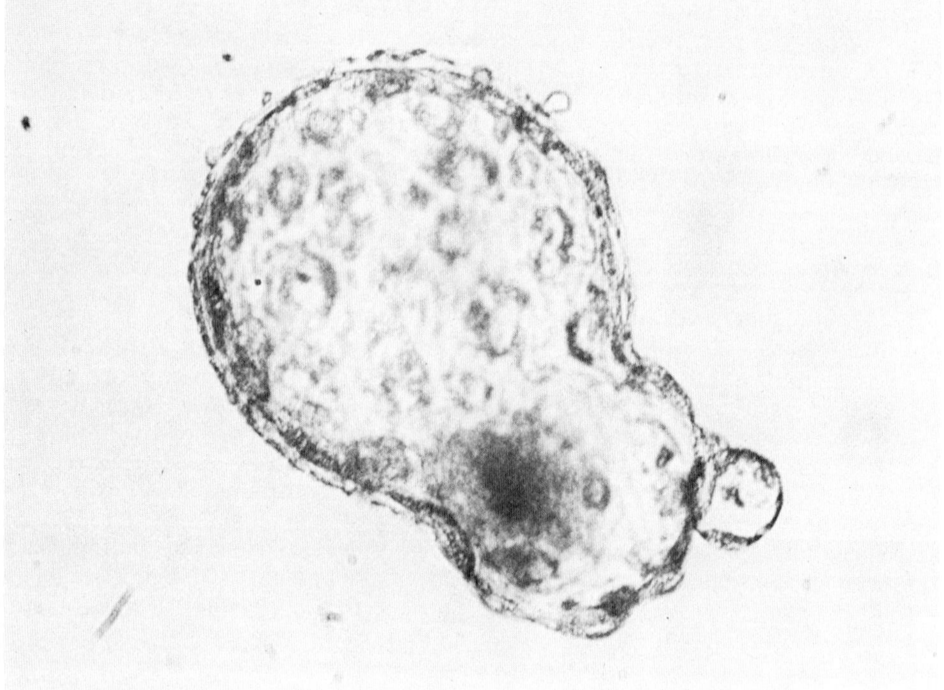

Figure 3. Hatching pig blastocyst, collected on day 5.5 to 6.0; approx. diameter 207 μm.

2.3. Uptake of Radiolabelled Steroid Hormones by Blastocysts

Prior to the uptake of radiolabelled steroid hormones, porcine morulae were cultured *in vitro* for 24 hr to the blastocyst stage. The culture medium was changed twice within 24 hr. Subsequently, embryos were randomly assigned to groups of 10 to 20 and further incubated in KRB containing 10% lamb serum (in two experiments pure KRB without BSA and lamb serum was used) and 10 nM or 20 nM (approximately 400,000 or 800,000 cpm/ml) [1,2,6,7-^{3}H] progesterone or [2,4,6,7,-^{3}H] estradiol-17β for 20, 120 or 360 min. Both steroids were obtained from New England Nuclear (radiochemical purity greater than 98%). In each experiment the uptake specificity was tested in parallel by incubation of embryos with radiolabelled steroids in the presence of a 100-fold excess (1 or 2 μM) of unlabelled progesterone or estradiol-17β. Each medium was standardized to a concentration of 0.1% ethanol. At the termination of each incubation, embryos were washed 5 times in KRB and transferred into vials containing 0.2 ml tissue solvent (Soluene® 350, Packard). Fifteen ml of scintillation fluid (Dimilume®) were added 5 hr later, and radioactivity in embryos was counted in a liquid scintillation spectrometer (Packard 2450 with a counting efficiency of 49%). Total activity in the incubation medium and background activity of the last washing medium were always controlled. The latter was subtracted from the activity determined in the embryos. The competitive inhibition was calculated as the difference

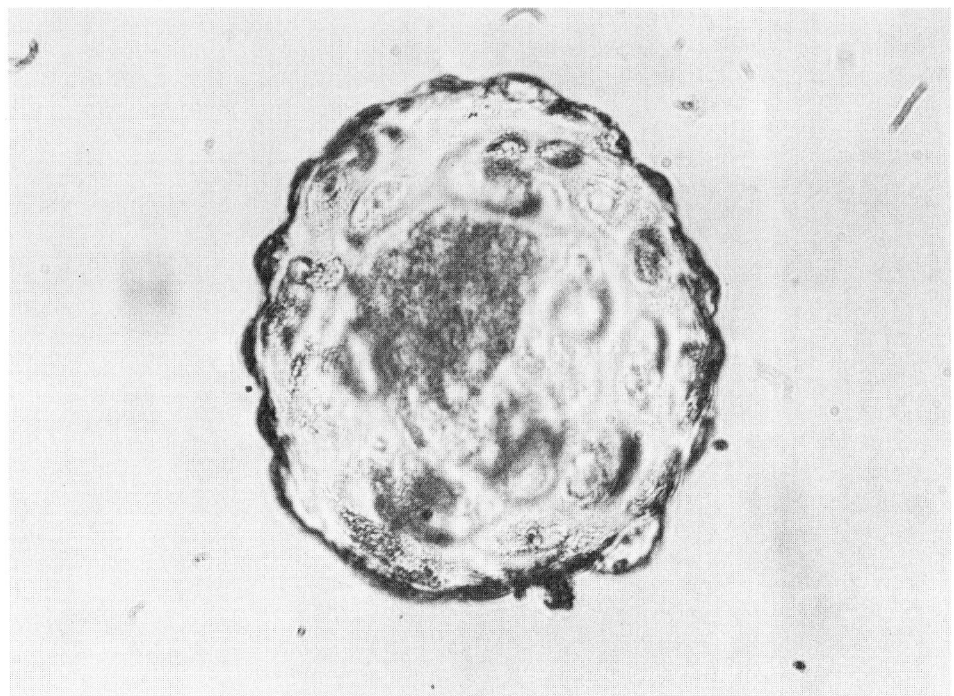

Figure 4. Hatched pig blastocyst, collected on day 6.0 to 6.5; diameter approx. 270 μm.

between total uptake of ^{3}H-progesterone or ^{3}H-estradiol-17β in the presence and in the absence of a 100-fold excess of unlabelled progesterone or estradiol-17β. The steroid uptake was always expressed as cpm per 10 blastocysts.

2.4. *In Vitro* Culture of Morulae in the Presence of Supplementary Steroids

Porcine morulae were cultured in medium supplemented with estradiol-17β (0.1 nM, 1 nM, 100 nM), progesterone (100 nM, 500 nM), cortisol (100 nM) or ethanol alone (controls). Estradiol-17β, progesterone and cortisol (all biochemical grade, Merck) were dissolved in ethanol and the appropriate amounts were dried down to give the desired concentration of steroids. Each medium was standardized to a concentration of 1 μl ethanol per ml.

2.5. *In Vitro* Culture of Morulae in the Presence of Charcoal-Stripped Lamb Serum

Culture media are usually supplemented with various blood sera or with BSA in order to stimulate *in vitro* development of mammalian embryos (see Wright and Bondioli, 1981). Both components are known to contain sex steroids apart from a number of other substances (Kane, 1978). In order to remove steroids, lamb serum was stirred gently at 4°C for at least 6 hr with 1% charcoal (Norit A, Serva No. 30890) and 0.1% dextran T 70, centrifuged and sterile filtered before use. Porcine morulae were incubated for 120 hr in either standard culture medium (controls), in KRB-solution supplemented with charcoal-stripped serum instead of normal lamb serum, or in KRB-solution supplemented with both charcoal-stripped serum and with 1 nM estradiol-17β.

2.6. *In Vitro* Culture of Morulae in the Presence of the Antiestrogen Nafoxidine

Antiestrogens prevent estrogens from expressing their full effects in estrogen target tissues (Katzenellenbogen *et al.*, 1979). The estrogen antagonist Nafoxidine is a nonsteroidal compound (n 1-2, P-3, 4 dihydromethoxy-2-phenyl-1-naphthyl-phenoxythyl) with a molecular weight of 462.0 (Nafoxidine-hydrochloride, Sigma Chemical Co., cat. no. 6632). It is known to compete with estradiol-17β for the same binding sites. Within the classic theory of steroid hormone action, the most likely mode of action of Nafoxidine is to inhibit replenishment of cytoplasmic estrogen receptors, which has been demonstrated in the immature rat uterus (Clark *et al.*, 1973, 1974; Katzenellenbogen and Ferguson, 1975; Ferguson and Katzenellenbogen, 1977). In our experiments Nafoxidine was dissolved in ethanol and prepared as a stock solution at a concentration of 1 μg/ml. Porcine morulae were incubated 24 hr in standard culture medium (controls), in medium supplemented with either Nafoxidine at concentrations of 3, 15, 30 or 60μg/ml, or with 15μg/ml Nafoxidine with different concentrations of estradiol-17β (1 nM, 100 nM, 100 μM), progesterone (100 nM) or cortisol (100 nM).

2.7. *In Vitro* Culture of Morulae in the Presence of an Estradiol-17 Beta Antiserum

The antiserum (E 12) had been raised against estradiol-17β in a rabbit and had been proven to be highly specific. The antiserum cross-reacted 0.6% with estrone, 0.1% with estriol and $<$ 0.1% with progesterone, testosterone and hydrocortisone (Elsaesser, 1980).

Porcine morulae were incubated for 120 hr in culture medium that was supplemented with either 5% normal rabbit serum (controls), 0.05, 0.5 or 5% estradiol-17β-antiserum, or 5% estradiol-17β antiserum and 1 μM or 100 μM estradiol-17β.

3. PROGESTERONE: EMBRYONIC UPTAKE AND ITS SPECIFICITY

After 6 hr of *in vitro* incubation, uptake of ^{3}H-progesterone amounted to 132 ± 18 cpm (mean ± SEM) per 10 blastocysts (Fig. 5). Addition of a 100-fold excess of unlabelled progesterone or estradiol-17β resulted in a small and nonsignificant reduction of the ^{3}H-progesterone uptake. The competitive inhibition was calculated to be 27.0% and 20.1% for progesterone and estradiol-17β, respectively. Thus, the progesterone uptake by the pig blastocyst apparently is nonspecific only.

Similar results were reported by Angle and Mead (1979) with rabbit embryos: a nonspecific uptake of progesterone into the blastocoel fluid was observed with maximum concentrations after 2 hr. Progesterone that accumu-

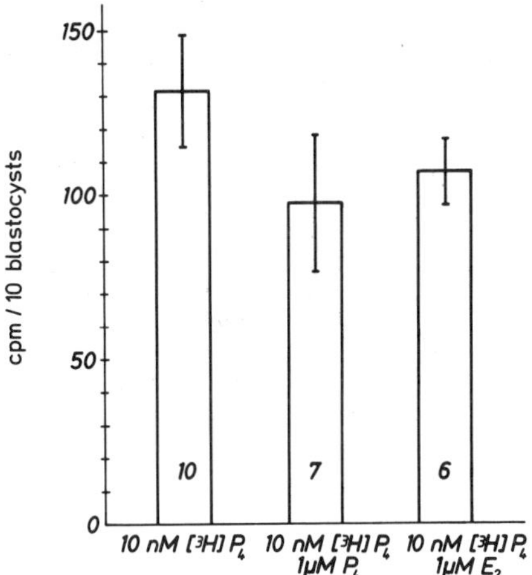

Figure 5. Specificity of uptake (mean ± SEM) of ^{3}H-progesterone by pig blastocysts after 6 hr of incubation. The number of experiments, each involving 5 to 10 embryos, is indicated at the bottom of the bars (data from Niemann and Elsaesser, 1984).

lates in rabbit blastocysts in the uterus probably originates from the uterine fluid (Borland *et al.*, 1977). It may well be that progesterone uptake into the blastocoel fluid of the pig blastocyst also occurs nonspecifically. Progesterone, which is necessary as a metabolic precursor of estradiol and estrone, may have importance for the pig embryo from day 11 to 12, since estrogens are needed as an embryonic signal for the maternal recognition of pregnancy (Perry *et al.*, 1973; Heap *et al.*, 1979). So far, however, no conclusive statements are possible regarding the kind of specific function progesterone has for the very early embryonic stages in the pig.

4. *IN VITRO* CULTURE OF MORULAE IN THE PRESENCE OF SUPPLEMENTARY STEROIDS

Neither inhibitory nor stimulatory effects on blastocyst formation were observed by morphological evaluation when embryos were incubated in culture medium supplemented with varying concentrations of either estradiol-17β, progesterone or cortisol. The percentage of blastocysts formed after 48 hr, as well as the number of hatching and hatched blastocysts after 120 hr, were identical in all groups, *i.e.*, 95-100% blastocysts, 65-75% hatching and hatched blastocysts (Niemann and Elsaesser, 1986b).

Thus, steroids added to the culture medium in concentrations simulating physiological or superovulatory conditions did not affect porcine early embryonic development up to the stage of the hatched blastocyst. Inhibitory effects following supplementation of the culture medium with progesterone have been reported in mice, rabbit and rat embryos, when progesterone was added in pharmacological doses, *i.e.*, > 4 μg/ml (Whitten, 1957; Daniel, 1964; Daniel and Levy, 1964; Daniel and Gowan, 1966; Pratt and Daniel, 1967; Dickmann, 1970; Kirkpatrick, 1971). In agreement with our own findings, lower concentrations of progesterone (< 4 μg/ml) did not affect early embryonic

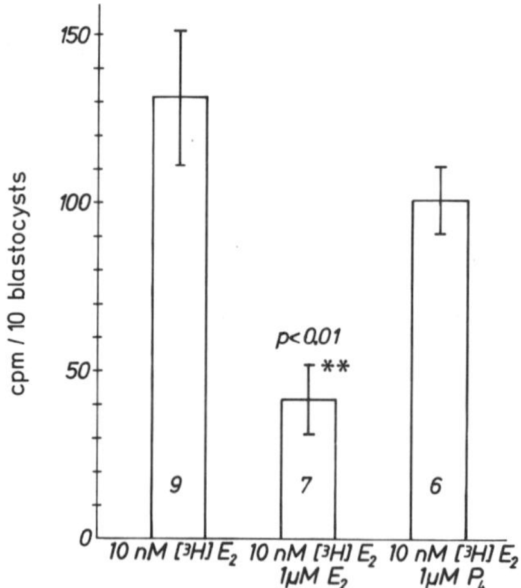

Figure 6. Specificity of uptake (mean ± SEM) of ^{3}H-estradiol-17β by pig blastocysts after 6 hr of incubation. The number of experiments, each involving 10 to 20 embryos, is indicated at the bottom of the bars (data from Niemann and Elsaesser, 1984).

development (Whitten, 1957; Cline *et al.*, 1977; Roblero and Izquierdo, 1976). A combination of progesterone and uterine proteins has even been shown to stimulate the development of mouse and rabbit early embryos (El-Banna and Daniel, 1972; Roblero and Izquierdo, 1976). In agreement with these results obtained from *in vitro* studies, we found that *in vivo* only excessively high dose levels of exogenous progesterone (3000 mg/day) disturbed early embryonic development in swine (Niemann and Elsaesser, 1984).

Regarding estrogens, supplementary estradiol-17β in the culture medium was found to stimulate the uptake of nucleic or amino acids in mouse early embryos in some experiments (Smith and Smith, 1971; Harrer and Lee, 1973), although other investigators did not find a stimulatory effect (Weitlauf, 1973).

5. ESTRADIOL 17 BETA: EMBRYONIC UPTAKE AND ITS SPECIFICITY

Similar to progesterone, the uptake of ^{3}H-estradiol-17β amounted to 134 ± 25 cpm per 10 blastocysts following 6 hr of incubation (Fig. 6). However, in contrast to progesterone, addition of a 100-fold excess of unlabelled estradiol-17β significantly ($p < 0.01$) reduced the uptake of ^{3}H-estradiol-17β by 67.7% to 43 ± 13 cpm per 10 blastocysts. In the presence of a 100-fold excess of unlabelled progesterone, a slight and nonsignificant (23.3%) reduction of the ^{3}H-estradiol-17β uptake was determined.

Further experiments with ^{3}H-estradiol-17β revealed that after 2 hr of incubation the uptake was not significantly less (82 ± 2 cpm) compared to a 6 hr incubation and the significant ($p < 0.001$) competitive inhibition by a 100-fold excess of unlabelled estradiol-17β was still maintained. After 20 min of incubation, however, a significantly ($p < 0.005$) reduced uptake of 42 ± 2 cpm was found when compared to an incubation period of 2 or 6 hr and no competitive inhibition could be determined. The uptake of ^{3}H-estradiol-17β was also significantly ($p < 0.001$) less (52 ± 5 cpm) in degenerate embryos and unfertilized ova recovered on day 4 when compared to viable blastocysts after 2 or 6 hr of incubation and no competitive inhibition was apparent (Fig. 7).

Furthermore, the uptake of ^{3}H-estradiol-17β was tested after 6 hr of culture in a protein-free culture medium. The amount of uptake was similar to that of viable embryos incubated in complete medium and the competitive inhibition was still maintained (Fig. 8).

Our results clearly indicate a specific uptake and binding of estradiol-17β by the pig blastocyst, which is supported by the significant competitive inhibition in the presence of a 100-fold excess of unlabelled estradiol-17β. Estradiol-17β apparently is bound to specific binding sites in the early pig embryo, which need to be characterized biochemically in subsequent experiments. Obviously, uptake and specificity of estradiol-17β are dependent on the viability of the blastocyst, which suggests a blastocyst controlled mechanism. Furthermore, uptake and specificity apparently are independent of proteins present in the environment. Recent experiments suggest that estrone is taken up by the early pig blastocyst in a saturable manner as well, and either that estradiol-17β and estrone compete for the same binding sites or that estrone is metabolized to estradiol-17β in the blastocyst (Niemann and Elsaesser, unpublished observation).

Based on the results of the above mentioned experiments, we suggested a specific physiological function of estradiol-17β for the development of the zona pellucida intact pig embryo. Thus, in a subsequent series of experiments we tried to substantiate the suggested presence of estrogenic binding sites and

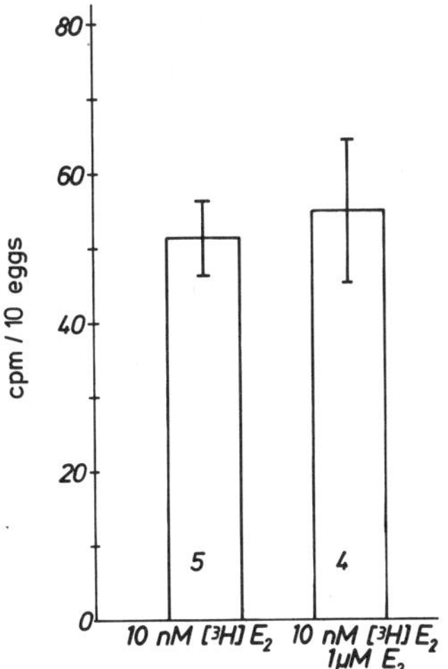

Figure 7. Specificity of uptake (mean ± SEM) of ^{3}H-estradiol-17β by degenerate embryos and unfertilized ova from day 4. The number of experiments, each involving 20 embryos, is indicated at the bottom of the bars (data from Niemann and Elsaesser, 1984).

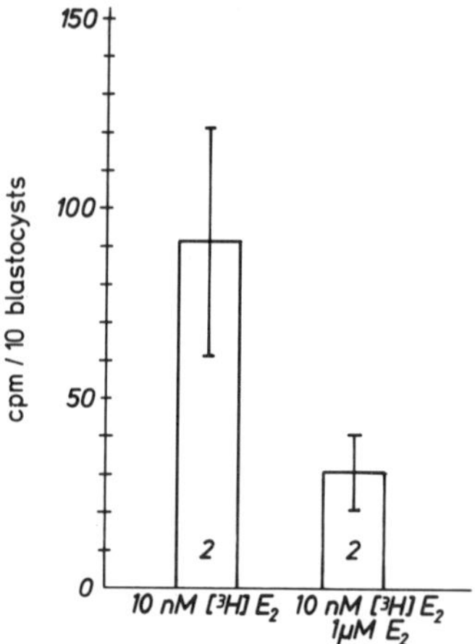

Figure 8. Specificity of uptake (mean ± SEM) of ^{3}H-estradiol-17β by pig blastocysts after 6 hr of incubation in protein-free culture medium. The number of experiments, each involving 20 embryos, is indicated at the bottom of the bars (data from Niemann and Elsaesser, 1984).

to study the role of estradiol-17β in the morula-blastocyst transformation in the pig.

6. ESTRADIOL WITHDRAWAL AND *IN VITRO* DEVELOPMENT OF MORULA STAGES

6.1. *In Vitro* Culture in the Presence of Charcoal-Stripped Lamb Serum

No treatment differences were observed when embryos were incubated for 48 hr in KRB-solution supplemented with charcoal-stripped lamb serum in the absence or presence of estradiol-17β. The percentage of morulae developing to blastocysts within 48 hr was 93.6% ± 4.7%, 93.1 ± 4.0% and 95.4% ± 2.0% for the controls, the "charcoal-group" and the "charcoal plus estradiol-17β group", respectively. Subsequent development was not affected, since the percentage of hatching and hatched blastocysts was similar (73-80%) in all groups after 120 hr of culture.

The lack of any detectable morphological effect in this experiment could be explained by the findings of Vignon *et al.* (1980) that incubation of serum with charcoal will only partially absorb conjugated steroids such as estrone-sulfate. Thus, during the process of *in vitro* incubation, hydrolysis might take place, providing free biologically active estrogens to the embryos, or alternatively, estrogen conjugates may be active *per se*. Thus, only very small amounts of exogenous estrogens could be necessary for undisturbed early embryonic development. On the other hand, the lack of any effect using charcoal-stripped lamb serum could be explained by embryonic synthesis of estrogens. Additional experiments involving enzyme inhibitors such as the aromatase inhibitor ATD are needed to elucidate this intriguing possibility.

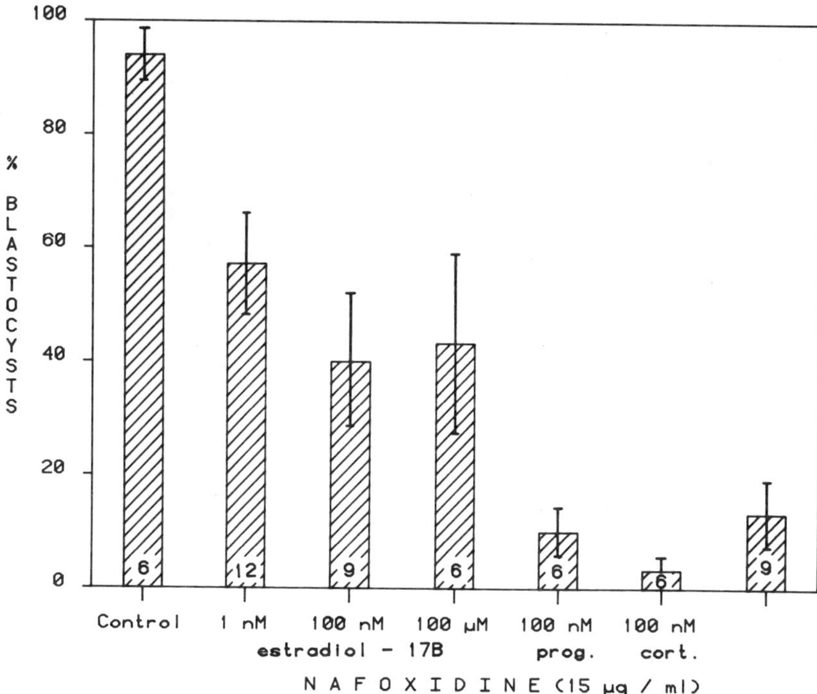

Figure 9. Specific reversal by estradiol-17β of inhibitory Nafoxidine (15 μg/ml) effects on development of porcine morulae. The number of experiments, each involving 5 embryos, is indicated at the bottom of the bars; prog. = progesterone, cort. = cortisol. (Data from Niemann and Elsaesser, 1986b).

6.2. *In Vitro* Culture in the Presence of the Antiestrogen Nafoxidine

Blastocyst formation rate was significantly ($p < 0.01$) reduced to 13.3 ± 5.8% at a Nafoxidine concentration of 15 μg/ml when compared to controls (93.3 ± 4.2%) (Fig. 9). Higher concentrations (30 μg/ml and 60 μg/ml) of Nafoxidine completely inhibited development, whereas lower concentrations did not seem to affect blastocyst formation (not shown in Fig. 9). The inhibitory effect of Nafoxidine (15 μg/ml) could be overcome by supplementation with 1 nM estradiol-17β since this significantly ($p < 0.01$) increased blastocyst formation rate to 57.2 ± 8.9%. Higher concentrations of estradiol-17β (100 nM, 100 μM) did not further enhance blastocyst formation. The stimulatory effects of estradiol-17β were specific, since Nafoxidine-reduced blastocyst formation rate remained low in the presence of 100 nM progesterone (10.0 ± 4.5%) or 100 nM cortisol (3.3 ± 3.3%). This is shown in Figure 9.

Results from this experiment clearly indicate that the physiological function of estradiol-17β is mediated through specific binding sites for estrogens, which was previously suggested based on the saturable accumulation of estradiol-17β in day 5 pig blastocysts. It has previously been suggested that receptors or specific binding sites for estrogens are involved in early embryonic development (Bhatt and Bullock, 1974; Holmes, 1976). Studies with rat embryos have shown that Nafoxidine inhibits morula-blastocyst transformation as well as blastocyst expansion and trophoblast maturation *in vitro*, and only the latter could be reversed by supplementation with estradiol-17β (Roy *et al.*, 1981, 1982). The incomplete reversal of the inhibitory effects of Nafoxidine by estradiol-17β is probably due to a reduced biological activity of estradiol-17β in the presence of the antiestrogen. A similar incomplete reversal has been observed with respect to rat blastocyst development (Roy *et al.*, 1982), and protein synthesis and weight response in the immature rat uterus (Katzenellenbogen and Ferguson, 1975).

6.3. *In Vitro* Culture in the Presence of an Estradiol-17 Beta Antiserum

Blastocyst formation rate was significantly ($p < 0.01$) reduced to 51.0 ± 6.7% in the presence of 5% estradiol-17β antiserum when compared to controls (93.1 ± 2.2%) (Fig. 10). The percentage of blastocysts was significantly ($p < 0.05$) increased to 79.7 ± 4.7% after supplementation with 100 μM estradiol-17β. A lower concentration of estradiol-17β (1 μM) resulted in a nonsignificant improvement of the blastocyst formation rate (65.8 ± 14.9%). Lower concentrations (0.05%, 0.5%) of the estradiol-17β antiserum had no inhibitory effects (not shown in Fig. 10). Furthermore, subsequent development of

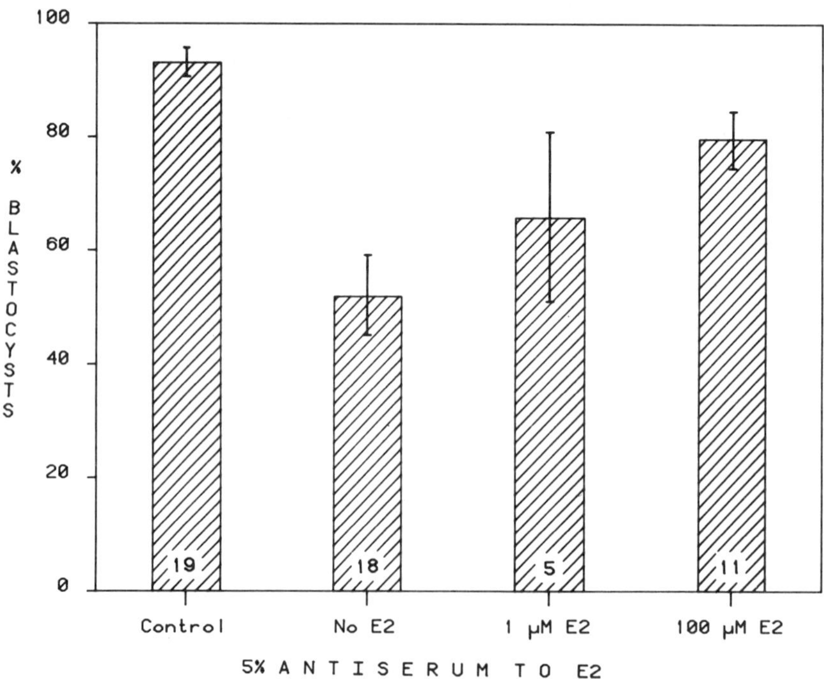

Figure 10. Inhibitory effects of an estradiol-17β antiserum on the development of porcine morulae and their reversal by estradiol-17β. The number of experiments, each involving 5 to 12 embryos, is indicated at the bottom of the bars (data from Niemann and Elsaesser, 1986b).

the blastocysts was not affected by the presence of antiserum since the percentage of hatching and hatched blastocysts was similar in all groups (Niemann and Elsaesser, 1986b).

The reversal of the withdrawal effects by supplementation with estradiol-17β indicates that the decreased blastocyst formation rate is due to an immunological inactivation of estradiol-17β molecules. Thus, less estradiol-17β would be available to form receptor complexes and consequently the response to estrogen was decreased. Since 50% of embryos still formed blastocysts in the presence of antiserum, the antibody-binding of estradiol-17β molecules may have been incomplete, leaving a sufficient number of estradiol-17β molecules to induce the estrogen response. If the pig morula is indeed able to synthesize its own estrogen, as suggested by Dickmann *et al.* (1976), it is conceivable that an estradiol-antiserum is not as effective in blocking estrogen response as are antiestrogens, which penetrate into the cells. At present, we cannot sufficiently explain why such high levels of estradiol were required to reverse the effects of the estradiol antiserum. Probably these amounts were necessary because of the considerable binding capacity of a 5% concentration of antiserum in the culture medium.

7. FURTHER CONSIDERATIONS

Any estrogen requirement for blastocyst formation may be met by the blastocyst itself or by an accumulation of estradiol-17β from the environment or both. The work of Perry *et al.* (1973) and Heap *et al.* (1979) suggests that porcine embryonic tissue does not acquire the enzymatic capacity to convert neutral steroids into estrogens before day 12 of pregnancy. However, Dickmann *et al.* (1976) provided at least limited evidence by histochemical methods that the zona pellucida intact pig embryo may not be different from rodent or rabbit embryos, which have been shown to synthesize estrogen. On the other hand, *in vivo* a supply of estrogens by the uterine environment is also conceivable, since uterine flushings from pigs already contain considerable amounts of estradiol-17β and estrone by day 6 of pregnancy (Zavy *et al.*, 1980).

Morula-blastocyst transformation is an important step in early embryonic development and results in two different cell types, those of the inner cell mass and of the trophoblast. The process coincides with a number of quantitative and qualitative changes in RNA, DNA and protein synthesis (Johnson, 1979). Since the embryonic genome is already active before blastocyst formation, as shown by RNA-synthesis in early mouse embryos (Warner and Hearn, 1977; Levey *et al.*, 1978), estrogenic function is probably the induction of genomic activity to start the process of blastocyst formation. The failure of any effect of supplementary estrogen on RNA-synthesis in mouse embryos (Warner and Tollefson, 1977, 1978) does not contradict this hypothesis. Supplementary estradiol-17β will not stimulate RNA-synthesis, if sufficient estrogen is supplied to the embryo by the culture medium or by the embryo itself.

In summary, the results of our experiments on pig embryos clearly demonstrate that estradiol-17β is required for the transformation of the compacted morula to the cavitated blastocyst stage (and possibly for the expansion of the blastocyst: Niemann and Elsaesser, unpublished) and most likely exerts its effect through specific binding sites for estrogens. The physiological function of estradiol-17β could be elucidated when estradiol-17β was biologically inactivated by the estrogen antagonist Nafoxidine or by an antiserum to estradiol-17β. Whether the estrogen requirement is met by an uptake of estrogens from the environment, or by embryonic estrogen synthesis, or both, is currently being determined. In subsequent experiments the role of other estrogens such as estrone or catecholestrogens for the development of zona pellucida intact pig embryos needs to be investigated.

ACKNOWLEDGMENTS

The expert technical assistance of Mr. H. H. Döpke is gratefully acknowledged.

8. REFERENCES

Angle, M.J., and Mead, R.A, 1979, The source of progesterone in preimplantation rabbit blastocysts, *Steroids* 33: 625-637.

Bhatt, B.M., and Bullock, D.W., 1974, Binding of oestradiol to rabbit blastocysts and its possible role in implantation, *J. Reprod. Fertil.* 39: 65-70.

Borland, R.M., Erickson, G.F., and Ducibella, T., 1977, Accumulation of steroids in rabbit preimplantation blastocysts, *J. Reprod. Fertil.* 49: 219-224.

Clark, J.H., Anderson, J.N., and Peck, E.J., Jr., 1973, Estrogen receptor antiestrogen complex: Atypical binding by uterine nuclei and effects on uterine growth, *Steroids* 22: 707-718.

Clark, J.H., Peck, E.J., Jr., and Anderson, J.N., 1974, Oestrogen receptors and antagonism of steroid hormone action, *Nature (London)* 251: 446-448.

Cline, E.M., Randall, B.A., and Oliphant, G., 1977, Hormone mediated oviductal influence on mouse embryo development, *Fertil. Steril.* 28: 766-771.

Daniel, J.C., Jr., 1964, Some effects of steroids on cleavage of rabbit eggs *in vitro*, *Endocrinology* 75: 706-710.

Daniel, J.C., Jr., and Gowan, M.C., 1966, Effects of some steroids in oral contraceptives on the cleavage of rabbit eggs *in vitro*, *J. Endocrinol.* 35: 155-160.

Daniel, J.C., Jr., and Levy, J.D., 1964, Action of progesterone as a cleavage inhibitor of rabbit ova *in vitro*, *J. Reprod. Fertil.* 7: 323-329.

Davis, D.L., and Day, B.N., 1978, Cleavage and blastocyst formation by pig eggs *in vitro*, *J. Anim. Sci.* 46: 1043-1053.

Dickmann, Z., 1970, Effects of progesterone on the development of the rat morula, *Fertil. Steril.* 21: 541-548.

Dickmann, Z., Dey, S.K., and Sengupta, J., 1976, A new concept: Control of early pregnancy by steroid hormones originating in the preimplantation embryo, *Vitam. Horm.* 34: 215-242.

Dziuk, P.J., 1971, Obtaining eggs and embryos from sheep and pigs, in: *Methods in Mammalian Embryology* (J.C. Daniel, Jr., ed.), Freeman and company, San Francisco, CA, pp. 76-85.

El-Banna, A.A., and Daniel, J.C., Jr., 1972, Stimulation of rabbit blastocysts *in vitro* by progesterone and uterine proteins in combination, *Fertil. Steril.* 23: 101-104.

Elsaesser, F., 1980, Effects of active immunization against oestradiol-17β, testosterone or progesterone on receptivity in the female rabbit and evaluation of specificity, *J. Reprod. Fertil.* 58: 213-218.

Ferguson, E.R., and Katzenellenbogen, B.S., 1977, A comparative study of antiestrogen action: Temporal patterns of antagonism of estrogen stimulated uterine growth and effects on estrogen receptor levels, *Endocrinology* 100: 1242-1251.

Harrer, J.A., and Lee, H.H., 1973, Differential effects of oestrogen on the uptake of nucleic acid precursors by mouse blastocysts *in vitro*, *J. Reprod. Fertil.* 33: 327-330.

Heap, R.B., Flint, A.P.F., Gadsby, J.E., and Rice, C., 1979, Hormones, the early embryo and the uterine environment, *J. Reprod. Fertil.* 55: 267-275.

Holmes, P.V., 1976, Evidence for cytosol receptor protein specific for estradiol in the preimplantation blastocyst, *VII. Intern. Congress on Anim. Reprod. and A.I. Cracow* Vol. 3: 290-293.

Johnson, M.H., 1979, Intrinsic and extrinsic factors in preimplantation development, *J. Reprod. Fertil.* 55: 255-265.

Kane, M.T., 1978, Culture of mammalian ova, in: *Control of Reproduction in the Cow* (J.M. Sreenan, ed.), Martinus Nijhoff, The Hague, pp. 383-397.

Katzenellenbogen, B.S., and Ferguson, E.R., 1975, Antiestrogen action in the uterus: Biological ineffectiveness of nuclear bound estradiol after antiestrogen, *Endocrinology* 97: 1-12.

Katzenellenbogen, B.S., Bhakoo, H.S., Ferguson, E.R., Lan, N.C., Tatee, T., Tsai, T.-S., and Katzenellenbogen, J.A., 1979, Estrogen and antiestrogen action in reproductive tissues and tumors, *Recent Prog. in Hormone Res.* 35: 259-292.

Kirkpatrick, J.F., 1971, Differential sensitivity of preimplantation mouse embryos *in vitro* to oestradiol and progesterone, *J. Reprod. Fertil.* 27: 283-285.

Levey, J.L., Stull, G.B., and Brinster, R.L., 1978, Poly (A) and synthesis of polyadenylated RNA in the preimplantation mouse embryo, *Dev. Biol.* 64: 140-148.

Niemann, H., and Elsaesser, F., 1984, Uptake and effects of ovarian steroids in the early pig embryo: *in vitro* and *in vivo* studies, *Theriogenology* 21: 84-102.

Niemann, H., and Elsaesser, F., 1985, Physiological significance of estrogens for the early embryonic development in the pig, in: *Early Pregnancy Factors* (F. Ellendorff, and E. Koch, eds.), Perinatology Press, New York, pp. 43-45.

Niemann, H., and Elsaesser, F., 1986a, Accumulation of sex steroids in preimplantation pig blastocysts and their importance for an undisturbed early embryonic development, in: *Embryonic Mortality in Farm Animals* (J.M. Sreenan, and M.G. Diskin, eds.), Martinus Nijhoff, The Hague, pp. 109-118.

Niemann, H., and Elsaesser, F., 1986b, Evidence for estrogen-dependent blastocyst formation in the pig, *Biol. Reprod.* 35: 10-16.

Niemann, H., Illera, M.J., and Dziuk, P.J., 1983, Developmental capacity, size and number of nuclei in pig embryos cultured *in vitro*, *Anim. Reprod. Sci.* 5: 311-321.

Paria, B.C., Sengupta, J., and Manchanda, S.K., 1984, Role of embryonic oestrogen in rabbit blastocyst development and metabolism, *J. Reprod. Fertil.* 70: 429-436.

Perry, J.S., Heap, R.B., and Amoroso, E.C., 1973, Steroid hormone production by pig blastocysts, *Nature (London)* 245: 45-47.

Pope, W.F., Maurer, R.R., and Stormshak, F., 1982, Intrauterine migration of the porcine embryo: Influence of oestradiol-17β and histamine, *Biol. Reprod.* 27: 575-579.

Pratt, R.M., and Daniel, J.C., Jr., 1967, Structural specificity of progesterone for inhibiting cleavage of the rabbit egg *in vitro*, *Steroids* 9: 257-262.

Roblero, L., and Izquierdo, L., 1976, Effect of progesterone on the cleavage rate of mouse embryos *in vitro*, *J. Reprod. Fertil.* 46: 475-476.

Roy, S.K., Sengupta, J., and Manchanda, S.K. 1981, Rat embryo development *in vitro*, *Acta Endocrinologica* 96: 546-551.

Roy, S.K., Sengupta, J., Paria, B.C., and Manchanda, S.K., 1982, *In vitro* inhibition of trophoblast maturation and expansion of early rat blastocysts by an oestrogen antagonist, *Acta Endocrinologica* 99: 129-135.

Sengupta., J., Dey, S.K., and Dickmann., Z., 1977, Evidence that "embryonic estrogen" is a factor which controls the development of the mouse preimplantation embryo, *Steroids* 29: 363-369.

Sengupta, J., Roy, S.K., and Manchanda, S.K., 1982, Effect of an oestrogen synthesis inhibitor, 1,4,6-androstatriene-3,17-dione, on mouse embryo development *in vitro*, *J. Reprod. Fertil.* 66: 63-66.

Sengupta, J., Paria, B.C., and Manchanda, S.K., 1983, Effect of an estrogen antagonist on development of blastocysts and implantation in the hamster, *J. Exp. Zool.* 225: 119-122.

Smith, D.M., and Smith, A.E.S., 1971, Uptake and incorporation of amino acids by cultured mouse embryos: Estrogen stimulation, *Biol. Reprod.* 4: 66-73.

Vignon, F., Terqui, M, Westley, B., Derocq, B., and Rochefort, H., 1980, Effects of plasma estrogen sulfates in mammary cancer cells, *Endocrinology* 106: 1079-1086.

Warner, C.M., and Hearn, T.F., 1977, The synthesis of RNA containing polyadenylic acid sequences in preimplantation mouse embryos, *J. Reprod. Fertil.* 50: 315-317.

Warner, C.M., and Tollefson, C.M., 1977, The effect of estradiol on RNA-synthesis in preimplantation mouse embryos cultured *in vitro*, *Biol. Reprod.* 16: 627-632.

Warner, C.M., and Tollefson, C.M., 1978, The effect of progesterone, estradiol and serum on RNA-synthesis in preimplantation mouse embryos cultured *in vitro*, *Biol. Reprod.* 19: 332-337.

Weitlauf, H.M., 1973, *In vitro* uptake and incorporation of amino acids by blastocysts from intact and ovariectomized mice, *J. Exp. Zool.* 183: 303-308.

Whitten, W.K., 1957, The effect of progesterone on the development of mouse eggs *in vitro*, *J. Endocrinol.* 16: 80-85.

Wright, R.W., Jr., and Bondioli, K.R., 1981, Aspects of *in vitro* fertilization and embryo culture in domestic animals, *J. Anim. Sci.* 53: 702-729.

Zavy, M.T., Bazer, F.W., Thatcher, W.W., and Wilcox, C.J., 1980, A study of prostaglandin $F_{2\alpha}$ as the luteolysin in swine. V: Comparison of prostaglandin F, progestins, estrone and estradiol in uterine flushings from pregnant and nonpregnant gilts, *Prostaglandins* 20: 837-851.

CHAPTER 7

GENETIC EXPRESSION DURING EARLY MOUSE DEVELOPMENT

TERRY MAGNUSON and CHARLES J. EPSTEIN

1. INTRODUCTION

As described in our earlier review (Magnuson and Epstein, 1981), preimplantation mammalian development is characterized by a complex but integrated continuum of changes in the synthesis of macromolecules (also recently reviewed by G. Schultz, 1986). As a result, embryos acquire properties that are different from those of embryos of earlier stages, and blastomeres within the same embryo eventually come to differ from one another. The developmental program that leads to the formation of the blastocyst, with its two differentiated cell types, the trophectoderm and inner cell mass, is actually initiated during the growth phase of oogenesis with the synthesis of significant quantities of RNA and protein (reviewed by Bachvarova, 1985; R. Schultz, 1986). Many of these macromolecules are carried through fertilization to the embryonic 2-cell stage, at which time the embryonic genome is activated in the mouse. The purpose of this chapter is to summarize the evidence concerning the gene expression that is necessary for early development. Our discussion, which updates our earlier review, will be restricted to oogenesis and to the first 5 days of post-fertilization mouse development, up to and including the time of initiation of implantation.

2. SYNTHESIS OF MATERNALLY-DERIVED PRODUCTS

The process of oogenesis is characterized by two important features.

Terry Magnuson Department of Developmental Genetics and Anatomy, School of Medicine, Case Western Reserve University, Cleveland, Ohio 44106, USA.
Charles J. Epstein Departments of Pediatrics and of Biochemistry and Biophysics, University of California, San Francisco, California 94143, USA.

One is oocyte growth, with the accumulation of materials that presumably constitute the maternal contribution to early development; the other is meiosis, which generates the haploid gamete. During the growth phase, the oocyte enlarges approximately 350-fold, and, on a per cell basis, the fully grown oocyte contains about 200 times more RNA, 60 times more protein, 1000 times more ribosomes, and 100 times more mitochondria than a typical mammalian somatic cell. The RNA composition includes 58-70% rRNA, less than 40% tRNA, and about 6-8% poly(A)-containing mRNA (reviewed by Bachvarova, 1985; G. Schultz, 1986; R. Schultz, 1986). After completing the growth phase, the oocyte is stimulated to resume meiosis I, at which point RNA synthesis declines to non-detectable levels. In spite of this transcriptional shutdown, the oocyte has stored a sufficient supply of rRNA, tRNA, mRNA, and ribosomes so that it is equipped to synthesize proteins through fertilization to the early 2-cell stage. In fact, many of the qualitative changes in proteins synthesized during this time have been shown to occur independently of transcriptional events, suggesting that a sequential activation and translation of selected maternally-derived mRNA subsets is occurring (Braude *et al.*, 1979; Petzoldt *et al.*, 1980; Cascio and Wassarman, 1982; Howlett and Bolton, 1985). Further evidence for stored message comes from *in vitro* translation experiments which have shown that mRNA encoding 2-cell stage specific polypeptides is present in the unfertilized egg (Braude *et al.*, 1979). In addition to selective activation of pre-existing mRNA, some of the changes in the protein pattern occurring during this time period also have been shown to be due to post-translational modifications of existing proteins (Van Blerkom, 1981, 1985; Howlett and Bolton, 1985).

3. ACTIVATION OF THE EMBRYONIC GENOME

One of the most interesting questions concerning early embryogenesis is the timing of onset of expression of the embryonic genome. It has been shown that low levels of heterogeneous RNA synthesis occurs in fertilized eggs. However, some of this incorporation is related to turnover of poly(A+) mRNA (Young and Sweeney, 1979; Clegg and Pikó, 1983a,b). Nevertheless, by the 2-cell stage, synthesis of all major classes of RNA is clearly detectable, indicating that their synthesis is dependent on embryonic transcription. The same inference can be drawn from an analysis of newly synthesized proteins, which has demonstrated the appearance at the 2-cell stage of a set of polypeptides the synthesis of which is sensitive to inhibitors of transcription (Flach *et al.*, 1982; Bolton *et al.*, 1984). These proteins, with relative molecular weights of 67 kD to 70 kD, are first detectable approximately 1 to 6 hr after the first cleavage, and some have been shown by peptide analysis to be heat shock proteins (Bensaude *et al.*, 1983). In addition, many of the other qualitative changes in the peptide pattern that subsequently occur between the early and the late 2-cell stages have also been shown to be dependent on embryonic transcription (Bolton *et al.*, 1984).

Since spermatozoa do not appear to contribute stored mRNA to the egg, the presence of a protein in the embryo that is coded for by the paternal genome also would constitute evidence that the embryonic genome is functional. The existence of several genetically determined protein variants provides a way for distinguishing maternal from paternal gene expression. For example, if females homozygous for one allele are mated with males homozygous for another, the time at which the paternally derived gene product is detected would be the latest time at which activation of the gene for that protein could have occurred. So far, the earliest times that paternal gene products have been detected are at the 2-cell stage (β_2-microglobulin: Sawicki *et al.*, 1982) and at the late 2- through 4-cell stage (β-glucuronidase: Wudl and Chapman, 1976).

4. ARE MATERNALLY-DERIVED PRODUCTS IMPORTANT FOR DEVELOPMENT BEYOND THE 2-CELL STAGE?

Although most of the maternally-derived poly(A+) RNA is degraded by the 2-cell stage (Bachvarova and DeLeon, 1980; Pikó and Clegg, 1982; Clegg and Pikó, 1983a; Giebelhaus *et al.*, 1983, 1985), it is difficult to determine whether any of this message persists and is translated at later times. However, it is known that the activities of several maternally-derived enzymes remain constant (in terms of amounts per embryo) until the 8- to 16-cell stage, after which they drop acutely (reviewed by Magnuson and Epstein, 1981). Furthermore, at least one maternally-derived protein product, glucose phosphate isomerase, is known to be stable throughout preimplantation development (West and Green, 1983; Gilbert and Solter, 1985). Nonetheless, in the absence of any clear-cut maternal effect mutations, it is difficult to determine the relative importance of these maternally-derived products on development.

5. BOTH THE MATERNAL AND PATERNAL GENOMES ARE REQUIRED FOR NORMAL DEVELOPMENT

Mouse embryos that have been artificially stimulated to develop in the complete absence of the male gamete (parthenogenesis) die during embryogenesis. If these embryos are haploid, they die sometime between early cleavage and implantation (Kaufman and Gardner, 1974). However, if diploid, they can develop to the 25 somite stage (Tarkowski *et al.*, 1970; Witkowska, 1973; Kaufman *et al.*, 1977). Parthenogenesis does not result in a cell-autonomous lethality, since artificially activated embryos can give rise to teratocarcinomas if transferred to ectopic sites (Graham, 1970; Iles *et al.*, 1975). Furthermore, they are able to combine with normal embryos to form viable chimeras in which many tissues, including functional germ cells, originate from the parthenogenetic embryo (Stevens, 1978).

It was originally postulated that parthenotes (parthenogenones) die because of expression of recessive lethal genes due to homozygosity, or, alternatively, because of a lack of an extragenetic contribution from spermatozoa (Graham, 1974). The latter hypothesis was supported by the observation that both diploidized androgenetic embryos (fertilized eggs containing only paternal chromosomes) and gynogenetic embryos (fertilized eggs containing only maternal chromosomes) appeared to be able to develop to term (Hoppe and Illmensee, 1977). However, a number of other investigators have attempted unsuccessfully to repeat these experiments (Modlinski, 1980; Markert, 1982; Surani and Barton, 1983; McGrath and Solter, 1984a; Surani *et al.*, 1984; Barton *et al.*, 1984). Instead, they found that both maternal and paternal pronuclei are necessary for normal development. Apparently, the mural trophoblast, ectoplacental cone, and the yolk sac develop poorly in gynogenetic embryos (Surani *et al.*, 1984), whereas the embryo proper of androgenetic embryos develops poorly even though the trophoblast develops relatively well compared to gynogenetic embryos (Barton *et al.*, 1984).

It has been suggested that Hoppe and Illmensee's results may have been due to incomplete removal of pronuclei (McGrath and Solter, 1984a; Surani *et al.*, 1984; Barton *et al.*, 1984), and it is now generally accepted that it is the genome and not the cytoplasmic contribution from the sperm that is required for successful development in mammals. This would suggest that the male and female genomic contributions to the zygote, although presumably identical in base sequence (except for polymorphisms), are not equivalent in a functional sense. In other words, one or the other set of chromosomes is in some way marked, as a result of which expression of particular loci is not the same in the zygote, embryo, or even adult.

The idea of differential transcription of the maternal and paternal genomes during development is also supported by data from other studies. For example, it has been known for a number of years that only the maternal X chromosome is active in extraembryonic membranes of mouse embryos (Takagi, 1978). Furthermore, maternal duplication/paternal deficiency or its reciprocal for the distal ends of chromosome 2 and 8 (Searle and Beechy, 1978), or the proximal end of chromosome 17 (Lyon and Glenister, 1977; McGrath and Solter, 1984b) is lethal, even though the individual is genetically balanced overall. Maternal disomy 6 (in which both chromosomes 6 are derived from the mother, again in a genetically balanced individual) is also lethal (Cattanach and Kirk, 1985), and maternal disomy 11 results in newborns that are consistently smaller than their normal littermates (Cattanach and Kirk, 1985). In contrast, paternal disomy 11 results in newborns that are larger than normal littermates. Although the above evidence clearly demonstrates a maternal/paternal genomic nonequivalency during development, not all of the genome is involved. For example, zygotes with maternal or paternal disomy for chromosomes 1, 4, 5, 9, 13, 14 and 15 survive normally (Lyon *et al.*, 1976; Cattanach and Kirk, 1985).

Two general mechanisms for maternal/paternal non-equivalency have been suggested. The first involves an as yet unidentified interaction between the maternal and paternal genomes (Markert, 1982). The second involves inheritance of a functional form of a gene(s) from only one parent (McGrath and Solter, 1984a; Barton *et al.*, 1984; Surani *et al.*, 1984). This would suggest that some form of imprinting or templating of the genome is involved. Recently, Groudine and Conklin (1985) found that certain genes which are expressed constitutively in somatic cells contain undermethylated sites in sperm. These sites corresponded to hypersensitive sites found in active copies of these genes in somatic cells. These investigators also described a *de novo* methylation process that occurs between the spermatogonial and primary spermatocyte stages of spermatogenesis and which specifically excludes DNA sequences included within nuclease hypersensitive sites in spermatogonial cells. These results are compatible with a model whereby *de novo* methylation could play a role in templating information in sperm DNA.

6. IS IT POSSIBLE TO CLONE MAMMALS BY NUCLEAR TRANSFER?

The ability of nuclei from various sources to support development when transplanted into enucleated zygotes has been a subject of interest for quite some time. Briggs and King (1952) originally observed that blastula nuclei transplanted into enucleated amphibian zygotes were capable of supporting normal development. Subsequently, Gurdon (1962a,b) demonstrated that nuclei obtained either from endodermal cells or from tadpole intestinal epithelium also were capable of supporting development when transplanted into irradiated eggs.

In the mouse, it has been reported that nuclei from inner cell mass cells could, when mechanically injected into enucleated zygotes, support development at least through the preimplantation stages and, in a small number of cases, to term (Illmensee and Hoppe, 1981; Hoppe and Illmensee, 1982). However, in direct contrast to these results, McGrath and Solter (1984b) found that nuclei from blastomeres of 4- and 8-cell embryos, as well as nuclei from inner cell mass cells, could not support development to the blastocyst stage when transplanted into enucleated zygotes using a karyoplast fusion technique instead of direct microinjection. A small fraction (19%) of zygotes receiving nuclei from 2-cell embryos were able to develop to the blastocyst stage. McGrath and Solter's (1984b) control experiments showed that neither cytoplasm nor the fusion process affected the developmental potential of the zygote. These investigators also observed the same lack of development when they used the mechanical injection technique of Illmensee and Hoppe (1981) and suggested that the discrepancy between the two approaches has to do with the method of enucleation. Illmensee and Hoppe (1981) achieved enucleation by the breakdown of pronuclei, possibly leaving behind pronuclear remnants, whereas McGrath and Solter (1984b) were able to remove an entire pronucleus by forming karyoplasts using an aspiration procedure.

It seems that, in the early mouse embryo, older nuclei require a cytoplasmic environment consistent with their nuclear function. The difference between amphibians and mice has been suggested to be due to differences in timing, in developmental terms, of activation of the embryonic genome (McGrath and Solter, 1984b). The mouse genome is activated by the 2-cell stage and there might not be enough developmental time to reprogram the transplanted nucleus. In contrast, the amphibian nucleus probably has sufficient time to reprogram since the embryonic genome is activated at the mid-blastula stage.

7. MUTATIONS AND CHROMOSOMAL ABNORMALITIES THAT AFFECT THE PREIMPLANTATION MOUSE EMBRYO

In the mouse, there are a number of genetic abnormalities that affect the development of the mouse embryo beginning as early as the 2-cell stage. These defects, which include conditions of chromosomal imbalance, recessive lethal mutations, and chromosomal deletions, provide absolute evidence that the embryonic genome is necessary very early in development. In the following paragraphs, those particular abnormalities that affect the mouse embryo during the preimplantation stages will be discussed briefly. More detailed reviews can be found in Magnuson and Epstein (1981) and Magnuson (1983, 1986).

7.1. c^{25H} (deletion)

The c^{25H} deletion is one of a series of overlapping chromosomal deficiencies studied primarily by S. Gluecksohn-Waelsch and her colleagues (reviewed by Gluecksohn-Waelsch, 1979). These deletions cover the albino locus located on mouse chromosome 7, and complementation tests have resulted in the classification of 6 of these deletions into 4 complementation groups. When homozygous, 4 of the deletions (c^{3H}, c^{112K}, c^{65K}, c^{14CoS}) are lethal at the time of birth. The remaining 2 deletions (c^{6H}, c^{25H}) result in death of the embryo during the early postimplantation and early preimplantation stages of development, respectively. In fact, the c^{25H} deletion represents one of the earliest acting genetic abnormalities that is known to affect the developing embryo. It is about 5 cM (7% of chromosome 7) in length (Miller *et al.*, 1974) and covers genes that are needed for normal development beginning at the 2-cell stage. The homozygous embryos stop dividing sometime between the 2- to 6-cell stage, with death occurring one to two days after this time (Lewis, 1978; Nadijcka *et al.*, 1979). The only obvious ultrastructural abnormality that has been found to be indicative of the lethal phenotype is the presence of aberrantly shaped nuclei.

7.2. Nullisomy

Embryos that are nullisomic for a chromosome have a greatly impaired developmental potential. For example, embryos nullisomic for chromosome 16 (Debrot and Epstein, 1986) or for the X chromosome (Morris, 1968; Burgoyne and Biggers, 1976; Luthardt, 1976; Tarkowski, 1977) die prior to the blastocyst stage. Haploid embryos that are also nullisomic for chromosomes 14 or 15 do not develop beyond the 2-cell stage (Kaufman and Sachs, 1975).

7.3. Autosomal Monosomy

Autosomal aneuploidy in the mouse is not compatible with post-natal survival (reviewed by Epstein, 1986). However, unlike autosomal trisomy, which is lethal sometime between mid-gestation and term, autosomal monosomy is lethal very early in development (Ford and Evans, 1973; Gropp *et al.*, 1974, 1975; Dyban and Baranov, 1978; Epstein and Travis, 1979; Baranov, 1983; Magnuson *et al.*, 1982, 1985). In fact, most monosomic embryos die during the pre- or peri-implantation period (8-cell through to implantation). This early lethality of monosomic embryos also appears to be true in man (Boué *et al.*, 1975; Hassold *et al.*, 1978). Although we do not understand the reasons for autosomal monosomy being lethal so much earlier in development than trisomy, the universal lethality does suggest that loci which are involved in dosage-dependent processes are scattered throughout the genome. The lethal phenotype associated with the various autosomal monosomies does not seem to be restricted to any stage-specific event, and one may therefore conclude that basic cellular processes or structures are affected (Magnuson *et al.*, 1985).

7.4. Haploidy

When a full set of chromosomes is missing, as is the case with haploid parthenogenetically-activated embryos (discussed above), the embryos die sometime between early cleavage and implantation (Kaufman and Gardner, 1974). In certain instances, parthenotes have been observed to develop at best to the 25 somite stage. However, these embryos have been diploid or of undetermined ploidy (Tarkowski *et al.*, 1970; Witkowska, 1973; Kaufman *et al.*, 1977). When haploid embryos have been produced by microsurgery, less than 10% developed to the blastocyst stage (Modlinski, 1975). Tarkowski and Rossant (1976) found that a higher percentage would develop to the blastocyst stage when haploid embryos were produced from bisected zygotes. In either case, the development of haploid embryos resembles the development of the monosomic embryos.

7.5. t^{12} and t^{w32} (t-complex)

These two haplotypes belong to the t-complex and are of the same complementation group. Both result in death of the embryo during the preimplantation stages of development. The t^{12} homozygotes arrest in development at the late morula stage (Smith, 1956), whereas the t^{w32} homozygotes arrest at the early morula stage (Hillman *et al.*, 1970). Even though death occurs at the morula stage, phenotypic effects have been detected as early as the 2-cell stage. Nuclear lipid droplets, excessive cytoplasmic lipid and binucleated cells have been found in presumptive homozygotes of both genotypes at this time (Hillman *et al.*, 1970; Hillman and Hillman, 1975). The two haplotypes differ in that nuclear fibrillo-granular bodies are observed in t^{12} but not t^{w32} homozygotes, and in that an early arrest of mitochondrial maturation with crystalline inclusions occurs in t^{w32} but not in t^{12} homozygotes. Ginsberg and Hillman (1975) have proposed an intrinsic defect of energy metabolism as being the basis for the developmental abnormalities and subsequent death associated with the two haplotypes. Some of our own work, which did not detect any differences in ATP levels, does not, however, support this hypothesis (Erickson *et al.*, 1974; Spielmann and Erickson, 1983).

An alternative suggestion is that the developmental defects are a consequence of altered surface antigens which result in abnormal cell-cell interactions and organizational problems within the embryo (Gluecksohn-Waelsch and Erickson, 1970; Bennett, 1975). However, no clear evidence exists to date that supports this hypothesis (reviewed by Magnuson, 1983). Thus, we still do not understand the mechanisms behind the lethality of these two recessive lethal mutations. In fact, even though a significant amount has been learned in the past few years about the organization and the molecular structure of the t-complex on the DNA level (reviewed by Silver *et al.*, 1984; Magnuson, 1986), we are still as ignorant as ever about the nature of the developmental defects attributed to any of the t-lethal genes. The cloning of

DNA sequences within the *t*-complex (Shin *et al.*, 1983, 1984; Rohme *et al.*, 1984; Fox *et al.*, 1984a,b) has not as yet yielded any information about the possible nature of the gene products affected by the *t*-lethal genes.

7.6. T^{hp} (deletion)

This deletion is located on chromosome 17 and results in a short-tailed mouse when present in heterozygous combination with a wild-type allele or in a tailless mouse when in combination with a *t*-haplotype (Johnson, 1974). T^{hp}/T^{hp} embryos have been reported to be unable to make the morula to blastocyst transition successfully (Babiarz, 1983). However, we have found (Magnuson and Epstein, unpublished results) that this is not the case. Using the *Rb(16.17)7Bnr* metacentric chromosome to mark the wild-type chromosome, about half of the homozygotes were observed to die at the morula stage whereas the remaining embryos were still viable at the late blastocyst stage.

Nothing is known about the defect that causes the lethality. However, an interesting aspect of this deletion is that when it is maternal in origin, $T^{hp}/+$ embryos die late in development (Johnson, 1974, 1975). In contrast, when the T^{hp} chromosome is sperm-derived, $T^{hp}/+$ adult progeny with hairpin tails are obtained. Thus, there is an apparent maternal effect associated with T^{hp}, the nature of which is not understood. When aggregation chimeras of T^{hp} (maternal in origin)/+ with +/+ embryos were made, the $T^{hp}/+$ component survived and was able to participate in normal development (Bennett, 1978). The germ cell defect, however, still persisted, and female chimeras failed to transmit the deletion to viable offspring. In recent nuclear transplant experiments, McGrath and Solter (1984c) have shown that the defect is nuclear in origin and not inherited through the maternal cytoplasm. They propose that some form of differential activity between the maternal and paternal genomes occurs during development, and a result of this is that the proximal portion of the maternally-derived chromosome 17 is necessary late in development (see Section 5 for other examples of maternal/paternal non-equivalency).

7.7. Ovum Mutant

This mutation represents the only possible maternal effect mutation so far described in mice. When females of the DDK strain are crossed with males of the same strain, perfectly normal mice are obtained. However, when these females are outcrossed, a significant loss of embryos occurs about the time of implantation. The affected embryos either fail to form blastocysts, or instead, form small blastocysts that induce a decidual reaction but fail to develop further (Wakasugi, 1973, 1974; Wakasugi *et al.*, 1967). A defect in trophectoderm formation has been suggested as a possible cause of death. It has been postulated that the developmental block is the result of an autosomal gene that codes for a factor in the egg which interacts specifically with a gene of sperm origin, and that this factor is incompatible with sperm of other strains.

7.8. Tail Short

Embryos homozygous for this mutation, located on chromosome 11, die at the blastocyst stage. These embryos are first detectable as small pale-staining morulae that divide at a slower rate than normal littermates

(Paterson, 1980). Heterozygous mice have skeletal abnormalities which include vertebral fusion, asymmetry of limb length, triphalangy, and an additional pair of ribs (Green, 1981).

7.9. Oligosyndactyly

This mutation is a radiation-induced mutation located on mouse chromosome 8. In heterozygous mice, the mutation results in syndactyly, diabetes insipidus (due to reduction in size of the kidney), and muscular anomalies (Grüneberg, 1956, 1961; Falconer *et al.*, 1964; Kadam, 1962). When present in the homozygous state, the mutation is lethal at about the time of implantation (Van Valen, 1966; Paterson, 1979). Although homozygous embryos are able to hatch from the zona pellucida and form blastocyst outgrowths (the *in vitro* equivalent to implantation), cells in mitosis begin to accumulate after the fifth to sixth cell division, *i.e.*, early blastocyst stage (Magnuson and Epstein, 1984). Even though the cytological appearance is one of mitotic cells treated with a microtubule inhibitor, the homozygous embryos do in fact have normal appearing intact spindles (Magnuson and Epstein, 1984). These results define the *Os* mutation as one that, in the homozygous state, prevents the movement of chromosomes from the metaphase plate of intact mitotic spindles. The homozygous embryos cannot be rescued when combined with wild-type embryos in an aggregation chimera, indicating that mutation results in a cell autonomous lethality.

The identification of the process of cell division as the target of effect of *Os* is an important point, since it demonstrates that a mutation which affects a basic cell process is still able to permit an embryo to progress fairly far in early development. The appearance of the metaphase arrest after 5 normal divisions may be due to the depletion of a maternally derived product that the mutant embryo is unable to synthesize. The relationship between the effects found in heterozygous mice with those in homozygous embryos remains unknown.

7.10. Agouti Locus

Among the alleles of the agouti locus, there are two, lethal non-agouti (a^x) and lethal yellow (A^y), that cause early lethality in homozygotes. Presumptive a^x/a^x embryos first appear abnormal at the mid-blastocyst stage (Papaioannou and Mardon, 1983). Nevertheless, these embryos are able to implant and form trophoblastic giant cells along with a primitive endoderm layer. However, by day 7.5 to 8.5, the putative homozygotes persist as disorganized clumps of embryonic tissue.

For A^y, it has been shown that homozygotes die sometime after implantation (Robertson, 1942a). Eaton and Green (1963) have attributed this death to defective differentiation of trophoblastic giant cells, and this conjecture has been indirectly supported by chimera experiments which suggest that homozygous inner cell mass cells are able to survive when combined with normal trophectoderm (Papaioannou and Gardner, 1979). Ovarian transplants have shown that the uterine environment is a contributing factor to the timing of lethality (Robertson, 1942b).

It was recently reported that the A^y mutation may have been caused by a retroviral integration (Copeland *et al.*, 1983). Such an event could mutagenize either by interrupting normal gene expression or by activating genes flanking the insertional site. Using the Moloney murine leukemia virus as a probe, we (Lovett *et al.*, 1985) have isolated molecular clones that span the entire proviral integration and include single-copy host DNA sequences flanking the provirus on both the 5' and 3' sides. These sequences have been mapped to a region on mouse chromosome 2 consistent with close linkage to the agouti locus and are transcribed in a mouse melanoma cell line and in spleens of mice heterozygous for A^y. The location of this transcription unit in close proximity to the viral transcriptional enhancer sequences and its expression in heterozygous tissues raises the possibility that one or more features of the dominant A^y phenotype (yellow hair pigmentation, obesity, increased susceptibility to the development of malignancy), and possibly the homozygous lethality, may be attributable to the proviral integration event and to inappropriate expression of a gene or genes in or near the agouti locus.

8. GENETIC CONTROL OF EARLY MAMMALIAN DEVELOPMENT: FUTURE APPROACHES

Clearly, the systems that would generate the most excitement in the analysis of early development are those that would have definite possibilities for relating phenotype to gene product and function. The reason for this is that, in spite of some detailed descriptions of the biological consequences associated with the above described genetic deficiencies, not a single instance can yet be cited where information also is known about the nature of a particular gene or its product. This inability to identify genes or gene products based on phenotypes is not surprising given the number and complexity of steps that are likely to exist between defective gene products and the detectable phenotypes.

There is much current interest in the possiblity of eliminating specific gene products and their functions by the use of either antibodies or complementary nucleotides. In fact, these approaches have already been used successfully in a few different instances. Antibodies to a gap junction protein were found to disrupt junction-mediated communication when injected into a specific blastomere of an 8-cell amphibian embryo (Warner *et al.*, 1984). As a result, subsequent developmental abnormalities occurred which included varying degrees of right/left asymmetry along with a small proportion of the embryos completely failing to form brains and eyes on either side.

In general, the use of antibodies would require prior preparation and characterization of a target protein, and for the mouse embryo, this approach may be applicable only during those stages of development that can be studied *in vitro*. In fact, investigations using antibodies with specificities directed against a cell surface molecule of embryonal carcinoma cells have led to the identification on 8-cell embryos of an exocellular molecule, uvomorulin, which is involved in compaction (Kemler *et al.*, 1977; Hyafil *et al.*, 1980, 1981; Peyrieras *et al.*, 1983).

The other method that has been used recently to neutralize specific gene products is one of blocking translation of specific messenger RNAs with complementary sequences termed "anti-sense RNAs". For example, experiments with the thymidine kinase gene (Izant and Weintraub, 1984, 1985) and the *E. coli* LacZ gene (Rubenstein *et al.*, 1984) have demonstrated that gene plasmids designed to produce anti-sense RNA can specifically diminish enzyme activity when co-transfected with the sense RNA (coding mRNA) into thymidine kinase minus or 3T3 cells, respectively. Moreover, this reduction in translation seems to require that only the 5' portion of the sense transcript be blocked. An alternative approach has been the direct injection of *in vitro* synthesized anti-sense RNA. For example, globin anti-sense RNA has been reported to block specifically the translation of globin message when both are co-injected into amphibian oocytes (Melton, 1985). However, an effective block occurred only if the anti-sense RNA was injected prior to the globin message or if the two were co-injected. If the globin message was injected first and then the anti-sense RNA, a low level of globin protein synthesis was detected. These results suggest that it may not be possible to inhibit completely protein synthesis from a message that is already being translated.

The potential use of anti-sense RNA in developmental studies has been demonstrated by Rosenberg *et al.* (1985). These investigators were able to produce phenocopies of the Krüppel mutation in *Drosophila* by reducing the level of the wild-type protein by injecting anti-sense RNA complementary to wild-type mRNA. Thus, the value of anti-sense transcription as a versatile tool for analysis of development in the mouse lies in the fact that it has the potential for analyzing cellular functions associated with a specific DNA sequence that may be expressed at a specific time and in discrete regions of the embryo.

9. SUMMARY

A considerable body of evidence in the mouse clearly indicates that the embryonic genome plays an essential role in development as early as the 2-cell stage. Although it seems clear now that both the maternal and paternal genomes are necessary for normal development, it is not at all understood how this differential genomic activity is regulated. Whether there is some kind of genomic imprinting that is established and carried through the germ line, or alternatively, some form of trans regulatory interaction that occurs, remains to be determined.

Although cells of the early cleavage-stage embryo are known to be pluripotent by the fact that they give rise to all cell types in the embryo, there is an immediate restriction in the ability of nuclei from early cleavage blastomeres to support development when transferred to an enucleated zygote. This may be due to irreversible genomic changes that occur beginning with fertilization, or it may simply be due to the fact that there is not enough developmental time for reprogramming to occur since the embryonic genome is active and needed as early as the 2-cell stage.

An intriguing question that remains to be answered is whether any of the maternally-derived message or its resulting product is necessary for development at later stages. The inability to generate maternal effect mutations in the mouse has made this difficult to answer.

To understand how development is initiated and regulated, it is important to distinguish genes involved in general cell structure or metabolism from genes that are associated with control of a developmental program. Traditionally, this has been approached by studying spontaneous or experimentally-induced mutations that affect development. In general, this approach has not been particularly fruitful for identifying gene products and functions. A more direct approach for identifying genes involved in developmental processes involves the elimination of specific products by the use of either antibodies or anti-sense RNA. These techniques would in theory allow production of phenotypes that are marked in such a way that biochemical and molecular experiments become possible.

ACKNOWLEDGEMENTS

The work described in this chapter which originated from our laboratories was supported by NIH Grants to C.J.E. (GM-24309, HD-03132, and HD-17001) and to T.M. (HD-19892). T.M. is a Pew Scholar in the Biomedical Sciences.

10. REFERENCES

Babiarz, B., 1983, Deletion mapping of the *T/t* complex: evidence for a second region of critical embryonic genes, *Dev. Biol.* 95: 342-351.

Bachvarova, R., 1985, Gene expression during oogenesis and oocyte development in mammals, in: *Developmental Biology. A Comprehensive Synthesis*, Volume 1 (L.W. Browder, ed.), Plenum Press, New York, pp. 453-524.

Bachvarova, R., and DeLeon, V., 1980, Polyadenylated RNA of mouse ova and loss of maternal RNA in early development, *Dev. Biol.* 74: 1-8.

Baranov, V.S., 1983, Chromosomal control of early embryonic development in mice. I. Experiments on embryos with autosomal monosomy, *Genetica* 61: 165-177.

Barton, S.C., Surani, M.A.H., and Norris, M.L., 1984, Role of paternal and maternal genomes in mouse development, *Nature (London)* 311: 374-376.

Bennett, D., 1975, The *T* locus of the mouse, *Cell* 6: 441-454.

Bennett, D., 1978, Rescue of a lethal *T/t* locus genotype by chimaerism with normal embryos, *Nature (London)* 272: 539.

Bensaude, O., Babinet, C., Morange, M., and Jacob, F., 1983, Heat shock proteins, first major products of zygotic gene activity in the mouse embryo, *Nature (London)* 305: 331-332.

Bolton, V.N., Oades, P.J., and Johnson, M., 1984, The relationship between cleavage, DNA replication, and gene expression in the 2-cell mouse embryo, *J. Embryol. Exp. Morphol.* 79: 139-163.

Boué, J., Boué, A., and Lazar, P., 1975, Retrospective and prospective epidemiological studies of 1500 karyotyped spontaneous human abortions, *Teratology* 12: 11-26.

Braude, P., Pelham, H., Flach, G., and Lobatto, R., 1979, Post-transcriptional control in the early mouse embryo, *Nature (London)* 282: 102-105.

Briggs, R., and King, T.J., 1952, Transplantation of living nuclei from blastula cells into enucleated frogs' eggs, *Proc. Natl. Acad. Sci. USA* 38: 455-463.

Burgoyne, P.S., and Biggers, J.D., 1976, The consequences of X-dosage deficiency in the germ line: Impaired development *in vitro* of preimplantation embryos from XO mice, *Dev. Biol.* 51: 109-117.

Cascio, S.M., and Wassarman, P.M., 1982, Program of early development in the mammal: Post-transcriptional control of a class of proteins synthesized by mouse oocytes and early embryos, *Dev. Biol.* 89: 397-408.

Cattanach, B.M., and Kirk, M., 1985, Differential activity of maternally and paternally derived chromosome regions in mice, *Nature (London)* 315: 496-498.

Clegg, K.B., and Pikó, L., 1983a, Poly(A) length, cytoplasmic adenylation and synthesis of poly A+ RNA in early mouse embryos, *Dev. Biol.* 95: 331-341.

Clegg, K.B., and Pikó, L., 1983b, Quantitative aspects of RNA synthesis and polyadenylation in 1-cell and 2-cell mouse embryos, *J. Embryol. Exp. Morphol.* 74: 169-182.

Copeland, N., Jenkins, N.A., and Lee, B.K., 1983, Association of the lethal yellow (A^y) coat color mutation with an ecotropic murine leukemia virus genome, *Proc. Natl. Acad. Sci. USA* 80: 247-249.

Debrot, S., and Epstein, C.J., 1986, Tetrasomy 16 in the mouse: a more severe condition than the corresponding trisomy, *J. Embryol. Exp. Morphol.* 91: 169-180.

Dyban, A.P., and Baranov, V.S., 1978, *The Cytogenetics of Mammalian Embryogenesis*, Nauka, Moscow.

Eaton, G.J., and Green, M.M., 1963, Giant cell differentiation and lethality of homozygous yellow mouse embryos, *Genetica* 34: 155-161.

Epstein, C.J., 1986, *The Consequences of Chromosome Imbalance: Principles, Mechanisms, and Models*, Cambridge University Press, New York, in press.

Epstein, C.J., and Travis, B., 1979, Preimplantation lethality of monosomy for mouse chromosome 19, *Nature (London)* 280: 144-145.

Erickson, R.P., Betlach, C.J., and Epstein, C.J., 1974, Ribonucleic acid and protein metabolism of t^{12}/t^{12} embryos and T/t^{12} spermatozoa, *Differentiation* 2: 203-209.

Falconer, D.S., Latsyzewski, M., and Isaacson, J.H., 1964, Diabetes insipidus associated with oligosyndactyly in the mouse, *Genet. Res.* 5: 473-488.

Flach, G., Johnson, M.H., Braude, P.R., Taylor, R.A.S., and Bolton, V.N., 1982, The transition from maternal to embryonic control in the 2-cell mouse embryo, *EMBO J.* 1: 681-686.

Ford, C.E., and Evans, E.P., 1973, Non-expression of genome unbalance in haplophase and early diplophase of the mouse and incidence of karyotypic abnormality in post-implantation embryos, in: *Proceedings of the Symposium on Chromosomal Errors in Relation to Reproductive Failure* (A. Boué, and C. Thibault, eds.), INSERM, Paris, pp. 271-285.

Fox, H.S., Martin, G.R., Lyon, M.F., Herrmann, B., Frischauf, A.M., Lehrach, H., and Silver, L.M., 1984a, Molecular probes define different regions of the mouse t-complex, *Cell* 40: 63-69.

Fox, H.S., Silver, L.M., and Martin, G.R., 1984b, An alpha globin pseudogene is located within the mouse t-complex, *Immunogenetics* 19: 125-130.

Giebelhaus, D.H., Heikkila, J.J., and Schultz, G.A., 1983, Changes in the quantity of histone and actin messenger RNA during the development of preimplantation mouse embryos, *Dev. Biol.* 98: 148-154.

Giebelhaus, D.H., Weitlauf, H.M., and Schultz, G.A., 1985, Actin mRNA content in normal and delayed implanting mouse embryos, *Dev. Biol.* 107: 407-413.

Gilbert, S.F., and Solter, D., 1985, Onset of paternal and maternal Gpi-1 expression in preimplantation mouse embryos, *Dev. Biol.* 109: 515-517.

Ginsberg, L., and Hillman, N., 1975, ATP metabolism in t^n/t^n mouse embryos, *J. Embryol. Exp. Morphol.* 33: 715-723.

Gluecksohn-Waelsch, S., 1979, Genetic control of morphogenetic and biochemical differentiation: lethal albino deletions in the mouse, *Cell* 16: 225-237.

Gluecksohn-Waelsch, S., and Erickson, R.P., 1970, The *T* locus of the mouse: implications for mechanisms of development, *Curr. Top. Dev. Biol.* 5: 281-316.

Graham, C.F., 1970, Parthenogenetic mouse blastocysts, *Nature (London)* 226: 165-167.

Graham, C.F., 1974, The production of parthenogenetic mammalian embryos and their use in biological research, *Biol. Rev.* 49: 399-422.

Green, E.L., 1981, *Genetic Variants and Strains of the Laboratory Mouse*, Oxford University Press, New York.

Gropp, A., Giers, D., and Kolbus, U., 1974, Trisomy in fetal backcross progeny of male and female metacentric heterozygotes of the mouse. I., *Cytogenet. Cell Genet.* 13: 511-535.

Gropp, A., Kolbus, U., and Giers, D., 1975, Systematic approach to the study of trisomy in the mouse, *Cytogenetics* 14: 42-62.

Groudine, M., and Conkin, K.F., 1985, Chromatin structure and *de novo* methylation of sperm DNA: implications for activation of the paternal genome, *Science* 228: 1061-1068.

Grüneberg, H., 1956, Genetical studies on the skeleton of the mouse. XVIII. Three genes for syndactylism, *J. Genet.* 54: 113-145.

Grüneberg, H., 1961, Genetical studies on the skeleton of the mouse. XXVII. The development of oligosyndactylism, *Genet. Res.* 2: 33-42.

Gurdon, J.B., 1962a, Adult frogs from the nuclei of single somatic cells, *Dev. Biol.* 4: 256-273.

Gurdon, J.B., 1962b, The developmental capacity of nuclei taken from intestinal epithelium cells of feeding tadpoles, *J. Embryol. Exp. Morphol.* 10: 622-640.

Hassold, T.J., Matsuymama, A., Newlands, J.M., Matsuura, J.S., Jacobs, P.A., Manuel, B., and Tsuei, J., 1978, A cytogenetic study of spontaneous abortions in Hawaii, *Ann. Hum. Genet.* 41: 443-454.

Hillman, N., and Hillman, R., 1975, Ultrastructural studies of t^{w32}/t^{w32} mouse embryos, *J. Embryol. Exp. Morphol.* 33: 685-695.

Hillman, N., Hillman, R., and Wileman, G., 1970, Ultrastructural studies of cleavage stage t^{12}/t^{12} mouse embryos, *J. Reprod. Fertil.* 33: 501-506.

Hoppe, P.C., and Illmensee, K., 1977, Microsurgically produced homozygous-diploid uniparental mice, *Proc. Natl. Acad. Sci. USA* 74: 5657-5661.

Hoppe, P.C., and Illmensee, K., 1982, Full-term development after transplantation of parthenogenetic embryonic nuclei into fertilized mouse eggs, *Proc. Natl. Acad. Sci. USA* 79: 1912-1916.

Howlett, S.K., and Bolton, V.N., 1985, Sequence and regulation of morphological and molecular events during the first cycle of mouse embryogenesis, *J. Embryol. Exp. Morphol.* 87: 175-206.

Hyafil, F., Morello, D., Babinet, D., and Jacob, F., 1980, A cell surface glycoprotein involved in the compaction of embryonal carcinoma cells and cleavage stage embryos, *Cell* 21: 927-934.

Hyafil, F., Babinet, C., and Jacob, F., 1981, Cell-cell interactions in early embryogenesis: a molecular approach to the role of calcium, *Cell* 26: 447-454.

Iles, S.A., McBurney, M.W., Bramwell, S.R., Deussen, Z.A., and Graham, C.F., 1975, Development of parthenogenetic and fertilized mouse embryos in the uterus and in extra-uterine sites, *J. Embryol. Exp. Morphol.* 34: 387-405.

Illmensee, K., and Hoppe, P.C., 1981, Nuclear transplantation in *Mus musculus*: developmental potential of nuclei from preimplantation embryos, *Cell* 23: 9-18.

Izant, J.G., and Weintraub, H., 1984, Inhibition of thymidine kinase gene expression by anti-sense RNA: a molecular approach to genetic analysis, *Cell* 36: 1007-1015.

Izant, J.G., and Weintraub, H., 1985, Constitutive and conditional suppression of exogenous and endogenous genes by anti-sense RNA, *Science* 228: 345-352.

Johnson, D.R., 1974, Hairpin-tail: A case of post-reductional gene action in the mouse egg?, *Genetics* 76: 795-805.

Johnson, D.R., 1975, Further observations on the hairpin-tail (T^{hp}) mutation in the mouse, *Genet. Res.* 24: 207-213.

Kadam, K.M., 1962, Genetical studies on the skeleton of the mouse. XXXI. The muscular anatomy of syndactylism and oligosyndactylism, *Genet. Res.* 3: 139-156.

Kaufman, M.H., and Gardner, R.L., 1974, Diploid and haploid mouse parthenogenetic development following *in vitro* activation and embryo transfer, *J. Embryol. Exp. Morphol.* 31: 635-642.

Kaufman, M.H., and Sachs, L., 1975, The early development of haploid and aneuploid parthenogenetic embryos, *J. Embryol. Exp. Morphol.* 34: 645-655.

Kaufman, M.H., Barton, S.C., and Surani, M.A.H., 1977, Normal postimplantation development of mouse parthenogenetic embryos to the forelimb bud stage, *Nature (London)* 265: 53-55.

Kemler, R., Babinet, C., and Jacob, F., 1977, Surface antigen in early differentiation, *Proc. Natl. Acad. Sci. USA* 74: 4449-4452.

Levey, I.L., Stull, G.B., and Brinster, R.L., 1978, Poly(A) and synthesis of polyadenylated RNA in the preimplantation mouse embryo, *Dev. Biol.* 64: 140-148.

Lewis, S.E., 1978, Developmental analysis of lethal effects of homozygosity for the c^{25H} deletion in the mouse, *Dev. Biol.* 65: 553-557.

Lovett, M., Yokoi, T., and Epstein, C.J., 1985, The lethal yellow mutation of the mouse: identification of a host transcription unit closely linked to a proviral integration site and homologous to a human DNA sequence, *Amer. J. Hum. Genet.* 37: A127 (abst).

Luthardt, F.W., 1976, Cytogenetic analysis of oocytes and early preimplantation embryos from XO mice, *Dev. Biol.* 54: 73-81.

Lyon, M.F., and Glenister, P.H., 1977, Factors affecting the observed number of young resulting from adjacent-2 disjunction in mice carrying a translocation, *Genet. Res.* 29: 83-92.

Lyon, M.F., Ward, H.C., Simpson, G.M., 1976, A genetic method for measuring non-disjunction in mice with Robertsonian translocations, *Genet. Res.* 26: 283-295.

Magnuson, T., 1983, Genetic abnormalities and early mammalian development, in: *Development in Mammals*, Volume 5 (M.H. Johnson, ed.), Elsevier Science Publishers, Amsterdam, pp. 209-249.

Magnuson, T., 1986, Mutations and chromosomal abnormalities: How are they useful for studying genetic control of early mammalian development?, in: *Experimental Approaches to Mammalian Embryonic Development* (J. Rossant, and R. Pedersen, eds.), Cambridge University Press, pp. 437-473.

Magnuson, T., and Epstein, C.J., 1981, Genetic control of very early mammalian development, *Biol. Rev.* 56: 369-408.

Magnuson, T., and Epstein, C.J., 1984, Oligosyndactyly: a lethal mutation in the mouse that results in mitotic arrest very early in development, *Cell* 38: 823-833.

Magnuson, T., Smith, S., and Epstein, C.J., 1982, The development of monosomy 19 mouse embryos, *J. Embryol. Exp. Morphol.* 69: 223-236.

Magnuson, T., Debrot, S., Dimpfl, J., Zweig, A., Zamora, T., and Epstein, C.J., 1985, The early lethality of autosomal monosomy in the mouse, *J. Exp. Zool.* 236: 353-360.

Markert, C.L., 1982, Parthenogenesis, homozygosity, and cloning in mammals, *J. Hered.* 73: 390-397.

McGrath, J., and Solter, D., 1984a, Completion of mouse embryogenesis requires both the maternal and paternal genomes, *Cell* 37: 179-183.

McGrath, J., and Solter, D., 1984b, Inability of mouse blastomere nuclei transferred to enucleated zygotes to support development *in vitro*, *Science* 226: 1317-1319.

McGrath, J., and Solter, D., 1984c, Maternal T^{hp} lethality in the mouse is a nuclear, not cytoplasmic, defect, *Nature (London)* 308: 550-551.

Melton, D., 1985, Injected anti-sense RNAs specifically block messenger RNA translation *in vivo*, *Proc. Natl. Acad. Sci. USA* 82: 144-148.

Miller, D.A., Dev, V.G., Tantravahi, R., Miller, O.J., Schiffman, M.B., Yates, R.A., and Gluecksohn-Waelsch, S., 1974, Cytological detection of the c^{25H} deletion involving the albino *(c)* locus on chromosome 7 in the mouse, *Genetics* 78: 905-910.

Modlinski, J.A., 1975, Haploid mouse embryos obtained by microsurgical removal of one pronucleus, *J. Embryol. Exp. Morphol.* 33: 897-905.

Modlinski, J.A., 1980, Preimplantation development of microsurgically obtained haploid and homozygous diploid mouse embryos and effect of pretreatment with cytochalasin B on enucleated eggs, *J. Embryol. Exp. Morphol.* 60: 153-161.

Morris, T., 1968, The XO and OY chromosome constitution in the mouse, *Genet. Res.* 12: 125-137.

Nadijcka, M.D., Hillman, N., and Gluecksohn-Waelsch, S., 1979, Ultrastructural studies of lethal c^{25H}/c^{25H} mouse embryos, *J. Embryol. Exp. Morphol.* 52: 1-11.

Papaioannou, V.E., and Gardner, R.L., 1979, Investigation of the lethal yellow A^y/A^y embryo using mouse chimeras, *J. Embryol. Exp. Morphol.* 52: 153-163.

Papaioannou, V.E., and Mardon, H., 1983, Lethal nonagouti (a^x): description of a second embryonic lethal at the agouti locus, *Dev. Genet.* 4: 21-29.

Paterson, H.F., 1979, *In vivo* and *in vitro* studies on the early embryonic lethal oligosyndactylism (*Os*) in the mouse, *J. Embryol. Exp. Morphol.* 52: 115-125.

Paterson, H.F., 1980, *In vivo* and *in vitro* studies on the early embryonic lethal tail-short (*Ts*) in the mouse, *J. Exp. Zool.* 211: 247-256.

Petzoldt, U., Hoppe, P.C., and Illmensee, K., 1980, Protein synthesis in enucleated fertilized and unfertilized mouse eggs, *Wilhelm Roux Arch. Dev. Biol.* 189: 215-219.

Peyrieras, N., Hyafil, F., Louvard, D., Ploegh, H.L., and Jacob, F., 1983, Uvomorulin: a non-integral membrane protein of early mouse embryo, *Proc. Natl. Acad. Sci. USA* 80: 6274-6277.

Pikó, L., and Clegg, K.B., 1982, Quantitative changes in total RNA, total poly(A), and ribosomes in early mouse embryos, *Dev. Biol.* 89: 362-378.

Robertson, G.G., 1942a, An analysis of the development of homozygous yellow mouse embryos, *J. Exp. Zool.* 89: 197-231.

Robertson, G.G., 1942b, Increased viability of homozygous yellow mouse embryos in new uterine environments, *Genetics* 27: 166-167.

Rohme, D., Fox, H.S., Herrmann, B., Frischauf, A.A., Edstrom, J.E., Mains, P., Silver, L.M., and Lehrach, H., 1984, Molecular clones of the mouse *t*-complex derived from microdissected metaphase chromosomes, *Cell* 36: 783-788.

Rosenberg, U.B., Preiss, A., Seifert, E., Jackle, H., and Knipple, D.C., 1985, Production of phenocopies by Krüppel antisense RNA injection into *Drosophila* embryos, *Nature (London)* 313: 703-706.

Rubenstein, J.R., Nicolas, J.R., and Jacob, F., 1984, Non-sense RNA (nsRNA): a tool to specifically inhibit gene expression *in vivo*, *C.R. Seances Acad. Sci.* 299: 271-274.

Sawicki, J.A., Magnuson, T., and Epstein, C.J., 1982, Evidence for expression of the paternal genome in the two-cell mouse embryo, *Nature (London)* 294: 450-451.

Schultz, G., 1986, Utilization of genetic information in the preimplantation mouse embryo, in: *Experimental Approaches to Mammalian Embryonic Development* (J. Rossant, and R. Pedersen, eds.), Cambridge University Press, pp. 239-265.

Schultz, R., 1986, Molecular aspects of mammalian oocyte growth and maturation, in: *Experimental Approaches to Mammalian Embryonic Development* (J. Rossant, and R. Pedersen, eds.), Cambridge University Press, pp. 195-237.

Searle, A.G., and Beechey, C.V., 1978, Complementation studies with mouse translocations, *Cytogenet. Cell Genet.* 20: 282-303.

Shin, H.S., Flaherty, L., Artzt, K., Bennett, D., and Ravetch, J., 1983, Inversion in the H-2 complex of *t* haplotypes in mice, *Nature (London)* 306: 380-383.

Shin, H.S., Bennett, D., and Artzt, K., 1984, Gene mapping within the *T/t* complex of the mouse. IV. The inverted MHC is intermingled with several *t*-lethal genes, *Cell* 39: 573-578.

Silver, L.M., Garrels, J.I., and Lehrach, H., 1984, Molecular studies of mouse chromosome 17 and the *t*-complex, in: *Genetic Engineering-Principles and Methods*, Volume 6 (J.K. Setlow, and A. Hollaender, eds.), Plenum Press, New York, pp. 141-156.

Smith, L.J., 1956, A morphological and histochemical investigation of a preimplantation lethal (t^{12}) in the house mouse, *J. Exp. Zool.* 132: 51-83.

Spielmann, H., and Erickson, R.P., 1983, Normal adenylate ribonucleotide content in mouse embryos homozygous for the t^{12} mutation, *J. Embryol. Exp. Morphol.* 78: 43-51.

Stevens, L.C., 1978, Totipotent cells of parthenogenetic origin in a chimeric mouse, *Nature (London)* 276: 266-267.

Surani, M.A.H., and Barton, S.C., 1983, Development of gynogenetic eggs in the mouse: implications for parthenogenetic embryos, *Science* 222: 1034-1036.

Surani, M.A.H., Barton, S.C., and Norris, M.L., 1984, Development of reconstituted mouse eggs suggests imprinting of the genome during gametogenesis, *Nature (London)* 308: 548-550.

Takagi, N., 1978, Preferential inactivation of the paternally derived X chromosome in mice, in: *Genetic Mosaics and Chimeras in Mammals* (L.B. Russell, ed.), Plenum Press, New York, pp. 341-360.

Tarkowski, A.K., 1977, *In vitro* development of haploid mouse embryos produced by bisection of one-cell fertilized eggs, *J. Embryol. Exp. Morphol.* 38: 187-202.

Tarkowski, A.K., and Rossant, J., 1976, Haploid mouse blastocysts developed from bisected zygotes, *Nature (London)* 259: 663-665.

Tarkowski, A.K., Witkowska, A., and Nowicka, J., 1970, Experimental parthenogenesis in the mouse, *Nature (London)* 226: 663-665.

Van Blerkom, J., 1981, Intrinsic and extrinsic patterns of molecular differentiation during oogenesis, embryogenesis, and organogenesis, in: *Cellular and Molecular Aspects of Implantation* (S.R. Glasser, and D.W. Bullock, eds.), Plenum Press, New York, pp. 155-176.

Van Blerkom, J., 1985, Post-translational regulation of early development in the mammal, in: *Differentiation and Proliferation* (C. Venizale, ed.), Van Nostrand Reinhold, New York, pp. 67-86.

Van Valen, P., 1966, Oligosyndactylism, an early embryonic lethal in the mouse, *J. Embryol. Exp. Morphol.* 15: 119-124.

Wakasugi, N., 1973, Studies on fertility of DDK mice: reciprocal crosses between DDK and C57BL/6J strains and experimental transplantation of the ovary, *J. Reprod. Fertil.* 33: 283-291.

Wakasugi, N., 1974, A genetically determined incompatibility system between spermatozoa and eggs leading to embryonic death in mice, *J. Reprod. Fertil.* 41: 85-96.

Wakasugi, N., Tomita, T., and Kondo, K., 1967, Differences of fertility in reciprocal crosses between inbred strains of mice: DDK, KK and NC, *J. Reprod. Fertil.* 13: 41-50.

Warner, A.E., Guthrie, S.C., and Gilula, N.B., 1984, Antibodies to gap-junctional protein selectively disrupt junctional communication in the early amphibian embryo, *Nature (London)* 311: 127-131.

West, J.D., and Green, J.F., 1983, The transition from oocyte-coded to embryo-coded glucose phosphate isomerase in the early mouse embryo, *J. Embryol. Exp. Morphol.* 78: 127-140.

Witkowska, A., 1973, Parthenogenetic development of mouse embryos *in vivo*. II. Postimplantation development, *J. Embryol. Exp. Morphol.* 30: 547-560.

Wudl, L., and Chapman, V., 1976, The expression of β-glucuronidase during preimplantation development of mouse embryos, *Dev. Biol.* 48: 104-109.

Young, R.J., and Sweeney, K., 1979, Adenylation and ADP-ribosylation in the mouse 1-cell embryo, *J. Embryol. Exp. Morphol.* 49: 139-152.

CHAPTER 8

DEFINING THE ROLES OF GROWTH FACTORS DURING EARLY MAMMALIAN DEVELOPMENT

ANGIE RIZZINO

1. INTRODUCTION

It is universally accepted that differential gene expression is central to the regulation of embryonic development. In reality, very little is known about the genes involved and how their expression is modulated. Although important regulatory genes are likely to be expressed during early development in response to environmental cues generated by the interactions of cells, even less is known about these events [see Chapter 7 (Ed.)]. Environmental cues that regulate cell behavior can take various forms, including: direct cell-cell contact, interactions with extracellular matrices, and interactions with diffusible factors. Diffusible factors can function over both short and long distances. Consequently, diffusible factors can be expected to affect the behavior of many cells and, in all probability, each factor affects the behavior of more than one cell type.

Evidence compiled during the past 30 years has provided convincing arguments that growth factors are important mediators of cell-cell interactions and that these factors are major regulators of both cell proliferation and differentiation (Gospodarowicz, 1981; Heldin and Westermark, 1984; James and Bradshaw, 1984). Currently, little is known about the roles played by growth factors during early mammalian development. This situation has not resulted from a lack of interest, but largely from the difficulties involved in working with embryos, especially with preimplantation and early postimplantation embryos of rodent species, which are composed of relatively few cells.

Undoubtedly, it will be a major undertaking to clarify the roles of growth factors during embryogenesis, but recent developments in the study of

Angie Rizzino Eppley Institute for Research in Cancer and Related Diseases, University of Nebraska Medical Center, 42nd and Dewey Avenue, Omaha, Nebraska 68105, USA.

growth factors have placed within our reach the capacity to understand at least some of their more important functions. This chapter outlines our current understanding of growth factors during early mammalian development and focuses on several recent advances, which strongly suggest that early embryonic cells respond to specific growth factors and that early postimplantation mouse embryos release endogenous growth factors. This discussion also illustrates strategies that can be used for furthering our understanding of the importance of growth factors during early development. An underlying theme of this chapter is that further progress in this area will be greatly influenced by improving the design of serum-free media for cultured blastocysts (64-cell stage). Consequently, the use and design of serum-free media for mouse blastocysts is discussed at length. To provide an appropriate frame of reference, the capacity of mouse embryos to develop in serum-supplemented media is briefly described in the next section.

This chapter does not discuss the production of steroids by early embryos or the roles of maternally-derived hormones. This is not intended to imply that these factors are of lesser importance; clearly, this is not the case. Instead, the decision to concentrate on growth factors reflects the need to increase our understanding of these factors during early development. Lastly, it should be noted that no distinction is made in this chapter between classical hormones, such as insulin, and classical growth factors, such as epidermal growth factor (EGF). In the case of factors produced by the embryo prior to development of the circulatory system, such a distinction has little meaning.

2. PREIMPLANTATION MOUSE EMBRYOS CULTURED IN SERUM-SUPPLEMENTED MEDIA

A complete review of the *in vitro* development of mouse embryos in serum-supplemented media is beyond the scope of this chapter. Therefore, only a brief overview of this topic is given here. For further details, the reader is referred to several key papers and reviews that address this subject in depth (Sherman and Wudl, 1976; New, 1978; Hsu, 1979; Wu *et al.*, 1981; Mintz, 1964; Cole and Paul, 1965; Menke and McLaren, 1970; Juurlink and Fedoroff, 1977).

The impressive work of several investigators has demonstrated that mouse embryos cultured from the 2-cell stage can develop *in vitro* to a stage containing somites and a beating heart, which is roughly equivalent to development on the eighth day of gestation *in utero*. [In this chapter, the day when the sperm plug is observed is considered the first day of gestation.] Development of preimplantation embryos to the early somite stage was first reported by Hsu (1973). Development from the blastocyst stage (fourth day of gestation) to the somite stage required 8 days in culture. It was made possible by frequent changes of the serum-supplemented medium and by replacement of bovine calf serum, first with fetal bovine serum and later with freshly prepared human cord serum (Hsu, 1973). More recently, Wu *et al.* (1981) have simplified these culture conditions and have reported that it is sufficient to culture blastocysts in medium containing 20% fetal bovine serum for the first 4 days of culture and in undiluted rat serum for an additional 4 days. Under these conditions, nearly 60% of the embryos reach the somite stage, which is comparable to the results obtained when embryos are cultured in undiluted human cord serum (Wu *et al.*, 1981). This is an important technical development, since rat serum is much easier to obtain than human cord serum.

The use of appropriate sera is also important during the earlier stages of development *in vitro*. Although mouse embryos from the 2-cell stage can develop to the blastocyst stage in the absence of serum, and can attach to the culture dish for a short period (Whitten, 1956; Brinster, 1965; Rizzino and Sherman, 1979), further development in culture requires serum. The addition of bovine calf serum to the culture medium supports stable attachment of blastocysts to the culture dish, but development is limited (Hsu, 1973). Further development to the egg cylinder stage (equivalent to the seventh day of gestation) occurs in medium supplemented with fetal bovine serum (Hsu, 1972, 1973; Wiley and Pedersen, 1977). The egg cylinder, which is surrounded by the mural trophoblast, parietal endoderm and ectoplacental cone, gives rise to the embryo proper and to accessory extraembryonic membranes. Wiley and Pedersen (1977) have shown that the structure of the egg cylinder formed by blastocysts *in vitro* is remarkably similar to that formed *in vivo*. In addition, each of the stages of development *in vitro* leading to the egg cylinder stage closely resembles the comparable stage *in vivo* (Wiley and Pedersen, 1977). However, some differences do exist between *in vitro* and *in vivo* development. The egg cylinder develops more slowly in culture than *in vivo*, by approximately 3 to 5 days (Wiley and Pedersen, 1977; Hsu, 1979; Sellens and Sherman, 1980), and the structures that are not part of the egg cylinder (parietal endoderm and ectoplacental cone) develop differently *in vitro* than *in vivo* (Wiley and Pedersen, 1977).

Other studies have shown that, despite the artificial conditions under which cultured embryos are grown, the trophectoderm also appears to differentiate in a relatively normal fashion, forming a monolayer of large polyploid cells (referred to as an outgrowth) on which the inner cell mass (ICM) rests (Fig. 1A). The trophectoderm *in vivo* differentiates into trophoblast cells, which exhibit several markers that can be readily monitored, including Δ^5-3β-hydroxysteroid dehydrogenase (3β-HSD). Chew and Sherman (1973) have shown that 3β-HSD appears *in utero* at approximately the ninth day of gestation and peaks during the 11th day of gestation. The same time course of 3β-HSD expression is observed with trophoblast cells formed in culture (Chew and Sherman, 1973, 1975). In addition, the trophoblast cells polyploidize, as they do *in vivo* (Barlow and Sherman, 1972). However, the development of trophoblast cells in culture, in particular the extent of their proliferation, is limited in comparison to that *in utero*.

These studies make two important points that are relevant to the roles of growth factors during early postimplantation development. First, early development can proceed outside the uterus, suggesting that the uterus does not provide unique signals for the early stages of mammalian development. Second, serum contains factors that are necessary and sufficient to support development, at least as far as the early somite stage. Given this situation, efforts to identify the serum factors required to support early postimplantation development *in vitro* should help identify at least some of the macromolecular requirements of developing embryos *in vivo* (supporting evidence for this point is discussed in Section 4). In addition, identification of the serum factors required for *in vitro* development could be used to improve the serum-free media for cultured blastocysts. This, in turn, would expedite the search

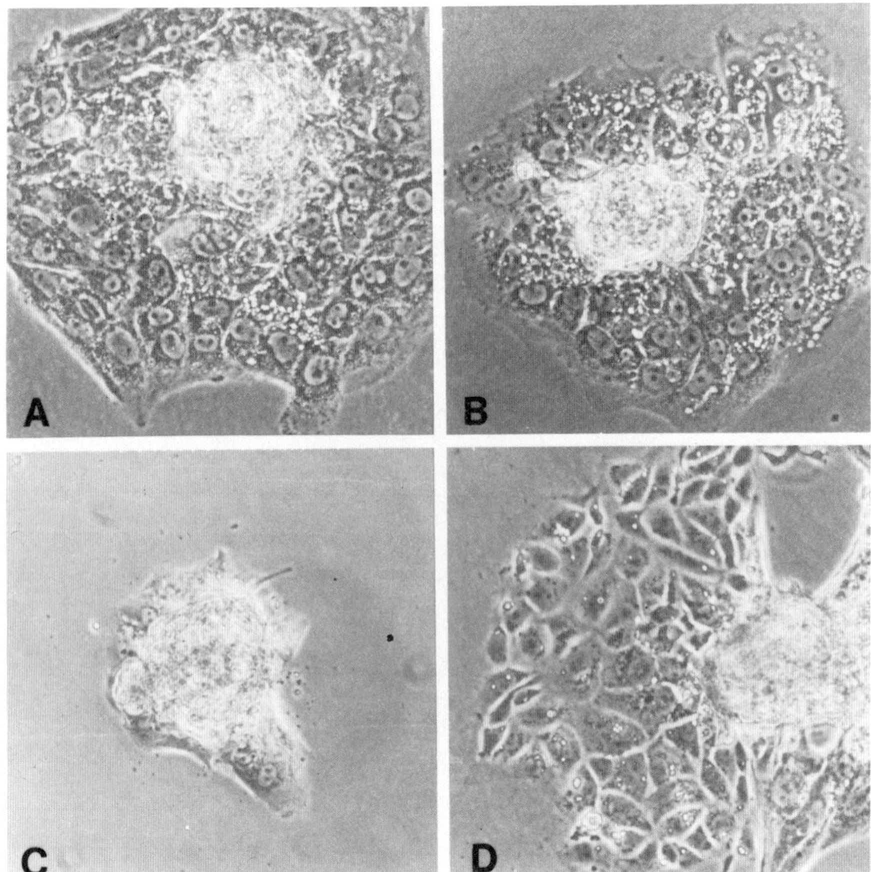

Figure 1. Morphology of cultured mouse embryos. Mouse blastocysts were cultured in (A) NCTC-109 plus 10% fetal bovine serum on regular tissue culture plastic, (B) in Modified EM-2, which is supplemented with insulin, transferrin and HDL on fibronectin-coated tissue culture plastic, (C) in Modified EM-2 from which HDL was omitted from the start of the experiment, or (D) in Modified EM-2 from which HDL was omitted on day 3. At the end of the 3rd day, the media were replaced with fresh media. The embryos were photographed on day 5. x 200 (optical magnification).

for growth factors produced by the embryo itself. As will be discussed in the next several sections, some progress is being made in the identification of factors required by cultured blastocysts, but more emphasis on this problem is needed.

3. MOUSE EMBRYOS CULTURED IN SERUM-FREE MEDIA

Numerous studies have demonstrated the advantages of undertaking cell culture in serum-free media supplemented with hormones and growth factors (reviewed in Rizzino *et al.*, 1979). Studies conducted in the absence of serum have not only helped identify the hormone, growth factor and attachment factor requirements of many different cell types, but have also facilitated the detection and subsequent isolation of growth factors released by cells (Rizzino *et al.*, 1979; Barnes and Sato, 1980; Rizzino, 1984a). Early attempts to culture mouse embryos in serum-free medium demonstrated that development to the blastocyst stage could occur in a relatively simple medium containing appropriate salts, glucose, lactate, pyruvate and bovine serum albumin (BSA).

In vitro development of embryos beyond the blastocyst stage was found to require more complex media, which contained amino acids (Gwatkin, 1966a; Spindle and Pedersen, 1973) and macromolecular components of serum (Mintz, 1964; Cole and Paul, 1965; Gwatkin 1966b; Menke and McLaren, 1970). Over the past 20 years, there have been sporadic reports of attempts to culture mouse blastocysts in serum-free media supplemented with purified or partially purified factors. Gwatkin (1966b) reported that partially purified fetuin promotes attachment and outgrowth of mouse blastocysts, as do several fractions of calf serum. Several years later, Jenkinson and Wilson (1973) reported that attachment and limited outgrowth can occur on collagen-coated surfaces in medium composed of a balanced salt solution, sugars and BSA. Neither study discussed the development of the ICM, but the ICM was unlikely to develop very far under the conditions used and, in all probability, did not survive for more than a few days.

Efforts to design serum-free media that support development beyond the blastocyst stage were continued by Rizzino and Sherman (1979). The approach taken was based in part on the finding that two embryonal carcinoma (EC) cell lines could be cultured in a serum-free medium, referred to as Embryonic Medium 1 or EM-1, which contains fetuin, insulin and transferrin (Rizzino and Sato, 1978). Given that EC cells appear to mimic specific stages of early mammalian development and share many properties with the undifferentiated ectoderm of the embryo (Graham, 1977; Martin, 1978, 1980; Solter and Damjanov, 1979), it seemed likely that early embryos and EC cells might share some of the same serum factor requirements.

Cultivation of blastocysts directly in EM-1 resulted in a high frequency of defective embryos. This suggested that the nutritional portion of EM-1, which is a 1:1 mixture of Dulbecco's Modified Eagle Medium and Ham's F-12, was not adequate. For this reason, several different culture media were tested, including NCTC-109 and the Standard Egg Culture Medium of Biggers. The best results were obtained with a 1:1 mixture of NCTC-109 and an enriched formulation of Biggers' Standard Egg Culture Medium. Standard Egg Culture Medium (Biggers *et al.*, 1971), which contains BSA, is enriched by adding vitamins and amino acids. The 1:1 mixture is supplemented with fetuin and is referred to as EM-2. Its formulation and preparation are given in the Appendix to this chapter (Section 9).

EM-2 supports hatching, attachment and outgrowth of mouse blastocysts at frequencies comparable to those observed in serum-supplemented media, but attachment and outgrowth are delayed by 12 to 15 hr and 5 to 8 hr, respectively. The trophoblast cells that form in EM-2 secrete plasminogen activator, have polyploid nuclei, and express 3β-HSD with the same time course as trophoblasts observed *in vivo* and in serum-supplemented medium (Chew and Sherman, 1973, 1975). The ICM also appears to undergo differentiation in EM-2, as demonstrated by the appearance of cells with the characteristic morphology of parietal endoderm and the ability to secrete plasminogen activator, but the ICM does not remain well organized (Rizzino and Sherman, 1979). In addition, the protein content and cell number of blastocysts cultured for 4 days in EM-2 are 2- and 3-fold lower than those of blastocysts cultured in medium supplemented with fetal bovine serum (Sellens and Sherman, 1980). For comparison, the conceptus at the eighth day of gestation *in utero* contains about 20 μg of protein and more than 16,000 cells, whereas blastocysts (day 4 embryos) cultured for 4 days in serum-supplemented medium contain only about 0.6 μg of protein and about 900 cells (see Sellens and Sherman, 1980). [See also Chapter 10 (Ed.).]

As an interesting aside, it has also been shown that serum-free media can be used to block *in vitro* development. Blastocysts cultured in the preimplantation culture medium of Goldstein *et al.* (1975) closely mimic the properties of embryos delayed *in utero* by ovariectomy. They exhibit similar cell numbers, mitotic indices, protein contents, and levels of marker enzymes (Sellens and Sherman, 1980).

Although EM-2 supports development beyond the blastocyst stage, it has several obvious drawbacks: the ICM appears to undergo only limited differentiation, the monolayer of trophoblast cells often deteriorates, and the medium itself is supplemented with fetuin, which is only partially purified and contains several impurities. Recently, some of these drawbacks have been addressed (Rizzino, 1985). Replacement of fetuin with the attachment factor fibronectin (or laminin) promotes the attachment of the embryo to the culture dish, but outgrowth does not occur (Fig. 1C). If high density lipoprotein (HDL) or low density lipoprotein (LDL) is also added to the medium, outgrowth occurs (Table I) and the trophoblast monolayer that forms is nearly as extensive as the monolayer of trophoblast cells observed in serum-supplemented medium (Fig. 1). In this serum-free medium, referred to as Modified EM-2, the trophoblast cells that form release plasminogen activator, as they do in serum-supplemented medium. However, other markers, such as 3β-HSD, have not yet been examined.

Modified EM-2 contains fibronectin (bound to the surface of the dish), insulin, HDL and BSA. BSA appears to be required for survival of the ICM. When BSA is omitted, the ICM deteriorates 3 to 4 days after embryo attachment and outgrowth occur (Rizzino, unpublished results). Similarly, it appears that HDL is necessary for promoting and maintaining trophoblast outgrowth. Removal of HDL from Modified EM-2 (3 days after blastocysts are placed in culture) results in deterioration of the trophoblast outgrowth, just as in EM-2. The removal of HDL on the third day of culture is also of interest because another cell type is observed under these conditions. These cells migrate away from the ICM and spread out on the surface of the culture dish in areas previously occupied by the trophoblast cells (this also occurs in EM-2). The cells secrete plasminogen activator and exhibit the characteristic morphology of extraembryonic endoderm. Under these conditions, one can determine the number of endoderm-like cells by direct observation with a phase-contrast microscope (Fig. 1D). Although the endoderm-like cells could be clearly distinguished in only 10% of the embryos, this observation suggests that the ICM differentiates in Modified EM-2, in much the same way it does in EM-2. [Presumably the endoderm-like cells form in a much higher percentage of the embryos, but they remain associated with the ICM and, therefore, cannot be distinguished morphologically.] Thus, the use of Modified EM-2 with removal of HDL on the third day provides a strategy for examining the effects of

Table I
Early Development of Mouse Blastocysts in Serum-Free Medium[a]

Factors added	No. of embryos	H	A	OG
None	20	15	3	0
FN	20	13	9	0
FN + HDL	18	18	18	17
FN + LDL	20	18	18	18
LM	23	22	18	0
LM + HDL	16	16	16	14
Fetuin	20	17	17	17
10% FBS	20	20	20	20

[a]Blastocysts were cultured in 2 ml of a 1:1 mixture of NCTC-109 and Enriched Standard Egg Culture Medium plus the factors indicated or in NCTC-109 alone when 10% fetal bovine serum (FBS) was used. Tissue culture dishes (35 mm) were coated with 20 μg of human fibronectin (FN) where indicated. Laminin (LM) and fetuin were added to the medium at 15 μg/ml and 500 μg/ml, respectively. LDL and HDL were used at 50 μg/ml and 300 μg/ml, respectively. On the third day, the numbers of embryos that had hatched (H), attached (A) and exhibited outgrowth (OG) were determined by observation under a dissecting microscope.

specific growth factors on the behavior of presumptive extraembryonic endoderm.

Recently, this system was used to examine the effects of platelet-derived growth factor (PDGF) on presumptive extraembryonic endoderm. PDGF increased the number of embryos with endoderm-like cells more than 2-fold and the number of endoderm-like cells per embryo more than 3-fold (Rizzino and Bowen-Pope, 1984). Although it remains to be determined whether the effect of PDGF is due to an increase in cell growth and/or other effects, such as increased cell migration, these findings argue that functional PDGF receptors appear relatively early during development. Viewed in general terms, these results indicate that serum-free media can be used to monitor the effects of specific growth factors on various early embryonic cell types. As further improvements are made in serum-free media for cultured blastocysts, it should be possible to focus on other embryonic cell types. However, as discussed in the next section, such improvements are likely to require a concentrated effort.

4. DESIGN OF SERUM-FREE MEDIA FOR MOUSE BLASTOCYSTS

A detailed discussion of the methodology used to design serum-free media is beyond the scope of this chapter. This topic has been discussed elsewhere for EC cell lines (Rizzino, 1984a) and other established cell lines (Rizzino *et al.*, 1979). Basically, 3 aspects of the culture medium must be considered: 1) selection of an appropriate attachment factor, 2) selection of the nutritional portion of the medium, and 3) identification of hormones, growth factors and serum accessory factors (*e.g.*, transferrin, HDL, BSA) that promote cell proliferation and maintain the organization of the ICM. The attachment requirement of mouse blastocysts appears to be satisfactorily met by using fibronectin-coated tissue culture dishes (Rizzino, 1985) or by adding laminin to the medium (Table I). Since Modified EM-2 supports extensive trophoblast differentiation and the ICM maintains a spherical shape, this medium, or one closely resembling it, appears to be a suitable starting point for developing improved serum-free media for cultured blastocysts. However, further study is needed to optimize the nutritional portion of this medium, and this study should be conducted before a serious effort is made to identify the hormone and growth factor requirements of developing embryos.

It is recognized that it will be a time-consuming process to identify the *in vitro* requirements of embryos. However, the information gained will not only aid in the development of improved serum-free media but is also likely to significantly broaden our understanding of the roles played by growth factors during early mammalian development. In this regard, results obtained thus far appear encouraging. Although formal proof is lacking, the factors that have been shown to improve development of cultured embryos are consistent with our current understanding of the needs of embryos *in vivo*.

In this connection, the production of laminin (Leivo *et al.*, 1980) and of fibronectin (Wartiovaara *et al.*, 1979) *in vivo* increase significantly at approximately the same time that implantation occurs. In culture, laminin and fibronectin promote attachment of the trophectoderm, which leads one to wonder whether these factors play an important role in implantation. [If this is the case, the factors are likely to be of maternal origin, since the trophoblast interacts with maternal tissues.] Consequently, the available data obtained *in vitro* and *in vivo* suggest that laminin and/or fibronectin begin to play an important role in cell-cell interactions by the end of the fourth day of gestation. The responses to LDL and HDL exhibited by trophoblast cells *in vitro* may also have a direct bearing on normal development. Recent studies have shown that maternal lipoproteins are the main source of cholesterol used by trophoblast cells to synthesize progesterone (Winkel *et al.*, 1980a,b, 1981). In this regard, it has been determined that trophoblast cells exhibit high affinity receptors for LDL and at least low affinity receptors for HDL (Winkel *et al.*, 1980b). Thus, the effects of LDL and HDL on trophoblast outgrowth appear to be mediated by physiological pathways. Lastly, the widespread use of BSA in preimplantation embryo culture media and the requirement for BSA in Modified EM-2 for survival of the ICM are consistent with the known presence of albumin in the environment of the embryo *in vivo*. However, the precise functions of albumin have not been determined and it is far from clear whether albumin serves the same functions *in vivo* and *in vitro*. Extrapolation of data from several systems suggests a variety of roles for albumin, including: carrier and source of fatty acids (Nilausen, 1978; Kane, 1979), chelator for toxic heavy metals (Abramczuk *et al.*, 1977), and modulator of proteins that interact with the embryo (Thorstensen and Romslo, 1984). [See also Chapter 10 (Ed.).]

5. PRODUCTION OF GROWTH FACTORS BY EARLY EMBRYOS

The studies discussed in the previous sections have focused on the importance of serum factors in embryonic culture media. However, an equally important picture is beginning to emerge: embryos at various stages of development produce growth factors that are likely candidates for regulating development. Epidermal growth factor (EGF) or, more likely, a closely related growth factor, is detectable by the 11th day of gestation in mouse embryos (Nexo *et al.*, 1980), by the 19th day of gestation in rat embryos (Matrisian *et al.*, 1982) and by the 8th day of gestation in chicken embryos (Mesiano *et al.*, 1985). Also, somatomedins are released by mouse embryonic explants from the 11th day of gestation (D'Ercole *et al.*, 1980). As one would expect, growth factors have been implicated in the control of development. Nerve growth factor (NGF) has long been believed to be intimately involved in the development of the nervous system (Cohen and Levi-Montalcini, 1956; Koroly and Young, 1981). Fibroblast growth factor (FGF) appears to influence myogenesis, at least in culture (discussed in Section 7), and EGF appears to play a role in the development of the lung (Catterton *et al.*, 1979; Sundell *et al.*, 1980) and the secondary palate (Hassell, 1975). In fact, the assay first used to monitor the purification of EGF depended on the ability of EGF to induce precocious eyelid opening and precocious eruption of the incisors in newborn mice (Cohen, 1962). More recently, embryos have been found to produce a new class of growth factors, which are referred to as transforming growth factors (TGFs). Acid-ethanol extracts prepared from mouse embryos at mid (12th to 13th day) and late (17th day) stages of gestation have been shown to contain factors (Proper *et al.*, 1982; Twardzik *et al.*, 1982) that are related, but not necessarily identical, to the two TGFs that have been purified thus far: TGF-α and TGF-β.

TGFs were first identified in medium conditioned by virally-transformed 3T3 cells and were originally referred to as sarcoma growth factor or SGF (De Larco and Todaro, 1978). Further study of SGF led to the finding that it was composed of two factors: TGF-α and TGF-β (Anzano *et al.*, 1983). TGF-α exhibits approximately 35-40% amino acid sequence homology with EGF (Marquardt *et al.*, 1984) and it competes with EGF for binding to membrane receptors. EGF and TGF-α bind to the same membrane receptor and they exert the same biological effects. They are, therefore, functionally equivalent. TGFs are usually assayed by their ability to induce the soft agar growth of non-transformed cells, such as normal rat kidney (NRK) cells. TGF-α and/or EGF are unable to induce NRK cells to form colonies in soft agar in the absence of TGF-β. TGF-β binds to a receptor distinct from that of EGF (Frolik *et al.*, 1984; Tucker *et al.*, 1984a) and is unable to promote the anchorage-independent growth of NRK cells in the absence of TGF-α or EGF (Roberts *et al.*, 1981; Rizzino, 1984b). More recently, it has been shown that TGF-β is closely related to the growth inhibitory factor discovered by Holley and his associates (Tucker *et al.*, 1984b) and TGF-β has been shown to inhibit the soft agar growth of some cells, *e.g.*, BSC-1 (Tucker *et al.*, 1984b) and A-549 (Roberts *et al.*, 1985). Although only TGF-α and TGF-β have been purified thus far, it is apparent that other TGFs exist (Halper and Moses, 1983; Yamaoka *et al.*, 1984; Sherwin *et al.*, 1983; Rizzino, 1983a). In addition, PDGF and FGF have been shown recently to induce the soft agar growth of several non-transformed cell lines (Rizzino *et al.*, 1986). Finally, it should be stressed that, despite their name, TGFs are produced by a wide range of normal cells (Roberts *et al.*, 1981) and, as we now know, this includes embryonic cells. Consequently, the term TGFs refers more to the method used to assay these factors than it does to their biological functions.

The presence of TGFs in extracts prepared from mid- and late-stage embryos raised a number of important questions, including: the cellular origin of these factors, their functions, whether TGFs are produced at earlier stages of development and whether these factors are released by the embryonic cells that produce them. The latter two questions have been examined recently and it is evident that mouse embryos cultured from the blastocyst stage produce and release one or more growth factors that promote the soft agar growth of NRK cells and thus exhibit the characteristic property of TGFs (Rizzino, 1985).

The production and release of a TGF-like growth factor(s) by early embryos was determined by co-culturing the embryos with NRK cells. The experiments were initiated by culturing several hundred mouse blastocysts in a single well of a Linbro dish (Flow Laboratories Inc.) for 3 days in the absence of NRK cells. During this period, attachment and outgrowth occurred and the ICMs formed 2-layered structures (indicative of the formation of extraembryonic endoderm) in at least 50% of the embryos. During the third day, the culture medium was removed and the embryos were overlaid with agar (0.5% Noble Agar in culture medium, which consisted of a 1:1 mixture of DME and Ham's F-12 supplemented with 10% bovine calf serum). This was followed by a second agar layer (0.3% in culture medium), which contained NRK cells as indicator cells. Under these conditions, the embryos and NRK cells are physically separated from one another and the soft agar growth of the NRK cells in the top agar layer is quantitated by direct observation with a phase-contrast microscope. It is worth noting that the colonies which formed under these conditions were relatively small, generally reaching only 6 to 10 cells per colony. Presumably, the size of the colonies would have been larger if more embryos had been used, but, given the number of embryos already employed to perform these assays, the use of more embryos is impractical.

These co-culture experiments were performed initially in medium supplemented with serum, which posed several problems. Firstly, the assay can not be performed in medium containing fetal bovine serum due to the high background soft agar growth of NRK cells induced by factors present in fetal sera. To minimize this problem, blastocysts were plated initially in medium supplemented with fetal bovine serum to provide good initial development. When the NRK cells were added on day 3, the medium supplemented with fetal bovine serum was replaced by medium supplemented with bovine calf serum from a carefully selected lot that minimized background growth of the NRK cells. Under these conditions, the ICM did not appear to grow significantly. Nonetheless, release of a TGF-like growth factor(s) was indicated by the soft agar growth of the NRK cells. Furthermore, it was determined that both isolated ICMs (prepared by immunosurgery) and ICM-free trophoblast monolayers (prepared by treating embryos with BUdR) release a TGF-like factor(s) (Rizzino, 1985).

Secondly, the use of serum-supplemented medium posed another problem. Since background growth of the NRK cells is not eliminated by even the best lots of calf serum, quantitation of the soft agar growth of the NRK cells was performed several days after the colonies first appeared (when the differential between the control [no embryos] and the experimental [with embryos] had increased). This made it difficult to determine when the TGF-like factors were first released. This problem was partially solved by the development of a serum-free assay for TGFs (Rizzino, 1984b). The assay is performed in a serum-free medium that contains insulin, transferrin, FGF and HDL in place of serum. Under these conditions, there is no background growth of the NRK cells. Readers interested in the details of this serum-free assay for TGFs and in TGF assays in general are referred to the original article (Rizzino, 1984b) and to a recent review (Rizzino, 1986).

The serum-free TGF assay was used to re-examine the release of a TGF-like factor(s) by blastocysts. To ensure the absence of serum, these experiments were initiated by culturing blastocysts in Modified EM-2. On day 3 of the experiment, the NRK cells were added and Modified EM-2 was replaced with the serum-free medium for NRK cells. In these particular experiments, EGF was added to the serum-free medium to maximize the response of the NRK cells to the TGF-like growth factors released by embryos. In agreement with the data from embryos in serum-supplemented medium, the embryos cultured in the absence of serum induced the soft agar growth of NRK cells (Rizzino 1984b). Moreover, the colonies that formed were readily counted only 7 days after the start of the experiment. This was only 4 days after the NRK cells were added and several days sooner than is possible in medium supplemented with serum. These results still do not allow one to pinpoint when these factors are first released by the embryo. However, since the blastocysts undergo only limited growth and development in these serum-free media, it is likely that the TGF-like growth factors are released by cells present in early postimplantation embryos.

Although these findings indicate that cultured embryos release a factor(s) that promotes the soft agar growth of NRK cells, the factor(s) involved has not yet been identified. To assist in identifying the TGF-like factor(s) released by early embryos, the growth factor requirements for promoting the soft agar growth of NRK cells in serum-free medium have been examined in detail (Rizzino, 1984b; Rizzino *et al.*, 1986). Four growth factors have significant effects on the soft agar growth of NRK cells: TGF-β, EGF (or other EGF-related growth factors), PDGF and FGF. When added together, TGF-β and EGF induce the soft agar growth of NRK cells and the response to these factors is augmented by PDGF and/or FGF. In addition, PDGF alone is sufficient to induce the soft agar growth of NRK cells in the serum-free medium, but this is not true for TGF-β, EGF or FGF individually.

The findings that soft agar growth of NRK cells is stimulated by 4 growth factors and also by cultured embryos suggests that one or more of these factors is released by early embryos. Since two of them, EGF and FGF, were present in the serum-free medium used to assay the embryos, it is likely that at least PDGF and/or TGF-β were released by the embryos in the serum-free experiments. However, this does not necessarily mean that the factors released are identical to either of these factors. In this regard, EC cells have been found to release a factor related to PDGF, but the evidence suggests that it is not identical with the PDGF present in platelets (Rizzino and Bowen-Pope, 1985). A similar situation exists for EGF. Embryos at the 11th to 12th day of gestation have been shown to produce a factor that competes with EGF in a radioreceptor assay, but this factor is not recognized by antibodies prepared against EGF (Nexo *et al.*, 1980). This suggests that this factor is an embryonic form of EGF. The EGF-related factor may be coded for by the gene for EGF but processed differently or it may be coded for by a different gene, as is the case for TGF-α (Derynck *et al.*, 1984). Finally, one cannot rule out the possibility that the TGF-like factor(s) released by early embryos is a completely new growth factor(s).

6. POSSIBLE ROLES OF GROWTH FACTORS DURING EARLY POST- IMPLANTATION DEVELOPMENT

The finding that early embryonic cells produce and respond to growth factors relatively early during development raises an important question: What are the functions of growth factors during early development? At present, one can only speculate on the answers. In doing so, it is relevant to discuss observations made in related systems, such as mouse EC cells, since this helps to formulate working hypotheses.

Early studies with EC cells in serum-free media led to the finding that EC cells condition their culture media (Rizzino and Crowley, 1980; Rizzino *et al.*, 1980). This, in turn, led to the finding that all EC cell lines examined (including F9, PC-13, OC-15-S1, 1003 and PSA-1) release TGF-like growth factors that promote the soft agar growth of NRK cells (Rizzino, 1982a; Rizzino *et al.*, 1983), whereas the endoderm-like derivatives of EC cells do not. The factors involved have not been characterized yet, but they appear to differ from TGF-α and TGF-β, since neither of these TGFs can be detected in either medium conditioned by EC cells or in cell extracts of EC cells (Rizzino, 1983a). It also remains to be determined whether the EC cells and early embryos produce the same TGF-like growth factors.

More recently, Gudas *et al.* (1983) and Rizzino and Bowen-Pope (1984, 1985) have determined that several EC cell lines release a factor that is related to PDGF. This phenomenon has been examined in depth with F9 and PC-13 EC cell lines. Both EC cell lines release a PDGF-like growth factor that competes with PDGF for binding to membrane receptors and is partially recognized by antisera prepared against PDGF (Rizzino and Bowen-Pope, 1985). Thus, the factor produced by EC cells is related, but not identical, to PDGF. However, the EC cell derived factor and PDGF are expected to be functionally equivalent, since they bind to the same receptor. It was also determined that EC cells lack detectable PDGF receptors and do not respond to exogenously added PDGF, whereas their endoderm-like derivatives, which do not produce this factor, express receptors for PDGF and respond to PDGF by increased growth (Rizzino and Bowen-Pope, 1984, 1985). Thus, EC cells release a factor that should be capable of binding to and stimulating the growth of some of their differentiated cells. This type of growth control is referred to as paracrine growth control (Dockray, 1979). As discussed above, PDGF can induce the soft agar growth of NRK cells in a serum-free TGF assay. This raises the possibility that PDGF is responsible, at least in part, for the TGF-like activity released by EC cells and early embryos. If it is, PDGF may exert a paracrine growth control function during early development and may influence the behavior of extraembryonic endoderm.

EC cells also release several other growth factors. If the observations made with EC cells reflect the situations in early embryos, one can anticipate that several growth factors are released during early development. The factors released by EC cells include: an insulin-like growth factor (Heath and Isacke, 1983), a factor that stimulates the growth of mouse fibroblasts (Heath and Isacke, 1984), and a factor(s) that affects the establishment of pluripotent cell lines from mouse embryos (Martin, 1981). The latter factor(s) is of particular interest because of its effects on mouse embryos, but it remains to be purified and characterized. In addition, it appears that PSA-1 EC cells produce a growth factor that stimulates their own growth (Jakobovits *et al.*, 1985), which is consistent with the finding that several EC cell lines condition their culture medium with factors that they require for proliferation at low cell densities in serum-free media (Rizzino and Sato, 1978; Rizzino and Crowley, 1980; Rizzino *et al.*, 1980).

Viewed as a whole, these studies suggest that growth factors play active roles during early development and that both autocrine and paracrine growth control are involved. This is a radical departure from the perception that early mammalian development is preprogramed, but this view is consistent with the events that occur during the fourth to eighth days of gestation. During this period, each of the three embryonic germ layers are formed and many new cell types appear. Consequently, during this period, growth factors are likely to begin coordinating the interactions between the newly formed embryonic cells. However, it is unclear precisely which cells are involved.

The results with EC cells suggest that the cells of the ICM (or cells formed relatively early by the ICM) produce a growth factor, possibly an embryonic form of PDGF, that affects the growth of extraembryonic endoderm. Other roles for growth factors produced by the ICM or its derivatives are more speculative, but it is worth bearing in mind that the ICM affects the trophoblast (reviewed by Ilgren, 1983) and growth factors could be involved. The roles played by growth factors produced by trophoblast cells are equally difficult to pin down. In this regard, a recent study strongly suggests that human cytotrophoblast cells from first trimester placenta produce a factor that is closely related to PDGF (Goustin *et al.*, 1985). Since PDGF is known to induce cell migration (Seppa *et al.*, 1982), it is possible that parietal endoderm migrates away from the developing egg cylinder (to occupy the inner surface of the trophoblast) in response to an embryonic form of PDGF released by trophoblast cells.

This recent study of PDGF production by human placenta (Goustin *et al.*, 1985) also demonstrated that cytotrophoblastic cell lines developed from first trimester placenta express receptors that bind PDGF and respond to it by increased thymidine incorporation. On the basis of these findings it was suggested that the PDGF-like growth factor released by the cytotrophoblast cells stimulates their own growth (autocrine growth control). However, the evidence is not conclusive since the presumptive cytotrophoblastic cell lines that respond to PDGF do not produce the PDGF-like factor. Although it is possible that these cells have lost the ability to produce this factor but not the ability to respond to PDGF, it is equally possible that there are two populations of cytotrophoblastic cells, one that responds to PDGF and one that produces the PDGF-like factor. In this regard, several recent reports that suggest autocrine growth control are subject to the same criticism, namely, the cells used may not be homogeneous. Hopefully, future studies will attempt to address this question by using cloned populations. Finally, even if the PDGF-like growth factor released by cytotrophoblastic cells does function in autocrine growth control, this factor is also likely to play a role in paracrine growth control. Besides its possible effects on extraembryonic endoderm, the PDGF-like growth factor is likely to play a role in the decidual reaction by affecting the uterine stroma.

7. POSSIBLE EFFECTS OF GROWTH FACTORS ON THE PROCESS OF DIFFERENTIATION

There is convincing evidence that growth factors can influence cell differentiation, but the mechanisms are far from clear. In some cases, it appears that growth factors influence differentiation by their effects on cell growth. One of the clearest examples is the effect of FGF on myogenesis. It has been apparent for a number of years that myoblasts fuse when the culture medium is depleted. More recently, it was shown that addition of FGF to the medium delays differentiation (Linkhart *et al.*, 1982). In this system, FGF appears to delay the onset of myogenesis by promoting cell growth. Similar observations have been made with EC cells. Two multipotent EC cell lines, OC-15-S1 and 1003, differentiate to endoderm-like cells when placed in a serum-free medium containing fibronectin, insulin, transferrin and HDL (Rizzino, 1983b). With the exception of fibronectin, these factors do not appear to induce differentiation. These findings suggest that serum factors block differentiation and that growth factors are involved. Thus, the decision of some embryonic cells to differentiate is very likely to depend on the availability of growth factors that block differentiation by maintaining the embryonic cells in a proliferative state (Rizzino, 1982b). In other cases, growth factors may stimulate cell proliferation and stimulate differentiation as a consequence. For example, EGF stimulates keratinization by stimulating the proliferation of keratinocytes (Cohen and Elliott, 1963).

Growth factors may exert other important regulatory effects that have far-reaching consequences, including effects on the production of extracellular matrices. In this connection, EGF is known to affect the production of fibronectin, hyaluronic acid and glycosaminoglycans by fibroblasts (Chen *et al.*, 1977; Lembach, 1976). Growth factors, such as EGF and FGF, may also modulate the release of other regulatory factors. For instance, human chorionic gonadotropin (hCG) may affect fetal development and EGF has been shown to increase the production of hCG by choriocarcinoma cells (Benveniste *et al.*, 1978). EGF is also known to decrease the release of growth hormone and increase the release of prolactin by the rat pituitary cell line GH3 (Johnson *et al.*, 1980). Likewise, FGF has recently been shown to increase the response of primary anterior pituitary cells to thyrotropin-releasing factor and to increase their release of prolactin and thyrotropin (Baird *et al.*, 1985). Finally, aberrant production of growth factors during early development can be expected to have very significant effects. Again, the action of EGF illustrates this point. Addition of EGF to chick embryo organ cultures can inhibit feather formation, apparently by keratinization of the subepidermal layer (Cohen, 1965). Thus, a better understanding of growth factors during development may provide a better understanding of certain birth defects.

8. CONCLUSIONS

Recent studies demonstrate that growth factors are produced by the embryo relatively early during development. Given the multitude of growth factor effects on cell proliferation and differentiation in other systems, one can anticipate that growth factors will be found to be major regulators of early embryonic cells. Formal proof of this expectation will require intense effort in several areas. The position taken in this chapter is that our understanding of growth factors during early development will advance most rapidly by studying embryos in serum-free media, since this facilitates both detection of growth factors released by embryonic cells and identification of their specific responses to growth factors. Consequently, the lack of serum-free media that support development of blastocysts to the egg cylinder stage is viewed as a major bottleneck. Unfortunately, solving this problem will not be a simple matter, since it will be necessary to refine the nutritional portion of the culture medium and to identify the factors in fetal bovine serum that support continued development of the ICM *in vitro*. However, development of an appropriate defined medium would greatly simplify and expedite our efforts to address several important questions, namely:

What are the identities and properties of the growth factors produced by early embryonic cells?
When do these factors first appear during development?
Which embryonic cells respond to these factors?
How do growth factors affect cell growth and differentiation?

Answering these questions will be a beginning, not an end point. As answers to these questions emerge, it will be possible to begin the study of genes that code for the embryonic growth factors and, ultimately, to determine how their expression is regulated during development. Such an understanding will shift the emphasis of early mammalian embryology from a primarily descriptive science, concerned with defining the course of cell-cell interactions, to a science heavily focused on the mechanisms by which embryonic cells interact and modulate the gene expression of one another.

9. APPENDIX

This section describes the composition and preparation of Modified EM-2, the serum-free medium used in the author's laboratory to culture mouse blastocysts. The synthetic portion of Modified EM-2 is a 1:1 mixture of NCTC-109 and an enriched formulation of the Standard Egg Culture Medium of Biggers. NCTC-109 (Whittaker M.A. BioProducts Inc., cat. no. 12-123B) is pretested for its ability to support development of blastocysts. When pretesting NCTC-109, it is supplemented with 10% fetal bovine serum. Several criteria are used to evaluate the quality of the medium, including the percentage of embryos that hatch, attach and outgrow in this medium (90% or more is considered acceptable). Other criteria are the percentage of embryos that form 2-layered ICMs (60% or more is considered acceptable) and the overall growth of the ICM. Once an acceptable lot of NCTC-109 has been identified, it is used for a period of two months. This pretesting ensures the quality of the medium, but it is time-consuming. Consequently, it is recommended that laboratories planning to use NCTC-109 over an extended period prepare it in-house from individual components (McQuilkin *et al.*, 1957; Morton, 1970). Once prepared in bulk and tested as outlined above, the powdered medium should be stable for at least one year when stored at 4°C in a desiccator.

The other half of the synthetic portion of the serum-free medium is referred to as Enriched Standard Egg Culture Medium. It is based on the Standard Egg Culture Medium of Biggers *et al.* (1971) and, in accordance with the findings of Spindle (1980), is enriched with amino acids and vitamins at twice the concentrations recommended by Spindle and Pedersen (1973). This medium (Table II) is prepared fresh each week from stock solutions. The stock solutions (Tables III and IV) are mixed one or two days prior to the start of each experiment. The Enriched Standard Egg Culture Medium is then mixed 1:1 with NCTC-109, the pH is adjusted to 7.2 to 7.4 and the medium is filter-sterilized (0.2 μm Sterivex filters, Millipore Corp.). This medium is supplemented with insulin (10 μg/ml), transferrin (5 μg/ml) and HDL (200-400 μg/ml). Fibronectin is a component of Modified EM-2, but it is not added directly to the medium. Instead, the tissue culture surface is coated with

Table II
Composition and Preparation of Enriched Standard Egg Culture Medium (ESECM)

Components	Quantities for 100 ml
100x Spindle-Pedersen amino acid mix[a]	1.0 ml
100x Tyrosine[a]	1.0 ml
25x Cystine[a]	4.0 ml
100x Glutamine[a]	1.0 ml
Spindle-Pedersen vitamins (100x MEM-vitamins from Gibco)	2.0 ml
Antibiotics (20,000 units penicillin G and 20,000 μg streptomycin sulfate/ml)	0.5 ml
10x ESECM salts[b]	10.0 ml
10x BSA (Pentex [Miles Laboratories]: 30 mg/ml in water; store at -20°C)	10.0 ml
10x Glucose (10 mg/ml; store at -20°C)	10.0 ml
Sodium pyruvate (28 mg/10 ml water; prepared fresh)	1.0 ml
Sodium lactate syrup [Sigma Chemical Co.] (warm to 37°C before pipetting)	0.37 ml
$NaHCO_3$	0.1892 g
Phenol red (0.5 mg/ml stock; store at -20°C)	0.5 ml
Water (highest purity possible)	58.0 ml
NaOH	to adjust pH (7.2 to 7.4)

[a]see Table III.
[b]see Table IV.

Table III
Amino Acid Stock Solutions Used in Enriched Standard Egg Culture Medium

Amino acids	Quantities (g/100 ml)	Final concentration in ESECM (mM)
100x Spindle-Pedersen amino acid mix[a]:		
Arg	0.21	0.2
His	2.16	2.06
Ile	0.26	0.4
Leu	1.31	2.0
Lys	3.65	4.0
Met	0.375	0.5
Phe	0.825	1.0
Thr	2.4	4.0
Trp	0.2	0.2
Val	1.175	2.0
Other stock solutions:		
100x Tyr[b]	0.181 in 0.1 N NaOH	0.2
25x Cys[b]	0.3 in 0.15 M HCl (does not dissolve readily)	1.0
100x L-gln[b]	2.92 in water (store frozen 5 ml/tube)	4.0

[a]Add one amino acid at a time (add lysine first, since it appears to be the least soluble amino acid used in the mix) to water that is prewarmed to 37°C. Filter the solution (prewashed 0.22μm Nalgene filter) and store at 4°C.
[b]Stable for approximately 3 months. The other amino acid solutions are stable for approximately 6 months.

fibronectin (2 μg/cm^2) as described previously (Rizzino and Crowley, 1980). The procedures for the preparation and storage of insulin and transferrin have been published (Rizzino and Crowley, 1980). HDL can be purchased from Meloy Laboratories. The density range of 1.087 to 1.21 has been found to give the most consistent results. Since HDL is expensive and its activity declines after 4 to 6 weeks, it may be preferable to prepare HDL in-house according to standard protocols (Havel *et al.*, 1955; Lindgren *et al.*, 1979). Once prepared, HDL should be stored at 4°C. It should not be frozen, since this significantly reduces its activity. As discussed in Section 3, LDL can be used in place of HDL. However, this may pose a technical problem. Many studies have shown that LDL can significantly inhibit the growth of cells cultured in serum-free media. Although the reasons are not clear, it appears that LDL is easily oxidized once prepared and, as a result, becomes cytotoxic for some cells, including EC cells (Rizzino, 1983b; Rizzino, unpublished results).

It should also be emphasized that the quality of the water employed will determine the success of experiments performed in serum-free media. Exper-

Table IV
10x Stock Solution of Salts Used in Enriched Standard Egg Culture Medium

Salts	Quantities (g/100 ml)
NaCl	4.14
KCl	0.356
KH_2PO_4	0.162
$MgSO_4.7H_2O$	0.294
$CaCl_2.2H_2O$	0.250

ience over a period of years at several different institutions has made it abundantly clear that one can expect variable and substandard results unless special care is given to obtaining water of the highest quality. Three times distilled water does not give consistent results. Water purified by carbon absorption and ion exchange (18 megohm), *e.g.*, using the Milli Q® (Millipore Corporation) or Nanopure II® (Barnstead) systems appears to provide water better suited to cell and embryo culture. Ultrafiltration, which removes pyrogens, is likely to further improve the quality of the water. However, even these steps should not be considered the final word and it is recommended that water from several sources be tested to determine the source that is most suitable for specific experiments. [See Appendix I at the end of this book (Ed.).]

Finally, the composition of Modified EM-2 should not be considered optimal. Its current complexity could probably be reduced and, thus far, very little time has been devoted to optimizing this medium. Consequently, further modifications of its composition are likely to result in media that support better development of cultured mouse blastocysts.

ACKNOWLEDGMENTS

Heather Rizzino is thanked for her editorial assistance and for her many helpful comments. Grant support from the National Cancer Institute (NIH Research Grant Number CA36727) and from the University of Nebraska Medical Center (22-271-732) for some of the work cited is acknowledged.

10. REFERENCES

Abramczuk, J., Solter, D., and Koprowski, H., 1977, The beneficial effect of EDTA on development of mouse one-cell embryos in chemically defined medium, *Dev. Biol.* 61: 378-383.

Anzano, M.A., Roberts, A.B., Smith, J.M., Sporn, M.B., and DeLarco, J.E., 1983, Sarcoma growth factor from conditioned medium of virally transformed cells is composed of both type α and type β transforming growth factors, *Proc. Natl. Acad. Sci. USA* 80: 6264-6268.

Baird, A., Mormede, P., Ying, S-Y., Wehrenberg, W.B., Veno, N., Ling, N., and Guillemin, R., 1985, A nonmitogenic pituitary function of fibroblast growth factor: Regulation of thyrotropin and prolactin secretion, *Proc. Natl. Acad. Sci. USA* 82: 5545-5549.

Barlow, P.W., and Sherman, M.I., 1972, The biochemistry of differentiation of mouse trophoblast: Studies on polyploidy, *J. Embryol. Exp. Morphol.* 27: 447-465.

Barnes, D., and Sato, G., 1980, Methods for growth of cultured cells in serum-free medium, *Anal. Biochem.* 102: 255-270.

Benveniste, R., Speeg, Jr. K.V., Carpenter, G., Cohen, S., Lindner, J., and Rabinowitz, D., 1978, Epidermal growth factor stimulates secretion of human chorionic gonadotropin by cultured human choriocarcinoma cells, *J. Clin. Endocrinol. Metab.* 46: 169-172.

Biggers, J.D., Whitten, W.K., and Whittingham, D.G., 1971, The culture of mouse embryos *in vitro*, in: *Methods in Mammalian Embryology* (J.C. Daniel, Jr., ed.), Freeman Press, San Francisco, pp. 86-116.

Brinster, R.L., 1965, Studies on the development of mouse embryos *in vitro*: I. The effect of osmolarity and hydrogen ion concentration, *J. Exp. Zool.* 158: 49-58.

Catterton, W.Z., Escobedo, M.B., Sexson, W.R., Gray, M.E., Sundell, H.W., and Stahlman, M.T., 1979, Effect of epidermal growth factor on lung maturation in fetal rabbits, *Pediat. Res.* 13: 104-108.

Chen, L.B., Gudor, R.C., Sun, T-T., Chen, A.B., and Mosesson, M.W., 1977, Control of a cell surface major glycoprotein by epidermal growth factor, *Science* 197: 776-778.

Chew, N.J., and Sherman, M.I., 1973, Δ^5-3β-hydroxysteroid dehydrogenase activity in mouse giant trophoblast cells *in vivo* and *in vitro*, *Biol. Reprod.* 9: 79-89.

Chew, N.J., and Sherman, M.I., 1975, Biochemistry of differentiation of mouse trophoblast: Δ^5-3β-hydroxysteroid dehydrogenase, *Biol. Reprod.* 12: 351-359.

Cohen, S., 1962, Isolation of a mouse submaxillary gland protein accelerating incisor eruption and eyelid opening in the new-born animal, *J. Biol. Chem.* 237: 1555-1562.

Cohen, S., 1965, The stimulation of epidermal proliferation by a specific protein (EGF), *Dev. Biol.* 12: 394-407.

Cohen, S., and Elliot, G.A., 1963, The stimulation of epidermal keratinization by a protein isolated from the submaxillary gland of the mouse, *J. Invest. Dermatol.* 40: 1-15.

Cohen, S., and Levi-Montalcini, R., 1956, A nerve growth-stimulating factor isolated from snake venom, *Proc. Natl. Acad. Sci. USA* 42: 571-574.

Cole, R.J., and Paul, J., 1965, Properties of cultured preimplantation mouse and rabbit embryos, and cell strains derived from them, in: *Preimplantation Stages of Pregnancy* (G.E.W. Wolstenholme, and M. O'Connor, eds.), Little, Brown and Co., Boston, pp. 82-122.

D'Ercole, A.J., Applewhite, G.T., and Underwood, L.E., 1980, Evidence that somatomedin is synthesized by multiple tissues in the fetus, *Dev. Biol.* 75: 315-328.

De Larco, J.E., and Todaro, G.J., 1978, Growth factors from murine sarcoma virus-transformed cells, *Proc. Natl. Acad. Sci. USA* 75: 4001-4005.

Derynck, R., Roberts, A.B., Winkler, M.E., Chen, E.Y., and Goeddel, D.V., 1984, Human transforming growth factor-α: Precursor structure and expression in *E. coli*, *Cell* 38: 287-297.

Dockray, G.J., 1979, Evolutionary relationships of the gut hormone, *Fed. Proc.* 38: 2295-2301.

Frolik, C.A., Wakefield, L.M., Smith, D.M., and Sporn, M.B., 1984, Characterization of a membrane receptor for transforming growth factor-β in normal rat kidney fibroblasts, *J. Biol. Chem.* 259: 10995-11000.

Goldstein, L.S., Spindle, A.I., and Pedersen, R.A., 1975, X-ray sensitivity of the preimplantation mouse embryo *in vitro*, *Rad. Res.* 62: 276-287.

Goustin, A.S., Betsholtz, C., Pfeifer-Ohlsson, S, Persson, H., Rydnert, J., Bywater, M., Holmgren, G., Heldin, C-H., Westermark, B., and Ohlsson, R., 1985, Coexpression of the *sis* and *myc* proto-oncogenes in developing human placenta suggests autocrine control of trophoblast growth, *Cell* 41: 301-312.

Graham, C.F., 1977, Teratocarcinoma cells and normal mouse embryogenesis, in: *Concepts in Mammalian Embryogenesis* (M.I. Sherman, ed.), The MIT Press, Cambridge, MA, pp. 315-394.

Gospodarowicz, D., 1981, Epidermal and nerve growth factors in mammalian development, *Ann. Rev. Physiol.* 43: 251-263.

Gudas, L.J., Singh, J.P., Stiles, C.D., 1983, Secretion of growth regulatory molecules by teratocarcinoma stem cells, in: *Teratocarcinoma Stem Cells* (L.M. Silver, G.R. Martin, and S. Strickland, eds.), Cold Spring Harbor Laboratory, Cold Spring Harbor, pp. 229-236.

Gwatkin, R.B.L., 1966a, Defined media and development of mammalian eggs *in vitro*, *Ann. N.Y. Acad. Sci.* 139: 79-90.

Gwatkin, R.B.L., 1966b, Amino acid requirements for attachment and outgrowth of the mouse blastocyst *in vitro*, *J. Cell Physiol.* 68: 335-344.

Halper, J., and Moses, H.L., 1983, Epithelial tissue-derived growth factor-like polypeptides, *Cancer Res.* 43: 1972-1979.

Hassell, J.R., 1975, The development of rat palatal shelves *in vitro*: An ultrastructural analysis of the inhibition of epithelial cell death and palate fusion by the epidermal growth factor, *Dev. Biol.* 45: 90-102.

Havel, R.J., Eder, H.A., and Bragdon, J.H., 1955, The distribution and chemical composition of ultracentrifugally separated lipoproteins in human serum, *J. Clin. Invest.* 34: 1345-1353.

Heath, J.K., and Isacke, C.M., 1983, Reciprocal control of teratocarcinoma proliferation, *Cell Biol. Intl. Rpts.* 7: 561-562.

Heath, J.K., and Isacke, C.M., 1984, PC13 embryonal carcinoma-derived growth factor, *EMBO J.* 3: 2957-2962.

Heldin, C-H., and Westermark, B., 1984, Growth factors: Mechanism of action and relation to oncogenes, *Cell* 37: 9-20.

Hsu, Y-C., 1972, Differentiation *in vitro* of mouse embryos beyond the implantation stage, *Nature (London)* 239: 200-202.

Hsu, Y-C., 1973, Differentiation *in vitro* of mouse embryos to the stage of early somite, *Dev. Biol.* 33: 403-411.

Hsu, Y-C., 1979, *In vitro* development of individually cultured whole mouse embryos from blastocyst to early somite stage, *Dev. Biol.* 68: 453-461.

Ilgren, E.B., 1983, Review article: Control of trophoblastic growth, *Placenta* 4: 307-328.

Jakobovits, A., Banda, M.J., and Martin, G.R., 1985, Embryonal carcinoma-derived growth factors: Specific growth-promoting and differentiation-inhibiting activities, in: *Growth Factors and Transformation* (J. Feramisco, B. Ozanne, and C. Stiles, eds.), Cold Spring Harbor Laboratory, Cold Spring Harbor, pp. 393-399.

James, R., and Bradshaw, R.A., 1984, Polypeptide growth factors, *Ann. Rev. Biochem.* 53: 259-292.

Jenkinson, E.J., and Wilson, I.B., 1973, *In vitro* studies on the control of trophoblast outgrowth in the mouse, *J. Embryol. Exp. Morphol.* 30: 21-30.

Johnson, L.K., Baxter, J.D., Vlodavsky, I., and Gospodarowicz, D., 1980, Epidermal growth factor and expression of specific genes: Effects on cultured rat pituitary cells are dissociable from the mitogenic response, *Proc. Natl. Acad. Sci. USA* 77: 394-398.

Juurlink, B.H.J., and Federoff, S., 1977, Effects of culture milieus on the development of mouse blastocysts *in vitro*, *In Vitro* 13: 790-798.

Kane, M.T., 1979, Fatty acids as energy sources for culture of one-cell rabbit ova to viable morulae, *Biol. Reprod.* 20: 323-332.

Koroly, M.J., and Young, M., 1981, Nerve growth factor, in: *Tissue Growth Factors* (R. Baserga, ed.), Springer-Verlag, New York, pp. 249-276.

Leivo, I., Vaheri, A., Timpl, R., and Wartiovaara, J., 1980, Appearance and distribution of collagens and laminin in the early mouse embryo, *Dev. Biol.* 76: 100-114.

Lembach, K.H., 1976, Enhanced synthesis and extracellular accumulation of hyaluronic acid during stimulation of quiescent human fibroblasts by mouse epidermal growth factor, *J. Cell Physiol.* 89: 277-288.

Lindgren, F.T., Jensen, L.C., and Hatch, F.T., 1979, The isolation and quantitative analysis of serum lipoproteins, in: *Blood Lipids and Lipoproteins: Quantitation, Composition and Metabolism* (G.J. Nelson, ed.), Robert E. Krieger Publishing Co., Huntington, NY, pp. 181-274.

Linkhart, T.A., Lim, R.W., and Hauschka, S.D., 1982, Regulation of normal and variant mouse myoblast proliferation and differentiation by specific growth factors, in: *Growth of Cells in Hormonally Defined Media*, Book B (G.H. Sato, A.B. Pardee, and D.A. Sirbasku, eds.), Cold Spring Harbor Laboratory, Cold Spring Harbor, pp. 867-876.

Marquardt, H., Hunkapiller, M.W., Hood, L.E., and Todaro, G.J., 1984, Rat transforming growth factor type 1: Structure and relation to epidermal growth factor, *Science* 223: 1079-1082.

Martin, G.R., 1978, Advantages and limitations of teratocarcinoma stem cells as models of development, in: *Development in Mammals*, Volume 3 (M.H. Johnston, ed.), North Holland Publishing, New York, pp. 225-266.

Martin, G.R., 1980, Teratocarcinomas and mammalian embryogenesis, *Science* 209: 768-776.

Martin, G.R., 1981, Isolation of a pluripotent cell line from early mouse embryos cultured in medium conditioned by teratocarcinoma stem cells, *Proc. Natl. Acad. Sci. USA* 78: 7634-7638.

Matrisian, L.M., Pathak, M., and Magun, B.E., 1982, Identification of an epidermal growth factor-related transforming growth factor from rat fetuses, *Biochem. Biophys. Res. Comm.* 107: 761-769.

McQuilkin, W.J., Evans, V.J., and Earle, W.R., 1957, The adaptation of additional lines of NCTC Clone 929 (strain L) cells to chemically defined protein-free medium NCTC 109, *J. Natl. Canc. Inst.* 19: 905-907.

Menke, T.M., and McLaren, A., 1970, Mouse blastocysts grown *in vivo* and *in vitro*: Carbon dioxide production and trophoblast outgrowth, *J. Reprod. Fertil.* 23: 117-127.

Mesiano, S., Browne, C.A., and Thorburn, G.D., 1985, Detection of endogenous epidermal growth factor-like activity in the developing chick embryo, *Dev. Biol.* 110: 23-28.

Mintz, B., 1964, Formation of genetically mosaic mouse embryos, and early development of "Lethal (t^{12}/t^{12})-Normal" mosaics, *J. Exp. Zool.* 157: 273-292.

Morton, H.C., 1970, A survey of commercially available tissue culture media, *In Vitro* 6: 89-108.

New, D.A.T., 1978, Whole-embryo culture and the study of mammalian embryos during organogenesis, *Biol. Rev.* 53: 81-122.

Nexo, E., Hollenberg, M.D., Figueroa, A., and Pratt, R.M., 1980, Detection of epidermal growth factor-urogastrone and its receptor during fetal mouse development, *Proc. Natl. Acad. Sci. USA* 77: 2782-2785.

Nilausen, K., 1978, Role of fatty acids in growth-promoting effect of serum albumin on hamster cells *in vitro*, *J. Cell Physiol.* 96: 1-14.

Proper, J.A., Bjornson, C.L., and Moses, H.L., 1982, Mouse embryos contain polypeptide growth factor(s) capable of inducing a reversible neoplastic phenotype in nontransformed cells in culture, *J. Cell Physiol.* 110: 169-174.

Rizzino, A., 1982a, Embryonal carcinoma cells release factors with transforming growth factor (TGF) activity, *J. Cell Biol.* 95 (2 Pt. 2): 181a.

Rizzino, A., 1982b, Growth and differentiation of embryonal carcinoma cells in defined media: The role of fibronectin, in: *Growth of Cells in Hormonally Defined Media* (G.H. Sato, A.B. Pardee, and D.A. Sirbasku, eds.), Cold Spring Harbor Laboratory, Cold Spring Harbor, pp. 209-218.

Rizzino, A., 1983a, Model systems for studying the differentiation of embryonal carcinoma cells, *Cell Biol. Intl. Rpts.* 7: 559-560.

Rizzino, A., 1983b, Two multipotent embryonal carcinoma cell lines irreversibly differentiate in defined media, *Dev. Biol.* 95: 126-136.

Rizzino, A., 1984a, Growth and differentiation of embryonal carcinoma cells in defined and serum-free media, in: *Methods for Serum-Free Culture of Epithelial and Fibroblastic Cells* (D. Barnes, D. Sirbasku, and G.H. Sato, eds.), Alan R. Liss Inc., New York, pp. 107-124.

Rizzino, A., 1984b, Behavior of transforming growth factors in serum-free media: An improved assay for transforming growth factors, *In Vitro* 20: 815-822.

Rizzino, A., 1985, Early mouse embryos produce and release factors with transforming growth factor activity, *In Vitro* 21: 531-536.

Rizzino, A., 1986, Soft agar growth assays for transforming growth factors and mitogenic peptides, in: *Methods in Enzymology - Peptide Growth Factors* (D. Barnes, and D. Sirbasku, eds.), (in press).

Rizzino, A., and Bowen-Pope, D., 1984, Production of and response to PDGF-like factors by early embryonic cells, *Fed. Proc.* 43: 519.

Rizzino, A., and Bowen-Pope, D.F., 1985, Production of PDGF-like growth factors by embryonal carcinoma cells and binding of PDGF to their endoderm-like differentiated cells, *Dev. Biol.* 110: 15-22.

Rizzino, A., and Crowley, C., 1980, Growth and differentiation of embryonal carcinoma cell line F9 in defined media, *Proc. Natl. Acad. Sci. USA* 77: 457-461.

Rizzino, A., and Sato, G., 1978, Growth of embryonal carcinoma cells in serum-free medium, *Proc. Natl. Acad. Sci. USA* 75: 1844-1848.

Rizzino, A., and Sherman, M.I., 1979, Development and differentiation of mouse blastocysts in serum-free medium, *Exp. Cell Res.* 121: 221-233.

Rizzino, A., Rizzino, H., and Sato, G., 1979, Defined media and the determination of nutritional and hormonal requirements of mammalian cells in culture, *Nutr. Rev.* 37: 369-378.

Rizzino, A., Terranova, V., Rohrbach, D., Crowley, C., and Rizzino, H., 1980, The effects of laminin on the growth and differentiation of embryonal carcinoma cells in defined media, *J. Supramol. Struct.* 13: 243-253.

Rizzino, A., Orme, L.S., and De Larco, J.E., 1983, Embryonal carcinoma cell growth and differentiation: Production of and response to molecules with transforming growth factor activity, *Exp. Cell Res.* 143: 143-152.

Rizzino, A., Ruff, E., and Rizzino, H., 1986, Induction and modulation of anchorage-independent growth by platelet-derived growth factor, fibroblast growth factor and transforming growth factor-β, *Cancer Res.* 46: 2816-2820.

Roberts, A.B., Anzano, M.A., Lamb, L.C., Smith, J.M., and Sporn, M.B., 1981, New class of transforming growth factors potentiated by epidermal growth factor: Isolation from non-neoplastic tissues, *Proc. Natl. Acad. Sci. USA* 78: 5339-5343.

Roberts, A.B., Anzano, M.A., Wakefield, L.M., Roche, N.S., Stern, D.F., and Sporn, M.B., 1985, Type β transforming growth factor: A bifunctional regulator of cellular growth, *Proc. Natl. Acad. Sci. USA* 82: 119-123.

Sellens, M.H., and Sherman, M.I., 1980, Effects of culture conditions on the developmental programme of mouse blastocysts, *J. Embryol. Exp. Morphol.* 56: 1-22.

Seppa, H., Grotendorst, G., Seppa, S. Schiffmann, E., and Martin, G.R., 1982, Platelet-derived growth factor is chemotactic for fibroblasts, *J. Cell Biol.* 92: 584-588.

Sherman, M.I., and Wudl, L.R., 1976, The implanting mouse blastocyst, in: *The Cell Surface in Animal Embryogenesis and Development* (G. Poste, and G.L. Nicolson, eds.), Elsevier/North-Holland Biomedical Press, Amsterdam, pp. 81-125.

Sherwin, S.A., Twardzik, D.R., Bohn, W.H., Cockley, K.D., and Todaro, G.J., 1983, High-molecular-weight transforming growth factor activity in the urine of patients with disseminated cancer, *Cancer Res.* 43: 403-407.

Solter, D., and Damjanov, I., 1979, Teratocarcinoma and the expression of oncodevelopmental genes, in: *Methods in Cancer Research*, Volume XVIII (W.H. Fishman, and H. Busch, eds.), Academic Press, New York, pp. 277-332.

Spindle, A., 1980, An improved culture medium for mouse blastocysts, *In Vitro* 16: 669-674.

Spindle, A.I., and Pedersen, R.A., 1973, Hatching, attachment, and outgrowth of mouse blastocysts *in vitro*: Fixed nitrogen requirements, *J. Exp. Zool.* 186: 305-318.

Sundell, H.W., Gray, M.E., Serenius, F.S., Escobedo, M.B., and Stahlman, M.T., 1980, Effects of epidermal growth factor on lung maturation in fetal lambs, *Amer. J. Pathol.* 100: 707-726.

Thorstensen, K., and Romslo, I., 1984, Albumin prevents non-specific transferrin binding and iron uptake by isolated hepatocytes, *Biochim. Biophys. Acta* 804: 393-397.

Tucker, R.F., Branum, E.L., Shipley, G.D., Ryan, R.J., and Moses, H.L., 1984a, Specific binding to cultured cells of ^{125}I-labeled type β transforming growth factor from human platelets, *Proc. Natl. Acad. Sci. USA* 81: 6757-6761.

Tucker, R.F., Shipley, G.D., Moses, H.L., and Holley, R.W., 1984b, Growth inhibitor from BSC-1 cells closely related to platelet type β transforming growth factor, *Science* 226: 705-707.

Twardzik, D.R., Ranchalis, J.E., and Todaro, G.J., 1982, Mouse embryonic transforming growth factors related to those isolated from tumor cells, *Cancer Res.* 42: 590-593.

Wartiovaara, J., Leivo, I., and Vaheri, A., 1979, Expression of the cell surface-associated glycoprotein, fibronectin, in the early mouse embryo, *Dev. Biol.* 69: 247-257.

Whitten, W.K., 1956, Culture of tubal mouse ova, *Nature (London)* 177:96.

Wiley, L.M., and Pedersen, R.A., 1977, Morphology of mouse egg cylinder development *in vitro*: a light and electron microscope study, *J. Exp. Zool.* 200: 389-402.

Winkel, C.A., Snyder, J.M., MacDonald, P.C., and Simpson, E.R., 1980a, Regulation of cholesterol and progesterone synthesis in human placental cells in culture by serum lipoproteins, *Endocrinology* 106: 1054-1060.

Winkel, C.A., Gilmore, J., MacDonald, P.C., and Simpson, E.R., 1980b, Uptake and degradation of lipoproteins by human trophoblastic cells in primary culture, *Endocrinology* 107: 1892-1898.

Winkel, C.A., MacDonald, P.C., Hemsell, P.G., and Simpson, E.R., 1981, Regulation of cholesterol metabolism by human trophoblastic cells in primary culture, *Endocrinology* 109: 1084-1090.

Wu, T-C., Wan, Y-J., and Damjanov, I., 1981, Rat serum promotes the *in vitro* development of mouse blastocysts during early somitic stages of embryogenesis, *J. Exp. Zool.* 217: 451-453.

Yamaoka, K., Hirai, R., Tsugita, A., and Mitsui, H., 1984, The purification of an acid- and heat-labile transforming growth factor from an avian sarcoma virus-transformed rat cell line, *J. Cell Physiol.* 119: 307-314.

CHAPTER 9

INTERACTION OF TROPHOBLASTIC VESICLES WITH BOVINE EMBRYOS DEVELOPING IN VITRO

YVES HEYMAN and YVES MÉNÉZO

1. INTRODUCTION

In many species, *in vitro* culture of fertilized embryos does not lead to normal cleavage and development as observed *in vivo*. When cultured *in vitro*, mouse eggs from outbred strains are blocked at the 2-cell stage unless they are transferred back into oviducts (Whittingham and Biggers, 1967). Hamster eggs also become blocked at this stage (Whittingham and Bavister, 1974) [see also Chapter 11 (Ed.)]. Most early bovine embryos (1- to 8-cell stage) arrest their development *in vitro* at the 8- to 12-cell stage (Thibault, 1966), in spite of a variety of culture media and conditions tested by several authors (reviewed by Wright and Bondioli, 1981).

Many factors can be involved in this block stage of embryos *in vitro*. Culture media are never similar to the tubal or uterine environment, apart from which the biophysical or biochemical conditions may sometimes be unsatisfactory. A lower oxygen tension in the culture medium is beneficial for sheep and cattle ova (Tervit *et al.*, 1972). The choice of energy source in the culture medium plays an important role in cleavage *in vitro*. Four-cell pig embryos cultured in Krebs-Ringer bicarbonate containing some or all of the following: glucose, pyruvate, lactate and BSA, developed at a higher rate in medium without lactate and pyruvate (Davis and Day, 1978).

In vivo it is possible that the oviduct environment compensates for some deficiencies of genomic expression in the embryo. The oviduct may secrete non-specific compounds such as lipids and growth factors that are necessary for the development of embryonic cell membranes, but it is difficult to include compounds such as lipids in culture media because they are generally non-soluble. During its migration through the oviduct, the embryo probably

Yves Heyman I.N.R.A., Station de Physiologie Animale, 78350 Jouy-en-Josas, France.
Yves Ménézo I.N.S.A., Laboratoire de Biologie, 69621 Villeurbanne Cedex, France.

induces the release of specific substances: catecholamines are found in the oviduct and seem to be related to embryo transport (Villalon *et al.*, 1982) and to embryo cleavage in invertebrates (Buznikov *et al.*, 1972). The partial ability of the rabbit oviduct to support development of embryos from other species (Boland, 1984) indicates that more than one specific metabolite is involved. Muggleton-Harris *et al.* (1982) demonstrated that in the mouse the 2-cell block *in vitro* can be overcome by injecting into blastomeres cytoplasm from other embryos which do not exhibit this block. Thus, there is some cytoplasmic control of preimplantation development *in vitro*. The block to development in the mouse does not seem to be detrimental since normal growth occurs once cleavage is induced.

Development of cattle embryos *in vitro* can be achieved from the morula to blastocyst stages (Renard *et al.*, 1976). After hatching and during the early elongation phase, cattle embryos can be dissected and cultured and it was shown that cultured trophoblastic vesicles were able to release an antiluteolytic and/or luteotrophic signal when transferred into the uterus of cycling recipients (Heyman *et al.*, 1984).

On the basis of these observations, and assuming either that metabolites can be provided by coculture with exogenous tissue or that a specific embryo cleavage signal exists, it may be supposed that cells from elongating blastocysts contain growth factors leading to an exponential multiplication of embryonic cells. [See also Chapter 8 (Ed.).] Taking into account the side effect of trophoblastic tissue in culture, we have set up a co-culture system of early stage bovine embryos together with trophoblastic vesicles in order to determine whether fragments of trophoblast were able to transmit compounds *via* the culture medium to 1- and 2-cell embryos and promote their further cleavage *in vitro*. This biological approach is different from the coculture systems developed by Kuzan and Wright (1982) or Allen and Wright (1984), who have demonstrated a positive effect of cell-cell contact on embryo development. [See also Chapter 12 (Ed.).]

The experiments reported here are just beginning and part of the data are still preliminary.

2. WORKING WITH TROPHOBLASTIC VESICLES

2.1. Preparation of Trophoblastic Vesicles

Embryo Collection. Day 14 bovine embryos were obtained from heifers superovulated by injections of pituitary extracts, *i.e.*, 32 mg FSH-P (Burns-Biotech) per animal, and inseminated at estrus (designated D0). Non-surgical recovery was performed by flushing the uterine horns with phosphate buffered saline (Tervit *et al.*, 1972) and embryos were rinsed twice and transferred into B2 medium (Api-System; see Ménézo, 1976; Ménézo *et al.*, 1984).

Dissection and Culture. Elongating blastocysts (0.4 to 40 mm) were used to prepare trophoblastic vesicles. Each embryo was cut into 2, 3 or more pieces to remove the embryonic disc as shown in Figure 1. Sectioning was performed in a glass dish using a scalpel under a binocular microscope or with a microscalpel and micromanipulators for spherical embryos. Just after

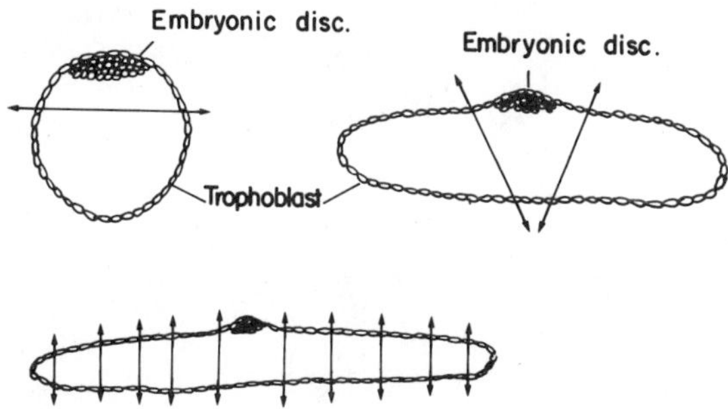

Figure 1. Diagrams showing planes of embryo sectioning to remove the embryonic disc and to prepare trophoblastic vesicles.

sectioning, each piece of tissue was cultured *in vitro* at 38°C in a multi-vial disposable box (Nunclon Delta) containing 500 µl B2 medium. The atmosphere was controlled (5% O_2, 5% CO_2, 90% N_2). For biochemical assay, some of the trophoblastic vesicles were cultured in medium without serum albumin. During culture, pieces of trophoblastic tissue changed into spherical trophoblastic vesicles (Fig. 2a). These were cultured for several days; the culture medium was changed every 24 hr and stored at -20°C until assayed.

In order to study the composition of the internal fluid of trophoblastic vesicles, at the end of culture, 50 to 100 vesicles were rinsed 3 times in physiological saline to eliminate traces of culture medium and opened by micromanipulation. The internal fluid was diluted in saline and passed through a Millipore filter (0.45 µm) to remove the cells.

In Vitro Development of Trophoblastic Vesicles. Cultures were observed daily and trophoblastic vesicle development was evaluated from the increase in their size. The diameter of trophoblastic vesicles was measured during the first 3 days of culture. More than 95% of trophoblastic vesicles developed *in vitro* within 48 hr in B2 medium, and 75% in the same medium without serum albumin. The mean diameter increase during 24-48 hr culture was 60% (from 1.0 mm to 1.6 mm). Some trophoblastic vesicles could be maintained in culture for periods as long as 2 or 3 weeks.

2.2. Structure of Trophoblastic Vesicles

Samples of trophoblastic vesicles before and after culture were fixed in 2% glutaraldehyde in 0.15 M cacodylate buffer, then in 1% OsO_4 in the same buffer for 1 hr each, dehydrated in ethanol and embedded in Epon. Semi-thin sections were stained with toluidine blue. Ultrathin sections were counterstained with uranyl acetate and lead citrate. Semi-thin sections revealed that trophoblastic vesicles were made of 2 layers of cells: one layer of trophoblastic cells lined by a thin layer of endoderm (Fig. 2b).

Scanning electron microscopy of a cultured trophoblastic vesicle showed numerous microvilli on the cell surface (Figs. 2c and 2d); ultrastructure of cultured trophoblastic vesicles (Fig. 2e) indicated that these vesicles may have an intensive secretory activity as they contained numerous inclusions.

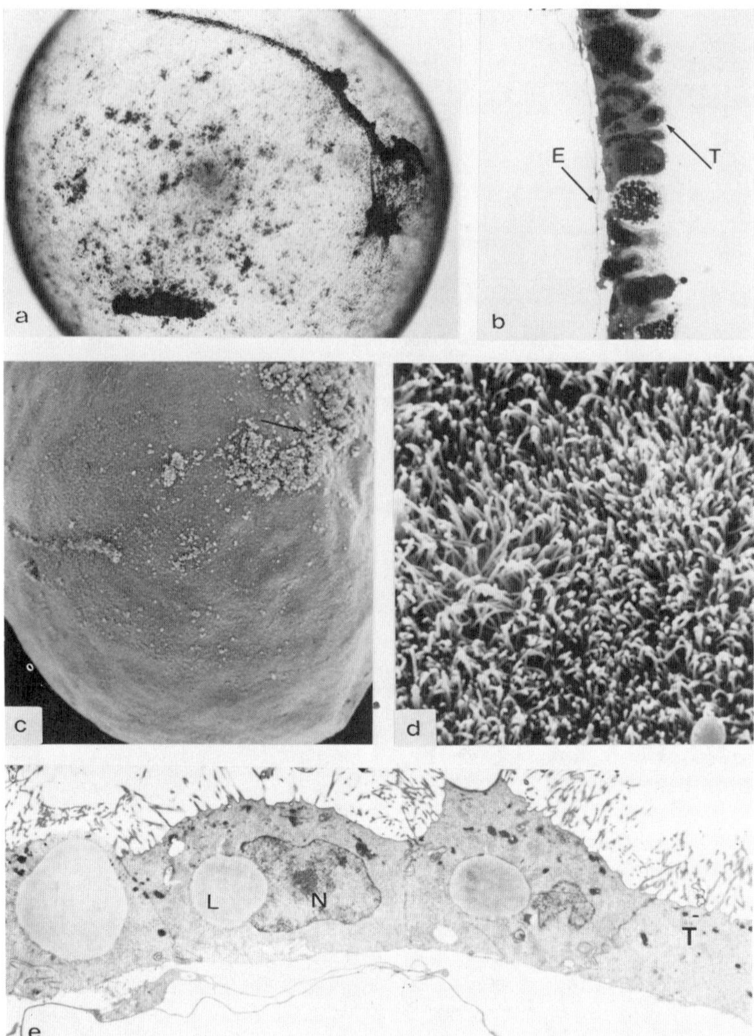

Figure 2. Bovine trophoblastic vesicles. (a) Trophoblastic vesicle after culture for 24 hr (x 70). (b) Semi-thin section of trophoblastic vesicle after culture; note the endoderm layer (E) and trophectoderm (T). (c) Scanning electron micrograph of a cultured trophoblastic vesicle. At low magnification, the surface appears smooth except on the scar (arrow). (Courtesy of Dr. Fléchon, I.N.R.A.) (x 120). (d) Higher magnification: surface of trophoblastic vesicle cells is densely covered with microvilli (x 3000). (e) Transmission electron micrograph of part of a trophoblastic vesicle showing the trophoblastic epithelium (T) containing numerous inclusions; nuclei (N) and lipid droplets (L). (Courtesy of Dr. Fléchon, I.N.R.A.) (x 2000). (Optical magnifications given).

2.3. Amino-Acid Uptake

Deproteinization was carried out by adding sulfosalicylic acid (30 mg/ml of sample). An internal staining reference was added at the same time (500 nM of norvaline, a synthetic amino-acid) to minimize dosage errors. Analyses were performed on a Kontron liquimat III auto-analyser.

Table I
Amino-Acid Uptake[a] by Bovine Trophoblastic Vesicles in Culture Medium According to Time in Culture

Culture period	Amino-acid[b]											
	Thre		Val		Leu		Phe		Lys		Arg	
	OSA	WSA	OSA	WSA	OSA	WSA	OSA	WSA	OSA	WSA	OSA	WSA
D1 - D2	12	10	63	47	34	18	15	10	34	20	111	78
D2 - D3	13	16	5	10	53	45	17	11	54	28	117	85
D6 - D7	0	16	6	80	20	48	ND	ND	ND	28	ND	21
D10 - D11	5	12	5	60	20	40	5	ND	15	24	ND	ND

[a]Mean value for 10 trophoblastic vesicles (ng/day).
[b]OSA: medium without serum albumin; WSA: medium with serum albumin; ND : not detected.

Amino Acid Uptake Variations Related to Culture Duration and Presence of Serum Albumin. Quantitative estimation was performed every day when culture medium was removed and changed. Table I shows amino acid uptake related to duration of culture in medium with or without albumin. Amino acid uptake was high even after 10 days of culture and, as indicated for *in vitro* development, the medium with serum albumin seemed to ensure better survival of the trophoblastic vesicles, especially over a 5-day culture period.

Amino Acids in the Internal Trophoblastic Fluid. At the end of culture, trophoblastic vesicles were opened and the internal fluid was analysed and compared to the external medium without serum albumin (Table II). The concentration ratio between internal and external medium was not similar for all amino acids. Basic amino acids (arginine, lysine) may be concentrated in the trophoblastic vesicle, but proline accumulated the most in internal fluid. In contrast, glycine, which was the most concentrated in the external medium, did not totally pass into the inner fluid.

Accordingly, passage of amino acids into the internal trophoblastic fluid depends upon the amino acid nature (charge), upon the external concentration and possibly also upon amino acid transport (Miller and Schultz, 1983; Hobbs and Kaye, 1985).

Biochemical parameters confirm the longterm viability of trophoblastic vesicles *in vitro* as shown by histological studies and by the beneficial effect of serum albumin in the medium on survival of trophoblastic vesicles.

2.4. Electrophoresis of Internal Fluid of Trophoblastic Vesicles

The internal fluid of trophoblastic vesicles cultured in medium with or without serum albumin was studied by electrophoresis on pAA 4/30 gradient gels. Before sample deposition, sucrose was added to the sample to increase viscosity. The buffer was Tris-glycine pH 8.6 and the migration period was 15 hr. Staining was performed as usual by Coomassie Blue in 7% acetic acid.

Table II
Relationship Between Amino-Acid Concentrations in Internal Fluid of Trophoblastic Vesicles and External Culture Medium

Amino-acid	Internal concentration[a] (nM)	External concentration (nM)	Ratio int./ext. concentrations
Thre	130	190	0.68
Ser	142	133	1.07
Pro	560	132	4.25
Gly	2218	6000	0.37
Val	530	546	0.97
Met	39	72	0.54
Ile	102	182	0.56
Leu	155	218	0.71
Tyr	114	132	0.86
Phe	152	152	1.00
Orn	127	109	1.17
Lys	412	234	1.76
His	233	171	1.36
Arg	540	297	1.82
Asn + glu + gln	1025	1250	0.82

[a]5 trials.

Destaining was done in 7% acetic acid with 1% glycerol. Electrophoresis (Fig. 3) showed that when serum albumin was added to the culture medium, a high amount of albumin was found in the trophoblastic vesicle internal fluid. This indicates that serum albumin can pass very easily through the trophoblastic layers. This transport has already been observed for mouse blastomeres (Glass and Hanson, 1975).

2.5. Qualitative Aspects of Peptide Secretions by Trophoblastic Vesicles

As was recently shown by Kane (1985), low molecular weight compounds bound to serum albumin stimulate rabbit blastocyst expansion. [See also Chapter 10 (Ed.).] Ménézo and Katchadourian (unpublished data) have observed that peptides can bind very tightly to serum albumin. Therefore, all the assays were performed in culture media with and without serum albumin to avoid any possible artefact due to peptide release from the serum albumin itself. Ion exchange chromatography showed a release from trophoblastic vesicles of at least 3 peptides: one acidic peptide exhibiting ionic chromatographic behavior similar to aspartic acid, another less acidic with migration similar to serine, and a neutral one migrating like valine. Analysis is now going on to isolate these compounds, which are also present in culture media of human trophoblasts (Ménézo, unpublished).

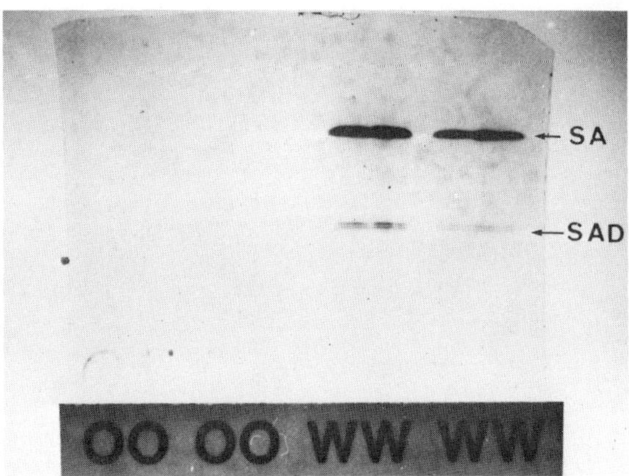

Figure 3. Electrophoresis of trophoblastic internal fluid. (WW) Internal fluid from trophoblastic vesicles cultured in medium with serum albumin. (OO) Internal fluid from trophoblastic vesicles cultured in medium without serum albumin. (SA) Serum albumin. (SAD) Serum albumin dimer. Note that when albumin is present in the medium (WW), the concentration is important (sample deposit 5 µl/10 µl).

3. EFFECT OF TROPHOBLASTIC VESICLES ON EARLY STAGE EMBRYO DEVELOPMENT *IN VITRO*

3.1. Collection and Culture of Early Stage Bovine Embryos

Embryos at the 1- to 8-cell stages (Figs. 4a and 4b) were recovered from superovulated heifers slaughtered on days 2, 2.5 or 3 (D0 = onset of estrus). Within a few minutes after slaughter, the genital tract was removed and each oviduct was isolated and flushed with 5 ml B2 medium. The embryos were classified according to their cell stage and allotted to different culture groups.

In vitro cultures were performed in a controlled humidified atmosphere incubator (5% O_2, 5% CO_2, 90% N_2) at 38°C. Vials for culture were 4-well culture plates (Nunclon-Delta). The basic medium used for culture was sometimes supplemented with 15% (vol./vol.) heat-treated fetal calf serum (FCS) prepared in our laboratory.

3.2. Advantage of a Coculture System with Trophoblastic Vesicles

Technique of Coculture. One trophoblastic vesicle (Fig. 2a) prepared as described earlier and exhibiting a diameter of about 1 mm, was simultaneously incubated with one to three early bovine embryos of the same initial cell stage, in 0.3 ml supplemented B2 medium for 3 to 4 days. Embryo development was evaluated morphologically by counting the blastomeres using an inverted microscope (Nikon-Diaphot) at 24 hr intervals. At the end of culture, the attainment of at least the 16-cell stage (Figs. 4d, 4e and 4f) indicated that the developmental block had been overcome. Treatment differences with controls were evaluated by χ^2 analyses.

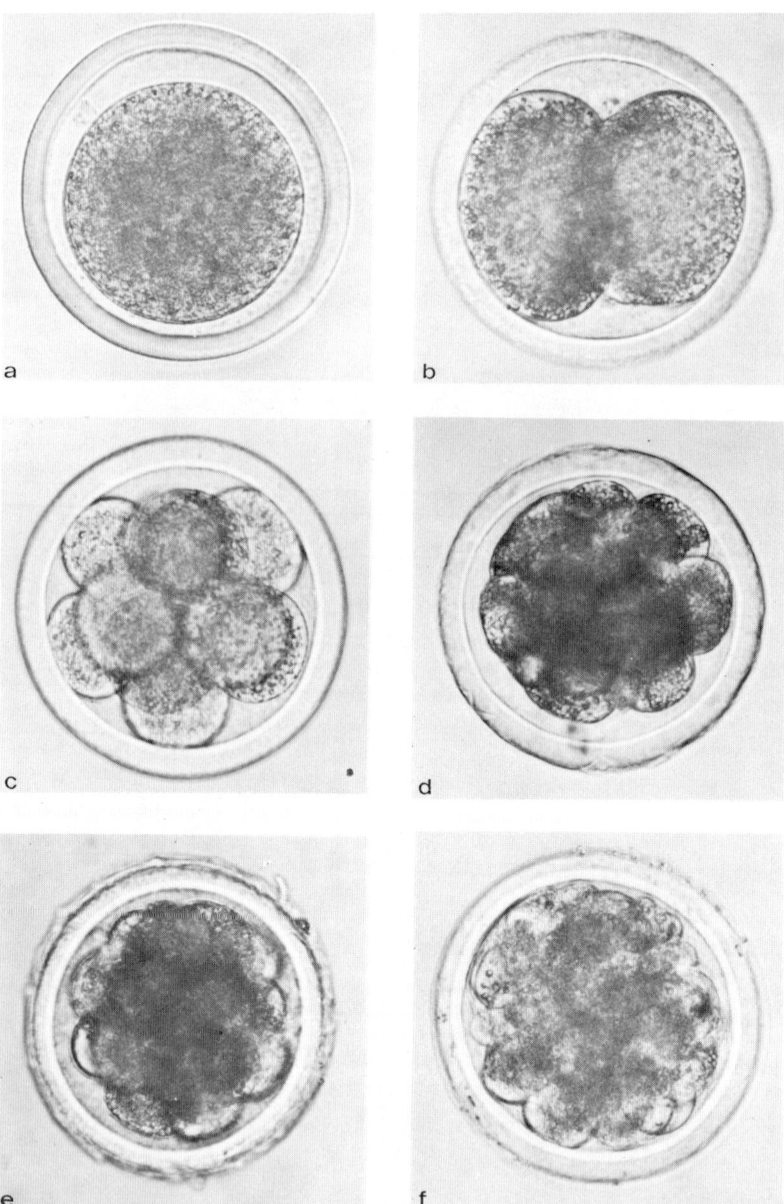

Figure 4. Early stage bovine embryos. (a) One-cell embryo before culture. (b) Two-cell embryo before culture. (c) Embryo blocked at the 8- to 10-cell stage *in vitro*. (d,e,f) Morulae (16- to 32-cells) obtained *in vitro* by coculture with trophoblastic vesicles. (x 500 optical magnification).

In Vitro Development of Early Stage Embryos in Coculture. A total of 87 bovine embryos (1- to 8-cell) out of 190 (46%) cocultured with trophoblastic vesicles in B2 medium supplemented with 15% FCS reached the morula stage

(≥ 16 cells) at the end of culture. In the control group without trophoblastic vesicles, this percentage was only 18% (19/106). The difference was highly significant ($p < 0.001$). This improved cleavage in coculture was observed irrespective of the initial cell stage (Table III). Forty-two % of the 1-cell cocultured embryos developed into morulae compared to 18% in controls ($p < 0.01$). For the 2-cell cocultured embryos, 38% reached the morula stage *vs.* 8% in controls ($p < 0.01$) and 70% of the 8-cell embryos developed past the block stage when cocultured, compared to 30% in the absence of trophoblastic vesicles ($p < 0.01$). The distribution of cell stages at the end of the culture period was significantly different in cocultured embryos and controls (Fig. 5). After coculture with trophoblastic vesicles, the cell stage with the highest percentage of embryos was always the morula stage, while in the control group (B2 + FCS) most embryos arrested their development at the 8-cell or at the 9- to 15-cell stage when their initial cell stage was 1- to 2-cell or 4- to 8-cells, respectively.

Thus, coculture with trophoblastic vesicles promotes *in vitro* cleavage of bovine embryos beyond the block stage; this effect is greater for 8-cell stage embryos than for 1- to 2-cell and 4-cell embryos. According to Seidel (1977), late 8-cell embryos should be easier to culture than those having just reached the 8-cell stage. However, early and late 8-cell embryos cannot be distinguished morphologically. In these experiments, after coculture with trophoblastic vesicles, the viability of early stage embryos was maintained although reduced: 4 pregnancies were initiated and one normal calf was born out of 10 morulae obtained *in vitro*, frozen and transferred to recipient heifers (Camous *et al.*, 1984). This was the first report of successful transfer after such a long culture period *in vitro*.

Table III
Effect of Coculture with Trophoblastic Vesicles on Development *in Vitro* of 1- to 8-Cell Bovine Embryos into Morulae (≥ 16 Cells)[a]

Initial stage of development	No. of embryos	Coculture with TV[b] (+)	Length of culture (days)	No. of embryos developing into morulae (%)[c]
1-cell	55	+	4	23 (42)[d]
	49	-	4	9 (18)
2-cell	80	+	3 or 4	30 (38)[d]
	25	-	3 or 4	2 (8)
4-cell	22	+	3	11 (50)
	12	-	3	2 (17)
8-cell	33	+	3	23 (70)[d]
	20	-	3	6 (30)

[a]Data from Camous *et al.* (1984).
[b]TV= trophoblastic vesicles.
[c]Values are compared with corresponding control group.
[d]Significantly different ($p < 0.01$) from controls (χ^2 analysis).

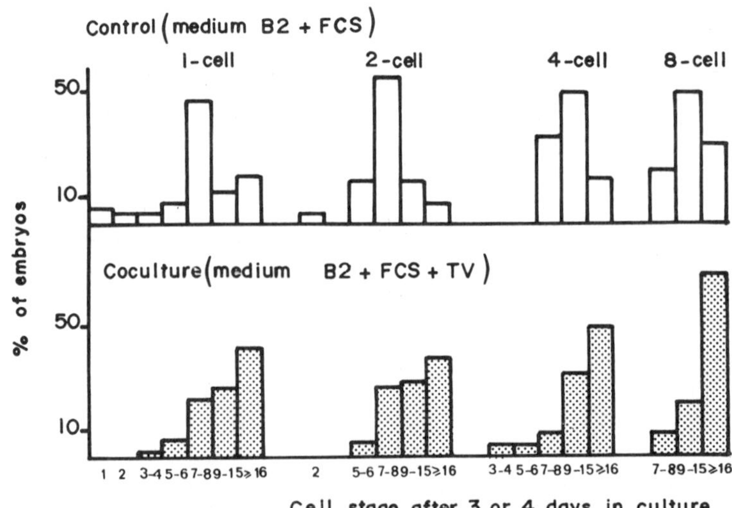

Figure 5. Distribution of cell stages after culture of 1- to 8-cell bovine embryos for 3 or 4 days in coculture with trophoblastic vesicles (medium B2 + FCS + trophoblastic vesicles, TV) or in control group (medium B2 + FCS).

Although a larger proportion of embryos developed into morulae after 3- or 4-day coculture with trophoblastic vesicles, in the present system the most advanced embryos exhibited a delay in development of about 24 hr compared with expected embryo development at the same age *in vivo*. According to Lindner and Wright (1983), *in vivo* most D5 embryos are at the 20- to 32-cell stage and most D6 embryos are compacted morulae.

In previous experiments, Camous *et al.* (1984) showed that addition of 15% FCS to a chemically defined medium (B2) had a beneficial effect on cleavage. One- to 8-cell bovine embryos were cultured in B2 alone or in B2 + 15% FCS. In B2 alone, 76% of the 1- to 2-cell embryos and 65% of the 4- to 8-cell embryos stopped cleaving at the 7- to 8-cell stage, while in B2 + FCS, 35% of the 1- to 2-cell embryos and 65% of the 4- to 8-cell embryos developed beyond the 8-cell stage. This indicates that serum addition to the culture medium partly meets the specific requirements of early bovine embryos *in vitro*. The beneficial effect of serum addition may be due to lipids: serum albumin can bind 1.6 mg/g of fatty acids (Ménézo *et al.*, 1982). [See also Chapter 10 (Ed.).] In the same way, Wright *et al.* (1976) obtained better *in vitro* development of early bovine embryos when using FCS as a supplement to Ham's F10 medium rather than BSA. However, later stage embryos (morulae) are able to reach the hatched blastocyst stage *in vitro* without any serum in B2 medium (Renard *et al.*, 1976).

To determine the respective roles of trophoblastic vesicles and serum during coculture, 1-cell bovine embryos were assigned to one of the 4 following culture conditions: a) B2 alone, b) B2 + 15% FCS, c) B2 + trophoblastic vesicles, d) B2 + FCS + trophoblastic vesicles. Results after 72 hr culture are indicated in Table IV. In B2 alone, only one embryo out of 22

(4.5%) cleaved past the 8-cell stage. Supplementation of the medium with FCS or trophoblastic vesicles increased this percentage (23.8% and 12.7%, respectively). When both FCS and trophoblastic vesicles were added to the B2 medium, cleavage was significantly higher ($p < 0.005$): 69% (38/55) of the initially 1-cell embryos cleaved past the 8-cell stage. Serum addition to B2 medium magnifies the effect of trophoblastic vesicles in the coculture system. We already observed that amino-acid uptake by trophoblastic vesicles was higher (Table I) when vesicles were cultured in medium with serum albumin (a serum fraction) than without. A synergistic effect could exist between trophoblastic vesicle signals and serum components. For example, peptides released by the trophoblastic vesicles might be bound and transported by albumin in the FCS. Lipids (LDL-HDL) are also more effectively transported by serum components.

However, one cannot completely exclude the hypothesis of a so-called "negative and indirect control" concept developed for cell proliferation by Soto and Sonnenschein (1984, 1985). According to this concept, serum, necessary for the supply of nutrients for embryo growth, should have an inhibitory effect on cleavage. In coculture, this inhibition should be neutralized by trophoblastic vesicle secretions.

3.3. Evidence for Release by Trophoblastic Vesicles of Factors into the Medium

In our experiments, the beneficial effect of trophoblastic vesicles on *in vitro* cleavage of early embryos during coculture was supposed to be acting *via* compounds released into the culture medium. Moreover, there was no direct contact between the early embryo and the trophoblastic vesicle cells during coculture. To test this hypothesis, we tried to culture 1- and 2-cell bovine embryos in media that had previously supported *in vitro* development of trophoblastic vesicles alone. Therefore, the culture medium from

Table IV
Effect of Serum and Trophoblastic Vesicles on *in Vitro* Cleavage of 1-Cell Bovine Embryos

Culture conditions[a]	No. of embryos	Stage of development (cell no.) at end of culture (72hr)						
		1	2	3-4	5-6	7-8	9-15	≥16
B2 alone	22	7	3	2	2	7	1	--
B2 + 15% FCS	67	9	2	5	5	30	7	9
Coculture B2 + TV	47	11	2	4	8	16	5	1
Coculture B2 + FCS + TV	55	--	--	1	4	12	15	23

[a]B2 = Ménézo's B2 medium (Ménézo, 1976); FCS= fetal calf serum; TV= trophoblastic vesicles.

trophoblastic vesicles was collected every 24 hr, Millipore-filtered (0.45 μm) and added to serum-supplemented B2 medium in a 1/1 vol. ratio. Each embryo was cultured in 500 μl of this "crossed-medium" without trophoblastic vesicles and the medium was changed every 24 hr during 3 days. Cleavage of 1- to 2-cell embryos was compared to that obtained during coculture with trophoblastic vesicles. Results are reported in Table V. A similar cleavage rate beyond the 16-cell stage was obtained when the embryos were cultured in crossed-medium (38.8%) or in coculture with trophoblastic vesicle (41.8%). This indicates that soluble components are released by trophoblastic vesicles into the culture medium and that these components are stable enough to remain active and promote the cleavage of early embryos *in vitro*. In order to characterize these released factors, different fractions of trophoblastic vesicle culture medium were separated and studied.

3.4. Molecular Weight Separation: Presumption that Active Compounds are not Macromolecular

Culture media, from 50 trophoblastic vesicles incubated 24 hr in B2 medium without serum albumin, were submitted to membrane ultrafiltration (UM10, Amicon Corp.). After washing 3 times with water, the macromolecular components (mol. wt > 10,000) were lyophilized, added to fresh B2 medium and tested for effects on *in vitro* cleavage of early embryos.

The ultrafiltrate was then concentrated and submitted to Sephadex G25 gel filtration. Elution was performed with 0.5% acetic acid. The molecular weight fraction 180-2500 was then collected. This fraction was also subjected to G15 Sephadex filtration (elution with 0.5% acetic acid in water) for more complete desalting. This low molecular weight fraction was lyophilized and its effect tested by addition to fresh B2 medium for *in vitro* culture of early stage embryos. After addition of these fractions to fresh B2 medium, the osmolarity was readjusted to 290 mOsmols using double distilled water.

Table V
In Vitro Development of 1- to 2-Cell Bovine Embryos

Culture conditions	No. of embryos	At end of culture (3-4 days)	
		No. (%) > 8-cell	No. (%) > 16-cell
Coculture with TV[a]	55	38 (69)	23 (41.8)
"Crossed medium"[b]	36	30 (83)	14 (38.8)
B2 + high mol. wt. fraction > 10,000	28	6 (21.4)	0 (0)
B2 + low mol. wt. fraction (180-2500)	21	13 (61.9)	5 (23.8)

[a]TV = trophoblastic vesicles.
[b]Crossed medium = 50% culture medium from TV + 50% serum-supplemented Ménézo's B2 (Ménézo, 1976).

A limited number of 1- and 2-cell bovine embryos has already been cultured in medium B2 (+ serum) supplemented either with high molecular weight fraction (n = 28 embryos) or low molecular weight fraction (n = 21 embryos) (Table V). After culture for 72 hr in medium supplemented with the high molecular weight fraction, only 21.4% of the embryos went beyond the 8-cell stage and none of them reached the 16-cell stage. In medium supplemented with the low molecular weight fraction, 61.9% of the embryos went beyond the 8-cell stage and 5 (23.8%) reached the morula stage *in vitro*.

These preliminary results indicate that active compounds released by trophoblastic vesicles into culture medium are of low molecular weight (180-2500). Small peptides might be involved as it was shown (see section 2.5) that at least 3 peptides were released by trophoblastic vesicles into the culture medium. However, this effect has still to be confirmed on a larger number of early embryos. Following this, further experiments need to be done using conventional biochemical techniques in order to isolate and determine the molecular structure of active compounds.

From these experiments it can be assumed that:

- coculture of early embryos with trophoblastic vesicles improves cleavage;
- this action is mediated *via* compounds released into the medium;
- serum addition to the culture medium increases efficiency.

It is not yet possible to determine whether the trophoblastic vesicle signal is a positive direct effect on early stage embryo cleavage or a "negative and indirect control" (Soto and Sonnenschein, 1984). But are embryo cleavage and cell multiplication analogous systems?

4. EVIDENCE FOR RELEASE BY TROPHOBLASTIC VESICLES OF SIGNALS FOR CORPUS LUTEUM FUNCTION

In cattle and in sheep, signals inhibiting luteolysis are undoubtedly produced by the embryo during its elongation phase, as shown by the experiments of embryo removal (Northey and French, 1980) or embryo transfer (Betteridge *et al.*, 1980). Using bovine and ovine trophoblastic vesicles, it was demonstrated that the embryonic disc is not necessary for this signal (Heyman *et al.*, 1984). Eleven- to 14-day-old embryos from which the embryonic disc was mechanically removed were capable of elongation *in utero* after transfer to cyclic cows and ewes. The life span of the corpus luteum was extended in 60% of the animals. This luteotrophic and/or antiluteolytic signal emitted by the trophoblastic vesicles may be due to released proteins since the same effect was obtained by Thatcher *et al.* (1985) who infused conceptus secretory proteins, isolated by dialysis from D15-D18 bovine conceptus culture media, into the uteri of 3 cyclic cows. These recipients exhibited cycles extended by 8 days compared to control cows.

This effect of trophoblastic vesicles on corpus luteum maintenance is presently used in our laboratory to try to improve embryo survival after transfer, especially in manipulated embryos (frozen or split) in which damage in membrane structure and function contribute to early embryonic mortality. To that end, a co-transfer experiment was made: one trophoblastic vesicle was added at the time of transfer to a frozen thawed whole blastocyst and both were deposited non-surgically into the uterine horn ipsilateral to the corpus luteum of recipient heifers. After transfer, pregnancy rate monitored by repeated rectal palpation was compared to that obtained after transfer of a single frozen thawed embryo without a trophoblastic vesicle. Preliminary results (Heyman, unpublished data) indicated that a 10% higher pregnancy rate was obtained by adding one trophoblastic vesicle to the frozen embryo at the time of transfer. After co-transfer, the confirmed pregnancy rate over 90 days was 56.6% *vs.* 44.4% in controls. It is interesting to note that D14 trophoblastic vesicles are active on D7 corpus luteum. This may be of economic interest for improving efficiency of embryo transfer in cattle.

The release of luteotrophic factors by bovine trophoblastic vesicles *in vivo* was also confirmed *in vitro* by Plante *et al.* (1985), who incubated trophoblastic tissue supernatant with rat granulosa cells and observed an increased progesterone production. This luteotrophic factor is not species-specific (Plante *et al.*, 1985; Martal *et al.*, 1984) and corpus luteum cells can be stimulated asynchronously by the trophoblastic vesicle signal. Moreover, these experiments point out the fact that subsequent viability after transfer of a manipulated embryo will depend upon its ability to emit various signals for pregnancy maintenance.

5. CONCLUSIONS

Coculture with trophoblastic vesicles or addition of trophoblastic vesicle culture medium has a beneficial effect on early bovine embryo cleavage. These observations seem different from those reported for other coculture systems.

Enhanced *in vitro* development of porcine embryos cocultured at various stages on porcine endometrial cell monolayers seems to be strictly dependent on a cell-embryo contact (Allen and Wright, 1984). When the embryos were isolated from the adjacent endometrial cells, there was no positive effect. Moreover, porcine endometrial cell supernatants were also ineffective. These authors, using the monolayer system, suggested that endometrial cell membrane projections needed to reach the surface of the embryonic cells, by penetrating the zona pellucida (see Chapter 12 [Ed.]). For Kuzan and Wright (1982), coculture of bovine morulae with bovine fibroblasts proved superior in supporting *in vitro* hatching. These authors used conditioned media to test the possibility that fibroblasts were either releasing a "factor" that promoted embryo hatching or were removing a toxic substance from the culture medium. The conditioned medium was ineffective.

These two papers do not exclude the possibility that an active factor, soluble but highly labile, may be released by cultured cell monolayers. Our experiments did not show the need for a cell-cell contact since conditioned media also improved early embryo cleavage. However, the necessity of adding serum in our coculture system makes it difficult to analyse the culture medium, in particular to determine the nature of the metabolic and/or specific effect.

Nevertheless, the possibility of culturing trophoblastic vesicles in a synthetic and defined medium, such as medium B2 without albumin, allows trophoblastic vesicle secretions to be examined quantitatively and qualitatively. However, this model is still not satisfactory:

- first of all, it includes two types of cells: trophoblastic and endodermic cells (the latter are probably necessary for continued growth of trophectoderm, as shown in the mouse by Ilgreen [1981]);

- secondly, this model cannot be compared with the whole embryo, since in trophoblastic vesicles the embryonic disc is removed and cannot exert any regulation on trophoblastic cells;
- thirdly, secretions of the trophoblastic vesicles *in vitro* may differ from those of an elongating blastocyst *in vivo* since *in vitro* the trophoblastic vesicle is not submitted to the same regulatory influences as *in utero*.

The trophoblastic vesicle is a complex system emitting various signals *in vitro* and *in vivo*. Trophoblastic vesicles should be used in the following sequence: *in vitro* fertilization, embryo culture and embryo transfer in bovine and perhaps in other species including human, for improving at least the last two steps. Further studies should be devoted to isolation, identification and determination of the activity of molecules involved in the observed interaction between trophoblastic vesicles and early embryos. This requires an ability to obtain *in vitro* pure trophoblastic cell lines.

ACKNOWLEDGEMENTS

We thank Annick Bouroche for text translation and Marie-Elisabeth Marmillod for typing the manuscript.

6. REFERENCES

Allen, R.L., and Wright, R.W., 1984, *In vitro* development of porcine embryos in coculture with endometrial cell monolayers or culture supernatants, *J. Anim. Sci.* 59: 1657-1661.

Betteridge, K.J., Eaglesome, M.D., Randall, G.C.B., and Mitchell, D., 1980, Collection, description and transfer of embryos from cattle 10-16 days after oestrus, *J. Reprod. Fertil.* 59: 205-216.

Boland, M., 1984, Use of the rabbit oviduct as a screening tool for the viability of mammalian eggs, *Theriogenology* 21: 126-137.

Buznikov, G.A., Sakharova, A.V., Manukhin, B.N., and Markova, L.N., 1972, The role of neurohumours in early embryogenesis. IV. fluorimetric and histochemical study of serotonin in cleaving eggs and larvae of sea urchins, *J. Embryol. Exp. Morphol.* 27: 339-351.

Camous, S., Heyman, Y., and Ménézo, Y., 1984, *In vitro* cleavage of cow embryos and further survival assessment *in vivo*, *Proceedings 10th Int. Congress on Anim. Reprod. and Artificial Insemination*, *Urbana*, Vol. II, brief communication no. 221.

Camous, S., Heyman, Y., Meziou, W., and Ménézo, Y., 1984, Cleavage beyond the block stage and survival after transfer of early bovine embryos cultured with trophoblastic vesicles, *J. Reprod. Fertil.* 72: 479-485.

Davis, D.L., and Day, B.N., 1978, Cleavage and blastocyst formation by pig eggs *in vitro*, *J. Anim. Sci.* 46: 1042-1053.

Glass, L.E., and Hanson, J.E., 1975, Transfer and localisation of maternal serum antigens by mouse preimplantation embryos and oviductal epithelium, *Differentiation* 4: 15-19.

Heyman, Y., Camous, S., Fèvre, J., Meziou, W., and Martal, J., 1984, Maintenance of the corpus luteum after uterine transfer of trophoblastic vesicles to cyclic cows and sheep, *J. Reprod. Fertil.* 70: 533-540.

Hobbs, J.G., and Kaye, P.L., 1985, Glycine transport in mouse eggs and preimplantation embryos, *J. Reprod. Fertil.* 74: 77-86.

Ilgreen, E.B., 1981, On the control of trophoblastic giant-cell transformation in the mouse: homotypic cellular interactions and polyploidy, *J. Embryol. Exp. Morphol.* 62: 183-202.

Kane, M.T., 1985, A low molecular weight extract of bovine serum albumin stimulates rabbit blastocyst cell division and expansion *in vitro*, *J. Reprod. Fertil.* 73: 147-150.

Kuzan, F., and Wright, R., 1982, Observations on the development of bovine morulae on various cellular and non-cellular substrata, *J. Anim. Sci.* 54: 811-816.

Lindner, G.M., and Wright, R.W., Jr., 1983, Bovine embryo morphology and evaluation, *Theriogenology* 20: 407-416.

Martal, J., Camous, S., Fèvre, J., Charlier, M., and Heyman, Y., 1984, Specificity of embryonic signals maintaining corpus luteum in early pregnancy in ruminants, *Proceedings 10th Int. Congress on Anim. Reprod. and Artificial Insemination, Urbana*, Vol. III, brief communication no. 510.

Ménézo, Y., 1976, Milieu synthétique pour la survie et la maturation des gamètes et pour la culture de l'oeuf fécondé, *C.R. Acad. Sci. D Paris* 282: 1967-1970.

Ménézo, Y., Renard, J.P., Delobel, B., and Pageaux, J.F., 1982, Kinetic study of fatty acid composition of Day 7 to Day 14 cow embryos, *Biol. Reprod.* 26: 787-790.

Ménézo, Y., Testard, J., Perrone, D., 1984, Serum is not necessary in human *in vitro* fertilization, early embryo culture, and transfer, *Fertil. Steril.* 42: 750-755.

Miller, J.G.O., and Schultz, G.A., 1983, Properties of amino-acid transport in rabbit preimplantation embryos, *J. Exp. Zool.* 228: 511-525.

Muggleton-Harris, A., Whittingham, D.G., and Wilson, L., 1982, Cytoplasmic control of preimplantation development *in vitro* in the mouse, *Nature (London)* 299: 460-462.

Northey, D.L., and French, L.R., 1980, Effect of embryo removal and intrauterine infusion of embryonic homogenates on the lifespan of the bovine corpus luteum, *J. Reprod. Fertil.* 12: 539-550.

Plante, C., Bousquet, D., Guay, P., Goff, A.K., and King., W.A., 1985, Luteotrophic factor secreted by bovine embryonic tissue, *Theriogenology* 23: 217 (abstr.).

Renard, J.P., du Mesnil du Buisson, F., Wintenberger-Torrès, S., and Ménézo, Y., 1976, *In vitro* culture of cow embryos from Day 6 and Day 7, in: *Egg Transfer in Cattle* (L.E.A. Rowson, ed.), Commission of the European Communities, Luxembourg, pp. 159-164.

Seidel, G.E., Jr., 1977, Short maintenance and culture of embryos, in: *Embryo Transfer in Farm Animals: A Review of Techniques and Applications* (K.J. Betteridge, ed.), Canada Dept. of Agriculture, Monograph 16, pp 20-24.

Soto, A.M., and Sonnenschein, C., 1984, Mechanism of estrogen action on cellular proliferation: evidence for indirect and negative control on cloned breast tumor cells, *Biophys. Biochem. Res. Commun.* 122: 1097-1103.

Soto, A.M., and Sonnenschein, C., 1985, The role of estrogens on the proliferation of human breast tumor cells (MCF7), *J. Steroid Biochem.* 23: 87-94.

Tervit, H.R., Whittingham, D.G., and Rowson, L.E.A., 1972, Successful culture *in vitro* of sheep and cattle ova, *J. Reprod. Fertil.* 30: 493-497.

Thatcher, W., Knickerbocker, J., Bartol, F., Bazer, F., Roberts, M., and Drost, M., 1985, Maternal recognition of pregnancy in relation to the survival of transferred embryos: Endocrine aspects, *Theriogenology* 23: 129-143.

Thibault, C., 1966, La culture *in vitro* de l'oeuf de vache, *Ann. Biol. Anim. Bioch. Biophys.* 6: 159-164.

Villalon, M., Ortiz, M.E., Aguayo, C., Munoz, J., and Croxatto, H.B., 1982, Differential transport of fertilized and unfertilized ova in the rat, *Biol. Reprod.* 26: 337-341.

Whittingham, D.G., and Bavister, B.D., 1974, Development of hamster eggs fertilized *in vitro* or *in vivo*, *J. Reprod. Fertil.* 38: 489-492.

Whittingham, D.G., and Biggers, J.D., 1967, Fallopian tube and early cleavage in the mouse, *Nature (London)* 213: 942-943.

Wright, R.W., Jr., and Bondioli, K.R., 1981, Aspects of *in vitro* fertilization and embryo culture in domestic animals, *J. Anim. Sci.* 53: 702-729.

Wright, R.W., Anderson, G.B., Cupps, P.T., and Drost, M., 1976, Blastocyst expansion and hatching of bovine ova cultured *in vitro*, *J. Anim. Sci.* 43: 170-174.

CHAPTER 10

IN VITRO GROWTH OF PREIMPLANTATION RABBIT EMBRYOS

MICHAEL T. KANE

1. INTRODUCTION

Culture of mammalian preimplantation embryos *in vitro* has for a long time been regarded as an important tool in the study of early embryonic development. Embryo culture itself and micromanipulation techniques dependent on it, such as embryo microdissection and chimaera production, have greatly increased our knowledge of developmental processes. Recent advances involving the microinjection of foreign DNA into mouse embryos and its subsequent incorporation and expression show exciting possibilities for elucidating the control of genes in mammalian embryos. This progress is awakening renewed interest in the culture of embryos from species other than the mouse.

Almost all early culture work was carried out on the rabbit embryo. These studies were facilitated by the fact that, before techniques for superovulation of prepubertal animals were developed, timing of egg stage and recovery was easier in an induced ovulator such as the rabbit than in a species with an estrous cycle such as the mouse. Research in this area began in Belgium with Brachet (1913), who cultured rabbit blastocysts. [See Chapter 1 (Ed.).]

The early work with rabbit embryos, before the advent of semi-defined and defined culture media (see below), was carried out with various "natural" media. One-cell and/or cleavage-stage rabbit embryos were cultured by Lewis and Gregory (1929) in chicken plasma and rabbit plasma, by Pincus (1930) in clotted rabbit plasma and chick embryo extract, by Pincus (1941a) in rabbit serum, and by Chang (1948, 1949) in blood sera of various animals. The degree of success attained in early culture work is not completely clear. However, it appears that 1-cell and early cleavage stages developed to the morula stage in plasma or serum (Lewis and Gregory, 1929) and that embryos collected at late

Michael T. Kane Physiology Department, University College, Galway, Ireland.

morula and early blastocyst stages continued to develop and blastocyst expansion occurred, although at a slower rate than *in vivo* (Pincus and Werthessen, 1938). No firm data reporting the growth of early cleavage stages to the blastocyst and expanding blastocyst stages were reported until Onuma *et al.* (1968) found that 2- to 8-cell embryos could be cultured to the expanding blastocyst stage in rabbit or bovine serum. Subsequently, Maurer *et al.* (1969) showed that 1-cell rabbit embryos could be cultured to the blastocyst stage in bovine serum. This was rapidly followed by the development of a complex semi-defined medium that allowed development of 1- to 4-cell embryos to the blastocyst stage (Kane, 1969; Kane and Foote, 1970a, 1971). [See also Chapter 1 (Ed.).] These results showed clearly that, unlike the case of the mouse and certain other species, there is no 2-cell or early cleavage block present in the cultured rabbit embryo.

A major goal of researchers in this area has been the development of strictly defined, reliable and reproducible culture media. Perhaps full attainment of this goal is impossible because most laboratory chemicals are contaminated with trace amounts of other chemical compounds. The implications of this problem for defined media are discussed by Ham (1981). In the present review, the term "defined medium" is restricted to media prepared from bench chemicals; macromolecular components (if present) consist of synthetic substances such as polyvinylalcohol (Bavister, 1981). The term "semi-defined medium" is used for media prepared from bench chemicals to which albumin or other protein has been added as a source of macromolecules. A medium containing any proportion of serum is not considered to be either defined or semi-defined.

The discovery by Whitten (1957) that 2-cell mouse embryos could be cultured to blastocysts in a semi-defined medium composed of a simple salt solution, with bovine serum albumin as the only macromolecule and lactate or pyruvate as an energy source, dramatically altered the culture of mammalian preimplantation embryos. The obvious advantages of a simple culture medium, together with the ease of superovulation techniques in prepubertal mice and the relative cheapness of the mouse as a source of ova, meant that most workers concentrated on the mouse embryo. This review will show why culture of the rabbit embryo is still relevant to the study of the control of preimplantation development and will outline the specific requirements for rabbit embryo culture *in vitro*. The culture of rabbit embryos has been specifically reviewed by Maurer (1978).

2. RELEVANCE OF RABBIT EMBRYO CULTURE TO STUDIES OF EMBRYONIC DEVELOPMENT

2.1. Comparison of Rabbit and Mouse Preimplantation Embryo Development

The major point to consider when comparing the development of the rabbit embryo with that of the mouse is the relatively small degree of growth of the mouse embryo that takes place before implantation in comparison with the much greater growth of the rabbit embryo.

This is true whether one looks at blastocyst cell count, diameter or protein content. At implantation, the mouse blastocyst comprises about a hundred cells (Bowman and McLaren, 1970), the total protein content is about

20 to 22 ng (Brinster, 1967a) and the diameter is < 200 μm. Remarkably, the mouse blastocyst has less total protein than the 1-cell embryo. In striking contrast, the rabbit blastocyst just before implantation has about 80,000 cells (Daniel, 1964), a diameter up to 7000 μm (Alliston and Pardee, 1973) and a protein content of about 200 μg (Lutwak-Mann, 1971). This enormous difference in the preimplantation growth pattern has several important consequences: (1) It explains many of the known differences in culture requirements between mouse and rabbit embryos, since a tissue which is gaining in mass must have increased requirements for development as compared with one that is decreasing in mass; (2) It explains some of the great differences in metabolism between rabbit and mouse blastocysts, *e.g.*, there is a 5500-fold increase in glucose oxidation from fertilization to implantation in the rabbit embryo as compared with only 60-fold for the mouse embryo (Brinster, 1967b, 1968); (3) It indicates that for some aspects of embryonic development the rabbit may be a more fruitful embryonic model than the mouse. This applies particularly to the possible involvement of endogenous or exogenous growth factors in early embryonic development (see Chapter 8 [Ed.]). It also means that the amounts of tissue available for biochemical analysis are much greater for rabbit than for mouse blastocysts.

2.2. Ease of Culture From the 1-Cell Stage

There is little difficulty in culturing 1-cell rabbit embryos to the morula or early blastocyst stages. There is no 1-cell block *in vitro* as there is for embryos of certain strains of mice and for other species (Maurer *et al.*, 1969; Kane, 1972; Kane and Foote, 1971). This should facilitate microinjection studies in which embryos are to be cultured subsequently.

3. HOW DOES GROWTH OF CULTURED RABBIT EMBRYOS COMPARE WITH GROWTH *IN VIVO*?

Before discussing the growth requirements of cultured rabbit embryos, it is useful to consider to what extent growth *in vitro* achievable by current culture techniques compares with normal growth *in vivo*. Figure 1 provides a comparison of the time-scale of embryo development *in vivo* and *in vitro* in terms of cell numbers.

3.1. Growth *in Vivo*

The following account of *in vivo* growth is summarized from reviews by Austin and Walton (1960) and Blandau (1961) and the work of Alliston and Pardee (1973). The paper of Alliston and Pardee is particularly useful in that it provides photographs of all stages from the 1-cell embryo to the fully expanded blastocyst. If the time of copulation is taken as 0 hr, then ovulation occurs at 9 1/2 to 10 hr and the first cleavage occurs at about 21 to 28 hr. The morula stage starts to form at about 48 hr and the blastocyst at about 70 hr. During its passage through the oviduct, a mucin coat is laid down around the zona pellucida and at the morula stage this coat is usually much thicker than the zona. At the blastocyst stage, the embryo enters the uterus and blastocyst expansion begins. The inner cell mass may be seen at 96 hr. The embryonic disc is visible at 5 1/2 days. At implantation, which starts at about 6 1/2 days,

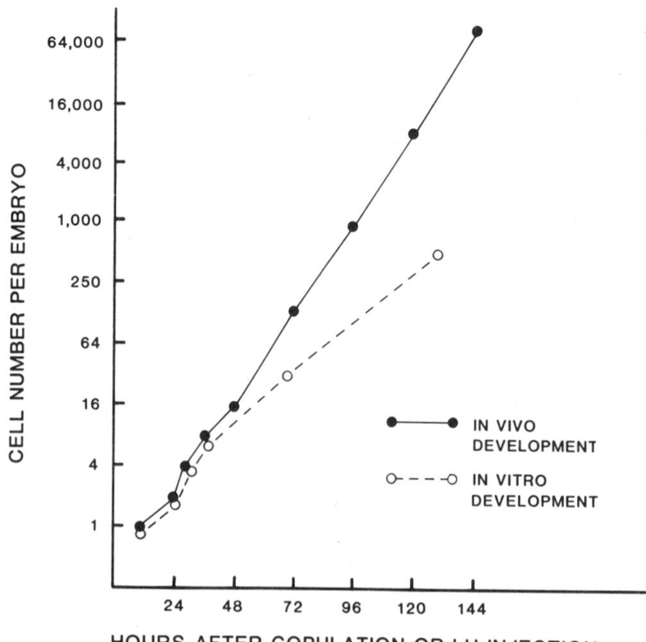

Figure 1. Relationship of cell number of rabbit embryos to time after copulation or LH injection and insemination, for embryos grown *in vivo* or cultured from the 1-cell stage *in vitro*. Information for *in vivo* stages taken from review by Blandau (1961) and from Daniel (1964). Data for *in vitro* culture from Kane (unpublished data). Graph shows (1) optimal development obtainable *in vitro*; generally, embryos cultured from the 1-cell stage do not develop much beyond about 500 cells in media presently available. Embryos cultured from morula stage can develop up to 2000 cells. (2) Comparative data for development *in vivo*.

the blastocyst may be 5 to 7 mm in diameter and contains about 80,000 cells (Daniel, 1964). Due to expansion of the blastocyst, the zona pellucida and the mucin coat are thinned even though there is a further mucin layer laid down in the uterus (the "gloiolemma" of Böving, 1957). There is also evidence of extensive remodelling of the embryonic coverings in the uterus (Denker and Gerdes, 1979). It is clear that blastocysts *in utero* do not hatch or shed the zona and mucin coat but that there is erosion of the blastocyst coverings (Enders, 1971; Denker, 1982).

3.2. Growth *in Vitro*

In vitro culture readily allows development from the 1-cell stage to the start of blastocyst expansion (Fig. 2), but further blastocyst growth is usually limited. An ultrastructural study of rabbit embryos grown *in vivo* and *in vitro* showed no observable differences over the first 4 days of development (Van Blerkom *et al.*, 1973). Embryos cultured from the 1-cell stage usually start blastocyst formation after about 3 days in culture and then start to expand. Expansion is usually limited with currently available techniques. In the case of embryos cultured from the 1-cell stage for 4 days (Kane, 1972), mean blastocyst diameter in the optimal medium was 483 μm; although individual

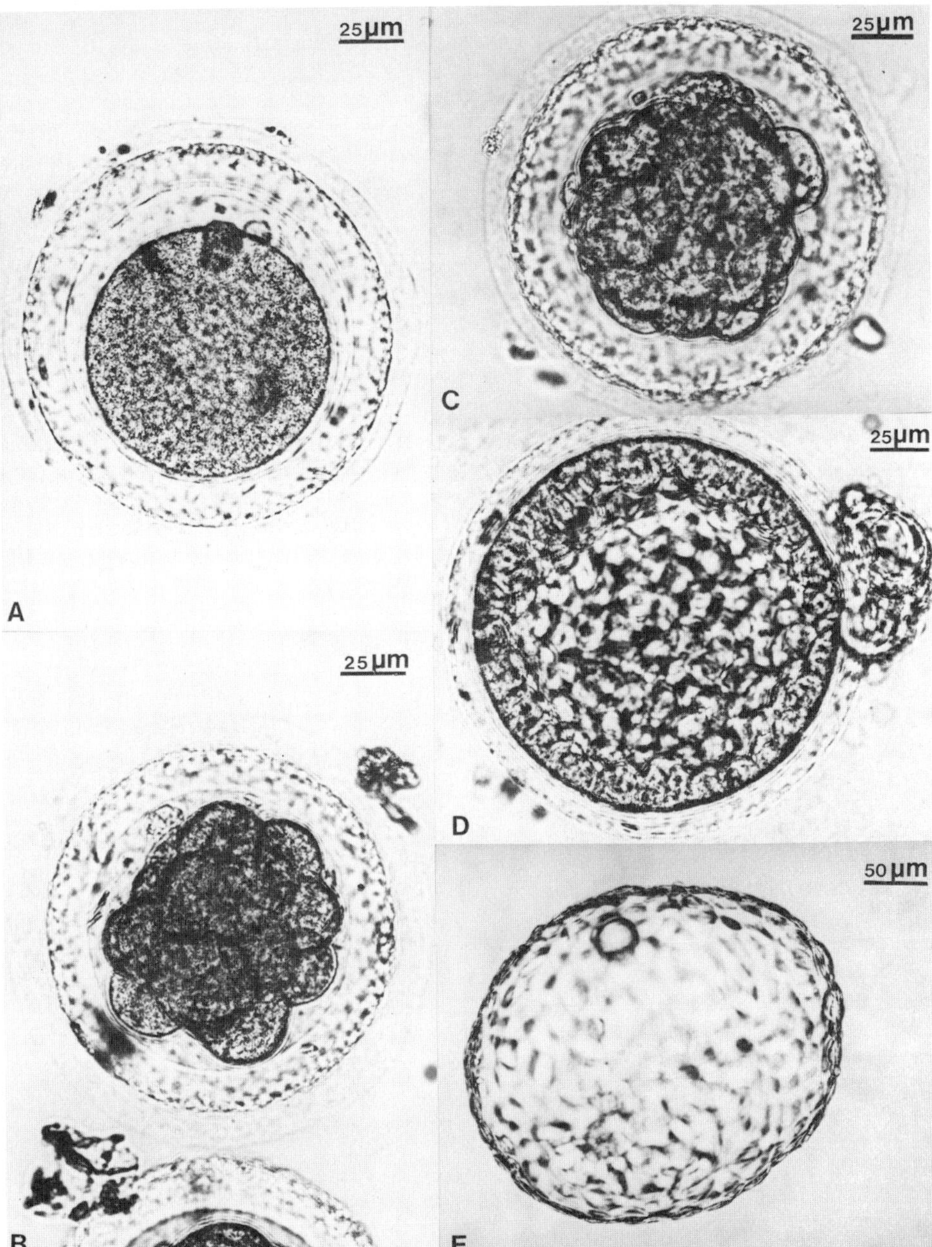

Figure 2. Rabbit embryos at different stages after culture *in vitro* for varying periods of time in a semi-defined medium containing 0.5% BSA (see Table V for composition of medium). Culture started at the 1-cell stage in each case. A: One-cell embryo at the start of culture (embryo collected 20 hr after LH injection). B: Sixteen cell embryo starting to compact to form an early morula after 24 hr culture. C: Late morula after 48 hr culture. D: Expanding blastocyst starting to hatch from the zona pellucida after 72 hr culture. E: Completely hatched blastocyst after 96 hr culture.

blastocysts attained diameters of 600 or 700 μm, this is still very much short of the development obtained *in vivo* (Fig. 1). Expanding blastocysts cultured from the 1-cell stage usually hatch at an early stage of expansion (Kane, 1972, 1975a). Kane (1975a) has provided evidence that this is due to the fact that blastocysts cultured from the 1-cell stage have only a very thin mucin coat. Blastocysts cultured in semi-defined medium from the morula stage (with a thick mucin coat) do hatch from the combined zona-mucin coat but only after a greater degree of expansion has occurred and in smaller numbers (Kane, 1975a). In general, blastocyst hatching is associated with an increase in cell number (Kane, 1983a) and is probably due to rupture of the blastocyst coats caused by blastocyst expansion (Kane, 1983b) and not to protease action. It is the author's experience that at this stage particularly, the strong zona-mucin coat complex, which (in cultured embryos) has not been softened by uterine enzymes, limits growth unless hatching takes place.

Blastocysts cultured from the morula stage in a semi-defined medium can show a considerable degree of expansion and increase in cell number. Kane (1985) has reported a mean blastocyst diameter of 457 μm and a cell count of 1128 after 4 days of culture in the optimal treatment. Individual blastocysts had diameters > 1200 μm with > 2000 cells. This is still greatly inferior to development *in vivo* but nevertheless indicates some progress in culturing blastocyst stages. One problem, however, is that *in vitro* growth depends very much on the quality of the bovine serum albumin (BSA) in the medium, which can vary considerably from batch to batch (Kane, 1983a). Until this problem is solved (see section 4.8, *An Embryotropic Factor from BSA*), culture at the blastocyst stage remains difficult. It is interesting in this regard that culture of 1-cell rabbit embryos in oviducal fluid collected at 8 to 9 days *post coitum* was very successful (Kille and Hamner, 1973): blastocysts frequently attained diameters of 3 to 4 mm.

Reports of transfer of embryos to recipient does, after long term culture in defined or semi-defined medium, are very sparse. Seidel *et al.* (1976) fertilized rabbit oocytes *in vitro* and cultured them in semi-defined medium for 72 hr to morulae or early blastocysts and then transferred them to recipient does; live young were born. However, similar embryos that were cultured for 84 hr did not result in birth of young. Maurer (1978) transferred embryos cultured from the 2- to 4-cell stage in semi-defined medium for 96 hr and found a small percentage of viable fetuses at 28 days.

One problem that arises in transfer studies is that it is very likely that hatched blastocysts which have completely shed the zona and mucin coat are not viable either in the oviduct or the uterus, possibly because of the action of the maternal immune system or of phagocytic cells. This is an area that needs investigation, particularly to find whether an artificial "mucin coat" could be formed to protect the embryo.

4. *IN VITRO* GROWTH REQUIREMENTS OF RABBIT EMBRYOS

The requirements for growth of rabbit embryos from 1-cell to morula are extremely simple, while requirements for development to the blastocyst stage and for blastocyst growth are much more complex. The known requirements for rabbit embryo growth are considered here under the headings of major inorganic salt constituents, pH, osmolarity, energy sources, amino acids, vitamins and cofactors, trace elements, nucleic acid precursors, hormonal factors, macromolecules and uterine factors.

4.1. Major Inorganic Salt Constituents, pH, and Osmolarity

There is a lack of information for the rabbit embryo regarding requirements for the major salt constituents Na, K, Ca, Mg, Cl, PO_4, and SO_4. The ionic composition of culture media in current use for rabbit embryos is based on Brinster's (1963) mouse medium with adjustments in osmolarity. It has been shown (Naglee *et al.*, 1969) that 2- to 4-cell rabbit embryos will develop to blastocysts in media with osmolarity varying from 230 to 339 mOsmols, with an optimum at about 290 mOsmols.

The main role of HCO_3^- in culture media is to function as part of a pH buffer system with CO_2, but it has a secondary role in 1-carbon metabolism and the formation of Krebs cycle metabolites. One-cell rabbit embryos can develop into morulae in the absence of HCO_3^- and CO_2 in a HEPES-buffered medium but development to the blastocyst stage requires HCO_3^- (Kane, 1975b), unlike the mouse embryo, in which cleavage is extremely limited in the absence of HCO_3^- (Quinn and Wales, 1973). This difference may be due to the larger mass of the rabbit embryo. The requirement for HCO_3^- at the blastocyst stage is probably due to an increased need for synthesis of Krebs cycle intermediates as the mass of the blastocyst increases.

In a bicarbonate-buffered medium, 1-cell rabbit embryos will develop to blastocysts over the pH range 6.64 to 7.91 (Kane, 1974). There is some indication that the optimum pH for rabbit blastocyst expansion (7.6) is higher than that for development to the blastocyst stage (7.3). It has been reported that the pH of rabbit uterine fluid may be as high as 7.69 (McLachlan *et al.*, 1970) or 7.9 (Vishwakarma, 1962), possibly due to high levels of HCO_3^-.

4.2. Energy Sources

The energy requirements for cleavage of the 1-cell mouse embryo can only be supplied by two substrates, pyruvate and oxaloacetate (Biggers *et al.*, 1967). Requirements for the rabbit are clearly different. One-cell rabbit embryos have been cultured to expanding blastocysts in a complex medium containing BSA, amino acids, vitamins and trace elements but without carbohydrate-type energy sources (Kane, 1972). However, the addition of pyruvate did increase blastocyst expansion. The development in the absence of carbohydrate-type energy sources was partly explained by the finding that fatty acids bound to BSA could act as energy sources for rabbit embryos (Kane, 1979; Kane and Headon, 1980). A wide range of fatty acids, both long- and short-chain, can fulfill this role for the rabbit embryo in the presence of defatted charcoal-treated BSA (Kane, 1979). Kane (unpublished data) recently examined the question of the energy substrates that will support cleavage of 1-cell rabbit embryos to the morula stage in a simple defined medium containing Krebs-Ringer bicarbonate-type salts supplemented with a synthetic macromolecule, polyvinylalcohol (PVA). The embryos were stained at the end of the culture period by the method of Grayson (1978) and the number of cells in each embryo was counted. The most remarkable finding was that, even in the complete absence of any exogenous energy source, up to 3 or 4 cleavage divisions took place; this is indicative of the substantial energy reserves of the rabbit embryo as compared with the mouse embryo. Of the substrates tested separately, only pyruvate, glucose, a group of 20 amino acids and ordinary non-defatted BSA, gave a significant increase in development and allowed growth to the early morula stage. Several other substrates, including oxaloacetate, malate, lactate, phosphoenolpyruvate and acetate, had smaller and non-significant effects on cleavage. These results are in contrast with a much earlier study by Daniel (1967a) who found that there was no growth of 1-cell embryos in Ham's F10 in the absence of either pyruvate, phosphoenolpyruvate or lactate. The discrepancy in these results may be due to the fact that the absence of a macromolecule from Daniel's culture medium may have restricted growth. [See Chapter 1 for formulation (Ed.).]

4.3. Amino Acids

Amino acids or a nitrogen source are not essential for cleavage of rabbit ova as I have found (Kane, unpublished data) that 1-cell rabbit embryos will cleave in a simple defined medium without amino acids and with PVA as the only macromolecule. This is in agreement with an earlier report (Kane, 1979) that 1-cell embryos cleaved to the morula stage in a semi-defined medium without amino acids but with charcoal-treated fatty acid-free BSA as the only macromolecule. These results are in contrast to the report of Daniel and Olson (1968) who found that certain amino acids were essential for early cleavage of rabbit ova. The difference in results may arise from the absence of macromolecules from the medium used by Daniel and Olson. In the absence of macromolecules, amino acids may have served to chelate traces of heavy metal ions.

There is, however, an absolute requirement for amino acids for the development of cleavage stage rabbit embryos to blastocysts (Kane and Foote, 1970a). When 2- to 4-cell rabbit embryos were cultured either in a complex medium based on Ham's F10 containing amino acids, vitamins, trace elements and nucleic acid precursors or in media with one of those nutrient groups omitted, the amino acid group was the only one whose omission completely abolished all blastocyst development (Table I). Of the 20 amino acids tested by Kane and Foote (1970a), the most important for blastocyst formation appeared to be methionine, serine, and threonine.

Daniel and Krishnan (1967) studied the effects of the amino acids of Ham's F10 on expansion of 5-day rabbit blastocysts and found that 10 amino acids, *i.e.*, arginine, lysine, histidine, tryptophane, methionine, phenylalanine, leucine, valine, threonine and serine were indispensable for growth; the rest were non-essential. It seems logical that in formulating culture media for rabbit blastocysts, a full complement of the usual 20 amino acids should be used, possibly together with some of the other amino acids found in the reproductive tract and used in Ménézo's B-2 medium (Ménézo, 1976).

4.4. Vitamins and Cofactors

Perhaps the earliest report of the beneficial use of a vitamin in embryo culture was the report of Pincus (1941b) showing that thiamine added to serum stimulated rabbit embryo growth. Although the vitamins are not necessary for

Table I
Effect of Omission of Either Amino Acids, Vitamins, Trace Elements or Nucleic Acid Precursors from a Complex Tissue Culture Medium on Growth of 2- to 4-Cell Rabbit Embryos to Blastocysts[a]

Treatment	No. of embryos	% Early blastocysts	% Expanding blastocysts
Basic medium[b]	57	83	42
No amino acids	57	0	0
No vitamins	57	54	5
No trace elements	57	79	39
No nucleic acid precursors	57	81	63

[a]Modified from Kane and Foote (1970a).
[b]The basic medium contained 1.5% BSA, Krebs-Ringer bicarbonate salts, glucose and the amino acids, vitamins, trace elements and nucleic acid precursors of Ham's F10 medium.

the development of early cleavage stage rabbit embryos up to the start of blastocyst formation, blastocyst expansion is extremely limited in the absence of vitamins (Kane and Foote, 1970a; and see Table I). Daniel (1967b) found that rabbit blastocysts collected at the 5-day stage required the following vitamins and cofactors for growth: inositol, pyridoxine, riboflavin, thiamine, niacinamide and folic acid. Choline chloride, vitamins B_{12}, calcium pantothenate, biotin, ascorbic acid, ergosterol and α-tocopherol were not required. It is of interest here to remember that Lutwak-Mann (1959) found that, of the B vitamins, nicotinic acid and B_{12} were present in greatest amounts in 6-day rabbit blastocysts and that thiamine, folinic acid and riboflavin were present in much smaller amounts. Inositol and choline were not detected. She also found that B_{12} and nicotinic acid were particularly rich in the endometrium of pregnant rabbits and that their concentration was under progestational control.

4.5. Trace Elements

There is very little information on trace element requirements of rabbit embryos. Kane and Foote (1970a) found that omission of Fe, Cu and Zn did not impair development of 2- and 4-cell embryos to the blastocyst stage. As mentioned already, 1-cell rabbit embryos will cleave to the morula stage in a very simple defined medium containing only PVA, the Krebs-Ringer bicarbonate salts and an energy source (Kane, unpublished data). However, Daniel and Millward (1969) reported that the omission of ferrous ions from Ham's F12 medium completely abolished cleavage of rabbit embryos and caused collapse of rabbit blastocysts. Again, the difference in results may be due to the use of macromolecule-free F12 by Daniel and Millward. The question of trace element requirements is complicated by the contamination of macromolecules

and even of the highest quality grades of salts (such as NaCl and KCl) with trace elements. It would seem reasonable that, at least in the case of rabbit blastocyst growth, in which tissue mass is increasing, that trace elements should be routinely added.

4.6. Nucleic Acid Precursors

There is evidently no requirement for nucleic acid precursors for cleavage of early stage rabbit embryos or for development to the blastocyst stage (Kane and Foote, 1970a; and see Table I). On the contrary, the omission of the nucleic acid precursors thymidine and hypoxanthine as a group improved blastocyst formation and expansion. It was later confirmed (Kane and Foote, 1971) that thymidine had a slight toxic effect on blastocyst formation but that hypoxanthine had no effect. However, Daniel (1967b) found that expansion of blastocysts collected at the 5-day stage was promoted by the addition of hypoxanthine.

4.7. Hormonal Factors

It is surprising that, given the multitude of reproductive hormones, it is very difficult to find a single well-documented direct growth promoting effect of a known hormone on embryos. A careful review by Warner (1977) found little evidence for beneficial direct effects of steroid hormones on preimplantation embryos. There was, however, evidence for deleterious direct effects at high concentrations of hormones. It seems that the beneficial effects of hormones such as progesterone on the embryo are exerted indirectly by their effects on the reproductive tract. The addition of a "cocktail" of hormones and growth factors including glucagon, thyrocalcitonin, transferrin, triiodothyronine, insulin, parathyroid hormone, liver cell growth factor, LRH, FSH and LH to the culture medium did not stimulate rabbit blastocyst growth (Kane, unpublished data). [See also Chapter 6 (Ed.).]

4.8. Macromolecules

General Role of Macromolecules. It is generally agreed that there is a macromolecular requirement for all embryo culture and this requirement is usually filled by the addition of protein, most commonly BSA. Deletion of BSA from a complex medium resulted in failure of 2- and 4-cell embryos to progress to the blastocyst stage (Kane and Foote, 1970b). It is not easy to dissect out the reasons for this requirement, which probably vary somewhat with stage and species of embryo and other medium constituents. One reason may be a presently ill-defined physical effect on the electrical surface charge or stickiness of embryos or of blastomeres. This effect may be related to the very great difficulty in handling embryos in macromolecule-free media due to their tendency to stick to glass and plastic surfaces.

Bavister (1981) introduced the use of a synthetic polymer, PVA, as a macromolecule for the culture of mammalian sperm. In our experience, PVA has proved to be extremely useful for the handling of embryos in protein-free medium. Embryos in medium containing PVA do not become sticky and hard to handle as they do in macromolecule-free media or in media containing other synthetic polymers such as polyvinylpyrrolidone or ficoll. PVA is also useful for culture of 1-cell rabbit embryos to early morulae but BSA is necessary for culture to the blastocyst stage.

A second reason for macromolecular requirements may be a protective effect produced by some types of macromolecules, most notably BSA with its great ligand-binding abilities against toxic components of the medium. These toxic effects could arise from normal medium components such as amino acids or vitamins at unfavourable levels; from contaminants present in the water or associated with other constituents of the medium; from the plastic culture dish; or from the paraffin oil used to cover droplets of medium. [See Appendix I (Ed.).] Hence changes in the method of culture or medium constituents could alter macromolecular requirements.

A third reason, which is potentially the most interesting, is that small molecules bound to BSA may have marked effects on embryonic growth. The small-molecule binding properties of albumin are well known (Goodman, 1958; Westphal, 1970). Kane (1979) showed that fatty acids bound to albumin could act as energy sources for growth of 1-cell embryos to the morula stage. This was later confirmed for blastocyst growth by Kane and Headon (1980), who also found that a BSA contaminant other than a fatty acid appeared to cause increased blastocyst expansion and blastocyst hatching. Efforts to extract this factor with chloroform were unsuccessful. The importance of variability in commercial samples of BSA due to contaminants was highlighted by the finding that, following the use of two batches of BSA from the same supplier, one batch resulted in a high proportion of completely hatched blastocysts whereas the second batch gave no hatched blastocysts and less than half the numbers of cells per blastocyst (Kane, 1983a).

An Embryotropic Factor from BSA. Decisive proof that the growth-promoting or embryotropic factor in BSA is a low molecular weight compound was provided by extraction of BSA with formic acid (Kane, 1985). BSA was dissolved in 5% formic acid, then filtered through a membrane filter with a 10,000 mol. wt. cut-off and freeze-dried (subsequent unpublished work has shown that the factor will pass through a 1,000 mol. wt. filter). The yield of extracted material was about 10 mg/g of BSA. Addition of this extract at a concentration of 0.2 mg/ml (Table II) to a complex semi-defined medium containing charcoal-treated BSA resulted in an 8-fold increase in cell number and more than a 2-fold increase in diameter in blastocysts cultured from the morula stage. There was some indication that either the extraction process did not extract all the embryotropic activity or (perhaps more likely) that some activity was destroyed by the extraction process. In subsequent work, extraction through a hollow fiber cartridge has given a higher yield of activity.

When the extract was chromatographed on a Sephadex G-10 column using 2.5% formic acid as the eluent, the embryotropic activity was eluted at about the total bed volume, suggesting the factor was of very low molecular weight. However, on using 1 M NH_4OH, the embryotropic activity eluted at about the void volume, indicating a much higher molecular weight. The explanation for this difference is that, while in principle G-10 separates on a mol. wt. basis (varying from 0-700), in practice this separation behaviour can be greatly modified by the presence of aromatic groups in the compounds

Table II
Effect of Low-Molecular Weight Extract of BSA on Rabbit Blastocyst Formation, Expansion and Cell Division[a]

		Basic medium containing:				
	Basic medium alone[b]	Low molecular weight extract (mg/ml)			Unextracted BSA (mg/ml)	
		0.04	0.2	1.0	1	5
No. of embryos	51	50	48	49	56	57
% Blastocysts	100	100	98	98	96	96
Embryo cell count (mean ± SEM)	99 ± 7	210 ± 29	807 ± 65	806 ± 55	316 ± 55	1128 ± 122
Embryo diameter (μm, mean ± SEM)	155 ± 7	191 ± 9	349 ± 13	328 ± 14	214 ± 12	457 ± 27

[a]Data modified from Kane (1985).
[b]The basic medium was a complex tissue culture medium containing Krebs-Ringer bicarbonate type salts, trace elements, amino acids, vitamins, pyruvate, glucose and 0.5% charcoal-treated BSA.

being separated. Under certain circumstances, which depend on the pH of the eluent, compounds with aromatic groups are retarded and appear late in the elution profile. Based on these results, we (Kane and Gray, unpublished data) concluded that the embryotropic factor(s) was acidic with an aromatic group. We then chromatographed the active G-10 fraction on a QAE-A25 Sephadex anion exchanger; on this column the activity was not eluted until the pH was reduced to 2.2, confirming the acidic nature of the material. This column gave two peaks of activity, indicating either the existence of two or more compounds with embryotropic activity or the presence of monomers and dimers. This material is currently being purified by HPLC with a view to its chemical identification. Analysis of the active G-10 fraction and the QAE-A25 fractions have shown the presence of peptidic material. However, prolonged digestion with pronase did not destroy the embryotropic activity, although this does not eliminate the possibility of the material being a somewhat unusual peptide.

These results are exciting because: (a) it seems that the factor is a *sine qua non* for any marked degree of rabbit blastocyst growth. It has not been possible to substitute for it the very complex tissue culture medium, MCDB 104. (b) If the factor is hormone-like, it would be the first evidence for a hormone having a direct beneficial effect on a preimplantation embryo. (c) It would seem that the factor has interspecies activity since it is extracted from a bovine protein and stimulates growth of rabbit blastocysts. Another point of potential interest is the possibility that growth factors for early embryos and for cancer cells may be related (see Chapter 8 [Ed.]).

Enzymes as Exogenous Factors. The role of uterine and blastocyst proteases in the erosion of the blastocyst coverings *in vivo* has been discussed by Denker and Gerdes (1979) and by Denker (1982). Onuma *et al.* (1968) showed that pretreatment of early cleavage-stage rabbit embryos with pronase followed by culture in pronase-free serum allowed hatching of embryos at the blastocyst stage and increased expansion. Kane (1983b) showed that the addition of as little as 20 ng/ml of trypsin to a semi-defined culture medium caused blastocyst expansion and hatching. However, these blastocysts rapidly degenerated after hatching. A survey of the effects of a range of commercially available proteases and carbohydrases on culture of morulae to blastocysts indicated that none had any growth promoting effects and just two (trypsin and pronase) caused blastocyst hatching (Kane, unpublished data). The effect on hatching appeared to be due to the weakening of the zona by the protease rather than to a growth promoting effect.

4.9. Uterine Factors

The influence of uterine factors on the culture of rabbit embryos has in the past been a controversial subject. The discovery of a distinctive protein from the rabbit uterine fluid, called "blastokinin" by Krishnan and Daniel (1967) and "uteroglobin" by Beier (1968), was pioneering work which evoked a great surge of interest in the role of uterine fluid in preimplantation embryonic growth. However, the culture evidence which led to the coining of the term blastokinin and the claim that blastokinin functioned as an endogenous regulator of blastocyst development (Krishnan and Daniel, 1967) is not convincing. A careful examination of the Krishnan and Daniel (1967) paper shows that maternal serum proteins also allowed cavitation (the start of blastulation) and some degree of blastocyst expansion. The beneficial effect of blastokinin on blastocysts then reduces to an effect on the degree of blastocyst expansion. It is clear from the photographic evidence (Krishnan and Daniel, 1967) that blastocysts cultured in either maternal serum proteins or uterine fluid components do not differ greatly in size and that in both cases, blastocysts are much smaller than those grown *in vivo*. Unfortunately, no quantitative information was given on blastocyst size, the numbers of embryos per treatment were very small and there was no statistical analysis of data. In later work, examination of the effects of uterine functions on quantifiable parameters such as blastocyst diameter and uptake of protein and nucleic acid precursors showed no statistically significant effect of blastokinin (Daniel, 1971a; El-Banna and Daniel, 1972). The growth of 1-cell (Kane, 1972), 2- to 4-cell (Naglee *et al.*, 1969) and morula-stage (Kane, 1975a, 1983a) embryos to expanded blastocysts in media without uterine components but with BSA has been demonstrated in quantitative terms.

A careful exploration of the effects of uterine proteins in promoting growth of rabbit blastocysts *in vitro* (Maurer and Beier, 1976) showed that unfractionated uterine protein had a greater stimulatory effect on growth than did any specific uterine protein fraction, and that BSA appeared to be only marginally less effective than the unfractionated uterine protein. Because of the relative large quantities of uteroglobin or blastokinin present in uterine fluid on days 5, 6 and 7, it seems unlikely that uteroglobin functions as a maternal signalling factor. Hence the suggestion (Mukherjee *et al.*, 1980, 1982) that uteroglobin and β_2-microglobulin are translinked by uterine transglutaminase to mask the antigenicity of the developing trophoblast and embryo is extremely interesting.

While much attention has been focused on the role of uterine proteins, the subject of low molecular weight growth factors in uterine fluid has been left unexplored. The effect of such factors on the rabbit embryo might well repay investigation (see section 4.8, *An Embryotropic Factor from BSA*).

5. METHODOLOGICAL CONSIDERATIONS

Culture of rabbit embryos entails a series of procedures: superovulation, embryo collection, the culture procedures themselves and evaluation of embryos after culture. Most of these procedures have been adequately documented in the literature and this review will merely cite these references and add some comments and details.

5.1. Superovulation Procedures

The optimal superovulatory procedure probably involves FSH/LH or FSH/hCG. The time and dose rate schedule for gonadotropin injection and insemination has been given (Kennelly and Foote, 1965; Varian *et al.*, 1967). There is however an error in the Kennelly and Foote paper with respect to the dose of FSH, which is corrected in the paper of Varian *et al.* (1967). Artificial insemination is most conveniently carried out at the time of LH injection (Adams, 1961; Kennelly and Foote, 1965). Although Dutch belted rabbits were used in these experiments, these dose rates also apply to the larger New Zealand breed. However, puberty in New Zealand rabbits is later than in the Dutch belted breed. Some adjustment of the FSH dose may be necessary depending on the batch of FSH. Frequent occurrence of premature ovulation, resulting in many unfertilized ova, can often be remedied by lowering the dose of FSH. The premature ovulation may be due to LH contamination of the FSH preparation (FSH-P, Burns-Biotech). At the present time, LH is very difficult to obtain and it can be replaced by hCG (human chorionic gonadotropin). An alternative superovulation schedule is a single injection of pregnant mare's serum gonadatropin, PMSG (150 IU intramuscular) followed 60 to 72 hr later by 100 IU hCG intravenous (or 2.5 mg LH) and artificial insemination or breeding. The PMSG/hCG treatment gives fewer embryos than the FSH/LH treatment and is reported to be unsuitable for repeated superovulation (Maurer *et al.*, 1968).

5.2. Collection of Embryos

Collection from the Oviduct. Embryos may be safely collected from the oviducts up to the morula stage at about 44 to 48 hr after the ovulating injection of LH. Almost all embryos obtained at about 18 to 21 hr after LH injection are at the 1-cell stage, and at 26 to 30 hr after LH almost entirely 2- to 4-cell stages are obtained. Early morulae may be obtained at about 44 to 48 hr. Morulae may also be collected from the oviduct at later times.

However, there is a possibility that transport of embryos to the uterus is faster in superovulated than in non-superovulated does. Embryos may be collected after killing the donor doe as described by Kennelly and Foote (1965) or from anaesthetized donors as described by Maurer *et al.* (1968). Rabbits can be anaesthetized with sodium pentobarbital but the margin between proper anaesthesia and death with this anaesthetic is very fine. The safest anaesthesia procedure and the one currently used routinely in the author's laboratory is a modification of that described by Green (1975). The rabbits are injected with Hypnorm (0.3 ml/kg intramuscular) followed by intravenous injection of Valium 20 (1 ml, 5 mg diazepam/ml). Hypnorm is a mixture of a narcotic (fentanyl) and a steroid (fluanisone). Fentanyl, being a narcotic, is subject to strict regulations in many countries.

Collection from the Uterus. Blastocyst stages are best collected from the uterus starting at about 72 hr after LH injection, although some embryos may be in the uterus at an earlier time. Again, blastocysts may be collected either after killing the donors or under anaesthesia. The simplest procedure in both cases is as follows: after making a midline ventral incision, a nick is made in the vagina near the cervix and a cannula with an internal diameter large enough to accommodate the blastocysts is gently threaded through one of the cervical openings and held in place with the fingers. This resembles the procedure of Staples (1971) for cervical transfer of blastocysts. A blunt 18 g needle is then used to puncture the uterus near the uterotubal junction and 10 ml of medium is flushed through this needle; the medium and blastocysts are gently forced out through the cannula by "milking" the uterine horn with the fingers. The procedure is then repeated for the contralateral horn. If the rabbit is to be saved, the vaginal nick may be sewn up with a single stitch, although if care is taken in making the nick this is not strictly necessary. This procedure works for up to 5 1/2 day blastocysts. A procedure for collecting blastocysts near the time of implantation (6 to 7 days) is described by Daniel (1971b).

5.3. Culture Procedures

Embryos can be handled at room temperature but care should be taken to avoid temperature shock, *e.g.*, embryos should not be cooled suddenly from body temperature, especially to temperatures lower than room temperature. A simple medium can be used for both collection and subsequent washing of embryos but it should be buffered with a non-bicarbonate buffer to avoid pH changes during handling of the embryos. The following is a simple HEPES-buffered medium that is suitable for embryo collection and washing: 126 mM NaCl, 4.78 mM KCl, 1.71 mM $CaCl_2.2H_2O$, 1.19 mM KH_2PO_4, 1.19 mM $MgSO_4.7H_2O$, 0.5 mM sodium pyruvate, 10 mM HEPES, 500 IU Penicillin G, 500 µg streptomycin sulfate per ml and 0.1% BSA or 0.1% PVA. This medium is neutralized to pH 7.4 with 1 M NaOH at the temperature at which it will be used and may then be stored frozen in suitable aliquots. The levels of penicillin and streptomycin are high but in this laboratory antibiotics are not included in the media used for culture.

The importance of water quality for the culture of mouse ova has been documented by Whittingham (1971). For culture of rabbit embryos in our laboratory, we routinely use water from a system in which it is first distilled and then "polished" in a Milli-Q system (Millipore Corporation) consisting of carbon filtration and ion exchange cartridges followed by membrane filtration. It is unlikely that this quality of water is necessary when the medium contains high levels of BSA (5 to 15 mg/ml). However, at low levels of BSA or of other macromolecules, water quality may become more important.

The culture systems that are currently used for rabbit embryo culture are of two main types:

(1) Microdrops of medium under paraffin oil in plastic tissue culture dishes, as first described for mouse ova by Brinster (1963). These dishes are 60 x 15 mm (Falcon no. 3002F) and are filled with 10 ml oil. This is the system the author still prefers but using 500 μl rather than 200 μl culture drops. The most important factor is perhaps the quality of the paraffin oil (liquid paraffin). Currently the author finds either Merck Paraffin Art. 7174 or BDH Liquid Paraffin Light Product 29436 to be the most suitable. The Merck 7174 is preferred to Merck Art. 7162 previously used in this laboratory because it is made to pharmaceutical standards (DAB and USP) and is therefore less likely to have toxic effects. Extraction of the paraffin oil with saline solution may also be useful (see Chapter 11 [Ed.]).

The number of embryos per unit volume of medium may affect development. The results of an experiment in which embryos were cultured in numbers varying from one to 16 per 0.5 ml drop of medium (Table III: Kane, unpublished data) indicate that development at low embryo density was poor compared with development at higher density. This suggests that rabbit embryos may produce factors that affect each other's growth.

(2) Microwells as in the 96-well microculture trays (various manufacturers) or in the strips of Lux SAS Multiplates (Miles Scientific). Well capacity usually varies from 0.1 to 0.2 ml. While the use of paraffin oil to prevent evaporation from these well-type culture systems is not essential if the culture chamber containing the culture dishes is saturated with water vapour, it is perhaps advisable.

Table III
Effect of Number of Embryos per Culture Drop on Growth of 1-Cell Rabbit Embryos to Hatched Blastocysts

	No. of embryos / drop[a,b]				
	1	2	4	8	16
Total no. of embryos[c]	78	76	76	80	79
% Hatched blastocysts	28	36	46	52	51

[a]There were significant ($p < 0.01$) linear and quadratic effects of number of embryos per drop on development to the hatched blastocyst stage.
[b]Drop volume was 0.5 ml.
[c]There were 5 replicates. Numbers of embryos per replicate are not exactly equal due to loss of some embryos.

Either 5% CO_2 in air or a 5% O_2, 5% CO_2 and 90% N_2 mixture can be used for culturing rabbit embryos. Oxygen in the gas phase is clearly required for culture of preimplantation rabbit embryos (Pincus 1941a) but there is little information on optimal concentrations. Daniel (1968) found that for 5-day and 7-day blastocysts a concentration of 10% oxygen was optimal (10% was the lowest level tested); 95% O_2 inhibited differentiation.

The role of CO_2 has already been discussed in section 4.1. There is no information available on the effect of different CO_2 concentrations on culture of rabbit embryos. [See Chapter 11 for effect of CO_2 on hamster embryos (Ed.).]

One point to note with the use of CO_2 is that some commercial samples may be contaminated with levels of carbon monoxide that could be toxic to cells (McLimans, 1972). A second point is that where culture work is carried out at high altitudes, the pO_2 and pCO_2 will be affected not just by the proportions of gases in the gas mixture but also by the lowered atmospheric pressure at high altitude. This will reduce the pCO_2 in solution, and thereby influence pH control.

The gas phase of 5% CO_2 in air over the culture dishes may be produced by a controlled CO_2 incubator but the author has found that an airtight module such as a McIntosh and Fildes anaerobic jar or a Flow Laboratories Modular Unit, flushed for 5 min with the desired gas phase, works very satisfactorily and is a cheaper alternative. Culture is usually carried out at a temperature of 37 to 38°C. Elliott *et al.* (1974) and Maurer (1978) found that increased ambient or total gas pressure improved rabbit embryo growth in culture but this idea does not appear to have been followed up by other workers.

The culture requirements of embryos at different stages have been discussed in Section 4. Development of 1-cell embryos to the early morula stage will take place in a very simple defined or semi-defined medium (Table IV). When development to the late morula stage or the start of blastocyst formation is desired, then Ham's F10 plus 5 to 15 mg/ml BSA is adequate. For development to the expanding blastocyst stage, the medium given in Table V is superior to F10.

6. EVALUATION OF EMBRYOS

6.1. Statistical Considerations

Because there is considerable variation between studies on culture of rabbit embryos, particularly blastocysts, it is essential that experiments be properly controlled and analysed statistically. It has been stated that "*In spite of rigid efforts to keep the conditions of the culture techniques constant, individual embryos will still respond differently to highly local variations in media, temperature, light, pH, oxygen concentration, condition of glassware,* etc. *The best one can hope to do is to demonstrate general tendencies which may not survive statistical scrutiny*" (Daniel, 1971a; and also El-Banna and Daniel, 1972). This statement is a contradiction in experimental logic. It is precisely because these variations exist that statistical analysis is necessary.

Table IV
Composition of a Medium Suitable for Culture of 1-Cell Rabbit Embryos to Early Morulae[a]

Component	mM	g/Liter
NaCl	108.0	6.31
KCl	4.78	0.356
$CaCl_2.2H_2O$	1.71	0.251
KH_2PO_4	1.19	0.162
$MgSO_4.7H_2O$	1.19	0.294
$NaHCO_3$	25.07	2.106
Sodium pyruvate	0.50	0.055
BSA or PVA	---	1.000

[a]This medium is suitable for culture of 1-cell stages during a 48 hr period. PVA is polyvinyl alcohol (Type II, Sigma Chemical Co., cat. no. P-8136). BSA should be either crystallized or Fraction V (Sigma); fatty acid-free BSA is also suitable.

Ideally, all embryos from one doe should be divided equally between all treatments and this should be done separately for each donor in order to separate treatment and donor effects. Because of variation in numbers of embryos per doe this may not be practical, and instead embryos collected on one day may be pooled and then allocated randomly to the various treatments. It is seldom useful to have more than 10 or 12 embryos per treatment drop or well. Treatments should in most cases include both a zero and a positive control under exactly the same conditions as the treatment groups. This should then be replicated (preferably) at least 3 times in all. It is not advisable to allocate embryos from one doe to one treatment and those from another doe to another treatment.

Data on proportions of embryos reaching different stages can be analysed by chi-square, which can be partitioned in analysis of variance form to examine factorial effects (Steel and Torrie, 1960). A more efficient technique may be that described for mouse embryos by Biggers and Brinster (1965). This requires equal numbers of embryos per drop and per replicate but this is harder to achieve with rabbit embryos because of variability in the superovulatory response, and the smaller numbers of embryos generally recovered. Embryo cell counts and blastocyst diameters can be analysed by ordinary analysis of variance either on the untransformed data (assumed to be normally distributed) or suitably transformed to take account of deviations from normality.

6.2. Cleavage Stages and Morulae

During culture it is convenient if microscopical evaluation can be carried out in the culture dish and for this purpose an inverted compound microscope is most useful. For such examination, ordinary brightfield illumin-

Table V
Composition of a Medium Suitable for Culture of Rabbit Embryos from Early Cleavage Stages to Expanding Blastocysts[a]

Components	mM	g/Liter	Components	μM	μg/Liter
Salts:			Vitamins:		
NaCl	108.00	6.31	Biotin	0.1	2.4
KCl	4.78	0.356	Calcium pantothenate	3.0	715.0
$CaCl_2.2H_2O$	1.71	0.251			
KH_2PO_4	1.19	0.162	Choline chloride	5.0	698.0
$MgSO_4.7H_2O$	1.19	0.294	Myo-inositol	3.0	541.0
$NaHCO_3$	25.07	2.106	Niacinamide	5.0	615.0
			Pyridoxine HCl	1.0	206.0
			Riboflavin	1.0	376.0
Energy substrates:			Thiamine HCl	3.0	1012.0
			Folic acid	3.0	1320.0
Sodium pyruvate	0.50	0.055	Vitamin B_{12}	1.0	1360.0
Glucose	1.00	0.180	Lipoic acid	1.0	200.0
Macromolecules:			Trace elements:		
BSA		5-15	$FeSO_4.7H_2O$	3.00	834.0
			$CuSO_4.5H_2O$	0.01	2.5
Amino acids:			$ZnSO_4.7H_2O$	0.10	28.8
as per Ham (1963) or Ménézo (1976)					

[a]The medium is based on Brinster's (1963) mouse medium and Ham's F10 medium (Ham, 1963). Ham's paper should be read for advice on preparation of stock solutions. In our laboratory concentrated stock solutions are made up of the different groups of components (with the exception of $NaHCO_3$ and BSA) and these are stored frozen in aliquots. The complete medium is then made up from the frozen aliquots as desired and the BSA and $NaHCO_3$ added dry. Alternatively, the whole medium (excluding $NaHCO_3$ and BSA) could be made up in a single stock solution and frozen in aliquots. BSA may be either crystallized or Fraction V BSA but not charcoal-treated (fatty acid-free) BSA.

ation with the condenser diaphragm stopped well down to increase contrast seems to work best. It must be stressed that this kind of microscopic examination is merely a guide to the progress of embryo culture. Because fragmentation is very difficult to distinguish from cleavage, it is not valid to use this method to decide that normal cleavage has occurred or that normal morulae have formed. Failure to appreciate this point vitiates the conclusions of a number of published papers. This is particularly true for situations in which development is limited.

Use of phase contrast optics will not solve this problem. While it is possible using good Nomarski optics to visualize the nuclei in blastomeres of cleavage-stage rabbit embryos, it may not be practicable to use this as a routine procedure for evaluation of living embryos because nuclei may not be visible at certain stages of the cell cycle.

To assess development during early cleavage or to the morula stage, either one of two approaches should be used. Development should be allowed to proceed to the early blastocyst stage, when formation of the blastocoel cavity is an unmistakable sign of development. If a minimal culture medium, which does not support blastocyst formation, is being used to evaluate requirements for early cleavage, then at the end of the culture period the embryos should be transferred to a more complete medium that will allow blastocyst formation in order to assess their viability and the normality of cleavage (Kane, 1979). Alternatively, the embryos can be stained and the number of cells with clearly visible nuclei counted. A simple procedure is the method of Grayson (1978) used by Kane (1983a) for cell counts of blastocysts. This procedure involves first adhering a group of embryos to a microscope slide previously coated lightly with glycerine-albumin; the preparation can then be treated as a standard histology slide for fixing, staining and mounting (Grayson, 1978). This method works reasonably well for cleavage-stage embryos cultured from the 1-cell stage. It does not work well for morulae recovered *in vivo*, which have a thick mucin coat. The ultimate test that a culture system allows normal development is, of course, the birth of live young following transfer to a host mother (*e.g.*, Seidel *et al.*, 1976).

6.3. Blastocysts

For blastocyst stages, the problem of mistaking fragmentation for cell division does not arise. The proportion of embryos reaching the blastocyst stage can therefore be used as an index of development. The formation of a blastocoel cavity, inner cell mass and trophoblast are readily distinguishable, as are blastocyst expansion and hatching. As stated in section 3.2, hatching of rabbit embryos is an artefact of the culture system, probably caused by the lack of a mucin coat in culture, but it is usually associated with active blastocyst growth. Blastocyst expansion can be quantitated with an eyepiece micrometer. The method of Grayson (1978) can be readily used to stain blastocysts cultured from the 1-cell or morula stages and cell numbers can then be counted. This method is only feasible up to about 1000 cells per blastocyst due to the overlapping of cells and to the tedium of the method with increasing cell number. There is need for development of a method in which later stage blastocysts could be dissociated by enzymatic treatment and an aliquot counted by haemocytometer or cell counter to give an estimate of the total number per blastocyst. Uptake of radioactive nucleic acid precursors or amino acids can also be used (El-Banna and Daniel, 1972).

Again, the ultimate test must be the birth of live young following transfer of cultured embryos to foster mothers.

7. CONCLUSIONS

It is clear that rabbit embryo culture provides a very useful alternative model system to mouse embryo culture for the study of mammalian preimplantation embryonic development. The ease and simplicity of culture from 1-cell to morula and the absence of an early cleavage block in the rabbit are major advantages. However, the potential that lies in the relatively great increase in preimplantation growth of the rabbit blastocyst as compared with the mouse will not be tapped until the problems of duplicating normal growth *in vitro* are solved. Overcoming these problems should of itself shed considerable light on the factors regulating embryonic growth.

ACKNOWLEDGMENTS

I thank Professor M.J.T. Fitzgerald, Dr. J.M. Sreenan and Dr. P. Morgan for constructive criticism of the manuscript. Grant support from the Medical Research Council of Ireland, the Irish National Board for Science and Technology and An Foras Taluntais is gratefully acknowledged.

8. REFERENCES

Adams, C.E., 1961, Artificial insemination in the rabbit, *J. Reprod. Fertil.* 2: 521-522.

Alliston, C.W., and Pardee, N.R., 1973, Variability of embryonic development in the rabbit at 19 to 168 hours after mating, *Lab. Anim. Sci.* 23: 665-670.

Austin, C.R., and Walton, A., 1960, Fertilization, in: *Marshall's Physiology of Reproduction*, Volume 1, Part 2 (A.S. Parkes, ed.), Longmans, London, pp. 310-416.

Bavister, B.D., 1981, Substitution of a synthetic polymer for protein in a mammalian gamete culture system, *J. Exp. Zool.* 217: 45-51.

Beier, H.M., 1968, Uteroglobin: A hormone-sensitive endometrial protein involved in blastocyst development, *Biochim. Biophys. Acta* 160: 289-291.

Biggers, J.D., and Brinster, R.L., 1965, Biometrical problems in the study of early mammalian embryos *in vitro*, *J. Exp. Zool.* 158: 39-48.

Biggers, J.D., Whittingham, D.G., and Donahue, R.P., 1967, The pattern of energy metabolism in the mouse oocyte and zygote, *Proc. Natl. Acad. Sci. USA* 58: 560-567.

Blandau, R.J., 1961, Biology of eggs and implantation, in: *Sex and Internal Secretions*, Volume II (W.C. Young, and G.W. Corner, eds), Williams and Wilkins, Baltimore, pp. 797-882.

Böving, B.G., 1957, Rabbit egg coverings, *Anat. Rec.* 127: 270.

Bowman, P., and McLaren, A., 1970, Cleavage rate of mouse embryos *in vivo* and *in vitro*, *J. Embryol. Exp. Morphol.* 24: 203-207.

Brachet, A., 1913, Récherches sur le déterminisme héréditaire de l'oeuf des mammifères. Développement *in vitro* de jeunes vésicules blastodermiques du lapin, *Arch. Biol. (Paris)* 28: 447-504.

Brinster, R.L., 1963, A method for the *in vitro* cultivation of mouse ova from two-cell to blastocyst, *Exp. Cell Res.* 32: 205-208.

Brinster, R.L., 1967a, Protein content of the mouse embryo during the first five days of development, *J. Reprod. Fertil.* 13: 413-420.

Brinster, R.L., 1967b, Carbon dioxide production from glucose by the preimplantation mouse embryo, *Exp. Cell Res.* 47: 271-277.

Brinster, R.L., 1968, Carbon dioxide production from glucose by the preimplantation rabbit embryo, *Exp. Cell Res.* 51: 330-334.

Chang, M.C., 1948, The effects of low temperature on fertilized rabbit ova *in vitro*, and the normal development of ova kept at low temperatures for several days, *J. Gen. Physiol.* 31: 385-410.

Chang, M.C., 1949, Effects of heterologous sera on fertilized rabbit ova, *J. Gen. Physiol.* 32: 291-300.

Daniel, J.C. Jr., 1964, Early growth of rabbit trophoblast, *Amer. Naturalist* 98: 85-97.

Daniel, J.C., Jr., 1967a, The pattern of utilization of respiratory metabolic intermediates by preimplantation rabbit embryos *in vitro*, *Exp. Cell Res.* 47: 619-624.

Daniel, J.C., Jr., 1967b, Vitamins and growth factors in the nutrition of rabbit blastocysts *in vitro*, *Growth* 31: 71-77.

Daniel, J.C., Jr., 1968, Oxygen concentrations for culture of rabbit blastocysts, *J. Reprod. Fertil.* 17: 187-190.

Daniel, J.C., Jr., 1971a, Uterine proteins and embryonic development, in: *Schering Symposium on Intrinsic and Extrinsic Factors in Early Mammalian Development, Advances in the Biosciences*, Vol. 6 (G. Raspé, ed.), Pergamon Press, Oxford, pp. 191-203.

Daniel, J.C. Jr., 1971b, Culture of the rabbit blastocyst across the implantation period, in: *Methods in Mammalian Embryology* (J.C. Daniel, Jr., ed.), W.H. Freeman, San Francisco, pp. 284-289.

Daniel, J.C., Jr., and Krishnan, R.S., 1967, Amino acid requirements for growth of the rabbit blastocyst *in vitro*, *J. Cell Physiol.* 70: 155-160.

Daniel, J.C., Jr., and Millward, J.T., 1969, Ferrous ion requirement for cleavage of the rabbit egg, *Exp. Cell Res.* 54: 135-136.

Daniel, J.C., Jr., and Olson, J.D., 1968, Amino acid requirements for cleavage of the rabbit ovum, *J. Reprod. Fertil.* 15: 453-455.

Denker, H.W., 1982, Proteases of the blastocyst and of the uterus, in: *Proteins and Steroids in Early Pregnancy* (H-M. Beier, and P. Karlson, eds.), Springer Verlag, Berlin, pp. 183-208.

Denker, H.W., and Gerdes, H.J., 1979, The dynamic structure of rabbit blastocyst coverings. I. Transformation during regular preimplantation development, *Anat. Embryol.* 157: 15-34.

El-Banna, A.A., and Daniel, J.C., Jr., 1972, The effect of protein fractions from rabbit uterine fluids on embryo growth and uptake of nucleic acid and protein precursors, *Fertil. Steril.* 23: 105-114.

Elliott, D.S., Maurer, R.R., and Staples, R.E., 1974, Development of mammalian embryos *in vitro* with increased atmospheric pressure, *Biol. Reprod.* 11: 162-167.

Enders, A.C., 1971, The fine structure of the blastocyst, in: *The Biology of the Blastocyst* (R.J. Blandau, ed.), The University of Chicago Press, Chicago, pp. 71-94.

Goodman, D.S., 1958, The interaction of human serum albumin with long-chain fatty acid anions, *J. Amer. Chem. Soc.* 80: 3892-3898.

Grayson, K., 1978, An improved method for staining mammalian oocytes, *Stain Technol.* 53: 115-116.

Green, C.J., 1975, Neuroleptanalgesic drug combinations in the anaesthetic management of small laboratory animals, *Lab. Animals* 9: 161-178.

Ham, R.G., 1963, An improved nutrient solution for diploid Chinese hamster and human cell lines, *Exp. Cell Res.* 29: 515-526.

Ham, R.G., 1981, Introduction: cell growth requirements - the challenge we face, in: *The Growth Requirements of Vertebrate Cells in Vitro* (C. Waymouth, R.G. Ham, and P.J. Chapple, eds.), Cambridge University Press, Cambridge, pp. 1-15.

Kane, M.T., 1969, *In vitro* culture of two- to four-cell rabbit embryos to expanding blastocysts in serum extracts and synthetic media, *Ph.D. Thesis*, Cornell University.

Kane, M.T., 1972, Energy substrates and culture of single cell rabbit ova to blastocysts, *Nature (London)* 238: 468-469.

Kane, M.T., 1974, The effects of pH on culture of one-cell rabbit ova to blastocysts in bicarbonate-buffered medium, *J. Reprod. Fertil.* 38: 477-480.

Kane, M.T., 1975a, Inhibition of zona shedding of rabbit blastocysts in culture by the presence of a mucin coat, *J. Reprod. Fertil.* 44: 539-542.

Kane, M.T., 1975b, Bicarbonate requirements for culture of one-cell rabbit ova to blastocysts, *Biol. Reprod.* 12: 552-555.

Kane, M.T., 1979, Fatty acids as energy sources for culture of one-cell rabbit ova to viable morulae, *Biol. Reprod.* 20: 323-332.

Kane, M.T., 1983a, Variability in different lots of commercial bovine serum albumin affects cell multiplication and hatching of rabbit blastocysts in culture, *J. Reprod. Fertil.* 69: 555-558.

Kane, M.T., 1983b, Evidence that protease action is not specifically involved in the hatching of rabbit blastocysts caused by commercial bovine serum albumin in culture, *J. Reprod. Fertil.* 68: 471-475.

Kane, M.T., 1985, A low molecular weight extract of bovine serum albumin stimulates rabbit blastocyst cell division and expansion *in vitro*, *J. Reprod. Fertil.* 73: 147-150.

Kane, M.T., and Foote, R.H., 1970a, Culture of two- and four-cell rabbit embryos to the expanding blastocyst stage in synthetic media, *Proc. Soc. Exp. Biol. Med.* 133: 921-925.

Kane, M.T., and Foote, R.H., 1970b, Fractionated serum dialysate and synthetic media for culturing two- and four-cell rabbit embryos, *Biol. Reprod.* 2: 356-362.

Kane, M.T., and Foote, R.H., 1971, Factors affecting blastocyst expansion of rabbit zygotes and young embryos in defined media, *Biol. Reprod.* 4: 41-47.

Kane, M.T., and Headon, D.R., 1980, The role of commercial bovine serum albumin preparations in culture of one-cell rabbit embryos to blastocysts, *J. Reprod. Fertil.* 60: 469-475.

Kennelly, J.J., and Foote, R.H., 1965, Superovulatory response of pre- and post-pubertal rabbits to commercially available gonadotrophins, *J. Reprod. Fertil.* 9: 177-188.

Kille, J.W., and Hamner, C.E., 1973, The influence of oviducal fluid on the development of one-cell rabbit embryos *in vitro*, *J. Reprod. Fertil.* 35: 415-423.

Krishnan, R.S., and Daniel, J.C., Jr., 1967, "Blastokinin": Inducer and regulator of blastocyst development in the rabbit uterus, *Science* 158: 490-492.

Lewis, W.H., and Gregory, P.W., 1929, Cinematographs of living developing rabbit eggs, *Science* 69: 226-229.

Lutwak-Mann, C., 1959, Biochemical approach to the study of ovum implantation in the rabbit, in: *Implantation of Ova, Memoirs of the Society for Endocrinology*, No. 6 (P. Eckstein, ed.), Cambridge University Press, pp. 35-46.

Lutwak-Mann, C., 1971, The rabbit blastocyst and its environment: Physiological and biochemical aspects, in: *The Biology of the Blastocyst* (R.J. Blandau, ed.), University of Chicago Press, Chicago, pp. 243-260.

Maurer, R.R., 1978, Advances in rabbit embryo culture, in: *Methods in Mammalian Reproduction* (J.C. Daniel, Jr., ed.), Academic Press, New York, pp. 259-272.

Maurer, R.R., and Beier, H.M., 1976, Uterine proteins and development *in vitro* of rabbit preimplantation embryos, *J. Reprod. Fertil.* 48: 33-41.

Maurer, R.R., Hunt, W.L., and Foote, R.H., 1968, Repeated superovulation following administration of exogenous gonadotrophins in Dutch-belted rabbits, *J. Reprod. Fertil.* 15: 93-102.

Maurer, R.H., Whitener, R.H., and Foote, R.H., 1969, Relationship of *in vivo* gamete aging and exogenous hormones to early embryo development in rabbits, *Proc. Soc. Exp. Biol. Med.* 131: 882-885.

Ménézo, M.Y., 1976, Milieu synthétique pour la survie et la maturation des gamètes et pour la culture de l'oeuf fécondé, *C.R. Acad. Sci. (Paris)* 282: 1967-1970.

McLachlan, J.A., Sieber, S.M., Cowherd, C.M., Straw, J.A., and Fabro, S., 1970, The pH values of the uterine secretions and preimplantation blastocyst of the rabbit, *Fertil. Steril.* 21: 84-87.

McLimans, W.F., 1972, The gaseous environment of the mammalian cell in culture, in: *Growth, Nutrition and Metabolism of Cells in Culture*, Volume I (G.H. Rothblat, and V.J. Cristofalo, eds.), Academic Press, New York, pp. 137-170.

Mukherjee, A.B., Laki, K., and Agrawal, A.K., 1980, Possible mechanism of success of an allotransplantation in nature: *Mammalian Pregnancy*, Med. Hypotheses 6: 1043-1051.

Mukherjee, A.B., Ulane, R.E, and Agrawal, A.K., 1982, Role of uteroglobin and transglutaminase in masking the antigenicity of implanting rabbit embryos, *Amer. J. Reprod. Immunol.* 2: 135-141.

Naglee, D.L., Maurer, R.R., and Foote, R.H., 1969, Effect of osmolarity on *in vitro* development of rabbit embryos in a chemically defined medium, *Exp. Cell Res.* 58: 331-333.

Onuma, H., Maurer, R.R., and Foote, R.H., 1968, *In vitro* culture of rabbit ova from early cleavage stages to the blastocyst stage, *J. Reprod. Fertil.* 16: 491-493.

Pincus, G., 1930, Observations on the living eggs of the rabbit, *Proc. Roy. Soc. (London) Ser. B* 107: 132-167.

Pincus, G., 1941a, Factors controlling the growth of rabbit blastocysts, *Amer. J. Physiol.* 133: 412-413.

Pincus, G., 1941b, The control of ovum growth, *Science* 93: 438-439.

Pincus, G., and Werthessen, N.T., 1938, The comparative behaviour of mammalian eggs *in vivo* and *in vitro*. III Factors controlling the growth of the rabbit blastocyst, *J. Exp. Zool.* 78: 1-19.

Quinn, P., and Wales, R.G., 1973, Growth and metabolism of preimplantation mouse embryos cultured in phosphate-buffered medium, *J. Reprod. Fertil.* 35: 289-300.

Seidel, G.E., Jr., Bowen, R.A., and Kane, M.T., 1976, *In vitro* fertilization, culture and transfer of rabbit ova, *Fertil. Steril.* 27: 862-870.

Staples, R.E., 1971, Blastocyst transplantation in the rabbit, in: *Methods in Mammalian Embryology* (J.C. Daniel, Jr., ed.), W.H. Freeman, San Francisco, pp. 290-304.

Steel, R.G.D., and Torrie, J.H., 1960, *Principles and Procedures of Statistics*, McGraw Hill Book Company, New York.

Van Blerkom, J., Manes, C., and Daniel, J.C., Jr., 1973, Development of preimplantation rabbit embryos *in vivo* and *in vitro* 1. An ultrastructural comparison, *Dev. Biol.* 35: 262-282.

Varian, N.B., Maurer, R.R., and Foote, R.H., 1967, Ovarian response and cleavage rate of ova in control and FSH-primed rabbits receiving varying levels of luteinizing hormone, *J. Reprod. Fertil.* 13: 67-73.

Vishwakarma, P., 1962, The pH and bicarbonate-ion content of the oviduct and uterine fluids, *Fertil. Steril.* 13: 481-485.

Warner, C.M., 1977, RNA polymerase activity in preimplantation mammalian embryos, in: *Development in Mammals*, Volume I (M.H. Johnson, ed.), North-Holland, Amsterdam, pp. 99-136.

Westphal, U., 1970, Corticosteroid binding globulin and other steroid hormone carriers in the blood stream, *J. Reprod. Fertil., Suppl.* 10: 15-38.

Whitten, W.K., 1957, Culture of tubal ova, *Nature (London)* 179: 1081-1082.

Whittingham, D.G., 1971, Culture of mouse ova, *J. Reprod. Fertil., Suppl.* 14: 7-21.

CHAPTER 11

STUDIES ON THE DEVELOPMENTAL BLOCKS IN CULTURED HAMSTER EMBRYOS

BARRY D. BAVISTER

1. INTRODUCTION

1.1. Rationale for Studies on Hamster Embryos

As mentioned in the Preface to this book, most of what we presently know about the regulation of preimplantation embryo development is derived from studies with mouse embryos under *in vitro* culture conditions. This information is of enormous value, yet we should be concerned about the relative lack of critical data available for other species (with the notable exception of the rabbit: see Chapters 1 and 10). Accepting that embryos of other species need to be intensively investigated, using experimental approaches similar to those applied to mouse embryos over the past 20 years or so, why choose the golden hamster as a model species? The scarcity of published reports on the culture of hamster embryos underscores the difficulty of growing these embryos *in vitro*; in fact, there has never been a report of successful culture of hamster 2-cell embryos. For some unknown reason, early cleavage stage embryos of the golden hamster are extraordinarily sensitive (compared to mouse embryos) to conventional culture conditions, and the "2-cell block" appears to be absolute (Yanagimachi and Chang, 1964; Whittingham and Bavister, 1974).

In my laboratory, we regard the refractoriness of hamster embryos to growth in culture as a challenge that offers an opportunity to gain information about the environmental factors which are important regulators of embryo development *in vitro*. Refractoriness to *in vitro* culture is not limited to hamster embryos. Even in the mouse, 1-cell embryos are highly sensitive to the culture environment (Biggers, 1971; Cross and Brinster, 1973; Spielmann

Barry D. Bavister Department of Veterinary Science, University of Wisconsin, Madison, Wisconsin 53706, USA.

et al., 1980) and embryos from outbred strains also exhibit a 2-cell block *in vitro* (Goddard and Pratt, 1983). Rat embryos usually will not develop beyond the 2- to 4-cell stage in culture (Whittingham, 1975). Other chapters in this book describe some of the problems encountered in attempts to sustain normal growth *in vitro* of preimplantation embryos from a variety of species. My experimental rationale is that, by intensively studying embryos which are difficult to grow *in vitro*, we should be able to elucidate reasons for retarded or blocked development; then, perhaps, this information might be successfully applied to the culture of refractory embryos from other species, such as the domesticated (farm) animals (see Chapter 12).

Apart from providing comparative data on embryo culture requirements, there are sound practical advantages of using the golden hamster to study embryogenesis *in vitro*. The females have a predictably regular 4-day estrous cycle; the stages of the cycle are characterized by well-defined vaginal exudates, which makes selection of animals for superovulation or for breeding very easy (Orsini, 1961). The animals are readily superovulated with pregnant mare's serum gonadotropin (PMSG), usually yielding 40 to 60 embryos after mating. The golden hamster is eminently suitable for *in vitro* fertilization (IVF) studies, and standardized protocols have been devised for this purpose (Bavister, 1987). Karyotyping is easily performed with hamster cells because of the polymorphic appearance of the chromosomes (Basler, 1978). Because each uterine horn in the golden hamster has a separate cervix, migration of transferred embryos between horns does not occur. This allows control and experimental embryos to be transferred to alternate uterine horns of the same animal, which simplifies experimental design and increases the power of statistical analysis, compared to the situation in which different recipients are used for the 2 classes of embryo.

At present, the value of the golden hamster for studies on preimplantation embryogenesis is quite limited because of the blocks to development *in vitro*. However, once we are able to grow 1-cell embryos into viable blastocysts in culture, this species should become very useful for studies on the regulation of early development.

1.2. Present Status of Hamster Embryo Culture Capabilities

Hamster 1-cell embryos will undergo the first cleavage division *in vitro* quite readily, but never accomplish the second division. This 2-cell block occurs in embryos that are fertilized *in vivo* or *in vitro*. Approximately 70 to 90% of *in vivo* fertilized zygotes will reach the 2-cell stage in culture (Whittingham and Bavister, 1974; Golden, unpublished). About 80% of ova inseminated *in vitro* are monospermically fertilized and two-thirds of the total ova are able to cleave once in culture (Juetten and Bavister, 1983a,b; Fig. 1). However, blocked embryos are not viable, as shown by embryo transfer (Sato and Yanagimachi, 1972; Whittingham and Bavister, 1974), and they are also morphologically abnormal (see section 3.1).

In vivo fertilized 4-cell embryos very rarely develop further *in vitro*. If they are collected just before the start of the third cleavage division, some embryos may reach the 5-, 6- or even 7-cell stage during culture for a few hours (Bavister, unpublished), but formation of 8-cell embryos *in vitro* almost never occurs. In one study, a total of 7 out of 185 cultured 4-cell embryos were able to develop into blastocysts *in vitro* (Bavister *et al.*, 1983a).

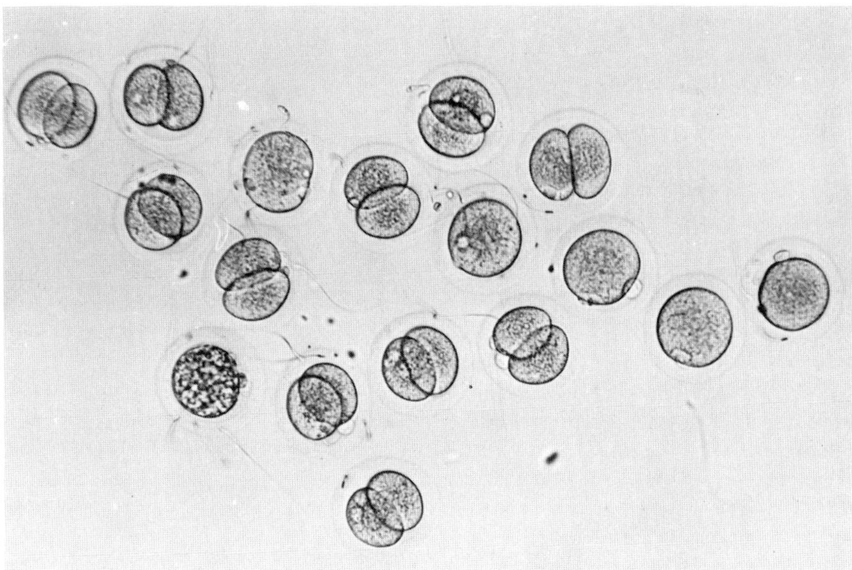

Figure 1. Group of *in vitro* fertilized hamster embryos, approx. 24 hr after insemination. One egg has degenerated and 5 failed to cleave; the remainder *appear* to be normal 2-cell embryos but the blastomere swelling effect is apparent (see Fig. 5). Phase-contrast, optical magnification x 125. From Bavister (unpublished data).

This response was independent of the culture treatments, and we have been unable to repeat it.

Eight-cell embryos collected from superovulated, mated hamsters very soon after the third cleavage division has occurred can be grown to the blastocyst stage in culture; however, they are extremely dependent on the presence of amino acids in the medium (Bavister *et al.*, 1983a). Recent developments in our laboratory have enabled a very high percentage of 8-cell hamster embryos to reach the blastocyst stage in culture (see section 4.2).

Three approaches are being used in our laboratory to overcome the developmental problems exhibited by hamster embryos *in vitro*. First, we are examining aspects of the physico-chemical environment provided by the culture medium to identify the cause of the very rapid loss of viability shown by 2-cell embryos exposed to culture media (Sato and Yanagimachi, 1972; Farrell and Bavister, 1984). This seems to be due to some trauma experienced by these embryos upon contact with the artificial environment. Secondly, in parallel experiments, 8-cell embryos are being cultured to the blastocyst stage. Early 8-cell embryos appear to share some of the sensitivity to culture conditions shown by earlier stages. As we improve the culture conditions for this stage of development, we anticipate that some 4-cell embryos can be persuaded to grow *in vitro*. By "working backwards" in this way, we should eventually be able to culture 2-cell embryos also (see Section 3). The third approach uses cross-species oviduct culture to by-pass blocks to development in hamster embryos (see section 3.3). In section 2, technical procedures are described for the collection, culture and transfer of hamster embryos.

2. TECHNICAL PROCEDURES.

2.1. Embryo Collection

Most of the animals used in our research program are colony-bred by us; this reduces the incidence of disease and lowers stress levels compared with commercially-bred animals. It is critically important to maintain animals on an automatic lighting schedule with a minimum of 14 hr light per 24 hr cycle; reproductive performance in rodents is exceedingly sensitive to lighting conditions. References to the timing of embryo collection (see below) relate to our colony schedule, in which lights come on at 06:00 hr. The temperature is maintained between 72° F and 76° F (22° to 24° C); relative humidity should be between 30 to 40%.

Females weighing about 90 to 120g are injected intraperitoneally with 15 IU PMSG by 10:00 hr on the morning of the post-estrus discharge; this discharge is easily recognized in the hamster by its viscosity and pungent odor (Orsini, 1961). We obtain consistently good superovulation responses with Organon® PMSG from Diosynth, Inc.; gonadotropin preparations from some other suppliers have not been satisfactory. The dose is quite critical: too much PMSG can interfere with the animals' cycles and/or with embryogenesis. No hCG (human chorionic gonadotropin) is used; an animal's endogenous LH is sufficient to ovulate large numbers of eggs (sometimes as many as 80). On the evening of the third day following PMSG injection (approx. 80 hr later), superovulated females are checked for estrus. A blunt toothpick is inserted into the vagina and withdrawn, pulling out a glistening strand of cervical mucus 15 to 30 cm long (Orsini, 1961). This "spinnbarkeit" is very characteristic of estrus in the hamster. Animals that fail this test usually will not mate. Females are placed into the cages of proven fertile males (never the other way round), with 2 females to each group of 2 or 3 males. Animals in estrus will stand rigidly in lordosis and mating will commence within a few minutes. Next morning, the females are removed and checked for the presence of a copulation plug in the vagina. Usually, we confirm fertile mating by examining vaginal smears for the presence of sperm.

One-cell embryos for culture are usually collected between 10:00 hr and 12:00 hr on the morning after mating (day 1); 2-cells 24 hr later (day 2); mostly 4-cell embryos between 24:00 hr and 08:00 hr on day 3; 8-cells from about 10:00 hr on day 3; and morulae or early blastocysts from about 21:00 hr on day 3. More advanced blastocysts are difficult to obtain because attachment to the uterine epithelium begins early on day 4. However, it possible to recover blastocysts from animals that have been artificially inseminated instead of being mated (Ortiz *et al.*, 1986). The complete developmental schedule for hamster embryos *in vivo* is given in Bavister *et al.* (1983a), although the timing of syngamy was stated incorrectly as 15 hr after egg activation (sperm penetration) instead of 18 to 20 hr. One-cell to 4-cell embryos are recovered by flushing excised oviducts into a sterile petri dish (about 0.2 ml of medium per oviduct), using a stainless steel 30g hypodermic needle (2.5 to 4.0 cm length) attached to a tuberculin (1 ml) syringe. The tip of the needle is blunted and polished, and it is helpful to bend the last 5 to 8 mm of the needle at an angle of 45° to facilitate penetration through the fimbrial opening into the ampulla. The ampulla is grasped tightly around the needle with fine forceps before flushing.

Eight-cell and later stage embryos are flushed using the same procedure except that the uterine horns are excised together with the oviducts, and 0.5 to 1.0 ml of flushing medium per oviduct/horn is used. Embryos are located as rapidly as possible using a bright-field dissecting microscope; one with an adjustable substage mirror is preferable. The embryos are then washed (usually only once) in the flushing medium and distributed among all the culture treatments (see section 2.3). The time between killing the donor animal and the placement of embryos in the culture medium should be minimised. Maintenance of sterility during these procedures is most important.

2.2. Media for Embryo Collection and Culture

Our standard culture medium is a modified Tyrode's solution containing additional bicarbonate, pyruvate, lactate and bovine serum albumin (BSA). This medium (TALP) was originally designed for hamster IVF (Bavister and Yanagimachi, 1977) but it is also suitable for culturing some stages of hamster, bovine and rhesus monkey embryos (Bavister *et al.*, 1983a,b; Hoppe and Bavister, 1983, 1984; and see Chapter 13). The composition of TALP and of a HEPES-buffered version is given in Chapter 13. Procedures using these media are described by Bavister (1987). For the beginner, TALP-HEPES is useful because its stable pH in air allows more time for manipulating embryos. However, there is evidence that exposure of hamster embryos to media lacking $NaHCO_3/CO_2$ is somewhat detrimental (see section 4.2: *CO_2, HCO_3^- and pH*), so we often use CO_2-equilibrated TALP for the embryo collection and washing. This requires an ability to work very rapidly to avoid the deleterious effects of elevated pH, which will occur within about 30 sec after exposure of TALP to air. The problem can be reduced substantially by using mineral or silicone oil to cover each embryo collection dish.

Media are usually prepared fresh, *i.e.*, salt solutions TL or TL-HEPES are used after storage for up to one week; BSA and pyruvate are added immediately before use. However, it is possible to store the media for at least 4 weeks at -20° C without detriment provided (i) BSA is present, or pyruvate is absent during storage, and (ii) the containers are tightly sealed (Stewart-Savage and Bavister, 1987). Containers for storing culture media should be borosilicate glass (*e.g.*, Pyrex®), which we treat with a siliconizing fluid (*e.g.*, Prosil 28®, SCM Specialty Chemicals); alternatively, tissue-culture grade flasks can be used, such as Falcon Plastics 200 ml flasks (cat. no. 3024). Culture media are sterilized with syringe filters. We prefer to use Millex-GV® filters (Millipore Corp., cat. no. SLGV025LS); alternatively, for bulk media, we use negative or positive pressure filters (respectively: Nalgene®, Nalge Co., cat. no. 4500020, 0.2 μm; and Sterivex-GS®, Millipore Corp., cat. no. SVGSB1010, 0.22 μm). Certain types of membrane filter are definitely cytotoxic; filters, like all culture apparatus, should be checked with a sensitive bioassay before using them. We routinely use a sperm motility bioassay for this purpose that has proved very useful in detecting sources of cytotoxic contaminants (Bavister and Andrews, 1987; and see Appendix I).

The inclusion of 4 amino acids (glutamine, methionine, phenylalanine and isoleucine) in the culture medium has a pronounced effect on development of 8-cell hamster embryos (see section 4.2: *Amino acids*) and may also be required for initial cleavage divisions (Bavister *et al.*, 1983a; Juetten and Bavister, 1983a). Accordingly, we routinely include these amino acids in our

TALP medium for embryo culture. A 100x stock solution containing all 4 amino acids (glutamine 14.6 mg, methionine 0.8 mg, phenylalanine 1.6 mg and isoleucine 2.6 mg, per ml) is stored in small aliquots at -20° C.

We have grown 8-cell hamster embryos to the blastocyst stage both in 3 ml TALP with 2 ml oil overlay in a 35 mm plastic petri dish (Falcon no. 1008) or in 100 μl drops in a 60 mm dish (Falcon no. 1007) under 8 ml oil. Results are the same with these two methods. We prefer the latter method because several different treatments (culture medium modifications) can be accommodated in the same dish. This is not only convenient for randomly assigning embryos to each treatment but also eliminates within replicate variability due to differences in the quality of individual culture dishes. Either mineral (paraffin) oil or silicone oil can be used, but equilibration with 0.9% saline is a good precaution to follow before use. An additional precaution that we take with mineral oil is to wash it with saline to extract water-soluble contaminants. About 250 ml of oil is swirled thoroughly (not *shaken*) with about twice the volume of saline in a 1 L separating funnel. The saline is drawn off from the bottom, and the extraction procedure is repeated twice more. The saline is prepared with tissue-culture grade water. Dishes containing TALP are equilibrated with 5% CO_2 in air at 37° C for several hours or even overnight before use. Exposure of equilibrated dishes to air, for addition or for examination of embryos, must be kept to a minimum to reduce detrimental effects of changes in pH (see also Appendix I on 'Incubators').

2.3. Experimental Design and Statistical Analysis of Data

Factorially-designed experiments are used whenever possible to facilitate analysis of factors important for regulating embryo development. This type of design is almost mandatory to avoid the striking between-female (probably also between-male) differences that we observe in terms of embryo growth *in vitro*. As much as possible, all embryos from one donor female are distributed randomly among all the treatments (including controls) being tested in any one replicate of an experiment. If possible, all treatments are represented in one 60 mm culture dish. This design considerably strengthens statistical analysis of data (see also Chapter 10 and Appendix I).

If only 2 treatments are being compared, then the proportions of embryos forming a blastocele are normalized using the arcsin transformation; Student's t-test can then be applied to find significant differences. If more than 2 treatments are involved, then the arcsin transformation followed by 2-way analysis of variance is used when the data are distributed normally. If the normal distribution assumption is invalid (usually the case with treatments that result in zero responses or that have few embryos per treatment), then we use a non-parametric test such as the Friedman test (Conover, 1980).

2.4. Endpoints for Evaluating *in Vitro* Embryo Development

It is likely that many experiments using cultured embryos have been inconclusive or even misleading because the chosen endpoints (or responses) were not sufficiently rigorous. Numerous factors found to be "unimportant" in various studies, such as amino acids, gas partial pressures, osmotic pressure, etc., may in fact have had a significant effect on the *viability* of cultured embryos, which is rarely assessed. [See also Chapter 14 and Appendix I.]

Evaluation of Embryos in Vitro. The most commonly used endpoint is the proportion of embryos reaching the blastocyst stage *in vitro* after a standardized period of culture. This endpoint is very simple and quantitative, and can provide useful data. The duration of culture can be adjusted to give more information about developmental characteristics, and a rough measure of the stage of blastocyst development attained (*e.g.*, early, middle, late) can be employed (Bavister *et al.*, 1983a). A disadvantage is that "late" blastocysts can vary considerably in size due to cycles of inflation and deflation (Fig. 2), apart from which, assignment to different size categories is rather subjective.

Blastocyst cell number is a very informative endpoint which should be examined, since there are often (usually?) substantial discrepancies in cell counts of embryos grown *in vivo* and *in vitro* (see Chapter 10). The time in culture needed to double the cell number is a useful quantitative parameter (Harlow and Quinn, 1982). Rodent blastocysts contain about 50 cells on day 4

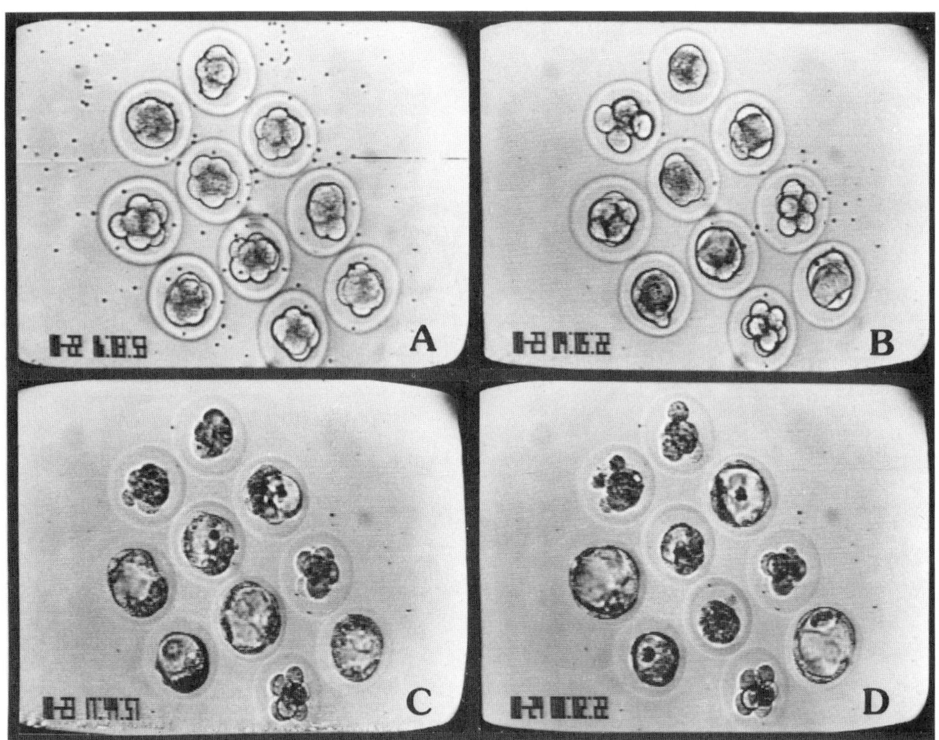

Figure 2. Time-lapse videomicrographs of successive stages of *in vitro* development of hamster 8-cell embryos. (A) Compacting 8-cells and early morulae. (B) Eight hr later, 3 embryos have "decompacted" and 4 are forming blastocele cavities. (C) After another 13.5 hr, 6 embryos have developed prominent blastoceles, 2 (bottom and right) would be classified as "non-starters" [see section 4.2: *CO_2, HCO_3^- and pH*] and 2 (at top) appear to have ceased development at the morula stage. (D) One embryo that had a large blastocele has collapsed (lower center); such embryos often reinflate. About 6 hr after previous frame. All videomicrographs made with Nomarski optics at 50 x magnification; embryos cultured continuously under 10% CO_2 in air on the microscope stage at 37° C (see Fig. 9). From Bavister and Carney (unpublished data).

(Harlow and Quinn, 1982), which can easily be counted by (for example) staining with the Hoechst fluorescent DNA-binding dye (Critser and First, 1986). An example of a Hoechst-stained hamster embryo is shown in Figure 3. A fluorescent vital dye, fluorescein diacetate (FDA), that gives information about cell membrane integrity has been used to assess embryo viability (Jackowski, 1977; Mohr and Trounson, 1980; Hutz *et al.*, 1985). However, this test provides only limited information about cell viability and the correlation of dye sequestration with embryo developmental capacity is influenced by whether the test is applied before or after embryo culture (Hoppe and Bavister, 1984; see also Chapter 14). The possible presence of ultrastructural defects should also be considered, since these can occur in cultured embryos that look quite normal under the light microscope (Chakraborty, 1981).

Continuous time-lapse videomicrography is a very useful technique for obtaining information about embryo development. We have cultured hamster 8-cell embryos on the stage of an inverted microscope (Nikon Diaphot) fitted with an environment control chamber to maintain a temperature of 37°C. Humidified CO_2 in air mixture is passed continuously over the culture dish so that pH is maintained at the desired level. Embryos cultured for 24 to 30 hr under these conditions will develop into blastocysts (Fig. 2). With longer culture periods, some embryos are able to hatch from their zonae pellucidae

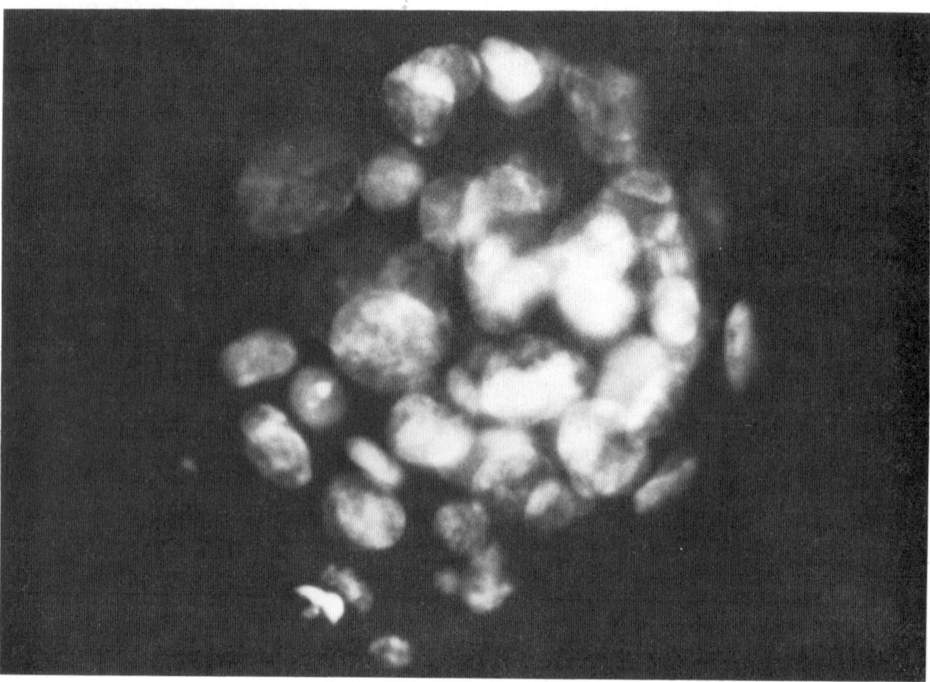

Figure 3. Use of Hoechst stain for cell counting. Hamster blastocyst cultured from 8-cell stage, stained with Hoechst 33342 and photographed under UV light. Light areas are cell nuclei, which can easily be counted (40 shown here). Optical magnification x 400. From Carney and Boatman (unpublished data).

(see section 4.2). Continuous time-lapse recording has provided information that we could not have obtained otherwise, for example on the cycles of inflation and deflation in cultured blastocysts (Fig. 2c,d) and on the "trophectodermal processes" appearing in blastocysts near to the time of hatching (see section 4.2: *Hatching of Hamster Blastocysts in Vitro*). Moreover, this technique has been successfully applied to IVF rhesus monkey embryos that were monitored for periods of several days during culture on the microscope stage (Boatman *et al.*, 1987; see also Chapter 13).

Embryo Transfer. The conclusive test of the efficacy of a culture treatment is the viability of embryos following transfer to pseudopregnant recipient females. For embryo transfers, we routinely use a drawn-out Pasteur pipet with a tip length of about 2.5 cm and an internal diameter of about 150 μm. For oviductal transfers, the tip of the pipet is rapidly passed through a flame to eliminate sharp edges; for uterine transfers, the tip is left quite sharp. Procedures for these two transfer approaches have been described (Bavister *et al.*, 1983a; Farrell and Bavister, 1984). In both approaches, correct placement of the pipet tip in the lumen of the reproductive tract is critically important; it should require almost no pressure (using a mouth-tube) to eject the embryos into the lumen.

Examination of embryo transfer recipients is usually carried out on day 9 or 10 after transfer, when fetal development is well advanced (Fig. 4).

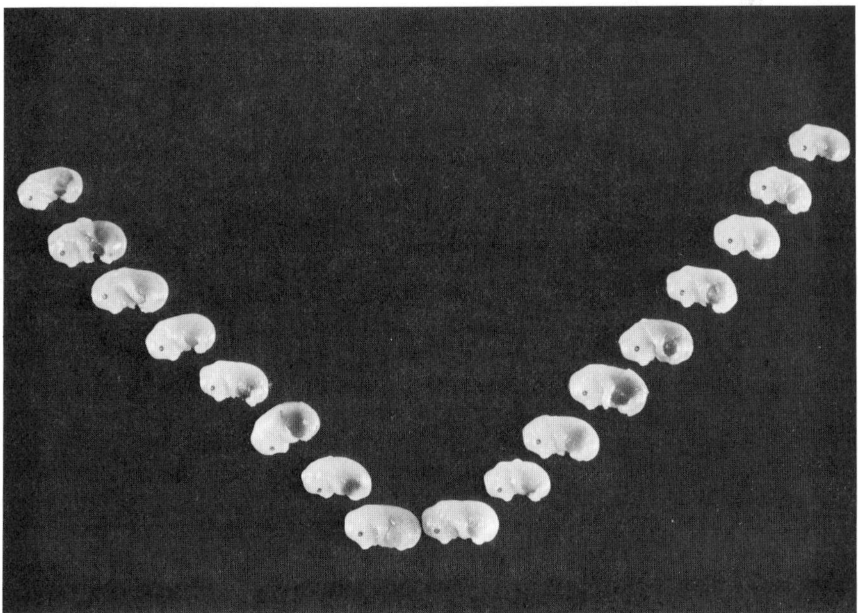

Figure 4. Results of hamster embryo transfer. Hamster fetuses dissected from uterine horns of recipient female 11 days after transfer (day 14 of gestation, 2 days before term) of 8-cell embryos exposed to fluorescein diacetate and UV light (left horn) and control embryos (right horn). There was no detectable difference between the treatments. From Hoppe and Bavister (1984), with permission.

Laparotomy is performed under Nembutal anesthesia and the numbers of fetuses and/or implantation sites in each uterine horn are counted (Orsini, 1962). Alternatively, uterine horns can be removed after injecting the animal with dye to reveal implantation sites (Orsini and Donovan, 1971). Some recipients should be allowed to carry their pregnancies to term (day 16) to confirm the viability and normality of fetuses (Sato and Yanagimachi, 1972; Bavister *et al.*, 1983a).

3. THE 2-CELL BLOCK TO DEVELOPMENT

3.1. Timing of the Block and Appearance of Blocked Embryos

Since early cleavage stage embryos of most other species will undergo at least some degree of further growth *in vitro*, it is puzzling why golden hamster 2-cell embryos are so refractory to culture. The block is encountered with both *in vitro* and *in vivo* fertilized embryos, so abnormalities resulting from IVF can be discounted as causative factors. *In vivo* fertilized embryos at the very late 2-cell stage block very quickly (in less than one hour) when placed in culture (Bavister, unpublished data), and their viability (judged by oviductal embryo transfer) is compromised within minutes of exposure to Medium 199 or to phosphate- or HEPES-buffered balanced salt solutions (Sato and Yanagimachi, 1972; Farrell and Bavister, 1984). The rapidity of these detrimental effects suggests that the 2-cell block is caused by some physico-chemical imbalance of the culture environment, and/or to the rapid loss from the embryos of some critically important factor normally present in the oviduct. Early cleavage stage embryos may require specific growth factors (see Chapters 8 and 9), but in the case of cultured hamster 2-cell embryos, such factors may be secondary to proper physico-chemical conditions in the culture environment (see section 3.2).

Problems induced by the culture medium environment are also manifested at the 1-cell stage: IVF hamster eggs, which look morphologically normal under the light microscope, nonetheless lack viability, as shown by results of transferring such embryos to recipient females within a few hours after insemination of eggs (Whittingham and Bavister, 1974). However, 1-cell hamster embryos, fertilized either *in vitro* or *in vivo*, will readily proceed through the first cleavage division *in vitro* (Yanagimachi and Chang, 1964; Whittingham and Bavister, 1974; Juetten and Bavister, 1983a,b). Consistently in these studies, about 60 to 80% of *in vitro* inseminated hamster eggs cleaved once in culture. Usually, at least 70% of *in vivo* fertilized eggs cleave to the 2-cell stage *in vitro* (Golden, unpublished data). In spite of this promising start to development *in vitro*, all of the 2-cell embryos are blocked.

A symptom of the 2-cell block has recently been found, consisting of a pronounced swelling of the blastomeres in 2-cell embryos cultured from the 1-cell stage (Fig. 5). This swelling phenomenon is also observed with *in vivo* fertilized embryos cultured from the 2-cell stage, being apparent within a few hours after exposure of the embryos to the culture medium. We have attempted to quantitate the degree of blastomere swelling but it is quite variable in embryos from different donors; in general, the swollen blastomeres contain about 130 to 150% of the volume of normal (freshly-recovered) 2-cell blastomeres (Bavister, unpublished).

Swelling of rat eggs produced by "experimental variations of the medium" was described by Defrise (1933) as one of several "lytic phenomena" but it is unclear from this report whether the author was referring to 1- or 2-cell embryos or which particular variations of the medium produced the

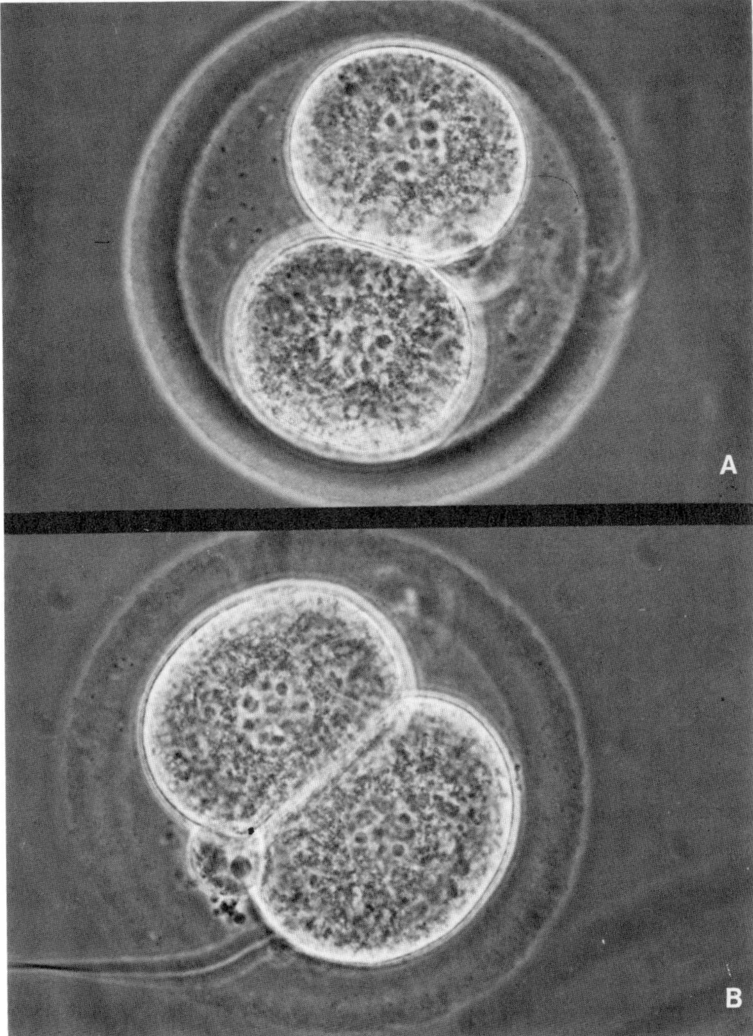

Figure 5. Blastomere swelling effect in cultured hamster embryos. Comparison of 2-cell embryos grown (A) *in vivo*, freshly collected, or (B) *in vitro* following IVF. Embryo (A) has small, spherical blastomeres with minimal contact area and large perivitelline space. Embryo (B) has swollen blastomeres (second polar body also visible) with greatly increased contact area and reduced perivitelline space (tail of accessory [non-fertilizing] sperm also visible). Two-cell embryos grown *in vivo* also usually undergo swelling as in (B) after exposure to culture media for a few hours. Phase-contrast microscopy of living embryos, both at same optical magnification (x 500). From Bavister (unpublished data).

swelling effect. In a detailed discussion of the anomalies present in blocked 2-cell mouse embryos in culture, Goddard and Pratt (1983) described the intercellular flattening between blastomeres found in the majority of cases as being analogous to compaction. This feature is remarkably similar in appearance to the flattening seen in blocked hamster embryos (Fig. 5). Perhaps the cause of the block is similar in these 2 species.

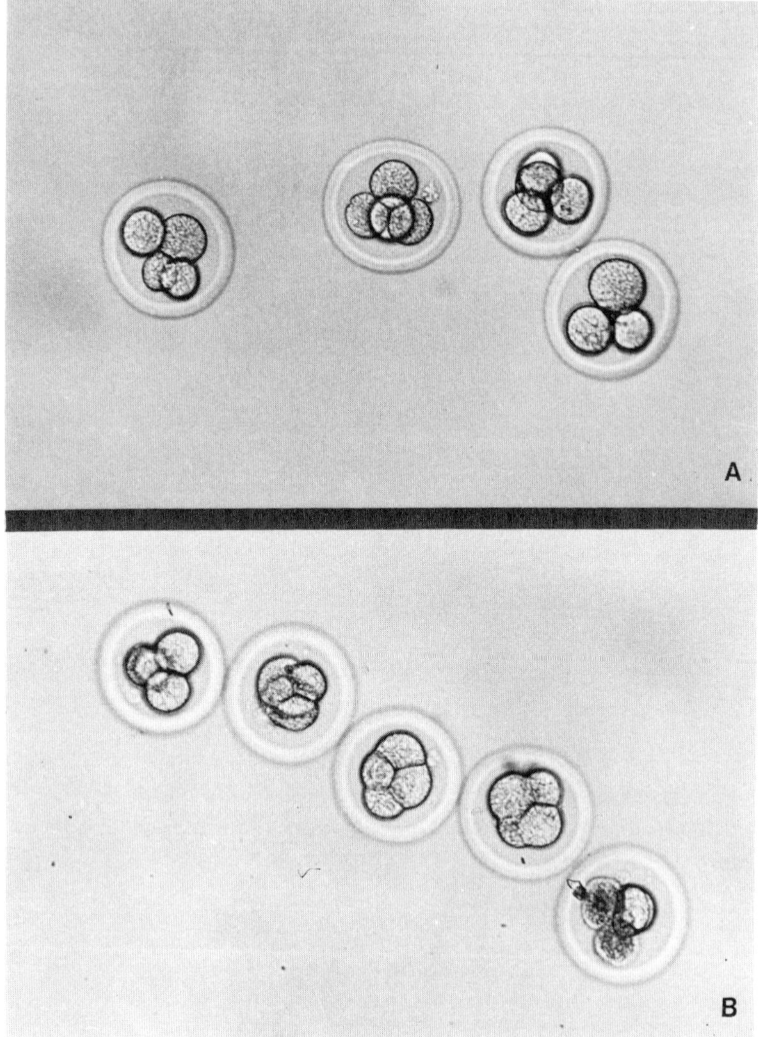

Figure 6. Reaction of 4-cell hamster embryos to culture conditions. (A) Three 4-cell embryos and one 3-cell, freshly collected; blastomeres are small with minimal contact area. (B) Four-cell embryos after several hr in culture: a change resembling premature compaction has occurred although there is no obvious swelling of the blastomeres. One embryo (at right) has degenerated. Phase-contrast, optical magnification x 125. From Bavister (unpublished data).

A different manifestation of the block to development is seen in cultured 4-cell hamster embryos. In this case, the blastomeres usually flatten against each other, again in a manner that superficially resembles compaction, but swelling is not observed (Fig. 6).

It is reasonable to suppose that the blastomere swelling effect is linked in some way with the loss of viability and developmental capacity in cultured hamster 2-cell embryos. We are using it as an indicator of damage to hamster embryos *in vitro*. We hope to eliminate it by modifying the culture medium; this may allow embryos to progress beyond the 2-cell stage in culture. When we discover the cause of the swelling effect, it may have some general relevance for understanding the reactions of cells to artificial environments. We may also find that studies on the swelling effect with hamster embryos are relevant to attempts at culturing embryos of other species. Blastomere swelling can be seen in cultured IVF rhesus monkey and bovine embryos, although no correlations were made with embryo viability in these cases (Bavister *et al.*, unpublished observations; E. Critser, personal communication).

3.2. Possible Causes of the Block

Is the Culture Environment Toxic to Embryos? Recognising the difficulties of identifying possible growth-inhibiting features of the culture medium using 2-cell hamster embryos, we have attempted to throw some light on the problem by obtaining comparative data using 2-cell embryos from C_3B_6/F_1 mice. These embryos will readily grow to the blastocyst stage *in vitro*, using (*e.g.*) BWW or a complex medium such as Ham's F10. Our hypothesis was that if some component(s) of our TALP medium was found to be actually harmful to early cleavage stage mouse embryos, this property might account for the block to development in cultured hamster embryos.

In the first series of experiments, one oviduct of a superovulated, mated mouse was flushed with TALP equilibrated with 5% CO_2 in air. Two-cell embryos were transferred rapidly to dishes of the same medium in the 37° C incubator. The embryos in the remaining oviduct were treated in the same way except that BWW was used for flushing and incubation. Embryos were examined for development to the blastocyst stage after incubation for 52 hr. A paired t-test was used for data analysis. This experiment was replicated 6 times, using one donor for each replicate, on two different days with freshly-prepared medium each time.

The results of these experiments (Table I) showed that medium TALP is not inhibitory to development of mouse 2-cell embryos *in vitro*. Presumably, hamster 2-cell embryos are much more sensitive than mouse embryos to some constituent(s) of the culture environment.

In most of our early experiments with hamster embryos, a HEPES-buffered version of TALP was used to collect embryos and to wash them briefly before transferring them to CO_2-buffered TALP for culture (Bavister *et al.*, 1983a). Conceivably, exposure to HEPES buffer might have been detrimental to embryo viability. We tested this idea with 2-cell mouse embryos. All the embryos from each donor were collected using TALP equilibrated with CO_2, then quickly partitioned into 3 groups: TALP (25 mM HCO_3^-, 5% CO_2 in air), TALP (10 mM HEPES, 25 mM HCO_3^-, 5% CO_2 in air), or TALP-HEPES (2.0 mM HCO_3^-, in air). The pH was 7.4 in all groups. Embryos were examined

Table I
Development of Mouse 2-Cell Embryos in TALP and BWW Culture Media[a]

	Culture medium	
	TALP[b]	BWW[c]
No. of 2-cell embryos cultured	64	82
No. of blastocysts	56	70
% Development[d]	87.5	85.4

[a]Data from Farrell (1983).
[b]TALP: Bavister *et al.* (1983a).
[c]BWW: Biggers *et al.* (1971).
[d]No significant difference between treatments.

Table II
Development of 2-Cell Mouse Embryos in Medium TALP (5% CO_2 in air), TALP-HEPES (5% CO_2 in air) or TALP-HEPES (in air)[a]

	Culture medium		
	TALP (5% CO_2 in air)	TALP-HEPES (5% CO_2 in air)	TALP-HEPES (in air)
No. embryos cultured	78	91	82
No. of blastocysts	45	58	0
% Development	58[b]	64[b]	0[c]

[a]Data from Farrell (1983).
[b,c]Numbers with different superscripts significantly different by 1-way ANOVA ($p < 0.001$).

after 52 hr culture at 37° C. Each donor animal constituted one replicate experiment and 9 replicates were done on 3 different days. Results were analyzed by one-way ANOVA.

Mouse embryos cultured in medium TALP containing 10 mM HEPES developed as well as the control embryos in normal TALP (with 25 mM HCO_3^- and 5% CO_2 in air in both cases), as shown in Table II. This result shows that HEPES itself is not harmful, at least to mouse embryos. As expected, embryos cultured in low HCO_3^- medium without CO_2 failed to develop.

We inferred from these experiments with mouse embryos that medium TALP and HEPES buffer probably were not exerting a toxic effect on hamster 2-cell embryos cultured *in vitro*. As a result, we began to look elsewhere for possible causes of the block to development.

Is There Some Imbalance in the Culture Medium? Some kind of imbalance in the physico-chemical environment seems the most likely explanation for the very rapid loss of viability in cultured 2-cell hamster embryos. Identifying the cause of the block is not easy. Reliable information on the composition of oviduct secretions is scarce and virtually no data are available for the golden hamster. However, judging from careful measurements on oviduct fluids from several species, it is likely that our standard "IVF" culture medium is unphysiological in some important respects. This medium, like most other balanced salt solutions used for cell culture (Bavister, 1981) is based on the ionic composition of blood plasma, which differs from oviduct secretions, particularly in concentrations of Na^+, K^+ and Cl^-. For example, the Na^+/K^+ ratio of TALP medium is about 8 times greater than that found in mouse oviduct fluid, which is 5.4 (Borland *et al.*, 1977). The Na^+/K^+ ratios of oviduct fluid in several other species range between about 10 and 20 (Bavister, 1981), but this is still considerably lower than the ratio in blood plasma, which is about 45. From basic knowledge of the effects of elevated extracellular Na^+ and/or reduced K^+ on somatic cell functions, it could be predicted that exposure of blastomeres of early cleavage stage embryos to these ionic conditions might influence developmental capabilities. Similarly, the Cl^- concentration in TALP medium is much higher than that of oviduct fluid from most species (Bavister, 1981). Adverse extracellular Cl^- levels might produce alterations in the transmembrane Cl^-/HCO_3^- balance (Donnan shift) with potential consequences for intracellular pH control. [The vulnerability of intracellular pH control to apparently innocuous culture conditions is strongly indicated by our results with hamster 8-cell embryos (section 4.2: *CO_2, HCO_3^- and pH*).]

Our initial approach to solving the 2-cell block has been based on this kind of logic. We incubated 1-cell embryos (collected about 8 hr after egg activation) in a variety of modifications of TALP medium. End-points were the percentage of embryos cleaving to the 2-cell stage after 24 hr, the proportion showing blastomere swelling, and the number of embryos able to reach the 4-cell stage after another 24 hr in culture. In a factorially-designed experiment, we found no effect of varying the Na^+/K^+ ratio (45, 15 and 5) at the same time as the Ca^{2+}/Mg^{2+} ratio (4 and 2). In all cases, development to the 2-cell stage was about 70%, all of the embryos showed the swelling effect and none reached the 4-cell stage (Golden and Bavister, unpublished). The same result was obtained when a wider range of Ca^{2+}/Mg^{2+} ratios was tested at a fixed Na^+/K^+ ratio (15). Apparently, the first cleavage division in hamster embryos is unaffected by wide variations in extracellular ion concentrations and ratios, and imbalances in the ions that we examined seem unlikely to be primary causes of the 2-cell block. Other ionic constituents could be important regulators of embryo development, but the primary cause of the block seems likely to reside with other physico-chemical factors.

One of the most obvious factors is osmotic pressure, since there is a pronounced swelling of the blastomeres in cultured hamster embryos (Fig. 5). However, raising the osmotic pressure to almost 400 mOsmols with Na^+ and K^+ salts did not eliminate the swelling effect in cultured 2-cell hamster embryos (Farrell, unpublished data). This leads us to consider the colloid osmotic pressure. Protein concentrations in the oviduct are generally lower than in blood plasma; for example, rat oviduct fluid contains about 20 mg/ml protein (Shalgi *et al.*, 1977), which we calculate would produce about 5 mOsmols colloid osmotic pressure. In contrast, we use 3 mg/ml BSA in our culture medium, which would generate negligible osmotic pressure. Placing early cleavage stage embryos into medium with a low protein concentration could subject them to considerable osmotic stress, sufficient to drive water into the cells and account for the observed swelling effect. One obvious remedy is to raise the BSA concentration in the medium. However, hamster embryos, at least from the 8-cell stage, do not grow as well *in vitro* when the BSA is raised much above 3 mg/ml (Bavister *et al.*, 1983a). Toxic effects of some BSA preparations have been reported (Caro and Trounson, 1984). Alternatives to BSA need to be examined for raising the colloid osmotic pressure of the culture medium.

Several other factors that may be implicated in the 2-cell block are oxygen toxicity, lack of usable energy substrates and disturbances of intracellular pH control. Any of these could be capable of producing the very rapid loss of viability that we have observed on exposure of hamster embryos to culture medium. Implications of O_2 toxicity and of inappropriate energy substrates for culture of mammalian embryos are discussed in Appendix I.

Apart from H^+ and CO_2, little attention seems to have been given to the possibility that some environmental constituents may have profound effects on intracellular pH of cultured embryos. Our results with 8-cell hamster embryos (see section 4.2) indicate that neither extracellular pH nor HCO_3^- concentration are critically important but the concentration of CO_2 and the presence of certain membrane-permeant organic acids may be very important for intracellular pH regulation. This possibility has not yet been sufficiently evaluated with the 2-cell stage embryos.

In our laboratory, slow progress is being made towards identifying the causative factor(s) of the 2-cell block using the empirical approach, *i.e.*, examining one or 2 constituents of the culture environment at a time. This approach is tedious, but it appears to be the most practical route, given the obvious difficulty of obtaining reliable data on conditions for embryo development *in vivo*.

3.3. Culture of Hamster Embryos in Explanted Oviducts

A potentially useful aid to the growth of hamster embryos *in vitro* is to culture them within an oviduct which is itself maintained in culture, so that the oviduct provides the environment that directly nourishes the embryos. This technique was used by Biggers *et al.* (1962) to support the first 3 cleavage divisions of mouse embryos, before culture media were devised for this purpose. Whittingham (1968) also used this method to bypass the 2-cell block in embryos from some strains of mice. We have attempted to utilize a modification of the "raft" culture system described by these authors to obtain development of hamster embryos *in vitro* beyond the 2-cell stage.

Excised oviducts were placed on sterile 25 mm diameter Millipore filters (0.22 μm) covering the center wells of organ culture dishes (Falcon Plastics, no. 3037) which contained 1 ml of TALP equilibrated with 5% CO_2 in air. Initially, hamster ampullae were cultured but these did not usually survive for 48 hr. Ampullae from Swiss albino mice could be maintained in these cultures for at least 48 hr, and so these were tested for ability to support development of hamster embryos. One-cell embryos were collected from superovulated, mated hamsters and inserted into explanted mouse oviducts from which the native eggs had been removed by flushing.

As shown in Table III, only 76 (9%) of a total of 830 hamster zygotes cultured in this way reached the 4- to 8-cell stage. This low rate of success should be weighed against the fact that none of many thousands of embryos incubated directly in culture medium has developed past the 2-cell stage (Bavister, unpublished; Sato and Yanagimachi, 1972; Farrell and Bavister, 1984). The internal environment provided by the cultured mouse oviducts was able to support development to the 4- to 8-cell stage in 94% of fertilized mouse eggs (Table III); 98% of these went on to form blastocysts after another 48 hr culture directly in TALP (data not shown).

The difference between the results with hamster and mouse embryos using the same culture technique is intriguing: the fact that a significant proportion of hamster embryos developed at all indicates that the oviductal milieu is not species-specific and that hamster embryos can develop after removal from their natural environment, while the large between-species difference again emphasizes the differential sensitivity of hamster and mouse embryos to culture conditions.

This cross-species culture technique provides a means of bypassing the 2-cell block so that, for example, *in vitro* fertilized hamster eggs could perhaps be cultured to the 8-cell stage or further. In Whittingham's (1968) experiments, mouse zygotes could develop into blastocysts within explanted oviducts. We might envisage IVF hamster zygotes growing to the blastocyst stage within mouse oviducts, then being transferred to recipient hamsters to test viability; alternatively, mouse oviducts might be used to bypass the 2-cell block, followed by culture of embryos directly in medium from the 8-cell to blastocyst stages before transferring the embryos. With these approaches, fertilization and early development in the hamster could be examined much more extensively than at present.

It may eventually be possible to use the cultured mouse oviduct to grow embryos of some other species that are normally difficult to maintain directly in culture medium, such as bovine or porcine early cleavage stage embryos (Wright and Bondioli, 1981; Davis and Day, 1978). There are numerous reports of successful cross-species preimplantation embryo growth *in vivo* (*e.g.*, Boland, 1984; Eyestone *et al.*, 1985). We might also envisage culturing IVF primate eggs in explanted mouse oviducts to increase the proportion of

Table III
Stage of Development of Hamster 1-Cell and Mouse 1- and 2-Cell Embryos after 48 hr in Cultured Mouse Oviducts[a,b]

Species	No.embryos cultured	No. morulae	No. 4- or 8-cells	No. 3-cells	No. blocked 2-cells	No. 1-cell or degenerate
Hamster	830[c]	0	76 (9)	17 (2)	458 (55)	279 (34)
Mouse	116[d]	75 (65)	34 (29)	0	3 (2.5)	4 (3.5)

[a]Data from Bavister and Minami (1986).
[b]Values in parentheses are percentages of total embryos cultured.
[c]Data from 40 oviduct cultures; embryos should be at 4- to 8-cell stage.
[d]Data from 8 oviduct cultures; embryos should be at 4-cell to morula stage.

embryos able to develop to the morula or early blastocyst stages; the likelihood of continued development of these stages after embryo transfer is greater than with early cleavage stages, on a per embryo basis.

4. GROWTH OF 8-CELL EMBRYOS *IN VITRO*

4.1. Indicators of Development *in Vitro*

In the hamster, the earliest developmental stage that will grow into a blastocyst *in vitro* is the 8-cell embryo collected about 54 hr after egg activation. This "early" 8-cell stage is considerably more difficult to grow in culture than the "late" 8-cell embryo collected only 8 hr later (Bavister *et al.*, 1983a). We hope that this differential sensitivity to culture conditions might indicate that the early 8-cell stage retains some vestige of the block to development observed in earlier cleavage stage embryos; in this case, modifications of the culture environment that would support good development of early 8-cell embryos might also sustain some 4-cell embryos. However, we have not found this to be the case so far: the 4-cell stage remains as refractory as ever to our culture conditions.

In our initial experiments with hamster 8-cell embryos, there is no doubt that many of the blastocysts grown *in vitro* were not normal. First, a large proportion of blastocysts were abnormally small (stages B1 to B2: Bavister *et al.*, 1983a). Very few expanded blastocysts with thinning zonae pellucidae were seen (stage B5). Secondly, the time needed for most blastocysts to grow from the early 8-cell stage *in vitro* was considerably longer (24 to 36 hr) than *in vivo* (about 18 hr: Sato and Yanagimachi, 1972). Thirdly, only a small proportion (10%) of the blastocysts grown in culture were able to develop normally to term following embryo transfer, although many more formed fetuses that were resorbing by day 5 or 6 after transfer (Bavister *et al.*, 1983a). Fourthly, none of the blastocysts was able to hatch from its zona pellucida; after prolonged culture, the embryos degenerated within the intact zonae.

Results using each of these indicators of development *in vitro* have been radically changed by various modifications that we have made to the culture environment. Overall, a very substantial improvement has been made in our ability to grow 8-cell hamster embryos into normal blastocysts *in vitro*, showing that this phase of preimplantation embryo development is very sensitive to changes in the culture environment.

4.2. Factors Regulating Growth of Embryos *in Vitro*

Amino Acids. In our early attempts to grow hamster 8-cell embryos *in vitro*, we found that about 20% of late 8-cell embryos would develop into blastocysts (stages B1 to B4), but virtually none of the early 8-cells would develop. Since Gwatkin and Haidri (1973) had described a requirement for several amino acids for maturation of hamster oocytes *in vitro*, we thought that preimplantation embryos might also show this requirement. This turned out to be true. The amino acids that we tested were glutamine, phenylalanine, methionine and isoleucine (1.0, 0.1, 0.05 and 0.2 mM, respectively). We found that inclusion of these amino acids in our standard TALP medium increased the proportion of early 8-cell hamster embryos that developed to the blastocyst stage within 25 hr from 2% to 36% (Bavister *et al.*, 1983a). The effect of the amino acids on growth of late 8-cell embryos was much less pronounced; the increase was from 22% to 66%. However, with both early and late 8-cell embryos, the effects of amino acids on growth was highly significant ($p < 0.0001$). These data show that late 8-cell embryos are more capable of developing to blastocysts *in vitro* than the early 8-cells, at least under these simple culture conditions, and that the need for exogenous amino acids is much greater with the early 8-cell stage embryos. This difference indicates that a substantial change in metabolic capabilities must be occurring during the 8-cell stage of development.

In order to investigate the amino acid requirements of hamster embryos more thoroughly, we studied the 4 amino acids individually. Early 8-cell embryos were cultured for 24 hr in TALP containing different combinations of 3 of these amino acids. Only the omission of glutamine resulted in a significant ($p < 0.05$) decrease in development to the blastocyst stage. Deletion of glutamine, phenylalanine, isoleucine and methionine resulted in 10%, 37%, 40% and 44% development, respectively (Carney and Bavister, 1985, 1987a). Moreover, when glutamine was used as the sole amino acid, development was supported as well as when all 4 amino acids were present. A critically important role for glutamine in growth of hamster 8-cell embryos is indicated from these results.

Since glutamine was found to be important for growth of hamster embryos, we examined the effect of using 20 amino acids together with glutamine in the culture medium, each at 0.1 mM (Carney and Bavister, 1987a). In this case, growth of 8-cell embryos was depressed compared to growth with glutamine as the only amino acid (39% *vs.* 51%, respectively). Some reduction of development was observed with these 20 amino acids even at 0.01 mM compared with results using glutamine alone. These data suggested that some of the amino acids might compete with essential amino acids for common transport mechanisms. To examine this possibility, 20 amino acids (0.1 mM each) were divided into 5 groups of 4, such that each group contained amino acids that do not compete with one another's transport mechanisms, based on data from studies with mouse embryos (Borland and Tasca, 1974; Keefer and Tasca, 1984). All groups contained glutamine (0.1 mM), which seems essential for hamster embryo development; therefore, we expected that other amino acids would either increase or decrease development over baseline, or produce no effect. One group, containing gln, gly, cys-SH, leu and lys, increased development over baseline results (41% *vs.* 23%, respectively). Two more groups, containing (1) gln, pro, cys, phe and arg, and (2) gln, ser, thr, val and glu, depressed development to about half that of baseline levels (Carney and Bavister, 1987a). Further work is clearly needed to segregate the beneficial amino acids from the detrimental ones, and to determine their optimal concentrations. However, it is apparent from our preliminary results that hamster 8-cell embryos have some rather specific amino acid requirements for growth *in vitro*, quite unlike mouse embryos.

Vitamins. In the same experiments in which we first examined amino acid requirements for growth of 8-cell embryos *in vitro*, we also investigated the effect of vitamins (Bavister *et al.*, 1983a). The vitamin supplement used in

Minimal Essential Medium (Grand Island Biological Co.) was used in TALP medium to culture early and late 8-cell embryos. There was no detectable response to the presence of vitamins in these experiments. After the culture conditions were markedly improved, there was still no effect of vitamins on development up to the late zonal blastocyst stage, although there was a marked effect on hatching (see "*Hatching of Hamster Blastocysts In Vitro*").

Energy Substrates. We have not conducted a thorough screening of substrates capable of supporting development of hamster 8-cell embryos *in vitro.* However, glutamine can support embryo growth to the blastocyst stage in the absence of all other energy substrates (Carney, unpublished data). This amino acid thus seems to be a key factor in development of hamster preimplantation embryos. Some evidence suggests that we may be mistaken in assuming that energy substrate needs are necessarily met because glucose, pyruvate and lactate are included in the culture medium (see Appendix I).

Oxygen Partial Pressure. This component of the culture environment has not been studied in a systematic manner. Our preliminary studies with "normal" (*i.e.*, atmospheric) O_2 *vs.* "low" (5%) O_2 in the gas phase did not show any differences in the proportion of 8-cell embryos reaching the blastocyst stage (Bavister, unpublished). However, these experiments were done at a time when our overall success rate with cultured hamster embryos was low, so any effect of O_2 may not have been rate-limiting. We propose to reexamine the importance of pO_2, now that we can routinely obtain >80% blastocyst development from hamster 8-cell embryos (see "*CO_2, HCO_3^- and pH*"). It would be remarkable if embryos, especially those as sensitive to culture as hamster embryos, were able to develop normally at a pO_2 that is 3 to 4 times the level that probably exists *in vivo* (see Appendix I). We may find that hyperoxia predominantly affects the *viability* of cultured embryos, *i.e.*, their ability to develop normally to term after embryo transfer, rather than their ability to produce apparently normal-looking blastocysts *in vitro* (see section 2.4).

Protein Source and Complex Culture Media. It might be argued that our initially low success rate in growing hamster 8-cell embryos to the blastocyst stage was due to use of a simple balanced salt solution that lacked many of the components needed for embryo development. We have found in fact that complex media are no better than our simple TALP medium in this respect, and in some circumstances they may inhibit embryo growth. In one series of experiments, Ham's F10 medium was compared to TALP, with either BSA (3 mg/ml, Fraction V, Sigma Chemical Co.) or fetal calf serum (FCS, 10%) as the protein source. Early 8-cell embryos from each donor hamster were assigned randomly to one of 4 treatments in a 2 x 2 factorial design; 4 donors were used for each of 3 replicate experiments. Embryos were cultured for 31 hr to allow maximal growth to take place, and the proportions of embryos reaching the blastocyst stage were recorded. The results are shown in Table IV.

No differences were found among 3 of the 4 treatment groups. However, there was a significant interaction ($p < 0.01$) between the Ham's F10 and BSA treatments; none of the embryos developed into blastocysts in this group. Similar results have been obtained recently with medium CMRL-1066 and BSA or FCS (Kraus, unpublished). No 4-cell hamster embryos cleaved in any of

Table IV
Development of Early 8-Cell Hamster Embryos in Simple (TLP) or Complex (Ham's F10) Media with BSA or FCS Protein Supplement[a]

Protein source	No. of embryos developing/no. cultured (mean ± SEM)[b,c] TLP	F10
BSA[d]	37/106 (35 ± 2.3)	0/102 (0.0)
FCS[e]	38/122 (31 ± 10.3)	27/78 (35 ± 2.9)

[a]Data from Farrell (1983).
[b]Experimental endpoint = blastocele formation.
[c]Interaction term significant at $p < 0.01$.
[d]BSA = Fraction V, Sigma Chemical Co. (3 mg/ml).
[e]FCS = Fetal calf serum, GIBCO (10% v/v).

these treatments. These experiments emphasize that problems with embryo development *in vitro* are not eliminated merely by using complex media to replace simple ones.

CO_2, HCO_3^- and pH. From results of our 2-cell embryo transfer experiments (Farrell and Bavister, 1984) we surmised that absence of HCO_3^- from the culture environment might partly be responsible for the observed rapid loss of viability (although in retrospect, the absence of CO_2 may have been a major factor). This idea was applied to culture of hamster 8-cell embryos. Embryos were flushed from reproductive tracts either with TALP-HEPES (air-buffered, with 2 mM $NaHCO_3$) or with TALP (5% CO_2-equilibrated, 25 mM $NaHCO_3$), using one medium for each uterine horn of each donor. The 2 groups of embryos were then cultured separately in 5% CO_2-equilibrated TALP for 24 hr. When late 8-cell embryos were used, there was no significant difference in responses to the 2 collection treatments. However, with early 8-cell embryos, there was a 37% increase in the proportion of embryos developing to the blastocyst stage when using CO_2-equilibrated medium for embryo collection compared to results with TALP-HEPES; this difference was significant, $p < 0.01$ (Carney and Bavister, 1985). Since we believed that HEPES is not toxic to embryos (Table II), the difference in response indicated that hamster embryos might be sensitive to HCO_3^- and/or CO_2 concentrations in the culture medium. Accordingly, a more elaborate 2 x 2 factorial experiment was designed in which 8-cell embryos were collected and cultured for 24hr in TALP containing either 25 mM or 50 mM $NaHCO_3$ equilibrated with 5% or 10% CO_2 (Table V). We found that using 10% CO_2 approximately doubled the proportion of embryos developing to the blastocyst stage in medium containing 25 mM $NaHCO_3$. However, increasing $NaHCO_3$ to 50 mM produced no improvement in response (at 5% CO_2) or it was less effective than 25 mM (at 10% CO_2). There was no critical effect of pH on embryo development over

Table V
Effect of Bicarbonate and CO_2 Concentration in Collection/ Culture Medium on Hamster Embryo Development *in Vitro*[a,b]

	5% CO_2 [$NaHCO_3$][e]		10% CO_2 [$NaHCO_3$][e]	
	25 mM[f]	50 mM	25 mM	50 mM
% Blastocysts (B1 - B5)[c]	44 (85/177)	43 (36/81)	88* (88/99)	74 (43/61)
% Late blastocysts (B3 - B5)[c]	30 (56/177)	34 (27/81)	70** (63/99)	58 (33/61)
% Non-starters (no development)	39 (62/177)	35 (27/81)	4 (4/99)	3 (7/61)
% Hatching[d] (% of total)	17 (39/177)	4 (3/81)	33 (31/99)	9 (6/61)
% Hatching[d] (% of blastocysts)	32 (39/85)	6 (3/36)	36 (31/88)	14 (6/43)
Approximate pH of medium	7.4	7.7	7.1	7.4

[a]Data adapted from Carney and Bavister (1987b).
[b]All embryos collected and cultured in TALP + glutamine + F10 vitamins.
[c]For definition of blastocyst stages B1 - B4, see Bavister *et al.* (1983a); stage B5 is an expanding blastocyst with stretched zona pellucida.
[d]Embryos with lysed zonae pellucidae at 48 hr in culture.
[e]All values expressed as percent development (no. in category/total no. embryos); percentages are calculated as mean of percentages for each replicate.
[f]Control (standard) culture condition.
*Significantly different from control, $p = 0.01$ (by Friedman test [Conover, 1980]).
**Significantly different from control, $p < 0.05$.

the range occurring in this experiment (7.1 to 7.7). The proportion of "non-starter" embryos (8-cell embryos showing no development) was dramatically reduced in 10% CO_2 compared to 5% CO_2 (25 mM $NaHCO_3$): nearly a 10-fold difference was found (4% and 39%, respectively). We consider the proportion of non-starter embryos to be an important indicator of the growth-supporting capabilities of culture media.

The importance of pH and $NaHCO_3$ concentration was evaluated more thoroughly in a second series of experiments. With 10% CO_2 in the gas phase, the $NaHCO_3$ content of the culture medium was varied from 6.25 mM to 50 mM, resulting in a pH range of 6.5 to 7.4. Over these ranges of pH and $NaHCO_3$ concentration, there were no differences in the proportions of embryos developing to the blastocyst stage or in the non-starter category (Fig. 7). These results confirmed the critical importance of pCO_2 for development of hamster 8-cell embryos *in vitro*.

From the data shown in Table V, it is clear that the predominant effect is of CO_2 concentration. Virtually all embryo culture experiments have used 5% CO_2 in the gas phase because this is the "physiological" concentration, which maintains pH at about 7.4 in conjunction with 25 mM $NaHCO_3$, which again is supposedly normal. These assumptions may not be true for all species. In cardiac venous blood of golden hamsters, pCO_2 was found to be 60 mm Hg,

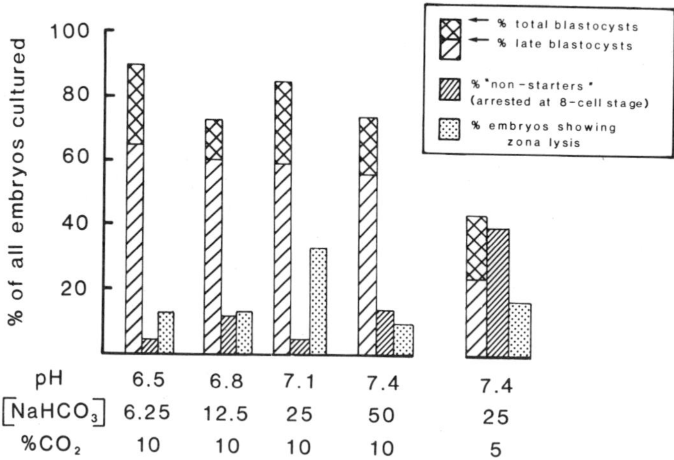

Figure 7. Effect of CO_2, HCO_3^- and pH on development of hamster 8-cell embryos to the blastocyst stage. Embryos cultured for 24 hr in TALP with 4 different $NaHCO_3$ concentrations under 10% CO_2 in air. Data isolated at right are responses of embryos to standard culture conditions (not obtained in same experiments). Adapted from Carney and Bavister (1987b).

corresponding to about 7.5% CO_2, rather than 40 mm, *i.e.*, 5% CO_2 (Lyman and Hastings, 1951). Blood pH was maintained at 7.4 by high levels of HCO_3^- (35 mM). Similar high pCO_2 levels were found by these authors in white rats. In view of these data, our use of 10% CO_2 for hamster embryo culture may not be so unphysiological as it first appears.

We hypothesized that CO_2 is acting as a highly membrane-permeant weak acid to maintain intracellular pH. This idea is predicated on the assumption that within the oviduct, hamster embryos are in equilibrium with CO_2 levels in excess of 5% and/or with extracellular weak organic acids. If the hypothesis is correct, then we should be able to replace the "excess" 5% CO_2 with one of several weak acids. This was tested by culturing 8-cell embryos in TALP containing different concentrations of acetic or lactic acids, using 3.0 mM $NaHCO_3$ under 5% CO_2 to achieve pH 6.5; this low pH reduces ionization of the weak acids and thus favors their passage across the cell membranes. To control against the possibility that these acids were simply being metabolized by the embryos, a non-metabolizable weak acid, 2,4 dimethyl-oxazolidine dione (DMO), was also tested. We found that acetic or lactic acids can indeed substitute for additional CO_2 in stimulating embryo development (Fig. 8). Embryo development in the presence of these weak acids was as good as in 10% CO_2. DMO was only partially effective; however, in later experiments, blastocyst development was as good in DMO with 5% CO_2 as in 10% CO_2 (Carney and Bavister, 1987b). As an additional control for any effect of low pH *per se* on development, some embryos were incubated at pH 6.5 in medium containing only standard (10 mM) lactate; the response in this treatment was hardly different from the normal pH (7.4) control (5% CO_2, 25 mM HCO_3^-). In these experiments, 25 mM lactate produced one of the highest responses; similar high lactate concentrations are found in some mouse embryo culture media (*e.g.*, Whitten's [1971] medium and BMOC-3 [Brinster, 1971, 1972]), which raises the possibility that lactate may be functioning

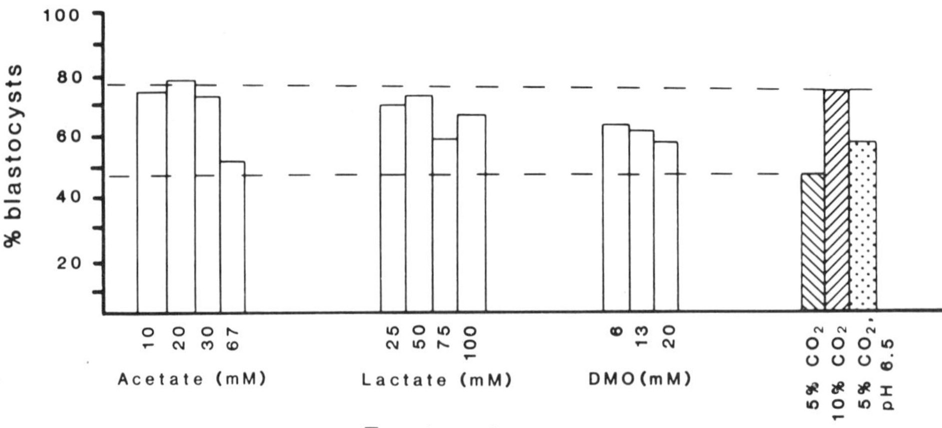

Figure 8. Effect of weak acids on development of hamster 8-cell embryos to blastocysts *in vitro*. All treatments contained 0.1 mM glutamine and F10 vitamins. TALP medium (10 mM lactate standard) was modified by lowering $NaHCO_3$ concentration to 3.0 mM (pH 6.5 under 5% CO_2 in air); acetate (Na salt) or extra lactate added to give final concentrations shown. DMO (2,4 dimethyloxazolidine dione) was also tested as a non-metabolizable weak acid, using 25 mM $NaHCO_3$, 5% CO_2, pH 7.4. Controls (all with 10 mM lactate) were TALP with (a) 25 mM $NaHCO_3$, 5% CO_2, pH 7.4); (b) 25 mM $NaHCO_3$, 10% CO_2, pH 7.1; (c) 3.0 mM $NaHCO_3$, 5% CO_2, pH 6.5. Adapted from Carney and Bavister (1987b).

to maintain intracellular pH in cultured mouse embryos. However, Brinster (1971) found no dose-response effect of CO_2 concentrations between 1 and 10% on development of cultured mouse embryos to the blastocyst stage. Our next experimental approach must be to measure intracellular pH in blastomeres of cultured hamster embryos, using pH-sensitive fluorochromes (*e.g.*, Schatten *et al.*, 1985), in order to directly assess the effect of culture medium composition on this parameter, and to correlate growth-stimulating effects of extracellular acidic components with intracellular pH levels. We have begun to examine viability of blastocysts grown under 10% CO_2 by transfer to pseudopregnant recipients. Results have been quite variable, probably in part because of difficulties with synchronization of recipients with embryos and in part because of mechanical problems with the transfer procedure. Of 21 cultured blastocysts that we recently transferred to 3 recipients, 12 had developed into fetuses when examined on day 9 of pregnancy, compared with 27 fetuses from 55 non-cultured (8-cell) embryos transferred into 3 recipients. Obviously, a much larger data-base is needed, but at least we can conclude from these preliminary results that culture of preimplantation embryos under 10% CO_2 is compatible with apparently normal post-implantation development.

Hatching of Hamster Blastocysts in Vitro. As mentioned in section 4.1, one of the correlates of poor developmental capacity of hamster embryos *in vitro* is that they completely fail to hatch from their zonae pellucidae. Although it is not clear whether rodent embryos normally "hatch" *in vivo* or if the zonae are dissolved by proteolytic enzymes, nevertheless the ability of embryos to escape from the zonae *in vitro* is one indication of their vigor, if not of their viability. Rabbit embryos exhibit some similarities to hamster embryos in that they have a requirement for amino acids and vitamins for blastocele development and hatching *in vitro* (Kane and Foote, 1970). Accordingly, we examined the effect of amino acids and vitamins on hatching of hamster embryos *in vitro*.

Late 8-cell embryos were recovered from donor females (30 to 60 per animal) by flushing reproductive tracts with TALP equilibrated with 5% CO_2 and cultured in the same medium (these experiments were conducted before we discovered the stimulatory effect of 10% CO_2). Embryos from each donor were distributed equally among treatment groups to provide one replicate experiment. Treatments consisted of supplementation with various amino acids and vitamins as used in Ham's F10 medium. After culture for 20 and 40 hr, embryos were examined with an inverted microscope for evidence of blastocele formation and for lysis of zonae pellucidae (lysis was used to denote both dissolution of zonae and splitting). As we had found previously (Bavister *et al.*, 1983a), amino acids (in this case, either glutamine or F10 amino acids) stimulated development of blastocysts, but few embryos showed zona lysis (Table VI). Vitamins alone did not stimulate blastocyst development and none of the embryos lysed their zonae. However, when vitamins and amino acids were used together, the majority of embryos reached at least the early blastocyst stage, and a substantial number showed zona lysis. In parallel experiments, hamster morulae were cultured using the same treatments (data not shown) and essentially the same results were obtained (Kane *et al.*, 1987).

Table VI
Effect of Amino Acid and Vitamin Supplementation on Development of Hamster 8-Cell Embryos *in Vitro*[a]

		20 hr in culture	40 hr in culture	
Supplement[b]	No. embryos cultured	% Blastocysts (no.)	% Zona lysis (no.)	Zona lysis as % of blastocysts
None (control)	34	24 (8)	0 (0)	0
L-glutamine	36	58 (21)*	6 (2)	10
F10 amino acids	35	71 (25)*	3 (1)	4
F10 vitamins	36	22 (8)	0 (0)	0
L-glutamine + F10 vitamins	35	77 (27)*	23 (8)**	30**
F10 amino acids + F10 vitamins	35	74 (26)*	29 (10)***	38***

[a]Data adapted from Kane *et al.* (1987); 7 replicates (embryos from 1 female = 1 replicate).
[b]All embryos cultured in a basic medium of TALP under 5% CO_2 in air, 37° C.
*Significantly different from control and F10 vitamins ($p < 0.001$).
**Significantly different from control and F10 vitamins ($p < 0.025$).
***Significantly different from control, F10 vitamins, F10 amino acids ($p < 0.001$) and L-glutamine ($p < 0.01$). All analyses by Friedman test (Conover, 1980).

We have also made observations on the mode of hatching of hamster embryos, using time-lapse videomicrography (Fig. 9). It should be noted that the time needed for hatching (or zona lysis) *in vitro* is abnormally long, so we cannot claim that the mechanism of hatching or the appearance of embryos are entirely normal. We observed that blastocysts were collapsed at the time they began to escape from their zonae, and numerous fine processes were visible projecting from the surface of embryos at this time (Fig. 9B, C). These processes were seen to be actively waving about (during normal-speed tape playback) and were probably responsible for the ability of hatched embryos to

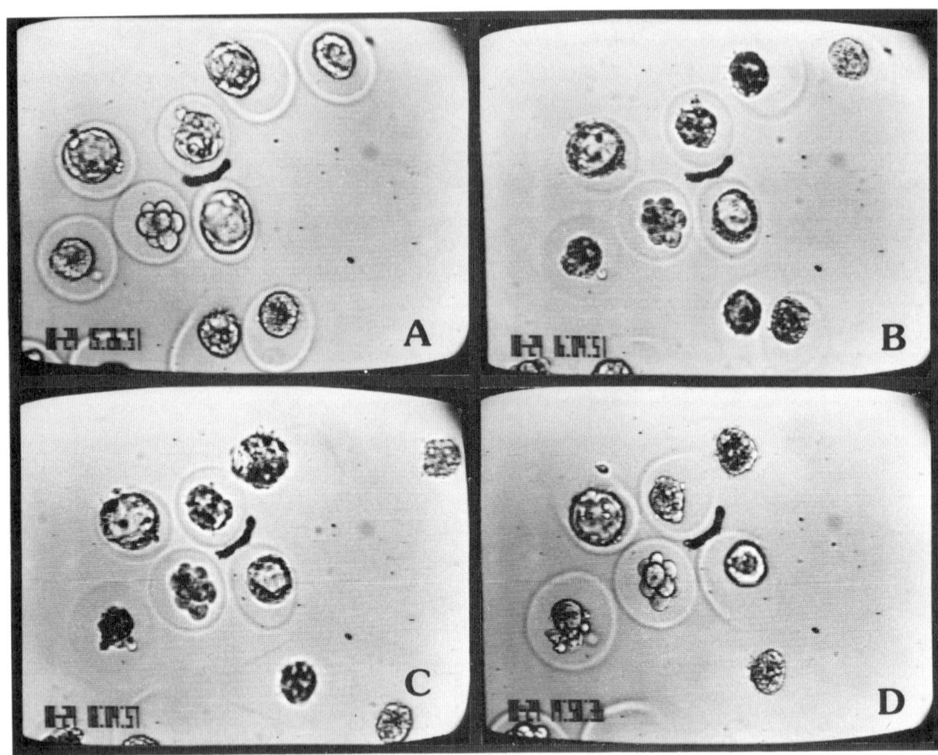

Figure 9. Time-lapse videomicrographs of successive stages in hatching of hamster embryos grown in culture from the 8-cell to the blastocyst stage. (A) Several zonae pellucidae are showing partial dissolution (lysis); one embryo (collapsed blastocyst) at top of group is emerging from a hole in its zona. Elapsed time since start of culture is approx. 48 hr. (B) Fine processes of trophectodermal origin can be seen protruding from the embryo at top of group, which has lysed about 50% of its zona, and from embryo at upper left, which is still enclosed in its zona. In the latter case, processes are immediately beneath area of thinning of zona. (38 min after previous frame). (C) Embryo at upper left has lysed its zona at one point (10 o'clock). Trophectodermal processes are clearly seen in the embryo at top, which has migrated to lie adjacent to another, zona-enclosed embryo. (2 hr after previous frame). (D) Zona pellucida of embryo at upper center has partially lysed adjacent to zona-free embryo at top. Other embryos have partially or completely lysed their zonae, but the zonae of the 8-cell (non-starter) embryo (lower center) and the degenerate embryo (bottom left) remained intact throughout the culture period. (1 hr 46 min after previous frame). Embryos videotaped continuously at 1/108 normal time with 50 x (Nomarski) optical magnification during culture under 10% CO_2 in air in the presence of glutamine and F10 vitamins. Data from Bavister and Carney (unpublished).

move about the culture dish. The processes may be similar to the trophectodermal processes described in guinea pig embryos by Blandau (1949); he suggested that these processes might be involved in hatching and/or in initial attachment to the uterine epithelium (or substratum in culture). We have seen similar processes in rhesus monkey embryos around the time of hatching *in vitro* (see Boatman *et al.*, 1987; and Chapter 13). The cultured hamster embryos did not split their zonae like mouse or monkey embryos; rather, at one side of each zona a region of thinning appeared, which eventually became a sizable hole through which the enclosed embryo seemed to crawl, probably by means of the surface processes (Fig. 9A, B). Of course, the collapsed state of the blastocysts may have distorted the sequence of events. In one case, an escaped embryo moved over to lie adjacent to another embryo that was still enclosed in its zona. The thinning of this zona and its eventual lysis took place directly next to the escaped embryo (Fig. 9C, D), which may indicate that the latter was still producing proteolytic enzymes. We intend to continue studying the development and hatching of hamster embryos in this way, and also to examine their interactions with substrates (cellular and non-cellular), to obtain information about the dynamics of development during the peri-implantation period.

5. CONCLUSIONS

Hamster embryos are very sensitive to artificial culture environments. It is difficult to sustain development of 8-cell embryos *in vitro*; however, examination of the environmental conditions that support growth is beginning to provide interesting data on the factors regulating embryogenesis in this species. No real progress has been made yet in elucidating reasons for the 2-cell block in cultured hamster embryos, which is absolute; however, we hope to find the answer by examining each of the components of the culture milieu in turn, which is not an overwhelming task with simple media.

To date, we have found that hamster 8-cell embryos are not very responsive to changes in external pH or HCO_3^- concentration, but their growth *in vitro* is surprisingly influenced by the pCO_2 in the gas phase. Indirect evidence suggests that this effect is exerted through intracellular pH regulation. This principle may also apply to embryos of other species. Unlike mouse embryos, hamster embryos are quite dependent on amino acids in the medium, with glutamine being of greatest importance, perhaps serving in part as an energy substrate. The presence of vitamins is necessary for escape of the hamster embryo from the zona pellucida *in vitro*, as in the rabbit (see Chapter 10). We still need to examine the possible influences of hormones and growth factors on hamster embryo development *in vitro* (see Chapters 6, 8 and 10).

Data obtained from studying growth requirements of cultured hamster embryos may be useful for overcoming blocks to development observed in cultured embryos of other species. Such blocks seem to be the rule rather than the exception in species other than rabbits and primates. These blocks severely limit the amount of basic information that we can obtain about mammalian preimplantation development in general. Moreover, the inability to culture embryos of some commercially-valuable animals (such as cattle) from the 1-cell to the morula stage hinders the exploitation of experimental embryological techniques for increasing the efficiency of food production. Additionally, when we can grow embryos of most species from the zygote into viable blastocysts *in vitro*, the wealth of new data that will result may be invaluable to efforts aimed at improving the success rate of human IVF and embryo culture procedures.

ACKNOWLEDGMENTS

I am grateful to my associates Dorothy Boatman, Edward Carney, Patrick Farrell, Meg Golden, Emily Kraus, and John Stewart-Savage for providing unpublished data described in this chapter; to Linda Endlich and Robert Dodsworth for their help in preparing the Figures; to the College of Agriculture and Life Sciences, and the Regional Primate Research Center, University of Wisconsin-Madison, for providing research and personal support; and to the NIH for funding some of the work described above (grant no. HD 14765).

6. REFERENCES

Basler, A., 1978, Timing of meiotic stages in oocytes of the Syrian hamster (*Mesocricetus auratus*) and analysis of induced chromosome aberrations, *Hum. Genet.* 42: 67-77.

Bavister, B.D., 1981, Analysis of culture media for *in vitro* fertilization and criteria for success, in: *Fertilization and Embryonic Development In Vitro* (L. Mastroianni, Jr., and J.D. Biggers, eds.), Plenum Press, New York, pp. 41-60.

Bavister, B.D., 1987, A consistently successful procedure for *in vitro* fertilization of golden hamster eggs, *Gamete Res.* (submitted).

Bavister, B.D., and Andrews, J.C., 1987, A rapid sperm motility bioassay procedure for quality control testing of water and culture media, *J. In Vitro Fertil. and Embryo Transfer* (submitted).

Bavister, B.D., and Minami, N., 1986, Use of cultured mouse oviducts to bypass *in vitro* development block in cleavage stage hamster embryos, *Biol. Reprod.* 34 (Suppl. 1): 191a.

Bavister, B.D., and Yanagimachi, R., 1977, The effects of sperm extracts and energy sources on the motility and acrosome reaction of hamster spermatozoa *in vitro*, *Biol. Reprod.* 16: 228-237.

Bavister, B.D., Leibfried, M.L., and Lieberman, G., 1983a, Development of preimplantation embryos of the golden hamster in a defined culture medium, *Biol. Reprod.* 28: 235-247.

Bavister, B.D., Boatman, D.E., Leibfried, L., Loose, M., and Vernon, M.W., 1983b, Fertilization and cleavage of rhesus monkey oocytes *in vitro*, *Biol. Reprod.* 28: 983-999.

Biggers, J.D., 1971, New observations on the nutrition of the mammalian oocyte and the preimplantation embryo, in: *The Biology of the Blastocyst* (R.J. Blandau, ed.), University of Chicago Press, Chicago, IL, pp. 319-327.

Biggers, J.D., Gwatkin, R.B.L., and Brinster, R.L., 1962, Development of mouse embryos in organ cultures of fallopian tubes on a chemically defined medium, *Nature (London)* 194: 747-749.

Blandau, R.J., 1949, Observations on implantation of the guinea pig ovum, *Anat. Rec.* 103: 19-47.

Boatman, D.E., Morgan, P.M., and Bavister, B.D., 1987, Culture of *in vitro* fertilized rhesus monkey oocytes to peri-implantation stages of development, *Biol. Reprod.* (submitted).

Boland, M.P., 1984, Use of the rabbit oviduct as a screening tool for the viability of mammalian eggs, *Theriogenology* 21: 126-137.

Borland, R.M., and Tasca, R.J., 1974, Activation of a Na^+-dependent amino acid transport system in preimplantation mouse embryos, *Dev. Biol.* 30: 169-182.

Borland, R.M., Hazra, S., Biggers, J.D., and Lechene, C.P., 1977, The elemental composition of the environments of the gametes and preimplantation embryo during the initiation of pregnancy, *Biol. Reprod.* 16: 147-157.

Brinster, R.L., 1971, *In vitro* culture of the embryo, in: *Pathways to Conception* (A. Sherman, ed.), Charles C. Thomas, Springfield, IL, pp. 245-277.

Brinster, R.L., 1972, Cultivation of the mammalian embryo, in: *Growth, Nutrition and Metabolism of Cells in Culture*, Vol. II (G.H. Rothblat, and V.J. Cristofalo, eds.), Academic Press, New York, pp. 251-286.

Carney, E.W., and Bavister, B.D., 1985, Development of hamster preimplantation embryos *in vitro*: effect of bicarbonate and amino acids, *Biol. Reprod.* 32 (Suppl. 1): 98a.

Carney, E.W., and Bavister, B.D., 1986, Increased atmospheric carbon dioxide stimulates hamster embryo development *in vitro*, *Biol. Reprod.* 34 (Suppl. 1): 199a.

Carney, E.W., and Bavister, B.D., 1987a, Stimulatory and inhibitory effects of amino acids on development of hamster 8-cell embryos *in vitro*, *J. In Vitro Fertil. and Embryo Transfer* (in press).

Carney, E.W., and Bavister, B.D., 1987b, Regulation of hamster embryo development *in vitro* by carbon dioxide, *Biol. Reprod.* (in press).

Caro, C.M., and Trounson, A., 1984, The effect of protein on preimplantation mouse embryo development *in vitro*, *J. In Vitro Fertil. and Embryo Transfer* 1: 183-187.

Chakraborty, J., 1981, Fine structural abnormalities in the developing mouse embryo, *Gamete Res.* 4: 535-545.

Conover, W.J., 1980, *Practical Nonparametric Statistics*, John Wiley and Sons, New York, pp. 299-305.

Critser, E.S., and First, N.L., 1986, Use of a fluorescent stain for visualization of nuclear material in living oocytes and early embryos, *Stain Technol.* 61: 1-5.

Cross, P.C., and Brinster, R.L., 1973, The sensitivity of one-cell mouse embryos to pyruvate and lactate, *Exp. Cell Res.* 77: 57-62.

Davis, D.L., and Day, B.N., 1978, Cleavage and blastocyst formation by pig eggs *in vitro*, *J. Anim. Sci.* 46: 1043-1053.

Defrise, A., 1933, Some observations on living eggs and blastulae of the albino rat, *Anat. Rec.* 57: 239-250.

Eyestone, W.H., Northey, D.L., and Leibfried-Rutledge, M.L., 1985, Culture of 1-cell bovine embryos in the sheep oviduct, *Biol. Reprod.* 32 (Suppl. 1): 100a.

Farrell, P.S., 1983, A comparative study of culture requirements for hamster and mouse preimplantation embryo development, *M.S. Thesis*, University of Wisconsin-Madison, Madison, WI.

Farrell, P.S., and Bavister, B.D., 1984, Short-term exposure of two-cell hamster embryos to collection media is detrimental to viability, *Biol. Reprod.* 31: 109-114.

Goddard, M.J., and Pratt, H.P.M., 1983, Control of events during early cleavage of the mouse embryo: an analysis of the '2-cell block', *J. Embryol. Exp. Morphol.* 73: 111-133.

Gwatkin, R.B.L., and Haidri, A.A., 1973, Requirements for the maturation of hamster oocytes *in vitro*, *Exp. Cell Res.* 76: 1-7.

Harlow, G.M., and Quinn, P., 1982, Development of preimplantation mouse embryos *in vivo* and *in vitro*, *Aust. J. Biol. Sci.* 35: 187-193.

Hoppe, R.W., and Bavister, B.D., 1983, Effect of removing the zona pellucida on development of hamster and bovine embryos *in vitro* and *in vivo*, *Theriogenology* 19: 391-405.

Hoppe, R.W., and Bavister, B.D., 1984, Evaluation of the fluorescein diacetate (FDA) vital dye viability test with hamster and bovine embryos, *Anim. Reprod. Sci.* 6: 323-335.

Hutz, R.J., DeMayo, F.J., and Dukelow, W.R., 1985, The use of vital dyes to assess embryonic viability in the hamster, *Mesocricetus auratus*, *Stain Technol.* 60: 163-167.

Jackowski, S.C., 1977, Physiological differences between fertilized and unfertilized mouse ova; glycerol permeability and freezing sensitivity. *Ph. D. Dissertation*, University of Tennessee, Knoxville, TN, pp. 15-16.

Juetten, J., and Bavister, B.D., 1983a, The effects of amino acids, cumulus cells, and bovine serum albumin on *in vitro* fertilization and first cleavage of hamster eggs, *J. Exp. Zool.* 227: 487-490.

Juetten, J., and Bavister, B.D., 1983b, Effects of egg aging on *in vitro* fertilization and first cleavage division in the hamster, *Gamete Res.* 8: 219-230.

Kane, M.T., and Foote, R.H., 1970, Culture of two- and four-cell rabbit embryos to the expanding blastocyst stage in synthetic media, *Proc. Soc. Exp. Biol. Med.* 133: 921-925.

Kane, M.T., Carney, E.W., and Bavister, B.D., 1987, Vitamins and amino acids stimulate hamster blastocysts to hatch *in vitro*, *J. Exp. Zool.* (in press).

Keefer, C.L., and Tasca, R.J., 1984, Modulation of amino acid transport in preimplantation mouse embryos by low concentrations of non-ionic and zwitterionic detergents, *J. Reprod. Fertil.* 70: 399-407.

Lyman, C.P., and Hastings, A.B., 1951, Total CO_2, plasma pH and pCO_2 of hamsters and ground squirrels during hibernation, *Amer. J. Physiol.* 167: 633-637.

Mohr, L.R., and Trounson, A.O., 1980, The use of fluorescein diacetate to assess embryo viability in the mouse, *J. Reprod. Fertil.* 58: 189-196.

Orsini, M.W., 1961, The external vaginal phenomena characterizing the stages of the estrous cycle, pregnancy, pseudopregnancy, lactation, and the anestrous hamster, *Mesocricetus auratus* Waterhouse, *Proc. Anim. Care Panel* 11: 193-206.

Orsini, M.W., 1962, Study of ovo-implantation in the hamster, rat, mouse, guinea-pig and rabbit in cleared uterine tracts, *J. Reprod. Fertil.* 3: 288-293.

Orsini, M.W., and Donovan, B.T., 1971, Implantation and induced decidualization of the uterus in the guinea pig, as indicated by Pontamine Blue, *Biol. Reprod.* 5: 270-281.

Ortiz, M.E., Bedregal, P., Carvajal, M.I., and Croxatto, H.B., 1986, Fertilized and unfertilized ova are transported at different rates by the hamster oviduct, *Biol. Reprod.* 34: 777-781.

Sato, A., and Yanagimachi, R., 1972, Transplantation of preimplantation hamster embryos, *J. Reprod. Fertil.* 30: 329-332.

Schatten, G., Bestor, T., Balczon, R., Henson, J., and Schatten H., 1985, Intracellular pH shift leads to microtubule assembly and microtubule-mediated motility during sea urchin fertilization, *Eur. J. Cell Biol.* 36: 116-127.

Shalgi, R., Kaplan, R., and Kraicer, P.F., 1977, Proteins of follicular, bursal and ampullar fluids of rats, *Biol. Reprod.* 17:333-338.

Spielmann, H., Eibs, H.G., and Jacob-Müller, U., 1980, *In vitro* methods for the study of the effect of teratogens on preimplantation embryos, *Acta Morphologica Acad. Sci. Hung.* 28: 105-115.

Stewart-Savage, J., and Bavister, B.D., 1987, Deterioration of stored culture media as monitored by a sperm motility bioassay, *Gamete Res.* (submitted).

Whitten, W.K., 1971, Nutrient requirements for the culture of preimplantation embryos *in vitro*, in: *Schering Symposium on Intrinsic and Extrinsic Factors in Early Mammalian Development, Advances in the Biosciences*, Vol. 6 (G. Raspé, ed.), Pergamon Press, Oxford, pp. 129-141.

Whittingham, D.G., 1968, Development of zygotes in cultured mouse oviducts, *J. Exp. Zool.* 169: 391-398.

Whittingham, D.G., 1975, Fertilization, early development and storage of mammalian ova *in vitro*, in: *The Early Development of Mammals* (M. Balls, and A.E. Wild, eds.), Cambridge University Press, Cambridge, U.K., pp. 1-24.

Whittingham, D.G., and Bavister, B.D., 1974, Development of hamster eggs fertilized *in vitro* or *in vivo*, *J. Reprod. Fertil.* 38: 489-492.

Wright, R.J., Jr., and Bondioli, K.R., 1981, Aspects of *in vitro* fertilization and embryo culture in domestic animals, *J. Anim. Sci.* 53: 702-728.

Yanagimachi, R., and Chang, M.C., 1964, *In vitro* fertilization of golden hamster ova, *J. Exp. Zool.* 156: 361-376.

CHAPTER 12

GROWTH OF DOMESTICATED ANIMAL EMBRYOS IN VITRO

RAYMOND W. WRIGHT, JR. and JAMES V. O'FALLON

1. INTRODUCTION

Early culture systems for sustaining the development of embryos from domesticated animals, *e.g.*, bovine, ovine and porcine, were of two kinds that were designed to meet different objectives. The first system was for long-term culture (days) whereby various media, atmospheres and embryo handling techniques could be studied. The second was a short-term culture system (hours) in which embryos could be held before being transferred to recipient females.

Over the last three decades, it has become possible to nurture embryonic development *in vitro* between the zygote and blastocyst stages, at least partially, in several mammals including the mouse (Brinster, 1965a-d; Chen and Hsu, 1982), rabbit (Maurer *et al.*, 1969; Kane and Foote, 1971; Ogawa *et al.*, 1971; Kane, 1972), hamster (Bavister *et al.*, 1983), ferret (Whittingham, 1975), cow (Wright *et al.*, 1976a,b), pig (Wright, 1977), sheep (Tervit *et al.*, 1972; Wright *et al.*, 1976c), and man (Edwards *et al.*, 1969, 1980). The following reviews, dealing with various aspects of embryo culture and storage in mammals, are recommended: Foote and Onuma (1970), Whittingham (1975), Maurer (1976), Seidel (1977), Anderson (1978), Kane (1978), Biggers (1979), Brinster and Troike (1979), Brackett (1981), and Wright and Bondioli (1981).

The problems remaining in embryo culture are as much a reflection of the current state of cell and organ culture in general (Ham, 1984) as that of embryo culture in particular. This chapter reviews various aspects concerning the *in vitro* growth and metabolism of mammalian embryos, especially of domesticated species, and alludes to areas in which increased knowledge is required for a complete understanding of embryo development.

Raymond W. Wright, Jr. and **James V. O'Fallon** Department of Animal Sciences, Washington State University, Pullman, Washington 99164-6332, USA.

Table I
Constituents of Complete Media Used for the Culture of Embryos from Farm Animals

Components	Ham's F-10 mg/liter	MEM[a] (Earle's salts) mg/liter	Medium-199[b] (Earle's salts) mg/liter	Modified Ham's F-10 mg/liter
Inorganic salts:				
$CaCl_2 \cdot 2H_2O$	44.10	200.00	200.00[c]	251.00
$CuSO_4 \cdot 5H_2O$	0.0025			
$FeSO_4 \cdot 7H_2O$	0.834			
KCl	285.00	400.00	400.00	356.00
KH_2PO_4	83.00			162.00
$Fe(NO_3)_3 \cdot 9H_2O$			0.72	
$MgSO_4$ (anhyd.)		97.67	97.67	
$MgSO_4 \cdot 7H_2O$	152.80			294.00
NaCl	7400.00	6800.00	6800.00	5696.00
$NaHCO_3$	1200.00		2200.00	2984.00
$NaH_2PO_4 \cdot H_2O$		140.00	140.00	
$Na_2HPO_4 \cdot 7H_2O$	290.00			
$ZnSO_4 \cdot 7H_2O$[c]	0.0288			
Other components:				
Glucose	1100.00	1000.00	1000.00	
Hypoxanthine	4.00		0.300	
Lipoic acid	0.20			0.20
Phenol red	1.20	10.00	20.00	
Sodium pyruvate	110.00			55.00
Thymidine	0.70			
Amino acids:				
L-alanine	9.00		50.00[d]	9.00
L-arginine HCl	211.00	126.00	70.00	211.00
L-asparagine H_2O	15.01			15.01
L-aspartic acid	13.00		60.00[d]	13.00
L-cysteine	25.00		20.00	25.00
L-cysteine HCl H_2O		31.29	0.110	
L-glutamic acid	14.70		150.00[d]	14.70
L-glutamine	146.00	292.00	100.00	146.00
Glycine	7.51		50.00	7.51
L-histidine HCl H_2O	23.00	42.00	21.88	23.00
L-isoleucine	2.60	52.00	40.00[d]	2.60
L-leucine	13.00	52.00	120.00[d]	13.00
L-lysine HCl	29.00	72.50	70.00	29.00
L-methionine	4.48	15.00	30.00[d]	4.48
L-phenylalanine	5.00	32.00	50.00[d]	5.00
L-proline	11.50		40.00	11.50
L-serine	10.50		50.00[d]	10.50
L-threonine	3.57	48.00	60.00[d]	3.57
L-tryptophan	0.60	10.00	20.00[d]	0.60
L-tyrosine	1.81	52.10	40.00	1.81
L-hydroxyproline			10.00	
L-valine	3.50	46.00	50.00[d]	3.50

(cont.)

Table I (cont.)
Constituents of Complete Media Used for the Culture of Embryos from Farm Animals

Components	Ham's F-10 mg/liter	MEM[a] (Earle's salts) mg/liter	Medium-199[b] (Earle's salts) mg/liter	Modified Ham's F-10 mg/liter
Vitamins:				
Biotin	0.024		0.010	0.024
D-Ca pantothenate	0.175	1.00	0.010	0.715
Choline chloride	0.698	1.00	0.500	0.698
Folic acid	1.320	1.00	0.010	1.320
i-Inositol	0.541	2.00	0.050	0.541
Niacinamide	0.615	1.00	0.025	0.615
Pyridoxine HCl	0.206	1.00	0.025	0.206
Riboflavin	0.376	0.10	0.010	0.376
Thiamine HCl	1.000	1.00	0.010	1.000
Vitamin B_{12}	1.360			1.360

[a]MEM = Minimum Essential Medium.
[b]Medium-199 contains numerous other components not found in the other 3 media.
[c]$CaCl_2$ anhydrous for Medium-199.
[d]DL amino acid mixtures for Medium-199.

2. MEDIA, SUPPLEMENTS AND ANTIBIOTICS

Embryos obtained from domestic animals have been cultured in a wide variety of defined and undefined media. Defined media are the choice when the objective is to study aspects of embryonic development. However, when the objective is to provide a system that supports *in vitro* embryo survival, then the appropriate medium is that which is effective. Several so-called complete and simple media have been used for flushing and culturing mammalian embryos, and a few studies have attempted to compare growth of bovine, ovine and porcine embryos in these media. The components of these media vary considerably; they are presented in Tables I and II.

2.1. Components of Synthetic Media

Almost nothing is known about the inorganic salt requirements for the culture of embryos from domestic animals. However, it is generally accepted that the principal function of NaCl is to regulate the osmolarity of the culture medium. Wales (1970) examined the influence of several inorganic salts on the development of preimplantation mouse embryos. In summary, he found that development occurred over a wide range of K^+ concentrations (0.6 mM to 48 mM) but was almost completely inhibited by the absence of K^+. Restall and Wales (1966) found the concentration of K^+ to be much higher in the ewe reproductive tract than in plasma. Two synthetic culture media formulated on the basis of oviduct and uterus secretions have reflected this high level of K^+ (Tervit *et al.*, 1972; Ménézo, 1976). The absence of Ca^{2+} ions in media inhibits cleavage and prevents the compaction of mouse morulae. However, apparently normal development has occurred over a relatively wide range (0.4 to 10 mM)

Table II
Constituents of Simple Media Commonly Used for the Culture of Embryos from Farm Animals

Constituent	Whitten's medium[a] g/liter	SOF[b] g/liter	BMOC-2[c] g/liter	BMOC-3[d] g/liter	Dulbecco's PBS[e] g/liter
Inorganic salts:					
$CaCl_2$		0.190	0.189	0.189	0.10
KCl	0.356	0.534	0.356	0.356	0.20
KH_2PO_4	0.162	0.161	0.162	0.162	0.20
$MgCl_2.6H_2O$		0.010			
$MgSO_4.7H_2O$	0.294		0.294	0.294	
NaCl	4.00	6.290	6.975	5.546	8.00
$NaHCO_3$	2.106	2.106	2.106	2.106	
$Na_2HPO_4.7H_2O$					2.16
Other components:					
Glucose	1.000	0.270		1.000	
Na pyruvate	0.036	0.036	0.028	0.056	
Ca lactate.$5H_2O$	0.527				
Na lactate	2.416	0.370	2.264	2.253	
Bovine serum albumin	1.00	32.00	1.00	5.00	

[a]Whitten (1971).
[b]Synthetic Oviduct Fluid (Tervit *et al.*, 1972).
[c]Brinster's Medium for Ovum Culture (Brinster, 1965d).
[d]Brinster (1971).
[e]Phosphate Buffered Solution (Dulbecco and Vogt, 1954).

of Ca^{2+} concentrations (Whitten, 1971; Ducibella and Anderson, 1975). The absence of PO_4, Mg^{2+} and SO_4 in media also appeared to have little effect on development of mouse embryos to the blastocyst stage *in vitro* (Wales, 1970).

An important role of HCO_3^- in culture medium is the regulation of pH. Studies by Brinster (1972) have shown that removal of HCO_3^- and CO_2 from the culture medium results in reduced development. Further investigations have shown that the mouse embryo fixes CO_2 starting at the 8-cell stage and continuing to the blastocyst stage, when overall embryo metabolism is highest (Wales *et al.*, 1969; Graves and Biggers, 1970; Quinn and Wales, 1971).

It is not possible to maintain bicarbonate-buffered media under a CO_2 atmosphere during current nonsurgical bovine and equine embryo collection procedures. This has led to a widespread replacement of HCO_3^- with phosphate-buffered media. This practice is of concern, particularly because Quinn and Wales (1973) found that development of 2- to 8-cell mouse embryos was significantly decreased when they were cultured in phosphate-buffered medium. Similarly, Trounson *et al.* (1976) cultured 6- to 7-day bovine embryos for 48 hr in phosphate-buffered saline (PBS) supplemented with 20% fetal calf serum (FCS) and found that only 13 of 26 embryos survived transfer to recipient females. No reports have evaluated bovine pre-morula stage embryos for their capacity to develop in phosphate *vs.* bicarbonate-buffered media.

2.2. Osmolarity and pH

Osmolarity and pH have seldom been studied as isolated treatments in systems used for the culture of embryos from farm animals. Results with laboratory animals, however, indicate that 2-cell mouse embryos develop into blastocysts in media ranging in osmolarity from 0.200 to 0.354 Osmols (Brinster, 1965a) and that 2-cell rabbit embryos develop to blastocysts in media ranging from 0.230 to 0.339 Osmols (Naglee *et al.*, 1969). Likewise, hamster embryos can develop over a wide range of osmolarity (Bavister *et al.*, 1983). A study in our laboratory showed a wide range in osmolarity among commonly used media and supplements (Table III). It appears that the effect of osmolarity of the medium on embryo development is minimal.

Most bicarbonate-buffered media are equilibrated with the gas in which the culture will be conducted, generally 5% CO_2 in air or 5% CO_2, 5% O_2, 90% N_2. Researchers at our laboratory measured the pH of the media and supplements commonly used for the culture of embryos from farm animals and found the range to be 7.1 to 7.4. Bovine serum albumin (BSA) has the tendency to lower the pH from a physiological range when used in concentrations similar to that (32 mg/ml) found in "synthetic oviduct fluid" (SOF: Tervit *et al.*, 1972). Phosphate buffered medium without supplements had a pH of 7.30. PBS with 1 or 30 mg/ml BSA or 10 or 20% FCS had pH levels of 7.21, 7.09, 7.25

Table III
Comparison of Measured Osmolarities of Complete and Simple Culture Media Used for the Culture of Embryos from Farm Animals[a,b]

		Medium supplemented with:			
		BSA, w/v[c]		Fetal calf serum, v/v	
Media	No supplement	1%	3%	10%	20%
Complete[d]					
Ham's F-10	293 ± 5	303 ± 6	314 ± 6	306 ± 5	301 ± 4
MEM	289 ± 5	294 ± 3	298 ± 4	285 ± 4	286 ± 3
TCM-199	289 ± 6	304 ± 4	304 ± 4	291 ± 4	284 ± 3
Modified Ham's F-10	270 ± 6	--	--	--	--
Simple[d]					
BMOC-3	316 ± 4	--	--	--	--
PBS	283 ± 3	284 ± 3	286 ± 4	281 ± 4	282 ± 3
SOF	276 ± 3	--	--	--	--
Whitten's	290 ± 3	--	--	--	--

[a]Values shown are means ± SD of 10 measurements of each medium in mOsmols determined with a vapor pressure osmometer.
[b]All media contained 1% (v/v) of antibiotic-antimycotic solution, Grand Island Biological Company.
[c]Bovine serum albumin, Fraction V, Sigma Chemical Co.
[d]Formulations as described in Table I or II.

and 7.15, respectively. No studies have been conducted with embryos of farm animals to define the optimal pH for growth in culture. However, studies with mice (Brinster, 1965a) and rabbits (Kane, 1974) indicated that development occurred over a wide range of pH (6.0 to 7.8). Thus, pH variation over a fairly wide range probably has a minimal effect on embryo development *in vitro*.

2.3. Energy Substrates

The early work of Whitten (1957) and Brinster (1965b) with mice and that of Kane (1976) with rabbits showed clearly that embryos require certain substrates at specific stages of development. No such studies have been conducted on the embryos of farm animals; however, Davis and Day (1978) have suggested that porcine embryos have a requirement for α-ketoglutarate and that lactate and pyruvate inhibit development. These results have not been confirmed by other laboratories, and satisfactory development has been achieved in media that contain lactate and pyruvate (Graves *et al.*, 1977; Wright, 1977; Lindner and Wright, 1978). The significance of these observations, as well as the interaction of energy substrates with other components of the culture systems, remains to be determined.

2.4. Amino Acids and Vitamins

Mouse embryos have developed to the blastocyst stage in a medium containing BSA or polyvinylpyrrolidone (PVP) without free amino acids (Brinster, 1965c; Cholewa and Whitten, 1970). However, amino acids are required for rabbit embryo development *in vitro* (Kane and Foote, 1970). Almost all media used for flushing or culturing embryos from farm animals contain amino acids, BSA or blood sera. It has been suggested that, unless careful dialysis is performed on BSA or serum supplements, free amino acids may be present (Wright *et al.*, 1978). Thus, probably few, if any, researchers have cultured pre-morula stage embryos of farm animals in amino acid-free medium.

Kane and Foote (1970) showed that the omission of a group of 11 water-soluble vitamins from a complex culture medium decreased the proportion of rabbit embryos forming blastocysts. However, vitamins do not seem to be necessary for blastocyst formation in the mouse (Kane, 1978), cattle (Tervit *et al.*, 1972; Kanagawa *et al.*, 1975), sheep (Tervit *et al.*, 1972; Trounson and Moore, 1974; Wright *et al.*, 1976c) and pig (Wright, 1977; Davis and Day, 1978; Lindner and Wright, 1978). No studies have examined the influence of specific vitamins on embryo development, except for one by Ménézo (1976), who used vitamin C to maintain the oxidation-reduction potential characteristic of the female tract.

2.5. Trace Elements

Trace elements are not required for blastocyst formation in the mouse or in the rabbit (Kane and Foote, 1970). However, Kane (1978) has suggested that analytical grade salts, which comprise the major salt constituents of the medium, may be contaminated with enough trace elements to supply the embryos' needs [see also Chapter 10 (Ed.)]. The role of trace elements in the culture of embryos from farm animals remains to be studied.

2.6. Macromolecules

Many studies have shown that the growth and development of mammalian embryos are enhanced when a macromolecular component is present in the culture medium. Cholewa and Whitten (1970) have demonstrated that BSA can be replaced with PVP for the culture of 2-cell mouse embryos to the blastocyst stage. However, BSA has been shown to be necessary for blastocyst formation in the rabbit (Kane and Foote, 1970). The total role of BSA in embryo development has not been fully explained. Commercially prepared BSA is a relatively impure protein that contains several low molecular weight proteins, fatty acids, and steroids (Kane, 1978). Undoubtedly, these contaminants may play an important role in embryo development *in vitro* [see Chapter 10 (Ed.)]. Lindner *et al.* (1979) demonstrated a beneficial effect of increasing BSA concentration on pre-morula ovine embryo development. However, Wright *et al.* (1976a) in a study with bovine embryos, and Wright *et al.* (1976c) with ovine pre-morula embryos, found no beneficial effect of increasing BSA concentrations from 1 to 4 mg/ml.

In addition to BSA, various sera have been used as media supplements. Generally, sera are heat-treated at 56°C for 30 min for the removal of compounds that have been shown to be toxic to embryos (Chang, 1949). As with BSA, the contribution of serum components to a culture system is not clear, but they generally exert a beneficial effect. Wright *et al.* (1976a) found that pre-morula bovine embryos developed slightly better in a variety of complete media supplemented with 10% rather than 20% heat-treated fetal calf serum (HTFCS). Generally, sera are diluted with some type of culture medium, but the amount and type of sera appropriate for each species and stage of development have not been determined. However, it appears that serum supplements of greater than 10% offer little advantage in the culture of embryos from farm animals.

Recently Allen *et al.* (1982) reported that the addition of 10% normal steer serum (NSS) was as effective a protein supplement as the more expensive fetal or newborn calf serum. This finding was confirmed (Canfield *et al.*, 1983) when NSS was shown to be effective for short term storage for bovine embryos in PBS-based medium.

2.7. Gas Atmospheres

Two gaseous atmospheres, 5% CO_2 in air and 5% CO_2, 5% O_2, 90% N_2, have been routinely used for the culture of mammalian embryos. The N_2 component of the gas mixture is considered to be inert. The 5% CO_2 component, in combination with 25 mM HCO_3^-, is used to regulate the pH at about 7.4. There is some evidence that bovine (Tervit *et al.*, 1972), ovine (Tervit *et al.*, 1972; Trounson and Moore, 1974; Wright *et al.*, 1976c) and porcine (Wright, 1977) pre-morula stage embryo development is superior under a 5% *vs.* a 20% oxygen atmosphere. However, Wright *et al.* (1976a), by culturing bovine embryos in microdrops of medium under paraffin oil, were unable to confirm the beneficial effect of the reduced oxygen atmosphere. These observations demonstrate the difficulty of comparing results obtained with different culture systems. Studies with mice suggest that the beneficial effect of a 5% O_2 atmosphere is most pronounced between the 1- and 2-cell stages (Whitten, 1971). Brinster and Troike (1979) have postulated that the oxygen effect is due to an alteration in the oxidation-reduction potential in the embryo, which at the early stages of development is most dependent on the $NAD^+/NADH$ ratio, and pyruvate and lactate (Brinster, 1965c).

3. GLUCOSE METABOLISM DURING EMBRYO DEVELOPMENT

3.1. Glucose Metabolism in General

The metabolism of glucose and its metabolites in the mammalian embryo has been the topic of many studies. An extensive review of this work is beyond the scope of this presentation and consequently the reader is referred to some prior reviews (Biggers and Stern, 1973; Brinster, 1973; Wales, 1975; Brackett, 1981; Pike, 1981). Only the following selected comments will be made.

It is now clear from experiments with mouse embryos that during preimplantation development there is a gradual change in energy substrate requirements (see Brinster, 1973). In the beginning, the oocyte needs pyruvate (or oxaloacetate) in the culture medium, but the 2-cell stage is able to develop when supplied with pyruvate, oxaloacetate, lactate or phosphoenolpyruvate.

Table IV

Metabolism of Glucose Labeled as [5-^{3}H], [1-^{14}C], or [6-^{14}C] by Preimplantation Mouse Embryos

Developmental stage	(A) [5-^{3}H] Glucose → ^3H_2O (pmol/4hr/embryo)[a]	(B) [1-^{14}C] Glucose → $^{14}CO_2$ (pmol/4hr/embryo)[a]	[6-^{14}C] Glucose → $^{14}CO_2$ (pmol/4hr/embryo)[a]	PPP (%)
2-Cell	2.12 ± 0.16 (31)	1.44 ± 0.05 (23)	1.05 ± 0.05 (21)	15.8
4-Cell	2.47 ± 0.16 (15)	1.03 ± 0.05 (13)	0.59 ± 0.03 (15)	9.5
8-Cell	2.52 ± 0.11 (14)	0.99 ± 0.08 (7)	0.36 ± 0.08 (7)	12.0
Compacted	5.94 ± 0.36 (18)	2.19 ± 0.10 (16)	0.47 ± 0.05 (11)	13.6
Early blastocyst	9.16 ± 0.54 (13)	2.76 ± 0.13 (14)	0.71 ± 0.06 (15)	9.5
Late blastocyst	20.05 ± 1.73 (7)	2.21 ± 0.15 (5)	0.39 ± 0.04 (5)	3.2
Compacted morula plus 50 μM DNP	9.79 ± 0.66 (9)	3.03 ± 0.47 (6)	2.86 ± 0.25 (6)	1.0
Compacted morula plus 25 μM PES	10.02 ± 0.80 (8)	8.84 ± 0.35 (9)	0.44 ± 0.05 (6)	71.0

[a]Values are mean ± S.E.M. for the number of individual embryo incubations given in parentheses.

After the 8-cell stage, the embryo is able to survive and develop using glucose or any of a number of other substrates including pyruvate, oxaloacetate, lactate, phosphoenolpyruvate, malate, α-ketoglutarate, acetate, or citrate (Brinster and Thomson, 1966). When the embryo reaches the blastocyst stage or at about the time of implantation, energy source requirements and energy metabolism are apparently similar to those of most adult cells. Some exceptions to the generalized findings with mouse and rabbit embryos (Brinster, 1965b; Daniel, 1967) have been reported. These include the inability of glucose to promote development of 8-cell sheep embryos or 16-cell cow embryos (Boone *et al.*, 1978) and the inhibition of development of 4-cell pig embryos by pyruvate (Davis and Day, 1978).

3.2. Pentose Phosphate Pathway Activity

The pentose phosphate pathway (PPP) of glucose metabolism plays a very important role in the development of an embryo. It generates both NADPH, which serves as a hydrogen and electron donor in reductive biosynthesis, and ribose-5-phosphate, which, along with its derivatives, is a constituent of ATP, coenzyme A, NAD^+, FAD^+, RNA, and DNA.

We have recently made the first quantitative determination of this pathway in preimplantation embryos (O'Fallon and Wright, 1986). The quantitative determination requires information on C-1, C-6, and total glucose metabolism, and the use of a mathematical formula proposed by Katz *et al.* (1966). Certain assumptions are required including the complete equilibration of the hexose phosphate pool and low activity of fructose-1, 6-diphosphatase and glucose-6-phosphatase. The formula can be represented as follows:

$$\text{Pentose phosphate pathway (PPP) activity} = \frac{S}{3-2S}$$

$$\text{where } S = \frac{G_1CO_2 - G_6CO_2}{1 - G_6CO_2}$$

and where GCO_2 is the specific yield of CO_2, defined as the fraction of the utilized glucose recovered as $^{14}CO_2$. Total glucose utilization is estimated from the rate of ^3H_2O production from [5-^{3}H] glucose (Katz and Rognstad, 1966; Neeley *et al.*, 1972; Fisher and Reicherter, 1984).

The production of CO_2 from glucose labeled in the first and sixth positions and ^3H_2O generated from [5-^{3}H] glucose by preimplantation mouse embryos is presented in Table IV. Pentose phosphate pathway activity varied with developmental stage, being highest in the 2-cell (15.8%) and lowest in the late blastocyst (3.2%) stages. This activity dropped as the embryo developed from the 2-cell to the 4-cell stage, then increased during development to the compacted morula stage. From there it gradually decreased again to the low value found at the late blastocyst stage.

Glucose-6-phosphate dehydrogenase, the rate-controlling enzyme of the PPP, is X-chromosome linked. It is difficult, however, to ascertain whether or not the observed PPP pattern has any correlation with X-chromosome activation and inactivation (Epstein *et al.*, 1978; Kratzer and Gartler, 1978).

Dinitrophenol (DNP), an uncoupler of oxidative phosphorylation, increased C-6-glucose metabolism 6 times while barely affecting that of C-1-glucose (O'Fallon and Wright, 1986). Phenazine ethosulfate (PES), an agent that oxidizes NADPH, increased C-1-glucose metabolism 4-fold but did not affect metabolism from C-6-glucose. The net effect of DNP was to decrease PPP activity from 12% to a very low level (1%), while that of PES was to increase the activity from 12% to 71%. The latter result indicates that the pentose phosphate cycle in preimplantation mouse embryos is dynamically controlled by the feedback inhibition of NADPH (Fabregat *et al.*, 1985). Furthermore, mouse embryos possess a tremendous reserve PPP potential, which is the case in certain other tissues, *e.g.*, brain and granular pneumocytes (Hothersall *et al.*, 1979; Fisher and Reicherter, 1984).

Because the PPP releases only C-1 as CO_2, while C-1 and C-6 are metabolized identically when processed through the Embden-Meyerhof and Krebs pathways, the C-1/C-6 ratio appears to be a feasible way to determine such an activity. However, no such correlation was obtained (Table IV).

An explanation of why C-1/C-6 is not an accurate indicator of PPP activity can be made as follows. From the above definition of PPP, 1/S is directly proportional to 1/PPP and thus as S increases so does PPP. However, S bears no relationship to the C-1/C-6 ratio. A low S (and hence low PPP) can be obtained whether the C-1/C-6 ratio is low or high. Likewise a high S (high PPP) is obtained whether the C-1/C-6 ratio is low or high. Furthermore, when G_1CO_2 is low (and hence G_6CO_2 is low also), the PPP is low (as must be the case by definition), although the ratio of G_1CO_2/G_6CO_2 (actually C-1/C-6) could be low or high. Thus, for these reasons, the C-1/C-6 ratio is not an accurate indicator of PPP activity.

3.3. Determining Metabolism in Individual Embryos

Since mammalian embryos can only be obtained in relatively small numbers, it is a goal of researchers to gather as much information as possible from the available embryos. We have recently developed an incubation chamber with which we can continuously monitor metabolism by single embryos of any radiolabeled compound catabolized to ^3H_2O or $^{14}CO_2$. The method employed is non-invasive and after metabolic sampling embryos can be successfully transferred to pseudopregnant recipients (O'Fallon and Wright, 1986).

The microvolume metabolic incubation chamber is depicted in Figure 1. This chamber possesses the following characteristics:

(1) The small (3 μl) incubation compartment allows:
   (a) the maintenance of a high specific activity of radiolabeled metabolic substrate at reasonable economic cost;
   (b) complete exchange of ^3H_2O produced by an embryo from a [^{3}H] substrate with unlabeled H_2O in the CHES trap; and
   (c) the maintenance of a high concentration of an expensive drug, hormone, or protein at minimal cost.

(2) The large (1.5 ml) CHES trap serves three purposes:
   (a) it provides sufficient water vapor pressure to maintain the 3 ml incubation medium against dehydration;

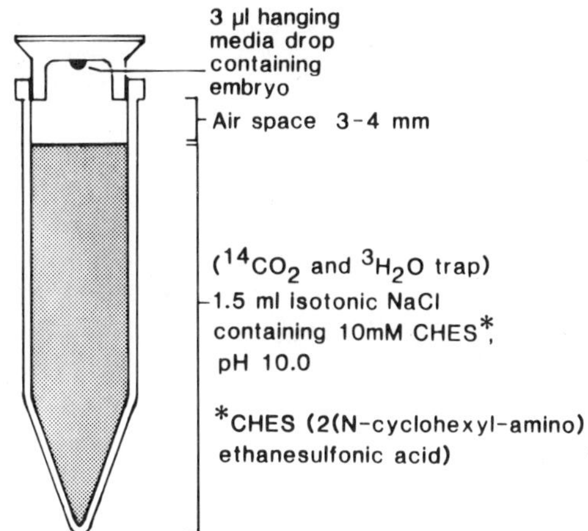

Figure 1. Microvolume metabolic incubation chamber (from O'Fallon and Wright, 1986, with permission).

(b) it allows the continuous entrapment of $^{14}CO_2$ produced by an embryo from a [^{14}C] substrate; and

(c) it permits the continuous collection of $^{3}H_2O$ from the incubation compartment. Thus, $^{3}H_2O$ is separated from [^{3}H] glucose as easily as $^{14}CO_2$ is from [^{14}C] glucose.

(3) In combination, the two compartments provide a system with which the metabolic activity of single embryos can be momentarily sampled without sacrificing the viability of the embryo. In fact, as soon as an incubation period is over, further studies with the developing embryo can ensue.

(4) The chamber is a closed system and thus different gaseous atmospheres can be used during the incubation. It should be noted that an acidic trap would be needed when a CO_2 atmosphere was used, thus eliminating $^{14}CO_2$ collection.

(5) It can be mentioned that the chamber will also be of value when radiolabeled substrates are not used. For example, because of its small size, the 3µl compartment will concentrate any products, *e.g.*, steroid hormones or proteins, that are released by the embryo, thus eliminating the need for extensive and sometimes destructive methods of concentrating such molecules.

As implied, both $^{3}H_2O$ and $^{14}CO_2$ can be collected simultaneously. In fact, we routinely conduct double label experiments with embryos in this chamber. In this respect, the chamber (Fig. 1) replaces the prototype previously described (O'Fallon and Wright, 1986).

Table V provides the results of several experiments in which individual mouse embryos were simultaneously incubated with both [1-^{14}C] and [5-^{3}H] glucose, and additional compounds. It should be noted that the ratio of $^{14}CO_2/^{3}H_2O$ so obtained provides a very precise profile of an individual

Table V
The [1-^{14}C]/[5-^{3}H] Ratio in Cultured Mouse Embryos Under Selected Experimental Conditions

Condition[a]	[1-^{14}C] Glucose → $^{14}CO_2$ (pmol/3hr/embryo)[b]	[5-^{3}H] Glucose → $^{3}H_2O$ (pmol/3hr/embryo)[b]	[1-^{14}C]/[5-^{3}H] Ratio
Control			
Early blastocyst (20)	0.77 ± 0.04	5.12 ± 0.45	0.15 ± 0.01
Late blastocyst (7)	1.22 ± 0.08	11.18 ± 0.96	0.11 ± 0.01
Pyruvate (2mM)			
Early blastocyst (16)	1.53 ± 0.06	5.26 ± 0.08	0.36 ± 0.05
Late blastocyst (5)	1.58 ± 0.15	19.93 ± 2.19	0.08 ± 0.01
Oxaloacetate (2mM)			
Early blastocyst (18)	1.44 ± 0.07	4.51 ± 0.35	0.37 ± 0.04
Late blastocyst (5)	1.55 ± 0.13	16.50 ± 2.23	0.10 ± 0.01
Acetaldehyde (2mM)			
Early blastocyst (9)	1.58 ± 0.08	3.36 ± 0.55	0.53 ± 0.05
Ethanol (20mM)			
Early blastocyst (10)	1.04 ± 0.09	6.77 ± 0.71	0.16 ± 0.01
Phenazine ethosulfate (10μM)			
Late blastocyst (8)	15.72 ± 1.04	22.41 ± 1.82	0.71 ± 0.02

[a]Individual embryos were incubated with 100 μM labeled glucose and additional compounds where indicated.
[b]Values are mean ± SEM for the number of individual embryo incubations given in parentheses.

embryo's metabolism. Moreover, the ratio concept allows the direct comparison of results obtained in different experiments. For example, if the $^{3}H_2O$ obtained from [5-^{3}H] glucose, which is a measure of glycolysis, is a function of the morphological stage of development of an embryo, then the metabolism of any ^{14}C-labeled compound can be normalized to glycolysis, allowing precise metabolic mapping of developing embryos.

In Table V it can be seen that three compounds with oxidative potential, namely, pyruvate, oxaloacetate, and acetaldehyde each specifically increases the [1-^{14}C]/[5-^{3}H] ratio in early blastocysts while ethanol has no such effect. It is interesting to note that late blastocysts apparently become refractory to stimulation by these compounds. However, late blastocysts can still be dramatically stimulated by PES (Table V).

4. EMBRYO CO-CULTURE

4.1. Co-culture of Porcine Preimplantation Embryos with Feeder Cell Monolayers or Culture Supernatants

The increased viability of certain cells cultured in the presence of a feeder layer of another cell type has been documented in numerous *in vitro* culture systems (*e.g.*, Kohler and Milstein, 1975; Martin, 1981). Enhanced *in vitro* development of embryos cultured in the presence of another cell type was first reported for the mouse embryo in coculture with irradiated HeLa cells (Cole and Paul, 1965). The specificity of the contribution of a feeder cell layer to embryonic development is unclear in view of the report that no differences were observed in the number of mouse embryos hatching from the zona pellucida when cultured on monolayers of liver, L, JLS-VII or teratocarcinoma cells (Glass *et al.*, 1979).

Murray *et al.* (1972) and Squire *et al.* (1972) were the first investigators to verify the presence of uterine specific proteins in the lumen of the porcine uterus. Chen and Bazer (1973) reported that the administration of an antiserum directed against one of these proteins (purple protein, acid phosphatase, Fraction IV) impaired normal placental function and fetal development, documenting the importance of uterine specific proteins in porcine embryonic development. Chen *et al.* (1975) utilized an antibody directed against the purple protein to establish its origin of synthesis as the endometrial cell. Basha *et al.* (1979) reported that the secretion of the purple protein by cultured porcine endometrial explants was dependent upon the stage of gestation of the donor.

Shaffer and Wright (1978) reported the attachment of swine blastocysts to a plastic substratum or collagen matrix with subsequent trophoblastic outgrowth. Kuzan and Wright (1982a) found that a greater number of hatched porcine blastocysts would attach to the substratum when cultured in the presence of bovine uterine fibroblasts compared with supernatants from bovine uterine fibroblast cell cultures.

Kuzan and Wright (1981) reported an increased incidence of attachment and trophoblastic outgrowth of porcine blastocysts to bovine uterine fibroblast monolayers while no such effect was apparent when blastocysts were cultured with bovine testicular fibroblasts. These results suggest that cell-embryo contact was necessary for the enhanced attachment and outgrowth, and further that there may be differences in embryo growth in culture depending on the source of fibroblast used.

4.2. Porcine Embryo Collection

Porcine embryos (4-cell to morulae) were collected from excised gilt reproductive tracts by flushing each uterine horn with 50 ml Minimal Essential Medium (MEM) as described by Shaffer and Wright (1978). Embryos were washed 3 times in MEM supplemented with 5% (v/v) heat-inactivated fetal calf serum, penicillin, streptomycin and fungizone (complete MEM). In previous experiments, complete MEM was found to provide the necessary constituents for the maintenance of porcine endometrial cells but only for moderate development of porcine embryos. The contribution to embryonic development from cell monolayers or from supernatants could then be distinguished (Allen and Wright, 1984).

4.3. Production of Endometrial Cell Monolayers

Porcine endometrial cell monolayers could be generated by either explant outgrowth and subsequent subculture, using 0.25% trypsin dispersal of endometrial slices, or by dispersal in collagenase as described for the rabbit by Riehl *et al.* (1983). Uteri that served as the endometrial cell source were generally not the same as those from which the embryos were recovered. Endometrial cells were cultured in 96-well microtiter plates as monolayers at 37°C under a 5% CO_2 in humidified air atmosphere. The medium used was complete MEM that was replenished by one-half replacement with fresh medium every 48 hr. The supernatants were saved and centrifuged to remove any contaminating endometrial cells and stored frozen at -20°C for later embryo culture. Endometrial monolayers were washed twice immediately before placing embryos in the co-culture system. Initial viability of the endometrial cell monolayers was determined by trypan blue exclusion and was usually greater than 80%.

4.4. Production of Ovarian Fibroblast Monolayers

Ovarian fibroblasts were generated by placing a 1 cm^2 section of ovarian tissue in 30 ml complete MEM in a 75 cm^2 tissue culture flask and allowing fibroblast-like cells to migrate from the tissue to form a monolayer. Two days before embryo collection, fibroblasts were removed from the flask by the addition of 0.25% trypsin, washed 3 times and subcultured into the wells of microtiter plates. Fibroblast monolayers were maintained at 37°C in a 5% CO_2 in humidified air atmosphere and washed twice immediately before embryo co-culture.

4.5. Production of Testicular Fibroblast Monolayers

Bovine testicular fibroblasts were obtained from aseptically collected calf testes by the method of Younger (1954). Cubes of testicular tissue, 1 to 2 mm^3, were plated in 25 cm^2 culture flasks and maintained at 37°C in a humidified gas atmosphere of 5% CO_2 in air. At 3-day intervals, half of the culture medium was replaced with fresh medium (MEM + 10% HTFCS). At 2-week intervals, cells were subjected to partial digestion by trypsin (0.25% solution, GIBCO Inc.) to release them from the plastic substratum and transferred to 75 cm^2 culture flasks. Fibroblasts were also plated in 24-well culture plates at a cell density of 40,000 cells per well for use in embryo culture. Conditioned medium collected from testicular fibroblasts was adjusted to pH 7.2, resterilized, and stored at 4°C for as long as 3 days prior to use in embryo culture.

4.6. Observations on Porcine Embryos Cultured With Feeder Cell Monolayers or Supernatant

Cultures were carried out in 96-well microtiter plates, 200 ml total volume, one embryo per well, at 37°C under a 5% CO_2 in humidified air atmosphere. The medium in all embryo cultures was complete MEM. Cultures were established using sterile technique under a laminar flow hood and all tissue culture reagents were obtained from GIBCO, Inc. Observations for stage of development were made every 12 hr by phase contrast microscopy (100x).

A chamber was assembled in which embryos could be maintained in culture with endometrial or fibroblast cells but without direct cell-embryo contact in order to establish the effect of culture supernatant dilution and freezing on embryonic development. Embryos were placed in the wells of a microtiter plate (1 embryo/well), with the embryos isolated from adjacent endometrial or fibroblast cells by a 0.22 μm membrane. This allowed for cell product diffusion into the well containing the embryo without physical cell-embryo contact.

The co-culture of porcine embryos at various stages of development on porcine endometrial cell monolayers resulted in enhanced embryo viability *in vitro* as assessed by progressive advancement in development (Allen and Wright, 1984; Table VI).

However, the specificity of the cell type used as the feeder monolayer presents an intriguing question when considering the findings that porcine ovarian fibroblasts appear to provide the same contribution to the support of embryo viability as endometrial cells. Kuzan and Wright (1982b) also reported that bovine fibroblast monolayers support increased development of bovine morulae when compared with fibroblast culture supernatants or endometrial explants. The failure of endometrial explants to support the *in vitro* development of bovine morulae (Kuzan and Wright, 1982b) may possibly be explained by the absence of cell-embryo contact that is present when embryos are cultured on cell monolayers.

Explanation of the mechanism(s) by which cell-embryo contact enhances embryo development is confounded by the presence of the zona pellucida, which represents an acellular barrier to physical communication between feeder cell and embryonic cell membranes. Three hypotheses can be advanced to explain these observations. First, it may be possible that fibroblast and (or) endometrial cell membrane projections penetrate the zona pellucida to reach the surface of the embryonic cells. In this regard, intercellular junctions have been observed between cumulus oophorus cells and the oocyte during oocyte maturation (Zamboni, 1970). Furthermore, Odor and Blandau (1969a,b) cultured mouse oogonia and oocytes in the presence of ovarian tissue and

Table VI
Mean Developmental Scores[a] of Porcine Embryos Cocultured with Selected Cell Monolayers or Supernatants

Treatment	Score
Medium only	0.63 ± 0.13[b]
Porcine endometrial cell monolayers	1.59 ± 0.22[c]
Porcine endometrial cell supernatants	0.48 ± 0.15[b]
Porcine ovarian fibroblast monolayers	1.44 ± 0.18[c]

[a]Mean developmental scores were determined by numerical coding of cell divisions as described by Allen *et al.* (1982).
[b,c]Values with different superscripts differ ($p < 0.05$).

reported a sequence of events wherein processes of follicular cells penetrate the zona pellucida and contact the oocyte. These events were followed by a breakdown of follicular cell processes, attachment of macrophages and subsequent penetration of the zona pellucida by macrophage projections. Our laboratory observed that porcine embryos removed from culture wells at the termination of co-culture experiments displayed endometrial cells attached to the zona pellucida (Allen and Wright, 1984). Second, at the time of implantation, embryonic cell processes could penetrate through the zona pellucida to contact endometrial cells. Such processes have been seen in the guinea pig (Blandau, 1949a,b), hamster and rhesus monkey (Boatman *et al.*, 1987; and see Chapters 11 and 13). Third, the possibility exists that while endometrial or fibroblast culture supernatants fail to provide the same factor(s) that cell contact provides, cell monolayers may produce substances (peptides, hormones, *etc.*) that are released in limited quantities and/or are metabolized rapidly by the adjacent embryo, making their detection in culture supernatants difficult. It is also possible that only certain cells of a monolayer produce such a factor(s), *i.e.*, embryos in contact with cells may release products that "trigger" endometrial or fibroblast factor production by only those cells in the vicinity of the embryo. If an embryo product and a feeder cell product positive feedback system should prove to be active, a model for the transfer of substances between endometrial cells and embryonic cells could be established.

5. REFERENCES

Allen, R.L., Bondioli, K.R., and Wright, R.W., Jr., 1982, The ability of fetal calf serum, newborn calf serum, and normal steer serum to promote *in vitro* development of bovine morulae, *Theriogenology* 18: 185-189.

Allen, R.L., and Wright, R.W., Jr., 1984, *In vitro* development of porcine embryos in coculture with endometrial cell monolayers or culture supernatants, *J. Anim. Sci.* 59: 1657-1661.

Anderson, G.B., 1978, Advances in large mammalian embryo culture, in: *Methods in Mammalian Reproduction* (J.C. Daniel, ed.), Academic Press, New York, pp. 273-283.

Basha, S.M.M., Bazer, F.W., and Roberts, R.M., 1979, The secretion of uterine specific, purple phosphatase by cultured explants of porcine endometrium. Dependency upon the state of pregnancy of the donor animal, *Biol. Reprod.* 20: 431-441.

Bavister, B.D., Leibfried, M.L., and Lieberman, G., 1983, Development of preimplantation embryos of the golden hamster in a defined culture medium, *Biol. Reprod.* 28: 235-237.

Biggers, J.D., 1979, Fertilization and blastocyst formation, in: *Animal Models for Research on Contraception and Fertility* (N.J. Alexander, ed.) Harper and Row, New York, pp. 223-252.

Biggers, J.D., and Stern, S., 1973, Metabolism of the preimplantation mammalian embryo, *Adv. Reprod. Physiol.* 6: 1-59.

Blandau, R.J., 1949a, Observations on implantation of the guinea pig ovum, *Anat. Rec.* 103: 19-47.

Blandau, R.J., 1949b, Embryo-endometrial interrelationship in the rat and guinea pig, *Anat. Rec.* 104: 331-359.

Boatman, D.E., Morgan, P.M., and Bavister, B.D., 1987, Culture of *in vitro* fertilized rhesus monkey oocytes to peri-implantation stages of embryo development, *Biol. Reprod.* (submitted).

Boone, W.R., Dickey, J.F., Luszcz, L.J., Dantzler, J.R., and Hill, J.R., 1978, Culture of ovine and bovine ova, *J. Anim. Sci.* 47: 908-913.

Brackett, B.G., 1981, *In vitro* culture of the zygote and embryo, in: *Fertilization and Embryonic Development In Vitro* (L. Mastroianni, Jr., and J.D. Biggers, eds.), Plenum Press, New York, pp. 61-79.

Brinster, R.L., 1965a, Studies on the development of mouse embryos *in vitro*. I. The effects of osmolarity and hydrogen ion concentration, *J. Exp. Zool.* 158: 49-58.

Brinster, R.L., 1965b, Studies on the development of mouse embryos *in vitro*. II. The effect of energy source, *J. Exp. Zool.* 158: 59-68.

Brinster, R.L., 1965c, Studies on the development of mouse embryos *in vitro*. III. The effect of fixed-nitrogen source, *J. Exp. Zool.* 158: 69-78.

Brinster, R.L., 1965d, Studies on the development of mouse embryos *in vitro*. IV. Interaction of energy sources, *J. Reprod. Fertil.* 10: 227-240.

Brinster, R.L., 1971, *In vitro* culture of the embryo, in: *Pathways to Conception: the Role of the Cervix and the Oviduct in Reproduction* (A.I. Sherman, ed.), Charles C. Thomas, Springfield, pp. 245-277.

Brinster, R.L., 1972, Cultivation of the mammalian embryo, in: *Growth, Nutrition and Metabolism of Cells in Culture*, Vol. II (G. Rothblat, and V. Cristofalo, eds.), Academic Press, New York, pp. 251-286.

Brinster, R.L., 1973, Nutrition and metabolism of the ovum, zygote, and blastocyst, in: *Handbook of Physiology-Endocrinology II*, section 7, part 2 (R.O. Greep, and E.A. Astwood, eds.), American Physiological Society, Washington, D.C., pp. 165-185.

Brinster, R.L., and Thomson, J.L., 1966, Development of 8-cell mouse embryos *in vitro*, *Exp. Cell Res.* 42: 308-315.

Brinster, R.L., and Troike, D.E., 1979, Requirements for blastocyst development *in vitro*, *J. Anim. Sci.* 49: 26-34.

Canfield, R.W., Gwazdauskas, F.C., Whittier, W.D., Lineweaver, J.A., Vinson, W.E., and Saacke, R.G., 1983, Influence of steer serum, bovine serum albumin, and uterine secretions from ovariectomized progesterone-estrogen treated cows on early bovine development, *J. Dairy Sci.* 66 (Suppl. 1): 237.

Chang, M.C., 1949, Effects of heterologous sera on fertilized rabbit ova, *J. Gen. Physiol.* 32: 291-300.

Chen, L.T., and Hsu, Y.C., 1982, Development of mouse embryos *in vitro*: preimplantation to the limb bud stage, *Science* 218: 66-68.

Chen, T.T., and Bazer, F.W., 1973, Effect of antiserum to porcine fraction IV protein on the conceptus, *J. Anim. Sci.* 37: 304 (abstr.).

Chen, T.T., Bazer, F.W., Beghardt, B.M., and Roberts, R.M., 1975, Uterine secretion in mammals: synthesis and transport of a purple acid phosphatase in pigs, *Biol. Reprod.* 13: 304-313.

Cholewa, J.A., and Whitten, W.K., 1970, Development of 2-cell mouse embryos in the absence of a fixed nitrogen source, *J. Reprod. Fertil.* 22: 553-555.

Cole, R.J., and Paul, J., 1965, Properties of cultured preimplantation mouse and rabbit embryos, and cell strains derived from them, in: *Preimplantation Stages of Pregnancy* (G.E.W. Wolstenholm, and M. O'Connor, eds.), Little, Brown and Company, Boston, pp. 82-112.

Daniel, J.C., 1967, The pattern of utilization of respiratory metabolic intermediates by preimplantation rabbit embryos *in vitro*, *Exp. Cell Res.* 47: 619-624.

Davis, D.L., and Day, B.N., 1978, Cleavage and blastocyst formation by pig eggs *in vitro*, *J. Anim. Sci.* 46: 1043-1053.

Ducibella, T., and Anderson, E., 1975, Cell shape and membrane changes in the 8-cell mouse embryo: Prerequisites for morphogenesis of the blastocyst, *Dev. Biol.* 47: 45-58.

Dulbecco, R., and Vogt, M., 1954, Plaque formation and isolation of pure lines with poliomyelitis viruses, *J. Exp. Med.* 99: 167-182.

Edwards, R.G., Bavister, B.D., and Steptoe, P.C., 1969, Early stages of fertilization *in vitro* of human oocytes matured *in vitro*, *Nature (London)* 221: 632-635.

Edwards, R.G., Steptoe, P.C., and Purdy, J.M., 1980, Establishing full-term pregnancies using cleaving embryos grown *in vitro*, *Brit. J. Obstet. Gynaecol.* 87: 737-756.

Epstein, C.J., Smith, S., Travis, B., and Tucker, G., 1978, Both X chromosomes function before X-chromosome inactivation in female mouse embryos, *Nature (London)* 274: 500-503.

Fabregat, I., Victorica, J., Satrustegui, J., and Machado, A., 1985, The pentose phosphate cycle is regulated by NADPH/NADP ratio in rat liver, *Arch. Biochem. Biophys.* 236: 110-118.

Fisher, A.B., and Reicherter, J., 1984, Pentose pathway of glucose metabolism in isolated granular pneumocytes: metabolic regulation and stimulation by paraquat, *Biochem. Pharmacol.* 33: 1349-1353.

Foote, R.H., and Onuma, H., 1970, Superovulation, ovum collection, culture and transfer: A review, *J. Dairy Sci.* 53: 1681-1692.

Glass, R.H., Spindle, A.I., and Pederson, R. A., 1979, Mouse embryo attachment to substratum and interactions of trophoblast with cultured cells, *J. Exp. Zool.* 208: 327-335.

Graves, W.M., and Biggers, J.D., 1970, Carbon dioxide fixation by mouse embryos prior to implantation, *Science* 167: 1506-1508.

Graves, W.M., Dickey, J.F., and McConell, J.C., 1977, *In vitro* development of porcine embryos, *J. Anim. Sci.* 45 (Suppl. 1): 164.

Ham, R.J., 1984, Formulation of basal nutrient media, in: *Cell Culture Method for Molecular and Cell Biology: Methods for Preparation of Media, Supplements, and Substrata for Serum-Free Animal Cell Culture*, Vol. 1 (D.W. Barnes, D.A. Sirbasku, and G.H. Sato, eds.), Alan R. Liss Inc., New York, pp. 3-21.

Hothersall, J.S., Baquer, N., Greenbaum, A.L., and McLean, P., 1979, Alternative pathways of glucose utilization in brain. Changes in the pattern of glucose utilization in brain during development and the effects of phenazine methosulfate on the integration of metabolic routes, *Arch. Biochem. Biophys.* 198: 478-492.

Kanagawa, H., Bedirian, K., Ringelberg, C., and Basrur, P.K., 1975, *In vitro* culture of bovine ova, in: *Proc. Eighth Ann. Meeting Society for the Study of Reprod.*, Fort Collins, CO (abstr. 74).

Kane, M.T., 1972, Energy substrates and culture of single cell rabbit ova to blastocysts, *Nature (London)* 238: 468.

Kane, M. T., 1974, The effects of pH on culture of one cell rabbit ova to blastocysts in bicarbonate-buffered medium, *J. Reprod. Fertil.* 38: 477-480.

Kane, M. T., 1976, Growth of fertilized one-cell rabbit ova to viable morulae in the presence of pyruvate or fatty acids, *J. Physiol.* 263: P235-P236.

Kane, M. T., 1978, Culture of mammalian ova, in: *Control of Reproduction in the Cow* (J.M. Sreenan, ed.), Martinus Nijhoff, The Hague, pp. 383-397.

Kane, M. T., and Foote, R. H., 1970, Culture of two- and four-cell rabbit embryos to the expanding blastocyst stage in synthetic media, *Proc. Soc. Exp. Biol. Med.* 133: 921-925.

Kane, M.T., and Foote, R.H., 1971, Factors affecting blastocyst expansion of rabbit zygotes and young embryos in defined media, *Biol. Reprod.* 4: 41-47.

Katz, J., and Rognstad, R., 1966, The metabolism of tritiated glucose by rat adipose tissue, *J. Biol. Chem.* 241: 3600-3610.

Katz, J., Landau, B.R., and Bartsch, G.E., 1966, The pentose cycle, triose phosphate isomerization, and lipogenesis in rat adipose tissue, *J. Biol. Chem.* 241: 727-740.

Kohler, G., and Milstein, C., 1975, Continuous cultures of fused cells secreting antibody of predefined specificity, *Nature (London)* 256: 495-497.

Kratzer, P.G., and Gartler, S.M., 1978, HGPRT activity changes in preimplantation mouse embryos, *Nature (London)* 274: 503-504.

Kuzan, F.B., and Wright, R.W., Jr., 1981, Attachment of porcine blastocysts to fibroblast monolayers *in vitro*, *Theriogenology* 16: 651-658.

Kuzan, F.B., and Wright, R.W., Jr., 1982a, Blastocyst expansion, hatching, and attachment of porcine embryos co-cultured with bovine fibroblasts *in vitro*, *Anim. Reprod. Sci.* 5: 57-63.

Kuzan, F.B., and Wright, R.W., Jr., 1982b, Observations on the development of bovine morulae on various cellular and non-cellular substrata, *J. Anim. Sci.* 54: 811-816.

Lindner, G.M., and Wright, R.W., Jr., 1978, Morphological and quantitative aspects of the development of swine embryos *in vitro*, *J. Anim. Sci.* 46: 711-716.

Lindner, G.M., Dickey, J.F., Hill, J.R., Jr., and Knickerbocker, J.J., 1979, Effect of bovine serum albumin concentration on the development of ovine embryos cultured in Brinster's and Whitten's medium, *J. Anim. Sci.* 49 (Suppl. 1): 314.

Martin, G.R., 1981, Isolation of a pluripotent cell line from early mouse embryos cultured in medium conditioned by teratocarcinoma stem cells, *Proc. Natl. Acad. Sci. USA* 78: 7634-7638.

Maurer, R.R., 1976, Storage of mammalian oocytes and embryos: A review, *Can. J. Anim. Sci.* 56: 131-145.

Maurer, R.R., Whitener, R.H., and Foote, R.H., 1969, Relationship of *in vivo* gamete aging and exogenous hormones to early embryo development in rabbits, *Proc. Soc. Exp. Biol. Med.* 131: 882-885.

Ménézo, M.Y., 1976, Milieu synthétique pour la survie et la maturation des gamètes et pour la culture de l'oeuf fécondé, *C. R. Acad. Sci. (Paris)* 282: 1967-1970.

Murray, F.A., Bazer, F.W., Wallace, H.D., and Warnick, H.C., 1972, Quantitative and qualitative variation in the secretion of protein by the porcine uterus during the estrous cycle, *Biol. Reprod.* 7: 314-320.

Naglee, D.L., Maurer, R.A., and Foote, R.H., 1969, Effect of osmolarity on *in vitro* development of rabbit embryos in a chemically defined medium, *Exp. Cell Res.* 58: 331-333.

Neeley, J.R., Denton, R.M., England, P.J., and Randle, P.J., 1972, The effects of increased heart work on the tricarboxylate cycle and its interactions with glycolysis in the perfused rat heart, *Biochem. J.* 128: 147-159.

Odor, D.L., and Blandau, R.J., 1969a, Ultrastructural studies on fetal and early postnatal mouse ovaries. I. Histogenesis and organogenesis, *Amer. J. Anat.* 124: 163-186.

Odor, D.L., and Blandau, R.J., 1969b, Ultrastructural studies on fetal and early postnatal mouse ovaries II. Cytodifferentiation, *Amer. J. Anat.* 125: 177-215.

O'Fallon, J.V., and Wright, R.W., Jr., 1986, Quantitative determination of the pentose phosphate pathway in preimplantation mouse embryos, *Biol. Reprod.* 34: 58-64.

Ogawa, S., Satoh, K., and Hashimoto, H., 1971, *In vitro* culture of rabbit ova from the single cell to the blastocyst stage, *Nature (London)* 233: 422-424.

Pike, I.L., 1981, Comparative studies of embryo metabolism in early pregnancy, *J. Reprod. Fertil.* (Suppl. 29): 203-213.

Quinn, P., and Wales, R.G., 1971, Fixation of carbon dioxide by preimplantation mouse embryos *in vitro* and the activities of enzymes involved in the process, *Aust. J. Biol. Sci.* 24: 1277-1290.

Quinn, P., and Wales, R.G., 1973, Growth and metabolism of preimplantation mouse embryos cultured in phosphate-buffered medium, *J. Reprod. Fertil.* 35: 289-300.

Restall, B.J., and Wales, R.G., 1966, The fallopian tube of the sheep. III. The chemical composition of the fluid from the fallopian tube, *Aust. J. Biol. Sci.* 19: 687-698.

Riehl, R.M., Pathak, R.K., and Harper, M.J.K., 1983, A reliable method for isolating endometrial epithelial cells from rabbits, and preliminary studies of prostaglandin uptake, *Biol. Reprod.* 28: 363-375.

Seidel, G.E., 1977, Short-term maintenance and culture of embryos, in: *Embryo Transfer in Farm Animals: A Review of Techniques and Applications* (K.J. Betteridge, ed.), Canada Dept. of Agriculture Monogr. 16, pp. 20-24.

Shaffer, S.J., and Wright, R.W., Jr., 1978, Attachment and trophoblastic outgrowth of swine blastocysts *in vitro*, *J. Anim. Sci.* 46: 1712-1716.

Squire, G.D., Bazer, F.W., and Murray, F.A., 1972, Electrophoretic patterns of porcine uterine protein secretions during the estrous cycle, *Biol. Reprod.* 7: 321-325.

Tervit, H.R., Whittingham, D.G., and Rowson, L.E.A., 1972, Successful culture *in vitro* of sheep and cattle ova, *J. Reprod. Fertil.* 30: 493-497.

Trounson, A.D., and Moore, N.W., 1974, Attempts to produce identical offspring in the sheep by mechanical division of the ovum, *Aust. J. Biol. Sci.* 27: 505-510.

Trounson, A.O., Willadsen, S.M., and Rowson, L.E.A., 1976, The influence of *in vitro* culture and cooling on the survival and development of cow embryos, *J. Reprod. Fertil.* 47: 367-370.

Wales, R.G., 1970, Effects of ions on the development of the preimplantation mouse embryo *in vitro*, *Aust. J. Biol. Sci.* 23: 421-429.

Wales, R.G., 1975, Maturation of the mammalian embryo: biochemical aspects, *Biol. Reprod.* 12: 66-81.

Wales, R.G., Quinn, P., and Murdoch, R.N., 1969, The fixation of carbon dioxide by the 8-cell mouse embryo, *J. Reprod. Fertil.* 20: 541-543.

Whitten, W.K., 1957, Culture of tubal ova, *Nature (London)* 179: 1081-1082.

Whitten, W.K., 1971, Nutrient requirements for the culture of preimplantation embryos *in vitro*, *in: Advances in The Biosciences*, Vol. 6 (G. Raspé, ed.) Pergamon Press, Oxford, pp. 129-141.

Whittingham, D.G., 1975, Fertilization, early development and storage of mammalian ova, in: *The Early Development of Mammals* (M. Balls, and A.E. Wild, eds.), Cambridge University Press, pp. 1-24.

Wright, R.W., 1977, Successful culture *in vitro* of swine embryos to the blastocyst stage, *J. Anim. Sci.* 44: 854-858.

Wright, R.W., Jr., and Bondioli, K. R., 1981, Aspects of *in vitro* fertilization and embryo culture in domestic animals, *J. Anim. Sci.* 53: 702-729.

Wright, R.W., Jr., Anderson, G.B., Cupps, P.T., and Drost, M., 1976a, Successful culture *in vitro* of bovine embryos to the blastocyst stage, *Biol. Reprod.* 14: 157-162.

Wright, R.W., Jr., Anderson, G.B., Cupps, P.T., and Drost, M., 1976b, Blastocyst expansion and hatching of bovine embryos cultured *in vitro*, *J. Anim. Sci.* 43: 170-174.

Wright, R.W., Jr., Anderson, G.B., Cupps, P.T., and Drost, M., 1976c, Culture of embryos from mature and prepuberal sheep, *J. Anim. Sci.* 42: 912-919.

Wright, R.W., Jr., Watson, J.G., and Chaykin, S., 1978, Factors influencing the *in vitro* hatching of mouse blastocysts, *Anim. Reprod. Sci.* 1: 181-188.

Younger, J.S., 1954, Monolayer tissue cultures I. Preparation and standardization of trypsin dispersed monkey kidney cells, *Proc. Soc. Exp. Biol. Med.* 85: 202-205.

Zamboni, L., 1970, Ultrastructure of mammalian oocytes and ova, *Biol. Reprod.* 2 (Suppl. 2): 44-63.

Chapter 13

IN VITRO GROWTH OF NON-HUMAN PRIMATE PRE- AND PERI- IMPLANTATION EMBRYOS

DOROTHY E. BOATMAN

1. INTRODUCTION

The close phylogenetic relationship between man and non-human primates has made these animals, especially the rhesus monkey, of particular interest for scientific research. However, since the early pioneer work (Hartman and Corner, 1941; Heuser and Streeter, 1941; Lewis and Hartman, 1941), only a few investigators have studied preimplantation embryology in non-human primates. A major reason for this situation is the expense of maintaining these long-lived animals (at least 30 years in captivity), which are largely monotocous and normally yield relatively few embryos in each reproductive season (albeit females may have as many as 18 reproductive seasons [years]). Additionally, some standard animal research procedures that would curtail the fertility of the animals (*e.g.*, the excision of oviducts for ova/embryo flushing) may be contraindicated by the goals of breeding programs that have been established to maintain adequate supplies of these animals. In the past, the scarcity of non-human primate embryos has tended to obviate their clear appropriateness as models for the study of early human development. Recent advances in the techniques of super-stimulation of follicular growth, in vitro fertilization (IVF) and embryo culture have increased the supply of non-human primate embryos (Bavister *et al.*, 1983a; Boatman and Bavister, 1984; Boatman *et al.*, 1986; Balmaceda *et al.*, 1984). These techniques, as well as improved procedures for superovulation (Schenken *et al.*, 1984; Stouffer, R.L., personal communication) and low trauma procedures for embryo collection (Pope *et al.*, 1980) and transfer (Bavister *et al.*, 1985) have the potential to significantly increase the utilization of non-human primate embryos for research into early embryonic

Dorothy E. Boatman Wisconsin Regional Primate Research Center, University of Wisconsin, Madison, Wisconsin 53715, USA.

development. However, up to the present time, procedures for obtaining predictable superovulation in non-human primates have not been routinely available. This limitation on superovulation has led several laboratories (including our own) to rely on IVF of oocytes aspirated from follicles of super-stimulated monkeys to provide information on fertilization and early embryonic development. For this reason, the principal focus of this chapter is on the development of rhesus monkey embryos *in vitro* subsequent to IVF of oocytes obtained from super-stimulated donors.

Increasing the supply of non-human primate embryos will permit research into fundamental problems using animal models that are more closely applicable to humans. For example, in most mammals, with the exception of the mouse, it is not known with certainty when the embryonic genome is activated during development (Geuskens and Alexandre, 1984; Tesarik *et al.*, 1986). The answer to this question in primates is of crucial importance for attempts to "multiply" the supply of primate embryos bearing genetic traits relevant to the investigation of human disease. It is also relevant to medical and ethical decisions concerning whether or not to replace human embryos after IVF. Additionally, little information is presently available concerning the nature, temporal onset and duration of early pregnancy signals produced by primate embryos (Chen *et al.*, 1985; O'Neill *et al.*, 1985). Knowledge concerning the nature and timing of these signals has importance for the control of human fertility. Human over-fecundity is rapidly reaching the status of a worldwide emergency that will affect the political future of all nations, regardless of their present economic status. There is a keen interest in devising better means of contraception that are long lasting, eventually reversible, cost effective and biochemically stable so that they may be administered far from urban amenities. Non-human primates are the most appropriate species for test trials on the efficacy and mode of action of any such contraceptives. On the individual's side of the equation, underfertility is of equal import. Timing of early embryonic pregnancy signals could well be a determining factor in embryo replacement following either embryo donation (Bustillo *et al.*, 1984) or IVF.

Greater availability of primate embryos and reliable methods for culturing them should benefit efforts to cryopreserve primate embryos, a matter with both practical and medical consequences. The ability to reliably freeze-store non-human primate embryos would benefit research in genetic manipulation of these embryos, since the supply of embryos and of synchronized recipients could be temporally dissociated. Prospects for species preservation, the "frozen zoo" concept, would become realistic rather than speculative. Humans would also be the beneficiaries of improvements in cryotechnology. They could conceive offspring early in their reproductive lifespan but defer childbearing until later in life. This would provide protection against the increased genetic anomalies associated with aging, as well as provide some assurance of having normal offspring for individuals engaged in high risk occupations, *e.g.*, handling radionucleotides. Since the present survival rate of cryopreserved primate (human and monkey) embryos following IVF is low (see section 6.4), radical modifications in freezing protocols (*e.g.*, vitrification: Rall and Fahy, 1984) may be the only means to substantially improve embryo viability. In addition, it would be most appropriate from an ethical standpoint to use non-human primate embryos to evaluate any new procedure prior to its application in human clinical practice.

2. TECHNICAL CONSIDERATIONS

2.1. Equipment

The laboratory for non-human primate IVF and embryo culture should be set up with care, utilizing basic principles of design appropriate for tissue culture laboratories (Paul, 1975). Minimal equipment requirements include a centrifuge, autoclave, water purification system, an incubator with reliable temperature and CO_2 controls, a dissecting microscope, a clean culture or laminar flow hood, and an inverted microscope. It is desirable to have Nomarski (or DIC) optics and an environmental chamber surrounding the microscope stage. It is also useful to have video recording capability (see also Chapter 11).

2.2. Water Quality

It is essential that the laboratory have access to extremely high quality water. This may be generated within the research laboratory by processing "institutional" distilled water through deionization cartridges, followed by Milli-Q® filtration; however, additional treatment (such as reverse osmosis) may be required (see Appendix I [Ed.]). Alternatively, any of several other trade-marked water purification systems may be suitable. The end product should be of at least 18 megohm resistivity and free of endotoxin contamination (see Appendix I). Glassware used in the IVF/embryo culture laboratory must be as scrupulously clean as the culture water (see review by Dandekar and Quigley [1984] on this aspect of laboratory management).

Water/culture medium quality should be monitored periodically using a suitable bioassay. Culture of mouse 2-cell embryos to hatched blastocysts within 96 hr is the most widely used test for human IVF culture conditions. However, this does not appear to be a very sensitive assay system for adverse water quality. It has been suggested that culture of the mouse 1-cell embryo is more rigorous as a bioassay since it is more difficult to culture the 1-cell stage to hatched blastocysts (Quinn *et al.*, 1985). Another check on the water/culture medium quality is provided by survival of ejaculated (homospecific) spermatozoa during prolonged culture (24 to 72 hr). A variation on this procedure is used routinely in our laboratory (see Appendix I). This rapid assay system (1 working day *vs.* 4 to 5 for the various mouse embryo culture assays) is verified periodically in our laboratory by culture of early 8-cell hamster embryos to hatching blastocysts.

2.3. Culture Media

The choice of the medium is itself an important experimental variable when beginning the culture of embryos from a new species. The approach that we have employed (described in more detail along with the experimental results in sections 5.2 and 5.3) is as follows. First, we looked for media known to support the development of embryos from a phylogenetically close species. Second, we decided to use the simplest medium formulation compatible with normal embryo development. Using simple formulations of media freshly-prepared in the laboratory with water and culture-ware of known quality has proven beneficial in many aspects of our work with gametes and embryos

(Bavister, 1981; Bavister *et al.*, 1983b). Since some of the components of complex media (*e.g.*, tryptophan, riboflavin, *etc.*) are quite unstable in solution and readily form toxic by-products such as peroxides (MacMichael, 1986), it seems wiser to include them only on a proven need basis.

There is no clear evidence for a sharp pH optimum *in vitro* in the range 7.1 to 7.6 for embryonic growth in the most characterized species, the mouse (Brinster, 1971). Without evidence to the contrary, it seems prudent to control pH of the medium at 7.4 for culture of non-human primate embryos. Prolonged handling of embryos in bicarbonate-based medium outside the CO_2 incubator should be avoided to minimize fluctuations in pH. Ova and embryos of many animal species will tolerate brief holding periods in a suitable air-buffered medium during experimental manipulations, such as ovum aspiration, micromanipulation, and embryo transfer (Brinster, 1971; Bavister *et al.*, 1983a, 1984). However, carbon dioxide and/or bicarbonate ion *per se* are important metabolic regulators of gamete and embryo functions (Brinster, 1971; Boatman and Bavister, 1983; Carney and Bavister, 1986, 1987; see also Chapter 11). Embryo culture should be performed under appropriate CO_2 gas tension (usually 5%) until or unless data show that the embryonic stage(s) can develop normally in the absence of CO_2.

Many investigators perform gamete handling and embryo culture in drops under an oil overlay in order to control pH and CO_2 tension during brief periods for manipulation and examination of embryos. This is particularly convenient for the culture of several treatments per 60 mm culture dish. If this procedure is used, it is imperative that the oil be free of toxic contaminants (see Appendix I at the end of this book). One should be wary of using pharmaceutical grade paraffin (mineral) oils since many of these contain stabilizers (such as α-tocopherol) which are deleterious to gametes and embryos (Brinster, 1971). An unfortunate choice of oil source could lead one to conclude erroneously that the medium itself was deficient for the culture of embryos. Aside from the "good *vs.* bad" oil consideration, oil overlays have the potential to extract lipid soluble substances from the culture medium that might be important cofactors for development (*e.g.*, fatty acids and steroids). This aspect of the culture environment must be considered together with the concerns over pH and CO_2 in making the decision to use (or not use) oil for embryo culture. For our own purposes, we have found silicone oil (Aldrich Chemical Co.) to be superior to paraffin oil. However, it is more gas permeable than paraffin oil and one must work rapidly when using it.

The formulations for our 3 basic media, TALP, TALP-HEPES, and modified CMRL-1066, are appended at the end of this chapter (Section 8).

3. SUPPLY OF *IN VIVO* FERTILIZED EMBRYOS

3.1. Reproductive Aspects

Natural Cycles. Under optimal conditions, embryos derived from *in vivo* conception would be the starting material of choice for examining embryo development after the first cleavage division (*i.e.*, subsequent to fertilization). By coordinating the time of embryo collection with known time of ovulation, it should be possible to collect embryos at specific stages of cleavage development. This stratagem works very well for many of the usual laboratory animal models, *e.g.*, mouse, hamster and rabbit (see Chapters 10, 11 [Ed.]). Precise schedules with embryo stages expected at specific times (hr/days) post-LH or post-hCG can be constructed (Bavister *et al.*, 1983b).

Ovulation in (old world) non-human primates occurs within a range of 24 to 56 hr after the estradiol (E_2) peak (Weick *et al.*, 1973). This is a much larger time range than occurs for many other species. With non-human primates, one can expect to obtain a range of embryo developmental stages even with timed flushing of the reproductive tract (Lewis and Hartman, 1941; Kraemer and Hendrickx, 1971; Pope *et al.*, 1983). Usually, embryos are obtained from natural (unstimulated) cycles. If several synchronized embryos are desired for simultaneous placement into various culture treatments, a fairly large colony of animals would be needed to provide them. One can predict the number of cycling females needed to ensure a specific number of synchronized embryos. The number of embryos desired (in this case, equals the number of females) is multiplied by the average cycle length of the animals: *e.g.*, 10 embryos x 28 days (average for rhesus monkeys: Catchpole and van Wagenen, 1975) = 280 cycling females. In actuality, embryos are recovered from only 1/4 to 1/2 of mated females (Enders *et al.*, 1982; Pope *et al.*, 1983), and cycle length can vary from 23 to more than 31 days in rhesus monkeys (Catchpole and van Wagenen, 1975). Thus, if 10 embryos at the same stage were needed at the same time for a specific experiment, a starting colony size of more than 560 cycling females would be needed. The large numbers of donor animals involved illustrate the impracticality of using multiple synchronous embryos in a single culture experiment.

Artificially Synchronized Cycles. Artificial synchronization of female cycles would have the potential to reduce the numbers of females needed for embryo collection or transfer. The basis of artificial synchronization is the promotion of premature luteolysis or regression of the corpus luteum. Reliable, routine procedures have not yet been reported for old world monkeys. In marmoset monkeys, a single injection of the prostaglandin $F_{2\alpha}$ analogue, cloprostenol, administered 8 to 17 days after ovulation, induced luteolysis and shortened the ovarian cycles. Matings during the post-injection cycle produced normal embryos (Summers *et al.*, 1985). The exact day of ovulation cannot be predicted with certainty following this type of synchronization because the length of the follicular phase varies from animal to animal. The range in time from injection to next day of ovulation was 9 to 13 days (Summers *et al.*, 1985).

Superovulation. This is a more promising approach for improving the prospects of obtaining a predictable, and larger, yield of embryos from non-human primates. Early attempts at superovulation often resulted in overly long retention of the mature ovum within the follicle. Frequently, follicles luteinized without ovulating, or ovulated very asynchronously (Simpson and van Wagenen, 1975; Batta *et al.*, 1978). Ovulated ova were aged and abnormal (Batta *et al.*, 1978). This problem of post-mature ovulation and premature luteinization may have been caused by an inappropriate balance of FSH and LH activities in the gonadotropin preparations used for superstimulation (Murphy *et al.*, 1984).

Some of these problems with superovulation may have been overcome by using different sources of gonadotropins of primate (human) origin. Recently, a small scale study using "pure" FSH, Urofollitropin®, on cycle days 1 to 11 plus hCG on day 12, yielded 12 ovulated ova from 3 monkeys (Schenken *et al.*, 1984). A twice daily regimen of Pergonal® (60 IU total/ day, FSH/LH ratio = 1) on cycle days 2 to 10 plus hCG (1000 IU) on day 11 has been reported to be equally successful in producing multiple ovulations (R. Stouffer, personal communication). In the Urofollitropin® study, oviducts were flushed 72 hr after hCG (Schenken *et al.*, 1984), and in the Pergonal® study, after 48 hr (Stouffer, personal communication). The timeliness and synchrony of the ovulations were consistent with those of normal ova, although no morphological characterizations of the ova were provided. There are limitations on the numbers of times that treatment with foreign (non-homologous) gonadotropins may be repeated to produce superovulation or super-stimulation preceding follicular aspiration and IVF (Section 4). Repeated exposure to foreign gonadotropins eventually leads to the development of non-precipitating antibodies to the hormones and refractoriness or non-response to their subsequent administration (Ottobre and Stouffer, 1985; Bavister *et al.*, 1986).

3.2. Recovery of Ova and Embryos from the Female Reproductive Tract

Surgical Embryo Recovery. Although embryos (and ova) may be recovered from excised oviducts and uteri, the scarcity of animals limits the usefulness of this approach. The procedures for flushing oviducts and uteri of non-human primates at laparotomy have been thoroughly described by Brackett (1978). Retrograde flushing of fallopian tubes and uteri at laparotomy can be repeated in the same animal from 4 to 6 times before excessive interference by adhesion formation (Hendrickx and Kraemer, 1971; Brackett, 1978). Cleavage stage embryos are normally found in the oviducts for 3 to 4 days after ovulation. After that time, 16-cell, morula, and blastocyst stages are found in the uterus.

Non-surgical Embryo Recovery. Transcervical (non-surgical) procedures are used to obtain uterine stage embryos, usually 16-cells, morulae and blastocysts, 4 or more days (not to exceed 9 for old world monkeys) after ovulation. This procedure does not cause adhesions and infertility in the animals. Transcervical recoveries in non-human primates are most readily performed in monkeys, such as the baboon, that have a relatively straight cervical canal through which instruments may be inserted into the uterine lumen (Pope *et al.*, 1980). Trans-cervical recoveries are more difficult in various macaque species, including the rhesus monkey, in which the cervical canal is tortuous and has several blind endings (Jaszczak and Hafez, 1972).

The most extensive recoveries using the transcervical method have been reported by Pope *et al.* (1983) in the baboon. An Isaacs endometrial cell sampler was modified by placing a smaller diameter delivery cannula inside the sampler. The cannula was connected to a syringe filled with flushing medium. This produced a bidirectional flow system. The flushing, containing the embryos, returned passively and was collected into a sterile centrifuge tube outside the vaginal *os*. The recovery rate with this procedure was 35 to 57.5% (Pope *et al.*, 1980).

In our laboratory, we have found that similar techniques may be devised for rhesus monkeys. In test trials, we have used a modified 18g intravenous catheter placement unit utilizing concentric inflow and outflow channels connected to a push-pull system. This apparatus could be used only in multiparous rhesus monkeys whose cervical canals were sufficiently dilated to permit passage of the device into the fundus.

4. *IN VITRO* FERTILIZED EMBRYOS

4.1. Rationale

The unreliability of super*ovulation* in macaque monkeys has necessitated (at least in our laboratory) a reliance on super*stimulation* to recruit the growth of multiple follicles, which are aspirated at laparoscopy at a set time(s) after the injection of hCG. The aspirated oocytes are placed into culture, where they complete their maturation prior to insemination *in vitro*. The advantage of this procedure is that all aspects contributing to the development of the embryo, *e.g.*, oocyte maturation, sperm capacitation, fertilization, and metabolic requirements of the embryos, may be examined. Since the time of insemination is known with certainty, the precise cleavage timings of the embryos may be determined. Many of the resultant embryos from a specific male/female pair will be well synchronized: this number of synchronized embryos can range from as few as 2 to as many as 20 (Boatman *et al.*, 1986). The larger yield from a single animal donor illustrates the potential superiority of this approach compared to flushing embryos from natural cycles (section 3.1). [There may be less of an advantage when compared to superovulation, but see section 3.1 for problems with superovulation.]

Follicular aspiration followed by IVF and embryo culture is most advantageous for providing early cleavage stage embryos for research. However, the prolonged culture times required to grow non-human primate embryos to fully expanded or implanting blastocysts (see section 5.3) may be a significant deterrent to using IVF for producing later stage embryos. Another potential disadvantage to research using embryos derived from IVF concerns criticism pertaining to their normalcy. Only a small percentage of primate embryos derived from IVF implant and produce live young after embryo transfer (Soules, 1985; Bavister *et al.*, 1984; Boatman *et al.*, 1986). It is likely that IVF derived embryos, even if "normal" by morphological criteria at the light microscope level, are not fully equivalent to those produced after mating in unstimulated cycles (Vanderhyden *et al.*, 1986). The potential cause of this putative non-equivalency may be due to gonadotropin recruitment of less viable ova, which has recently been documented in superovulation studies in swine (Koenig *et al.*, 1986).

4.2. Procedures

Oocyte Maturity and IVF. Our procedures for follicular stimulation, semen preparation, and sperm capacitation have been published (Bavister *et al.*, 1983a; Boatman and Bavister, 1984; Boatman *et al.*, 1986). We have subsequently modified our procedures for evaluating egg maturity. As previously (Bavister *et al.*, 1983a), the initial assessment of the aspirated egg cumulus complex (ECC) is performed under the dissecting microscope and the appearance of the cumulus cells is used to assign a numerical grade. Then, ECC's are transferred to drops of equilibrated TALP under oil. Using sterile needles, the ECC's are stretched out and attached to the bottom of Falcon no. 1008 dishes (Becton Dickinson Co.). The ECC's are examined with a Nikon inverted microscope using DIC optics and a fully controlled environmental chamber (37° C, 5% CO_2; see section 2.1). The appearance of the corona radiata cells (condensed *vs.* elongated [radiant]), [in]visibility of the zona pellucida, presence or absence of a perivitelline space, spherical or eccentric vitellus, as well as definitive presence of the first polar body (PB1) are used to re-grade the ECC's as immature, nearly mature, mature or atretic (see Fig. 1A-D, atretic not shown). Eggs may have been classified as mature even if the PB1 was not visible due to optical interference from cumulus and corona cells (Fig. 1B). [Cumulus and corona cells are left undisturbed during the period of co-incubation with sperm to facilitate sperm capacitation and to act as

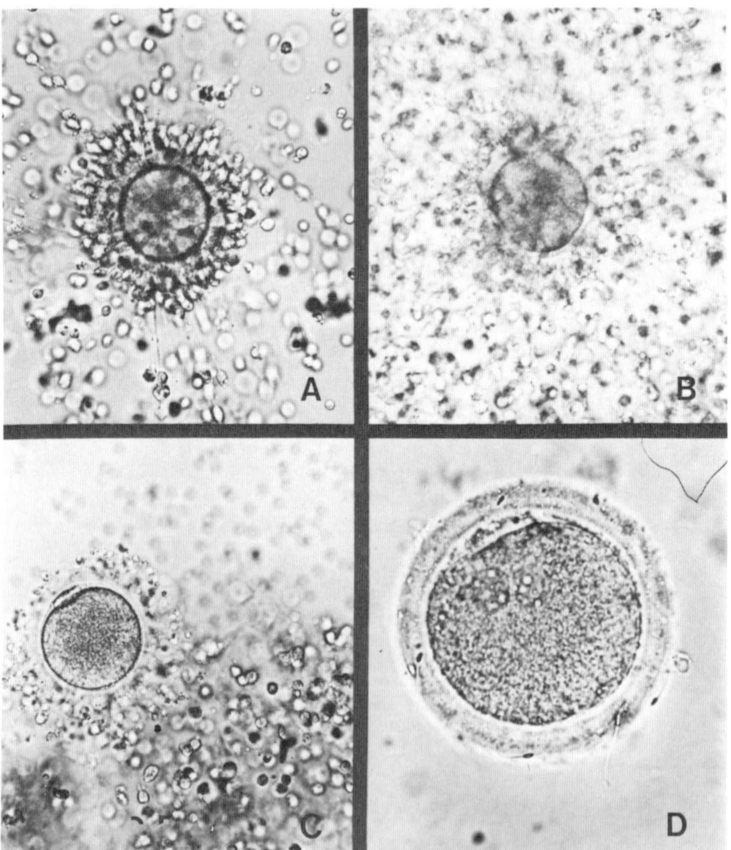

Figure 1. (A) Immature egg with unexpanded corona radiata. (B) Mature egg with expanded cumulus oophorus and corona radiata. (C) Mature egg with first polar body. (D) Fertilized egg (13 hr post-insemination [PI]) with two pronuclei near syngamy and two polar bodies (slightly out of focal plane). Optical magnification: A to C x 50, D x 100.

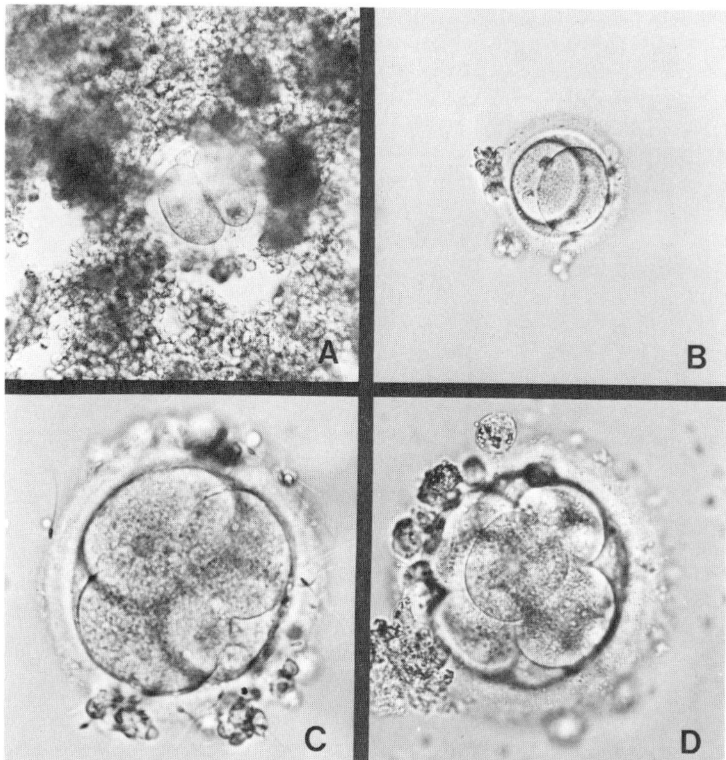

Figure 2. (A) Two-cell embryo (16 hr PI) with nucleated blastomeres and two polar bodies, in fertilization drop. (B) Same embryo as in A (19 hr PI) after removal of excess sperm and corona cells. Optical magnification A and B x 50. (C) Four-cell embryo (same embryo as in A and B, 34 hr PI). (D) Eight-cell embryo (42 hr PI). Optical magnification C and D x 100. Nucleated blastomeres can be seen in both embryos.

additional barriers to polyspermy.] Therefore, the maturational status of the egg (presence of PB1) must be re-confirmed after removal of cumulus and corona cells and excess sperm. This is done 12 to 15 hr after insemination.

Evaluation of Fertilization. At 12 to 15 hr after insemination, eggs are examined for the occurrence of normal fertilization. If sperm capacitation and egg insemination have been performed according to published procedures (Bavister *et al.*, 1983a; Boatman and Bavister, 1984), the penetrated egg(s) should have 2 well developed pronuclei and 2 polar bodies (Fig. 1D). If polyspermy occurs, it can readily be detected by the presence of more than 2 pronuclei. Eighteen hr or more after insemination, pronuclei may no longer be visible. By 24 hr, most zygotes have cleaved (Fig. 2A, B). At this point, polyploidy can no longer be detected in the living embryos. Since poly-pronucleate monkey (and human) embryos can continue development to at least the morula/early blastocyst stage (Fig. 3A shows one of these rhesus embryos at the 4-cell stage; see also Van Blerkom *et al.*, 1984), cleavage to the 2-cell stage is not a sufficient indicator that normal fertilization has occurred unless an additional assessment was made at the pre-syngamy stage.

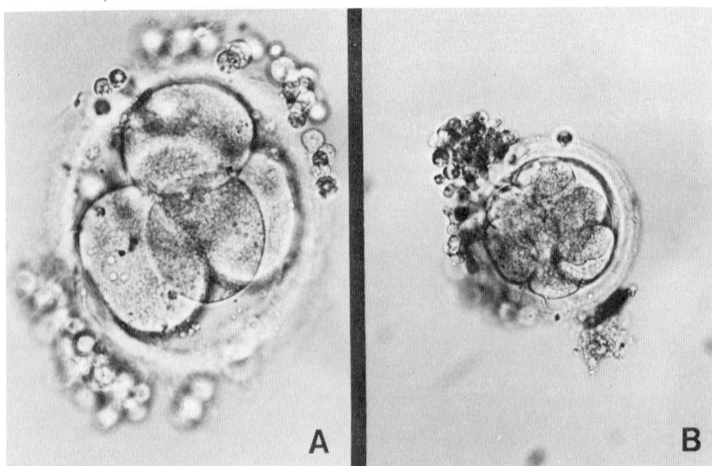

Figure 3. (A) Five-cell embryo (34 hr PI) derived from zygote with multiple (4) pronuclei; optical magnification x 100. (B) Four-cell embryo (39 hr PI) which gave rise to male offspring *Orwell*; numerous blebs observed in this embryo. Optical magnification x 50.

It is not uncommon to find anucleate fragments ("blebs") pinched off from the rest of the nucleated embryo during various cleavage divisions. The presence of a few blebs is compatible with normal embryonic and fetal development. Our second "test-tube" monkey (*Orwell*) resulted from the transfer of a single embryo which had some of these blebs (Fig. 3B). Nonetheless, since blebs occur with some frequency, and involve varying proportions of the total egg cytoplasm, it is important to have a microscope capable of resolving the presence or absence of nuclei in apparent blastomeres.

5. EMBRYO GROWTH CHARACTERISTICS

5.1. Oocyte Maturity and Consequences for Embryonic Development

Until recently, the study of mammalian oocyte maturation was concerned almost exclusively with maturation of the nucleus. It was assumed that the nucleus, activated at the completion of the follicular cycle by the LH surge, would express maturation competence only after all other aspects of oocyte cytoplasmic processing had been completed. The application of IVF as a research tool has revealed that the normal sequence of oocyte maturation can be temporally reordered, and that the egg nucleus can complete maturation in an oocyte which lacks the cellular machinery necessary for the completion of fertilization and/or normal embryonic development (Moor *et al.*, 1982; Leibfried and Bavister, 1983; Szöllösi and Gerard, 1983). A disturbance in the temporal order of oocyte maturation events is often expressed as the sequelae of premature harvesting of oocytes from follicles prior to the natural LH surge. The question then arises, whether defective cytoplasmic maturation occurs spontaneously in ova and could lead to interference with normal reproductive efficiency, or, alternatively, whether it is only an iatrogenic (*in vitro*) artifact, albeit one with considerable import for clinical IVF and for animal science.

There are some indications that naturally occurring defects in oocyte maturation could contribute to reproductive inefficiency in primates. There is considerable evidence for a high level of conception/embryonic wastage in old world primates, including macaques and humans. Estimates of pre-clinical pregnancy conception/embryonic losses range from 25% to as high as 50% (Hertig *et al.*, 1956; Hurst *et al.*, 1980; Hendrickx and Binkerd, 1980; Enders *et al.*, 1982). Abnormal preimplantation embryos have been flushed from the reproductive tracts of mated rhesus monkeys. Defects noted in the embryos included retarded cleavage timing as well as abnormal morphology (Hurst *et al.*, 1980; Enders *et al.*, 1982). Relevant to this, we have found oocytes with defective cytoplasm that were aspirated on the day of the E_2 surge from immediately preovulatory follicles of spontaneously cycling rhesus monkeys (P. Morgan, unpublished observations). Similar oocytes have been recovered from dominant follicles of cynomolgus monkeys (Kreitmann *et al.*, 1982). This unexpected variability of preovulatory primate oocytes compared to those of laboratory rodents suggests that spontaneously occurring deficiencies in oocyte (cytoplasmic) maturation may contribute to the high early pregnancy losses observed in primates. This inference is supported by our work on *in vitro* maturation of rhesus oocytes (Boatman *et al.*, 1986; Morgan *et al.*, 1986). We have shown that events regulating folliculogenesis and oocyte maturity may have a significant effect on embryo development *in vitro* after fertilization has occurred (Boatman *et al.*, 1986; Morgan *et al.*, 1986).

A large percentage of oocytes aspirated following superstimulation protocols were classified as non-ovulatory: 53.7% were immature (Table I). The immature oocytes would appear to be significantly less fertilizable than

Table I
Relationship of Initial Classification of Oocytes to Subsequent Fertilization and Development

Class	No. oocytes recovered (% total)	No. oocytes fertilized (% inseminated)	No. ≥6- to 8-cells *in vitro* (% of inseminated oocytes)
Mature[a]	66 (40.2)	40/61[c,e] (65.6)	35/58[d] (60.3)
Immature	88 (53.7)	35/83[c,e] (42.2)	21/83 (25.3)
Atretic[b]	10 (6.1)	NA NA	NA NA

[a]Incubated <8 hr prior to insemination, polar body (PB1) confirmed after cumulus dispersal (see section 4.2).
[b]PB1 but no cumulus oophorus or corona radiata at time of recovery, zona pellucida often thick or ragged, occasionally defective cytoplasm.
[c]Some oocytes not inseminated.
[d]Excludes 2 embryos transferred (1 live birth [Boatman *et al.*, 1986] and 1 lost prior to the 6- to 8-cell stage).
[e]χ^2, $p \leq 0.01$.

Table II
Comparison of Fertilization and Development of Ova Initially Mature *vs.* Ova Matured *In Vitro*

Class	No. ova fertilized/ no. inseminated (%)	No. ≥6- to 8-cells *in vitro* (% of ova fertilized)
Mature[a]	40/61 (65.6)	35/37[c,d] (94.6)
Maturation completed *in vitro*[b]	35/49 (71.4)	21/35[d] (60.0)

[a]Incubated <8 hr prior to insemination, first polar body (PB1) confirmed after cumulus dispersal.
[b]Incubated >8 hr prior to insemination, PB1 confirmed after cumulus dispersal.
[c]Excludes 2 embryos transferred (1 live birth [Boatman *et al.*, 1986] and 1 lost prior to 6- to 8-cell stage).
[d]χ^2, $p \leq 0.001$.

those of the immediately preovulatory oocytes (Table I). This result was not unexpected. However, when the results were re-analyzed for only those oocytes that matured (extruded PBI), a different result was obtained. Oocytes that had completed maturation *in vitro* were not less fertilizable than those which were initially mature (71.4% *vs.* 65.6%, respectively, Table II). Oocytes matured *in vitro* did not differ in the percentage that cleaved to the 2-cell stage (data not shown). However, the zygotes obtained from these two types of mature oocytes differed in their developmental abilities: 95% of zygotes from initially mature oocytes reached the 6- to 8- cell stage *in vitro* compared to only 60% for those derived from initially immature oocytes (Table II). In a separate study, we have shown that only 18% of the embryos derived from immature oocytes had normal cleavage timing at the 6- to 8- cell stage (see Fig. 4 illustrating *in vitro* vs. *in vivo* cleavage timing), compared to 62% for those from mature oocytes (Morgan *et al.*, 1986). We conclude that consideration of oocyte maturity is critical when embryo development is examined subsequent to follicular aspiration and IVF. We suggest that our findings on retarded embryo cleavage timing *in vitro* subsequent to disturbed oocyte maturation may have a bearing on the etiology of embryonic losses *in vivo*.

5.2. *In Vitro* Development of Oviductal Stage Embryos

Inasmuch as primate embryos from the fertilized egg up to the 8-cell stage (approximately) would be found normally in the oviduct, these cleavage stages will be considered "oviductal" for the purposes of this section regardless of their original source, *i.e.*, *in vivo* or *in vitro*.

The majority of embryos that arise from IVF of primate oocytes are transferred to recipients or are frozen-stored for future transfer at or before the 8-cell stage is attained (see sections 6.3 and 6.4). Additionally, these are the embryo stages most likely to be utilized for nuclear and/or genetic

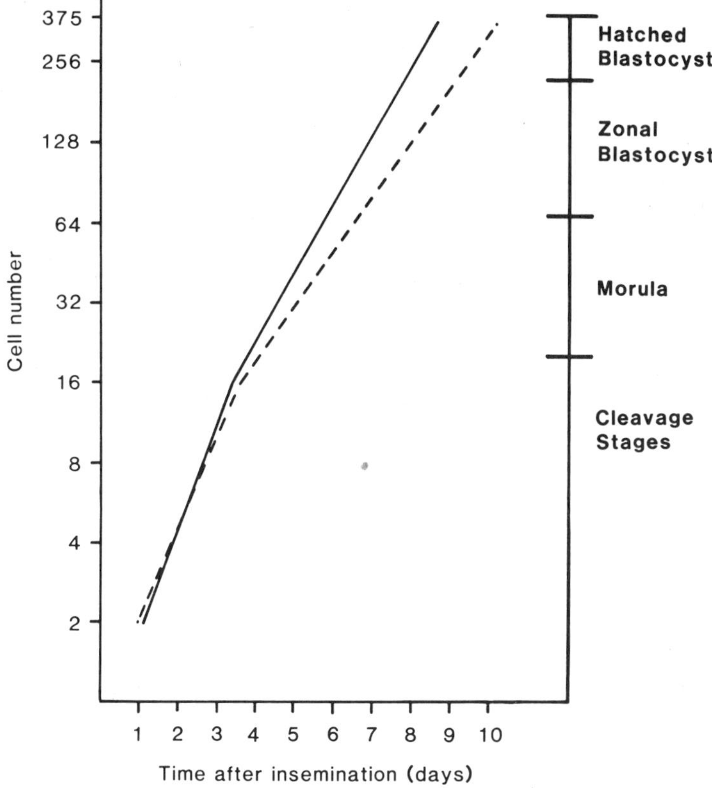

Figure 4. Cleavage timing of rhesus monkey embryos *in vivo* (solid lines) and *in vitro* (dashed lines). Phase 1 *in vivo* (2- to 16-cells) derived from mid-points of cleavage timings reported by Lewis and Hartman (1941). Phase 2 *in vivo* (16-cell to hatched blastocyst) derived from estimated fertilization age or midpoint of interval of rhesus and baboon embryos reported by Lewis and Hartman (1941), Heuser and Streeter (1941), Kraemer and Hendrickx (1971), Hendrickx and Kraemer (1968). Phase 1 *in vitro* obtained from cleavage timings of 100 rhesus embryos (Boatman, Morgan and Bavister, unpublished data). Phase 2 was obtained by assigning the same mean cell number to *in vitro* hatched blastocysts (Boatman *et al.*, unpublished data) as reported for *in vivo* embryos while using the mean observed hatching time.

manipulation experiments. Thus, culture requirements and growth characteristics for oviductal primate embryos are of particular importance for research and biomedical application.

The first consideration in culturing embryos is the choice of culture medium. We made certain assumptions when initiating our studies on fertilization and embryo culture (see also section 2.1). In preference to complex media such as Ham's F10, we used the simple medium TALP, which was devised by Bavister for IVF and embryo culture (EC) in the hamster (Bavister and Yanagimachi, 1977); an earlier version of TALP (Bavister, 1969) was used in some of the first experiments on human IVF and EC (Edwards *et al.*, 1969, 1980). There is some evidence that, under certain conditions, complex media can be detrimental to the growth of embryos *in vitro* (Farrell, Bavister and Kraus, see Chapter 11). We also assumed that the motility and viability of ejaculated rhesus spermatozoa in the medium of choice could be used as one index of the suitability of that medium. Since capacitated spermatozoa and the ovulated ovum/zygote share the same physiological site, the oviductal ampulla, during fertilization, they must both be able to survive in the chosen culture medium. Our preliminary tests showed that rhesus sperm maintained high viability and motility in medium TALP (Boatman and Bavister, 1984) for at least 48 hr. For embryo culture, we supplemented TALP with the amino acids glutamine, methionine, isoleucine, and phenylalanine, which were known to be beneficial for hamster oocyte maturation and embryo culture (Gwatkin and Haidri, 1973; Juetten and Bavister, 1983; Bavister *et al.*, 1983a, 1983b). For protein supplement, 3 mg/ml BSA (the standard concentration used in our hamster IVF and EC) was included at all stages of culture. A small amount (2%) of heat-inactivated rhesus blood serum was present only during the fertilization period.

In these early studies of rhesus embryo development *in vitro*, the use of the simple modified TALP medium without serum seemed to be validated. In TALP, 87% of IVF rhesus embryos cleaved to at least the 8-cell stage at 54 hr (mean time) after egg insemination (Table III; Bavister *et al.*, 1983a). The cleavage timings of the embryos were consistent with those expected for embryos *in vivo* (Bavister *et al.*, 1983a; and Fig. 4). The viability of embryos was tested by transfer to surrogate recipients. Twenty-two embryos cultured to the 2- to 8- cell stage (mean cell number 5.45) in the modified TALP medium were transferred to 11 recipients and two pregnancies were established; one of these resulted in the birth in August, 1983, of *Petri*, the first genetically-confirmed non-human primate birth following IVF and ET (Bavister *et al.*, 1984). We concluded that culture of rhesus embryos in simple

Table III

Comparison of Embryo Development in Two Culture Media

Culture medium	2-cells No. (%)[c]	2-cells Time[d]	≥6- to 8-cells No. (%)[e]	≥6- to 8-cells Time[d]	≥ morula No. (%)[e]	≥ morula Time[d]
TALP[a]	23/29 (79)	26 ± 5	20/23 (87)	54 ± 14	3/23 (13)	88 ± 11
CMRL (modified)[b]	75/110 (68)	23 ± 5.5	56/72[f] (78)	59 ± 19	46/68[f] (68)	111 ± 16

[a]Modified Tyrode's solution with supplements including 3 mg/ml BSA (Bavister *et al.*, 1983a; see also Section 8).
[b]CMRL-1066 with supplements including 20% fetal (human, calf) or adult (rhesus, horse) serum (Boatman *et al.*, 1986; see also Section 8). Oocytes inseminated in TALP, then transferred to CMRL at 12 to 18 hr post-insemination (PI).
[c]Percent of inseminated oocytes.
[d]Time (hr) PI; mean ± S.D.
[e]Percent of fertilized ova.
[f]Numbers of embryos reduced due to transfers to recipients.

medium was compatible with embryo viability. However, only 48% of the embryos reached the 16-cell stage (data not shown), and only 13% formed morulae (Table III; Bavister *et al.*, 1983a). This could be an indication that the metabolic requirements of the embryos were changing sometime after the 8-cell stage had been reached, and that the medium was inadequate for supporting these more advanced stages of development. The metabolic requirements of mouse embryos alter at the 8-cell stage (Kaye *et al.*, 1982; Rosenblum *et al.*, 1986), and hamster embryos also appear to change their nutrient requirements at this time (Bavister *et al.*, 1983b).

The success of Pope *et al.* (1982) in culturing naturally conceived (predominantly uterine stage) baboon embryos in the complex medium CMRL-1066 encouraged us to use this medium for rhesus embryos after fertilization had occurred in the TALP medium (Morgan *et al.*, 1984; Boatman *et al.*, 1986). We inseminated 151 rhesus oocytes in our standard TALP medium, then transferred the resulting zygotes and 2-cell embryos at 12 to 24 hr post-insemination to CMRL-1066 plus 20% fetal (human, calf) or adult (rhesus, horse) serum for continued culture. Seventy-two of the 75 embryos were maintained until either the 6- to 8- cell stage had been reached, or embryonic arrest had occurred. The cleavage timings of these embryos in complex medium, CMRL-1066 plus 20% serum, did not differ from those obtained in the earlier study using the simple medium TALP with 3mg/ml BSA (Table III). Although a lower percentage of the fertilized zygotes reached the 6- to 8-cell stage in CMRL than in the earlier study using TALP (78% *vs.* 87%, respectively: Table III), this was attributed to the higher proportion of initially immature oocytes used in the second study (Table II; see also section 4.1). One of the embryos grown in CMRL + serum was transferred at the 4-cell stage (Fig. 3B) to a surrogate recipient (on the day of its sex-skin breakdown), resulting in the birth of a male infant (*Orwell*) in June, 1984 (Boatman *et al.*, 1986). Although there does not *appear* to be any difference in viability of embryos cultured in these two types of medium (Table III), the numbers are too limited to draw any definitive conclusions based on embryo transfer.

5.3. *In Vitro* Development of Uterine Stage Embryos

For the purposes of discussion in this section, embryos having 10 or more cells are considered to be uterine stages because the majority of embryos that have been obtained by uterine flushing of rhesus and baboon monkeys had 10 or more cells (Hendrickx and Kraemer, 1968; Hurst *et al.*, 1976; Pope *et al.*, 1983). A small proportion of these embryos were at the 2- to 6- cell stages (Pope *et al.*, 1983). These younger stages may have been slow or arrested in their development, even though at least one of them was shown to be viable (Pope *et al.*, 1983).

Pre-hatching Development. The use of the more complex medium, CMRL-1066, plus 20% fetal human or calf serum, appeared to be beneficial in our culture system for increasing the survival of IVF embryos to the morula stage of development: 5 times as many survived in CMRL than in TALP (68% *vs.* 13%, Table III).

As stated in section 5.2, cleavage timing of IVF 2- to 8-cell embryos *in vitro* did not differ from that expected for embryos recovered from the female reproductive tract (doubling times 19 and 18.4 hr, respectively: Fig. 4).

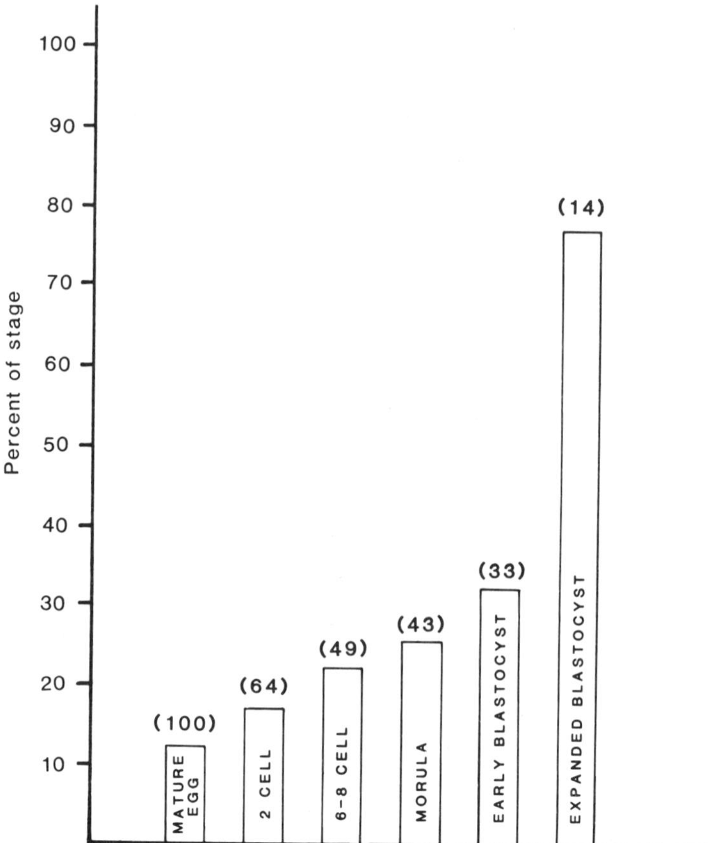

Figure 5. Probability at each stage of development for embryo to achieve hatched blastocyst status *in vitro*; values in parentheses = number of embryos at each stage. All data from transferred embryos excluded.

However, after the 8-cell stage, the doubling time of the cultured rhesus embryos slowed to approximately 34 hr (Fig. 4). It would appear from published cell counts of rhesus and baboon embryos flushed from reproductive tracts that they also have a slower rate of development after the 8-cell stage has been reached (estimated doubling time 27.4 hr, Fig. 4). Even so, the *in vitro* cultured embryos appeared to be approximately 25% slower than their *in vivo* counterparts. As a result, almost all of the "*in vitro* lag" expressed by the cultured rhesus embryos occurred after the 8-cell stage was reached. On average, the cultured embryos were 2 to 3 days behind schedule at the time of zona shedding (Fig. 4).

The greatest loss of *in vitro* cultured embryos occurred during their transit as uterine stages prior to hatching (Fig. 5). The decline in viability was linearly related to time in culture, with about a 9% loss per day (Fig 6). Since the embryos took a much longer time in developing from the 8-cell stage to hatched blastocyst (184 hr average) than from the time of insemination to the 8-cell stage (59 hr average), this resulted in the larger percentage loss for the

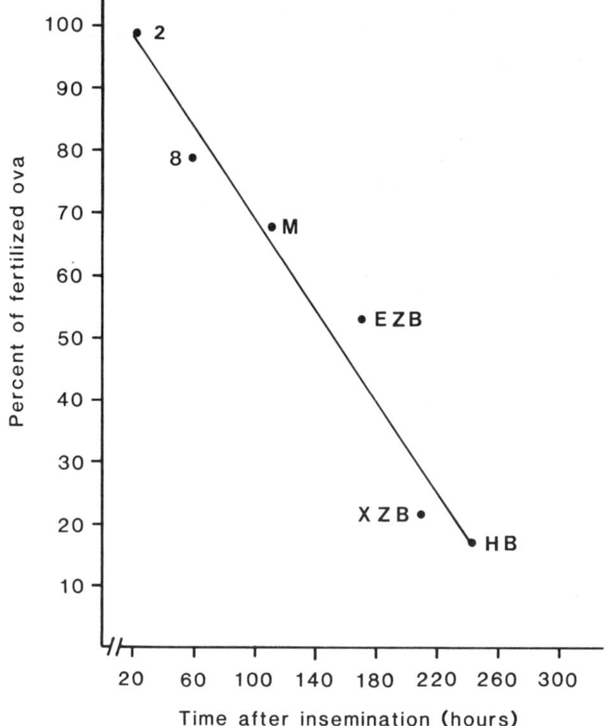

Figure 6. Percentage of fertilized ova reaching key developmental stages (2-cell, 8-cell, morula [M], early zonal blastocyst [EZB], expanded zonal blastocyst [XZB] and hatching blastocyst [HB]) as function of mean time to reach stage *in vitro*.

uterine stage of development. Overall, only 11% of starting mature ova survived to become hatched blastocysts *in vitro*, whereas 33% of early blastocysts hatched (Fig. 5).

Losses during culture do not appear to be restricted to primate embryos derived from IVF. In one study, 60 embryos flushed from the uterus were cultured (Pope *et al.*, 1982). The percentage survival (to hatching and/or beyond) of embryos that contained 12 or more cells at the time of recovery was greater than we have reported here (50% *vs.* 25%, respectively; Pope *et al.*, 1982); however, younger stages showed only 11% survival, a lower success rate than we report for similar stages (17 to 22%: Fig. 5; Pope *et al.*, 1982).

Blastocyst Expansion and Zona Shedding. The blastocoel cavity in cultured rhesus embryos, viewed with DIC optics, began as one or more small clear vesicles in the morula (not shown). At this time (and slightly preceding it) the morulae contained large, dark inclusion granules (Fig. 7B) which appeared similar to those described in mouse morulae (Mintz, 1964). Initially, the small clear vesicles appeared to be *within* cells rather than *between* them (not shown). They gradually increased in diameter and became confluent (Fig. 7C shows a blastocyst with non-confluent cavities); the first vesicle was noted at

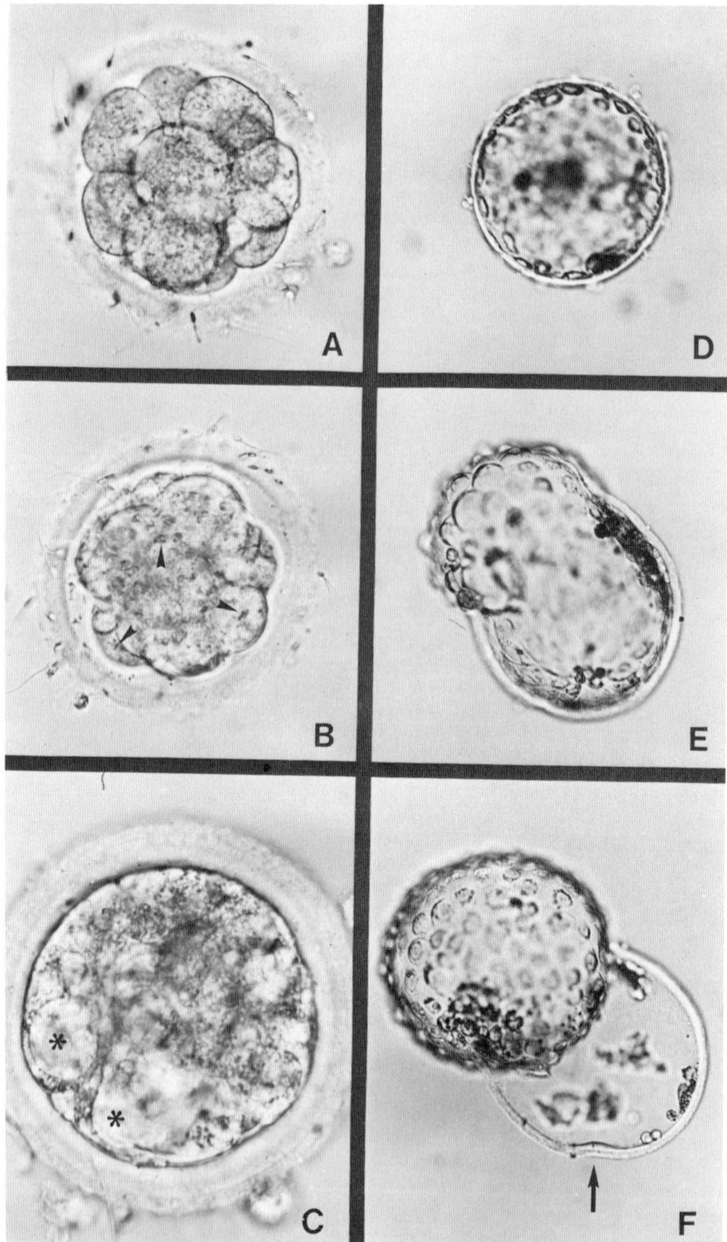

Figure 7. (A) Sixteen-cell embryo (same embryo as in Fig. 3 A-C, 87 hr PI). (B) Morula (same as A, 118 hr PI) showing compaction and dark granules (arrowheads). Optical magnification A and B x 100. (C) Zonal blastocyst (168 hr PI) with unfused blastocoel cavities (asterisks); optical magnification x 132. (D) Fully expanded zonal blastocyst (same embryo as C, 253 hr PI) showing numerous trophectodermal projections through the zona. (E) Early stage of hatching (same embryo as C and D, 265 hr PI). (F) Fully hatched embryo (same embryo as C to E, 268 hr PI), trophectodermal projection still tethered to zona (arrow). Optical magnification D to F x 50. E and F from Bavister (1986).

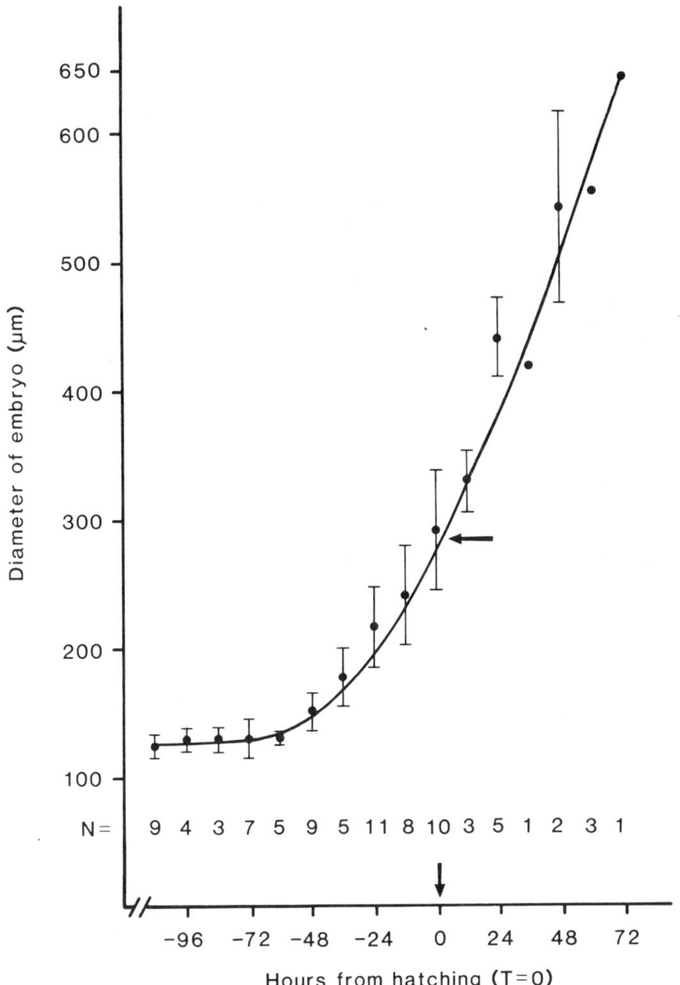

Figure 8. Kinetics of blastocoel expansion of 11 embryos that hatched *in vitro* normalized to the onset time of zona rupture (arrows). Measurements exclude the zona pellucida.

129 hr (mean time) post-insemination, and the time of expanded blastocyst formation (when all of the vesicles had fused) was 210 hr (Fig. 7D shows an expanded blastocyst). The prominent inclusion granules in the morulae became less discernable with time and were no longer seen in expanded blastocysts.

When embryo expansion was normalized to the time of zona rupture, the expansion of the blastocoel cavity was described by a polynomial function or exponential growth curve (Fig. 8). Maximal expansion of volume began about 24 hr preceding zona rupture and continued to 96 hr post-hatching. Individual embryos initiated blastocoel expansion at widely variant times after insemination (Fig. 9). One possible cause for this variation could be differing amounts of cell death. Longer times would then be needed for some embryos to produce enough viable cells to generate sufficient blastocoel fluid so that expansion would reach the critical hatching diameter (Figs. 8 and 9).

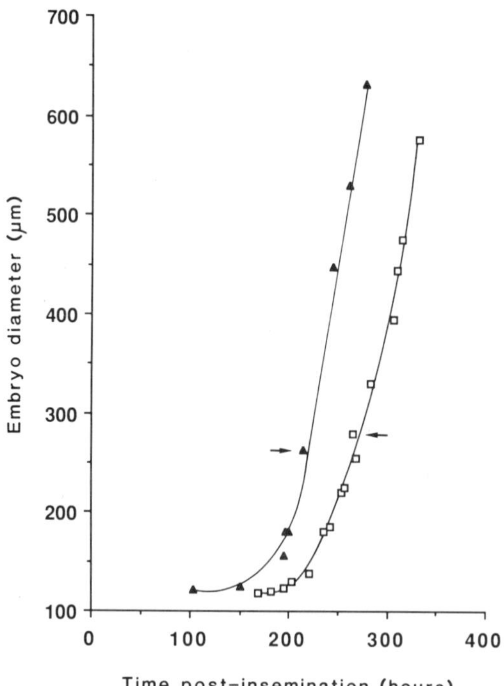

Figure 9. Variability in time taken to achieve hatching diameter (arrows); there is a 52 hr difference in hatching time for these two embryos.

Knowledge of the source(s) of energy used (required) for the metabolically demanding stage of blastocoel expansion is critically needed, since the largest loss of embryos occurred during this period of development. Fifty-eight percent of early blastocysts failed to complete expansion, and 67% failed to hatch (Fig. 5). Additionally, it needs to be determined if the energy sources are pre-packaged within the embryos at some particular stage of development, or if they are exogenously supplied at or near the time they are used for formation of the blastocoel fluid. It has been suggested by Wiley (see Chapter 4) that lipid granules stored in mouse embryos serve as energy sources for the pumping of blastocoel fluid as well as providing metabolic water for this purpose. This is of particular interest in light of the large inclusion granules that we observed in rhesus morulae, which seemed to disappear with progressive blastocoel formation. It would be interesting to find out if the inclusion granules seen by us are rich in lipids.

As pointed out by Enders and Hendrickx (1980), information concerning loss of the zona pellucida in primates is limited, although data on rhesus and human embryos suggest that the zona is lost prior to implantation. We have monitored the loss of the zona pellucida in 14 cultured embryos either discontinuously (recorded by still photography) or continuously (recorded by time-lapse videomicrography). *In vitro*, the embryos escaped by rupturing the stretched zonae after reaching a critical diameter (approximately 280 μm, Fig. 8) through expansion of the blastocoel (Fig. 7D-F). Escape of embryos may also be assisted by trophectodermal processes, which we observed projecting through the zona pellucida 17 hr (mean time) prior to zona rupture (Fig. 7D, F). [See also Chapter 11 (Ed.).]

Post-Hatching Development. After rupturing the zona pellucida, the embryos continued to expand and floated free from the zona pellucida. The embryos continued to float in the medium containing 20% serum for another 2 to 3 days. The maximum (spherical) diameter recorded was 644 μm. As the embryos started to deflate, they became more dense, assuming a more discoid shape, and began to settle. Only after considerable contraction did they contact the substrate and initiate attachment (Fig. 10A). In our initial

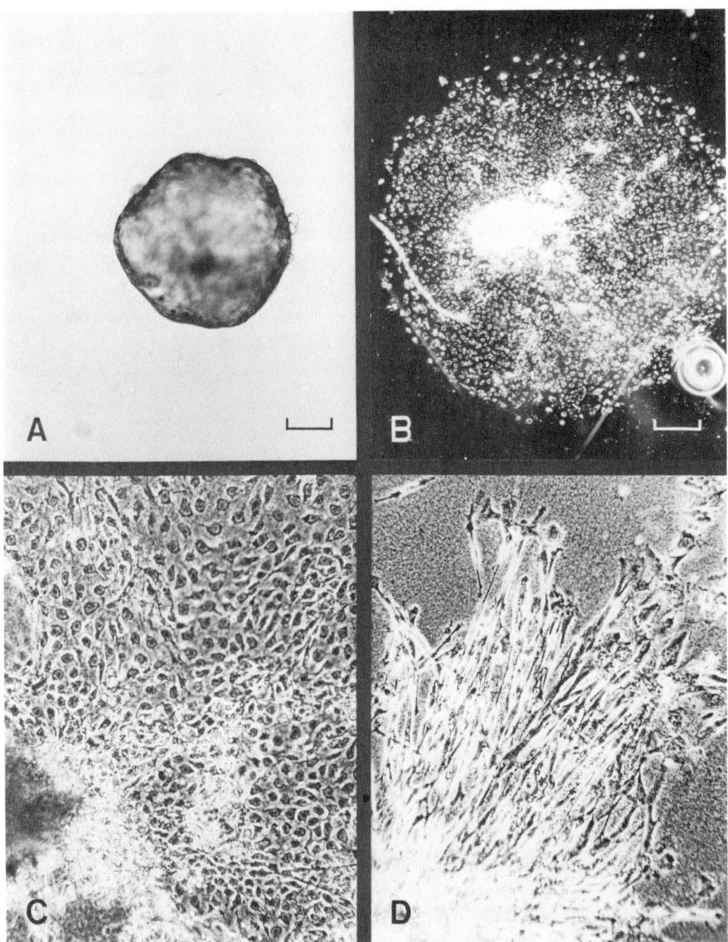

Figure 10. (A) Azonal blastocyst (same as Fig. 4C to F, 372 hr PI) at earliest stage of attachment to substrate (12 hr after deflation and settling); bar = 102 μm. (B) Trophoblast outgrowth on collagen gel (17.5 days PI); embryonic mass is bright central area. Bar = 333 μm. (C) Polygonal, confluent trophoblastic cells grown in calf serum on collagen (same embryo as B, 17.5 days PI). (D) Fusiform trophoblastic cells grown in human cord serum on collagen (20.5 days PI). Optical magnification C and D x 25.

experiments, hatched blastocysts were cultured in Falcon no. 1008 dishes. Blastocysts made only temporary attachment to this surface. In later experiments, embryos were grown in collagen-coated wells (on glass). After deflating, they were able to adhere firmly to this substrate and to proliferate trophoblastic outgrowths (Fig. 10B-D).

Embryos cultured for 9 to 13 days after insemination were sectioned for electron microscopic examination. They were found to have undergone some differentiation but had varying degrees of disorganization of the embryonic components of the inner cell mass (Enders *et al.*, 1987). The trophoblast had differentiated into cyto- and syncitio-trophoblast. In other embryos that were grown on collagen substrate, we found that the morphological appearance (under the light microscope) of the outgrowing trophoblastic cells was influenced by the choice of serum supplement: Hyclone® calf serum produced trophoblastic cells with polygonal confluent form (Fig. 10C, 17.5 days PI) whereas human fetal cord serum produced cells with a fusiform shape (Fig. 10D, 20.5 days PI).

In vivo conceived baboon embryos recovered at the 16-cell stage or later and then cultured were similar in appearance to the IVF cultured rhesus embryos that we have described, with some exceptions. In the published photographs (Pope *et al.*, 1982), the region of the inner cell mass of the hatched baboon embryos 5 to 7 days after hatching (10 to 12 days of culture) seemed to be better defined. If the inner cell mass was indeed better conserved than in our experiments, this could have been due either to conditioning in the uterine environment, or alternatively, the absence of an oil overlay in the culture system used by Pope *et al.* (1982) may have been beneficial. The morphological types of trophoblastic outgrowth cells that we have noted (polygonal and fusiform) were not specifically mentioned (however, outgrowing cells with poorly defined boundaries can be seen in the published photographs of Pope *et al.* [1982]). The sheets of cells that we have described emanating from the blastocyst mass proper became more prominent only 8 to 9 days after hatching. Moreover, the morphology of these cells, as well as that of the attaching blastocyst, may be determined in part by the type of substrate to which they attach, *i.e.*, various types of plastic, glass or collagen.

6. ASSESSMENT OF EMBRYONIC NORMALCY AFTER *IN VITRO* CULTURE

6.1. Morphological Criteria

Embryos may be subjected to repeated evaluations during their growth in culture using the light microscope to assess normal appearance. This is a major means of assessment in our work on primate embryos as discussed in preceding sections of this chapter (see also Figs. 1-3, 7, 10 and 11). Embryos may be preserved in fixatives at set times of development and the numbers of cells compared to those expected at an equivalent development time or stage *in vivo* (Wiley *et al.*, 1986; and see Chapter 10). Additionally, they may be fixed and sectioned for electron microscopy so that the distribution of organelles and cellular fine structure may be compared to embryos grown *in vivo* (discussed in section 5.3).

6.2. Cleavage Timing

Along with morphological appearance, the time taken to reach a particular stage, especially with reference to the "normal" time taken *in vivo*, is a valuable indicator of normal embryo development, one that we have relied upon heavily in our work (see sections 4.2 and 5, also Figs. 4 and 6).

6.3. Embryo Transfer

The ultimate test of viability after embryo culture is the birth of live young following the transfer of embryos to synchronous recipients. Our live birth rate for cultured IVF rhesus embryos is approximately 6% per *transferred embryo* (Table IV; Bavister *et al.*, 1984; Boatman *et al.*, 1986), similar to the viability rate of human embryos transferred after IVF (Soules, 1985). The pregnancy rate per *recipient* is influenced by the number of embryos transferred to each recipient: the rate is 8% for one IVF rhesus embryo and 15% for two (Table IV; Bavister *et al.*, 1984; Boatman *et al.*, 1986).

The testing of embryo viability in non-human primates by embryo transfer has some features that may be peculiar to non-human primate research. The follicular stimulation protocol (PMSG/hCG) that has given us our most reliable and highest yield of embryos (Boatman *et al.*, 1986) results in a post-aspiration (luteal phase) elevation of estradiol and progesterone to

Table IV
Effect of Transfer Method and Number of Embryos Transferred per Recipient on Pregnancy Outcome

	No. of transfers (= no. of recipients)				
	No. of embryos/ recipient			Total no.	No. of term
	1	2	≥ 3	recipients	pregnancies
Method of transfer:					
Non-surgical (uterine)	11	11	2	24	2
Surgical (oviductal)[a]	2	2	-	4	1
Total no. recipients	13	13	2	28	
Total no. embryos	13	26	9		
No. term pregnancies	1	2	0		3

[a]At laparoscopy

Table V
Summary of Embryo Transfers by Day of Ovulation of Recipients and Developmental Stage of the Most Advanced Embryo

	Estimated day of ovulation	No. of cells in most advanced embryo transferred to each recipient										Total recipients (by day of ovulation)	Term pregnancies
		3	4	5	6	7	8	10-11	16	M[c]	EZB[d]		
Day of sex-skin breakdown:													
-5	(-4, -3)			1			1					2	
-1	(+1)[a]		1									1	1
0	(+1, +2)		2	1	1	1	3			1		9	2
+1	(+2, +4)[b]	1	4		2		1	1	1			10	
+2	(+3, +4)		1					1			1	3	
+4	(+5, +6)				1						1	2	
+6	(+7, +8)										1	1	
Total recipients (by cell no. of most advanced embryo)		1	8	2	4	1	5	2	1	1	3	28	
Term pregnancies			2		1								3

[a]Confirmed by estradiol assay (sex-skin estimation 0, +1).
[b]One recipient confirmed by estradiol assay as +4 (sex-skin estimation +2, +3).
[c]M = morula.
[d]EZB = early zonal blastocyst.

supra-physiological levels in the donor animal. Therefore, we transfer IVF embryos to surrogate (synchronous) recipients. The problems of inducing artificial synchrony in monkeys has been discussed in section 3.1. Instead, we rely on the availability of a large colony of naturally cycling female monkeys. The most accurate method of determining synchrony is a daily rapid assay for estradiol. For practical reasons, only a relatively small sub-group of animals can be monitored with daily E_2 assays. The pool of animals for E_2 assay needs to be at least 2 to 3 times larger than the minimum anticipated number of embryos or transfers desired. The animals are selected for synchrony of the first day of their menstrual cycles, but not all of them will have a synchronous E_2 peak (or ovulation). Similarly, the embryos to be transferred will not be synchronous with respect to developmental stage (as a result of differing oocyte maturation/insemination times). Not infrequently, variability in the donor animal's response to the stimulation protocol (as detected by daily E_2 levels and ultrasonography) results in a last-minute rescheduling of the day of oocyte recovery. All of these sources of uncertainty can significantly reduce the number of known synchronous recipients (as deduced from E_2 levels). For these reasons, the majority of our embryo transfers have been to recipients selected on the basis of sex-skin coloration and swelling (Bavister *et al.*, 1984, 1985; Boatman *et al.*, 1986). In the rhesus monkey, the perineal or "sexual" skin shows color changes under the influence of estrogens that peak around midcycle. A significant decrease in color, "sex-skin breakdown", is a useful indicator that ovulation has occurred (Czaja *et al.*, 1977). In 80% of cases, sex-skin breakdown occurs 2 or 3 days after maximal (peak) E_2 (that is, 1 or 2 days after ovulation: Weick *et al.*, 1973), but in 18% of cases, it occurs 3 or more days after peak E_2 (Boatman *et al.*, 1986). Obviously, many of the embryos transferred to recipients selected by this method will not be perfectly synchronized with the recipient's uterus, and this will undoubtedly influence the transfer outcome. The importance of correct synchrony can be seen in our transfer results (Table V). Our 3 IVF births have resulted from transfer to recipients one to two days after estimated ovulation time (Table V). Additionally, cell stage at the time of embryo transfer appears to have been an important variable: 20% of the transfers in which the most advanced embryo had 3 to 6 cells resulted in a term pregnancy, whereas no pregnancy occurred when the transferred embryos had 7 or more cells (Table V). It remains to be determined whether this was a problem related to synchrony or to decreasing viability of embryos with increasing time in culture (see Fig. 6).

6.4. Cryopreservation

Freeze storage of cleavage stage embryos following IVF represents one means of increasing the numbers of synchronized recipients available for embryo transfer. At present, approximately 50 to 70% of frozen stored primate (human, monkey) IVF embryos survive freezing with 50% or more of the initial blastomeres intact (Mohr *et al.*, 1985; Balmaceda *et al.*, 1986). The pregnancy rate of the surviving embryos after transfer is 12 to 16% *per embryo* (Mohr *et al.*, 1985; Balmaceda *et al.*, 1986). Both glycerol and dimethylsulfoxide have been used as cryoprotectants for primate embryos (Trounson and Mohr, 1983; Pope *et al.*, 1984). [The IVF rhesus embryo illustrated in Figure 11 was dehydrated in 6 steps and cryopreserved in 1.5 M glycerol. After thawing, rehydration occurred by dropwise dilution of sucrose (Pope *et al.*, 1984).] Significantly more cryopreservation research using primate embryos must occur before genetic banking for humans or for endangered primates (the "frozen zoo") will result in a reliable and successful outcome.

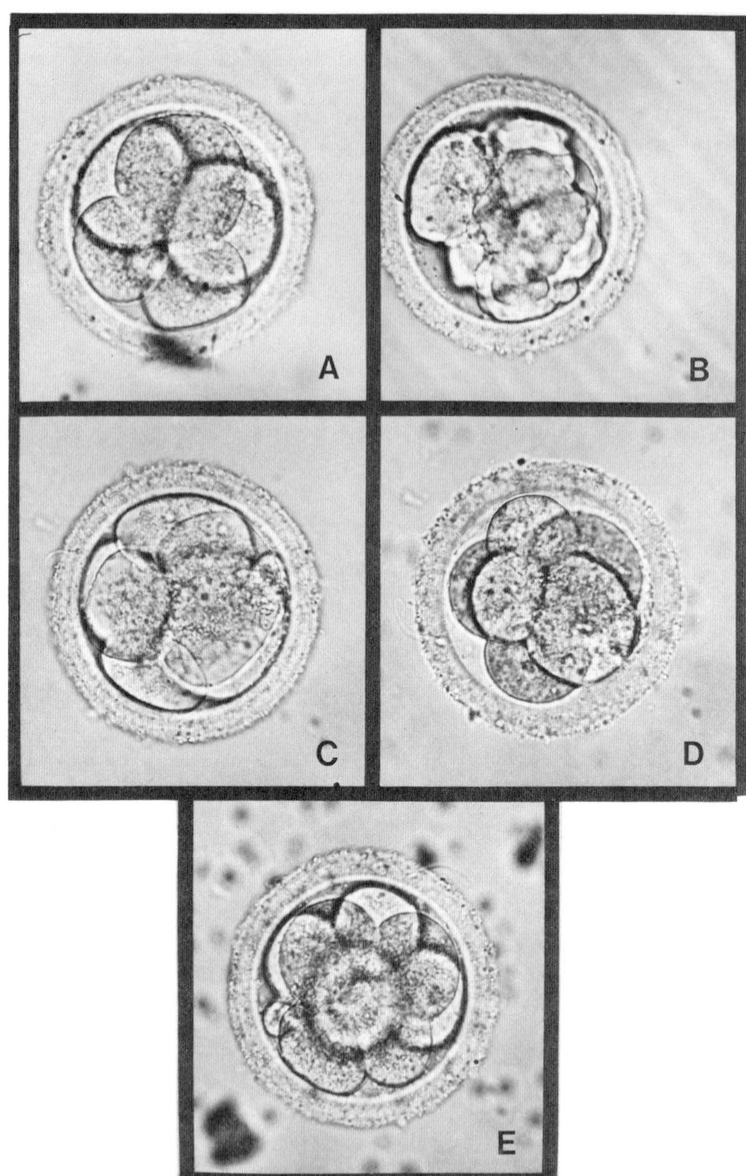

Figure 11. (A) Five-cell embryo (57 hr PI) prior to processing through steps of cryoprotectant. (B) Same embryo immediately after thawing, in 1.5M glycerol + PBS. (C) Same embryo (now 6-cell) 75 min post-thaw following sucrose dilution steps. (D) Same embryo (6-cell) 165 min post-thaw following 1.5 hr in culture (37°C, 5% CO_2 in air). (E) Same embryo (8-cell) 25 hr post-thaw. Optical magnification A to E x 100.

7. CONCLUSIONS

A major conclusion derived from evidence described in this chapter is that embryos from non-human primates can be cultured successfully *in vitro*. However, the components of the culture medium environment that contribute positively or negatively to the accomplishment of primate embryo growth *in vitro* are largely unknown. For example, IVF rhesus embryos can develop normally and with little loss of viability to at least the late 8-cell stage in culture (see section 5.2; Bavister *et al.*, 1983a). After this time, increasing loss of viability occurs. Relatedly, baboon embryos recovered from the uterus after the 10-cell stage can be cultured with a higher rate of success to hatching and peri-implantation stages than IVF embryos, while embryos with fewer cells at the time of embryo recovery are not as successfully cultured (see section 5.3; Pope *et al.*, 1982). From these observations, it would appear that the maternal environment provides component(s) as yet unknown that are essential for an important transition in early primate embryo development and which are either lacking or deficient in the culture medium. This is but one of numerous fundamental questions that remain to be answered which can be addressed by experiments on non-human primate embryos. Other important questions that need to be studied concern timing of the activation of the embryonic genome, the timing of compaction (gap junction formation), the nature of embryo/maternal signaling, and epigenetic regulation of pre-implantation embryo development. Additionally, we are far from possessing sufficient information concerning cellular events in mammalian fertilization, early embryonic development, and maternal *vs.* paternal inheritance to deduce a general mammalian "plan" (if indeed one exists). The ability to sustain normal development of early primate embryos *in vitro* will enable data derived from them, representing the most highly evolved mammals, to contribute to the general conceptual models that will be developed in the near future.

8. APPENDIX: CULTURE MEDIA

8.1. Media for Rhesus Embryo Culture

Two kinds of bicarbonate-CO_2 buffered culture media are used in our laboratory for IVF and culture of rhesus monkey embryos. A simple balanced salt solution (modified Tyrode's solution) is used for IVF of oocytes, and for sperm capacitation (Bavister *et al.*, 1983a; Boatman and Bavister, 1984). For culture of embryos up to the 8-cell stage, this medium is supplemented with serum and amino acids and the pyruvate concentration is increased to 0.5 mM (Bavister *et al.*, 1983a). Alternatively, oocytes are removed from the Tyrode's medium 12 to 18 hr after insemination and cultured in a modified CMRL-1066 medium; use of supplemented, complex medium is mandatory for culture of embryos beyond the 8-cell stage (Table III; Bavister *et al.*, 1983a; Morgan *et al.*, 1984). A low-bicarbonate, HEPES-buffered version of TALP is used for sperm washing, for handling oocytes during collection and for transporting embryos immediately before transfer to recipients.

For completion of oocyte maturation *in vitro* prior to insemination, TALP medium is supplemented with 20% serum. However, for IVF, oocytes are transferred to fresh drops of TALP either with 2% rhesus serum (Bavister *et al.*, 1983a) or (more recently) without serum. Supplementation of culture media with blood serum appears to be beneficial for development of rhesus monkey embryos beyond the 8-cell stage; however, there is no evidence that serum exerts any beneficial effect on earlier stages (Table III; and see Appendix I of this book). Sera used for embryo culture are always heat-inactivated (56 °C for 30 min); we prefer to do this immediately before use, without further passage of the serum through membrane filters. Midcycle rhesus monkey serum, human fetal cord serum and commercial sera have all been used successfully in our laboratory for the culture of rhesus embryos up to the expanded blastocyst stage or beyond. Rhesus serum is derived from blood drawn on the day of oocyte recovery for inclusion in the first change of medium (12 to 18 hr after insemination), or on appropriate days thereafter. Serum is filtered (Millex-GV®, Millipore Corp., cat. no. SLGV025LS) and stored in small aliquots at -20 °C if not used immediately. We have discontinued use of human cord serum because it is very variable in quality and difficult for us to obtain in reasonable quantities. We have found Hyclone® sera to be suitable for embryo culture; horse serum supports development of rhesus embryos to the blastocyst stage, but we now routinely use "characterized calf serum", which supports considerable post-hatching development (section 5.3 and Fig. 10B, C).

Embryos are cultured individually in 30 mm plastic petri dishes (Falcon no. 1008) in 100 μl drops of medium under 4 ml of saline-equilibrated, sterile silicone oil (Aldrich Chemical Co.) in a 5% CO_2 atmosphere. The medium is changed every 24 hr.

8.2. Preparation of Culture Media

Medium TALP. A stock solution, consisting of modified Tyrode's solution containing lactate (= TL), is prepared from pure salts and ultrapure water (Table VI). This solution is sterilized using a Millex-GV® filter and can be kept for 1 week at 4 °C or for 2 to 3 months at -20 °C. Immediately before use, the following ingredients are added: sodium pyruvate (0.1 mM for sperm preincubation, 0.5 mM for IVF and embryo culture); amino acids ([omitted from sperm preincubation medium] see below for preparation of stock solution); gentamycin sulfate (Garamycin®, Schering Corp.), 50 μg/ml; bovine serum albumin (Fraction V, Sigma Chemical Co.), 3 mg/ml (for sperm washing and IVF; also for embryo culture, unless serum is substituted).

Amino acid stock solution (100x) is prepared containing glutamine (100 mM), methionine (5.0 mM), isoleucine (20.0 mM) and phenylalanine (10.0 mM). This is stored at -20° C.

TALP-HEPES. The composition of the basic components of this medium is given in Table VI. Pyruvate (0.1 mM) and BSA (3 mg/ml) are added just before use. Heparin (5 USP units/ml) is added for oocyte recovery. Usually, about 1 liter of this medium is prepared, and sterilized with either Nalgene negative pressure filters (Nalge Co., cat. no. 4500020, 0.2 μm) or Sterivex-GS® positive pressure filters (Millipore Corp., no. SVGSB1010, 0.22 μm) into sterile borosilicate glass culture bottles (Corning no. 99448) siliconized with Prosil

Table VI
Composition and Preparation of Solutions TL and TL-HEPES

Component	Stock solution[a] (mM)	Stock solution[a] (g/100 ml)	Final solution (mM)	mls of stock solution[b] TL[f]	mls of stock solution[b] TL-HEPES[f]
HEPES[c]	-		10.0	-	*240 mg*
NaCl	157.0	0.92	114.0	to 100 ml	to 100 ml
KCl	166.0	1.24	3.16	1.9	1.9
$CaCl_2$	120.0	1.76	2.0	1.7	1.7
$MgCl_2.6H_2O$	120.0	2.44	0.5	0.41	0.41
Na lactate[d]	150.0	-	10.0	6.7	6.7
Water	-		-	-	7.1
$NaH_2PO_4.H_2O$[e]	20.5	-	0.35 }	1.7	1.7
Glucose[e]	295.0	5.31	5.0 }		
$NaHCO_3$	167.0	1.40	25.0 (TL) 2.0 (TL-HEPES)	15.0	1.2

Penicillin G *(10,000 units/100 ml)* and phenol red *(1 mg/100 ml)*

[a]Ultrapure water must be used to prepare stocks and media (see section 2.2). Osmotic pressure of each stock should be 285-290 mOsmols.
[b]Start medium preparation by placing HEPES (for TL-HEPES only, see 'c'), phenol red and penicillin G into a 100 ml volumetric flask, then add 50 ml of NaCl stock solution. Add remaining stock solutions in order listed with constant stirring. The final addition (after adjusting TL-HEPES to pH 7.4 with 1.0 M NaOH [approx. 0.4 ml/100 ml]) is NaCl stock solution to bring to final volume.
[c]N-2-Hydroxyethylpiperazine-N'-2-ethanesulfonic acid, free acid form (Sigma Chemical Co., no. H 3375).
[d]Sodium lactate stock solution is conveniently prepared by diluting 60% syrup (Sigma Chemical Co., no. L-1375) 1:35 vol/vol with water containing 1 mg/ml phenol red indicator. Adjust pH to 7.6 with 1M NaOH, then sterilize by Millipore filtration and store at 4°C for maximum of 1 week.
[e]Sodium phosphate (28 mg) is dissolved in 10 ml glucose stock solution. This phosphate-glucose stock is sterilized by Millipore filtration and stored at 4°C for 1 week maximum. Add phosphate-glucose stock solution dropwise over 1-2 min while stirring mixture continuously.
[f]Both solutions should be 285 to 290 mOsmols; millipore filter and store at 4°C. Sodium pyruvate (0.1 or 0.5 mM) and BSA (3 mg/ml) are added just before use to produce TALP and TALP-HEPES media; for culture of eggs and embryos, amino acids are added from stock solution (see text).

28® (SCM Specialty Chemicals) and capped with Teflon®-lined caps (Corning no. 9826).

Osmotic Pressure Adjustment. To adjust the osmotic pressure (OP) of a medium when it is too low (*e.g.*, 280 mOsmols; values much lower than this indicate a significant error in medium preparation), add solid NaCl as follows:

[157 mM NaCl = 9.176 mg/ml in pure water = 290 mOsmols/L]

Deficit OP (mOsmols/L) =
Desired OP (mOsmols/L) - Measured OP (mOsmols/L)

Deficit OP (mOsmols/L) ÷ 290 mOsmols/L x 9.176 mg/ml x [initial medium volume (ml) - OP sample volume (ml)] = mg NaCl to be added to remaining medium.

To adjust the OP of a hypertonic medium, see under "*Modified CMRL*".

Modified CMRL. Medium CMRL-1066 is purchased as a 10x solution without bicarbonate or glutamine (GIBCO, catalog no. 330-1540). We divide this stock solution into 10 ml aliquots in sterile, siliconized, glass screw-top test-tubes and store it at -20 °C. The working solution (1x) is prepared just before use. Addition of ingredients shown in Table VII raises the osmotic pressure of the medium to 320 or more, so adjustment of osmotic pressure is necessary. To adjust osmotic pressure to 290 ± 5 mOsmols:

a) remove 1 ml of medium and measure osmotic pressure (we use a Wescor vapor pressure osmometer).

b) osmotic pressure will be > 300 mOsmols; adjust by removing a volume of medium from the volumetric flask. This volume (V ml) is:

$$\frac{\text{initial reading (mOsmols/L)} - 290}{\text{initial reading}} \times 100 \text{ ml} = \text{V ml}$$

c) replace V ml of medium with V ml of water.

$$\text{Example: } \frac{320 - 290}{320} \times 100 \text{ ml} = 9.375 \text{ ml} = \text{V ml}$$

but 1 ml was already removed for osmotic pressure measurement, so remove another 8.375 ml of medium and replace with 9.375 ml water. Re-check osmotic pressure, which should now read 290 mOsmols.

The working solution is filter-sterilized using Millex-GV® filters. These filters do not alter the quality of culture media, based on bioassay data from our laboratory. The solution is stored in sterile 100 ml bottles at 4 °C. The final culture medium is prepared just before use by adding glutamine from a sterile 100x stock solution (100 mM, stored at -20° C), sodium pyruvate and serum (20%, see section 8.1). The culture medium is stored at 4 °C for 2 or 3 days and one aliquot is used every 24 hr to prepare fresh drops for embryo culture.

Table VII
Modification of CMRL-1066 for Culture of Rhesus Monkey Embryos[a]

Stock solution	Amount added[b]
Penicillin G	10,000 units
Gentamycin sulfate (10 mg/ml)	0.5 ml
CMRL-1066 (10 x stock)	10.0 ml
$NaHCO_3$	218.0 mg
Na lactate (290 mOsmols stock)	6.7 ml
Water	to 100.0 ml

[a]For culture, add glutamine, pyruvate and serum (see text).
[b]Mix well and adjust to 290 ± 5 mOsmols as explained in text.

ACKNOWLEDGMENTS

I am grateful to Barry Bavister, Patricia Morgan and Emily Kraus for providing unpublished data described in this chapter; to Linda Endlich and Robert Dodsworth for their help in preparing the figures; to the Regional Primate Research Center, University of Wisconsin-Madison, for providing skilled personnel and research support (NIH grant no. RR 00167); and to the NIH for funding (grant no. HD 14765).

9. REFERENCES

Balmaceda, J.P., Pool, T.B., Arana, J.B., Heitman, T.S., and Asch, R.H., 1984, Successful *in vitro* fertilization and embryo transfer in cynomolgus monkeys, *Fertil. Steril.* 42: 791-795.

Balmaceda, J.P., Heitman, T.O., Garcia, M.R., Pauerstein, C.J., and Pool, T.B., 1986, Embryo cryopreservation in cynomolgus monkeys, *Fertil. Steril.* 45: 403-406.

Batta, S.K., Stark, R.A., and Brackett, B.G., 1978, Ovulation induction by gonadotropin and prostaglandin treatments of rhesus monkeys and observations of the ova, *Biol. Reprod.* 18: 264-278.

Bavister, B.D., 1969, Environmental factors important for *in vitro* fertilization in the hamster, *J. Reprod. Fertil.* 18: 544-545.

Bavister, B.D., 1981, Substitution of a synthetic polymer for protein in a mammalian gamete culture system, *J. Exp. Zool.* 217: 45-51.

Bavister, B.D., 1986, Animal *in vitro* fertilization and embryonic development, in: *Manipulation of Mammalian Development, Developmental Biology,* Vol. 4 (R.B.L. Gwatkin, ed.), Plenum Press, New York, pp. 81-148.

Bavister, B.D., and Yanagimachi, R., 1977, The effects of sperm extracts and energy sources on the motility and acrosome reaction of hamster spermatozoa *in vitro*, *Biol. Reprod.* 16: 228-237.

Bavister, B.D., Boatman, D.E., Leibfried, M.L., Loose, M., and Vernon, M.W., 1983a, Fertilization and cleavage of rhesus monkey oocytes *in vitro*, *Biol. Reprod.* 28: 983-999.

Bavister, B.D., Leibfried, M.L., and Lieberman, G., 1983b, Development of preimplantation embryos of the golden hamster in a defined culture medium, *Biol. Reprod.* 28: 235-247.

Bavister, B.D., Boatman, D.E., Collins, K., Dierschke, D.J., and Eisele, S.G., 1984, Birth of rhesus monkey infant following *in vitro* fertilization and non-surgical embryo transfer, *Proc. Natl. Acad. Sci. USA* 81: 2218-2222.

Bavister, B.D., Collins, K., and Eisele, S., 1985, Non-surgical embryo transfer in the Rhesus monkey, *Theriogenology* 23: 177a.

Bavister, B.D., Dees, H.C., and Schultz, R.D., 1986, Refractoriness of Rhesus monkeys to repeated ovarian stimulation by exogenous gonadotropins is caused by formation of non-precipitating antibodies, *Amer. J. Reprod. Immunol. Microbiol.* 11: 11-16.

Boatman, D.E., and Bavister, B.D., 1983, Regulation of hamster sperm capacitation by bicarbonate ion-carbon dioxide, *J. Cell Biol.* 97: 12a.

Boatman, D.E., and Bavister, B.D., 1984, Stimulation of rhesus monkey sperm capacitation by cyclic nucleotide mediators, *J. Reprod. Fertil.* 71: 357-366.

Boatman, D.E., Morgan, P.M., and Bavister, B.D., 1986, Variables affecting the yield and developmental potential of embryos following superstimulation and *in vitro* fertilization in rhesus monkeys, *Gamete Res.* 13: 327-338.

Brackett, B.G., 1978, Experimentation involving primate embryos, in: *Methods in Mammalian Reproduction* (J.C. Daniel, Jr., ed.), Academic Press, New York, pp. 333-357.

Brinster, R.L., 1971, *In vitro* culture of the embryo, in: *Pathways to Conception* (A. Sherman, ed.), Charles C. Thomas, Springfield, IL, pp. 245-277.

Bustillo, M., Buster, J.E., Cohen, S.W., Thorneycroft, I.H., Simon, J.A., Boyers, S.P., Marshall, J.R., Seed, R.W., Louw, J.A., and Seed, R.G., 1984, Nonsurgical ovum transfer as a treatment in infertile women, *J. Amer. Med. Assoc.* 251: 1171-1173.

Carney, E.W., and Bavister, B.D., 1986, Increased atmospheric carbon dioxide stimulates hamster embryo development *in vitro*, *Biol. Reprod.* 34 (Suppl. 1): 199 (abs. no. 300).

Carney, E.W., and Bavister, B.D., 1987, Regulation of hamster embryo development *in vitro* by carbon dioxide, *Biol. Reprod.* (in press).

Catchpole, H.R., and van Wagenen, G., 1975, Reproduction in the rhesus monkey, *Macaca mulatta*, in: *The Rhesus Monkey*, Vol. II, *Management, Reproduction, and Pathology* (G.H. Bourne, ed.), Academic Press, New York, pp. 117-140.

Chen, C., Jones, W.R., Bastin, F., and Forde, C., 1985, Early pregnancy factor, in: *In Vitro Fertilization and Embryo Transfer, Ann. N. Y. Acad. Sci.*, Vol. 442 (M. Seppälä, and R.G. Edwards, eds.), New York Academy of Sciences, New York, pp. 420-428.

Czaja, J.A., Robinson, J.A., Eisele, S.G., Scheffler, G., and Goy, R.W., 1977, Relationship between sexual skin colour of female rhesus monkeys and midcycle plasma levels of oestradiol and progesterone, *J. Reprod. Fertil.* 49: 147-150.

Dandekar, P.V., and Quigley, M.M., 1984, Laboratory setup for human *in vitro* fertilization, *Fertil. Steril.* 42: 1-11.

Edwards, R.G., Bavister, B.D., and Steptoe, P.C., 1969, Early stages of fertilization *in vitro* of human oocytes matured *in vitro*, *Nature (London)* 221: 632-635.

Edwards, R.G., Steptoe, P.C., and Purdy, J.M., 1980, Establishing full-term human pregnancies using cleaving embryos grown *in vitro*, *Brit. J. Obstet. Gynaecol.* 87: 737-756.

Enders, A.C., and Hendrickx, A.G., 1980, Implantation in nonhuman primates: I. A comparison of morphological events, in: *Non-Human Primate Models for Study of Human Reproduction* (T.C.A. Kumar, ed.), S. Karger, Basel, pp. 99-108.

Enders, A.C., Hendrickx, A.G., and Binkerd, P.A., 1982, Abnormal development of blastocysts and blastomeres in the rhesus monkey, *Biol. Reprod.* 26: 353-366.

Enders, A.C., Boatman, D.E., Morgan, P.M., Schlafke, S.G., and Bavister, B.D., 1987, Differentiation of blastocysts derived from *in vitro* fertilized rhesus monkey ova, (in preparation).

Geuskens, M., and Alexandre, H., 1984, Ultrastructural and autoradiographic studies of nucleolar development and rDNA transcription in preimplantation mouse embryos, *Cell Different.* 14: 125-134.

Gwatkin, R.B.L., and Haidri, A.A., 1973, Requirements for the maturation of hamster oocytes *in vitro*, *Exp. Cell Res.* 76: 1-7.

Hartman, C.G., and Corner, G.W., 1941, The first maturation division of the macaque ovum, *Contrib. Embryol.* 29: 1-6.

Hendrickx, A.G., and Binkerd, P.E., 1980, Fetal deaths in nonhuman primates, in: *Human Embryonic and Fetal Death* (I.H. Porter, and E.B. Hook, eds.), Academic Press, New York, pp. 45-69.

Hendrickx, A.G., and Kraemer, D.C., 1968, Preimplantation stages of baboon embryos (*Papio* sp.), *Anat. Rec.* 162: 111-120.

Hendrickx, A.G., and Kraemer, D.C., 1971, Methods, in: *Embryology of the Baboon* (A.G. Hendrickx, ed.), University of Chicago Press, Chicago, IL, pp. 31-44.

Hertig, A.L., Rock, J., and Adams, E.C., 1956, A description of 34 human ova within the first 17 days of development, *Amer. J. Anat.* 98: 435-493.

Heuser, C.H., and Streeter, G.L., 1941, Development of the macaque embryo, *(Contrib. Embryol. 181)*, in: *Embryology of the Rhesus Monkey (Macaca mulatta)*, Carnegie Institution, Washington, D.C., Pub. no. 538, pp. 17-65.

Hurst, P.R., Jeffries, K., Eckstein, P., and Wheeler, A.G., 1976, Recovery of uterine embryos in rhesus monkeys, *Biol. Reprod.* 15: 429-434.

Hurst, P.R., Wheeler, A.G., and Eckstein, P., 1980, A study of uterine embryos recovered from rhesus monkeys fitted with intrauterine devices, *Fertil. Steril.* 33: 69-76.

Jaszczak, S., and Hafez, E.S.E., 1972, The cervix uteri and sperm transport in female macaques, in: *Medical Primatology (Proc. 3rd Conf. Exp. Med. Surg. Primates)*, Part I (E.I. Goldsmith, and J. Moor-Jankowski, eds.), S. Karger, Basel, pp. 263-270.

Juetten, J., and Bavister, B.D., 1983, The effects of amino acids, cumulus cells, and bovine serum albumin on *in vitro* fertilization and first cleavage of hamster eggs, *J. Exp. Zool*, 227: 487-490.

Kaye, P.L., Schultz, G.A., Johnson, M.H., Pratt, H.P.M., and Church, R.B., 1982, Amino acid transport and exchange in preimplantation mouse embryos, *J. Reprod. Fertil.* 65: 367-380.

Koenig, J.L.F., Zimmerman, D.R., Eldridge, F.E., and Kopf, J.D., 1986, Cytogenetic analysis of swine ova: the effect of superovulation and selection for high ovulation rate, *Biol. Reprod.* 34 (Suppl. 1): 59 (abs. no. 20).

Kraemer, D.C., and Hendrickx, A.G., 1971, Descriptions of stages I, II, and III, in: *Embryology of the Baboon* (A.G. Hendrickx, ed.), University of Chicago Press, Chicago, IL, pp. 45-52.

Kreitmann, O., Lynch, A., Nixon, W.E., and Hodgen, G.D., 1982, Ovum collection, induced luteal function, *in vitro* fertilization, embryo transfer and low tubal ovum transfer in primates, in: *In Vitro Fertilization and Embryo Transfer* (E.S.E. Hafez, and K. Semm, eds.), M.T.P. Press Ltd., Lancaster, U.K., pp. 303-324.

Leibfried, M.L., and Bavister, B.D., 1983, Fertilizability of *in vitro* matured oocytes from golden hamsters, *J. Exp. Zool.* 226: 481-485.

Lewis, W.H., and Hartman, C.G., 1941, Tubal ova of the rhesus monkey, *(Contrib. Embryol. 180)*, in: *Embryology of the Rhesus Monkey (Macaca mulatta)*, Carnegie Institution, Washington, D.C., Pub. no. 538, pp. 9-15.

MacMichael, G., 1986, The adverse effects of UV and short-wavelength visible radiation on tissue culture, *Amer. Biotechnol. Lab.* 4 (no. 4): 30-31.

Mintz, B., 1964, Gene expression in the morula stage of mouse embryos, as observed during development of t^{12}/t^{12} lethal mutants *in vitro*, *J. Exp. Zool.* 157: 267-272.

Mohr, L.R., Trounson, A., and Freeman, L., 1985, Deep-freezing and transfer of human embryos, *J. In Vitro Fertil. and Embryo Transfer* 2: 1-10.

Moor, R.M., Osborn, J.C., and Crosby, I.M., 1982, Cell interactions and oocyte regulation in mammals, in: *Follicular Maturation and Ovulation, Proc. IVth Reiner de Graaf Symp., Nijmegen, 1981* (R. Rolland, E.V. van Hall, S.G. Hillier, K.P. McNatty, and J. Schoemaker, eds.), Excerpta Medica, Amsterdam, pp. 249-264.

Morgan, P.M., Boatman, D.E., Collins, K., and Bavister, B.D., 1984, Complete preimplantation development in culture of *in vitro* fertilized rhesus monkey oocytes, *Biol. Reprod.* 30 (Suppl. 1): 96 (abs. no. 131).

Morgan, P.M., Boatman, D.E., and Kraus, E.M., 1986, Relationship between follicular fluid steroid hormone concentrations and *in vitro* development of rhesus monkey embryos, *Biol. Reprod.* 34 (Suppl. 1): 96 (abs. no. 94).

Murphy, B.D., Mapletoft, R.J., Manns, J., and Humphrey, W.D., 1984, Variability in gonadotrophin preparations as a factor in the superovulatory response, *Theriogenology* 21: 117-125.

O'Neill, C., Pike, I.L., Porter, R.N., Gidley-Baird, A.A., Sinosich, M.J., and Saunders, D.M., 1985, Maternal recognition of pregnancy prior to implantation: methods for monitoring embryonic viability *in vitro* and *in vivo*, in: *In Vitro Fertilization and Embryo Transfer, Ann. N. Y. Acad. Sci.*, Vol. 442 (M. Seppälä, and R.G. Edwards, eds.), New York Academy of Sciences, New York, pp. 429-439.

Ottobre, J.S., and Stouffer, R.L., 1985, Antibody production in rhesus monkeys following prolonged administration of human chorionic gonadotropin, *Fertil. Steril.* 43: 122-128.

Paul, J., 1975, Design and equipment of a tissue culture laboratory, in: *Cell and Tissue Culture* (J. Paul), 5th ed., Churchill Livingstone, Edinburgh, U.K., pp. 162-174.

Pope, C.E., Pope, V.Z., and Beck, L.R., 1980, Nonsurgical recovery of uterine embryos in the baboon, *Biol. Reprod.* 23: 657-662.

Pope, C.E., Pope, V.Z., and Beck, L.R., 1982, Development of baboon preimplantation embryos to post-implantation stages *in vitro*, *Biol. Reprod.* 27: 915-923.

Pope, C.E., Pope, V.Z., and Beck, L.R., 1984, Live birth following cryopreservation and transfer of a baboon embryo, *Fertil. Steril.* 42: 143-145.

Pope, V.Z., Pope, C.E., and Beck, L.R., 1983, A 4-year summary of the nonsurgical recovery of baboon embryos: a report on 498 eggs, *Amer. J. Primatol.* 5: 357-364.

Quinn, P., Warnes, G.M., Kerin, J.F., and Kirby, C., 1985, Culture factors affecting the success rate of *in vitro* fertilization and embryo transfer, in: *In Vitro Fertilization and Embryo Transfer, Ann N. Y. Acad. Sci.*, Vol. 442 (M. Seppälä, and R.G. Edwards, eds.), New York Academy of Sciences, New York, pp. 195-204.

Rall, W.F., and Fahy, G.M., 1985, Ice-free cryopreservation of mouse embryos at -196°C by vitrification, *Nature (London)* 313: 573-575.

Rosenblum, I.Y., Mattson, B.A., and Heyner, S., 1986, Stage-specific insulin binding in mouse preimplantation embryos, *Dev. Biol.* 116: 261-263.

Schenken, R.S., Williams, R.F., and Hodgen, G.D., 1984, Ovulation induction using "pure" follicle-stimulating hormone in monkeys, *Fertil. Steril.* 41: 629-634.

Simpson, M.E., and van Wagenen, G., 1962, Induction of ovulation with human urinary gonadotrophins in the monkey, *Fertil. Steril.* 13: 140-152.

Soules, M.R., 1985, The *in vitro* fertilization pregnancy rate: let's be honest with one another, *Fertil. Steril.* 43: 511-513.

Summers, P.M., Wennink, C.J., and Hodges, J.K., 1985, Cloprostenol-induced luteolysis in the marmoset monkey *(Callithrix jacchus)*, *J. Reprod. Fertil.* 73: 133-138.

Szöllösi, D., and Gerard, M., 1983, Cytoplasmic changes in the mammalian oocytes during the preovulatory period, in: *Fertilization of the Human Egg In Vitro* (H.M. Beier, and H.R. Lindner, eds.), Springer-Verlag, Berlin, pp. 35-55.

Tesarik, J., Kopecny, V., Plachot, M., Mandelbaum, J., Da Lage, C., and Fléchon, J.E., 1986, Nucleologenesis in the human embryo developing *in vitro*: ultrastructural and autoradiographic analysis, *Dev. Biol.* 115: 193-203.

Trounson, A., and Mohr, L., 1983, Human pregnancy following cryopreservation, thawing and transfer of an eight-cell embryo, *Nature (London)* 305: 707-709.

Van Blerkom, J., Henry, G., Porreco, R., 1984, Preimplantation human embryonic development from polypronuclear eggs after *in vitro* fertilization, *Fertil. Steril.* 41: 686-696.

Vanderhyden, B.C., Rouleau, A., Walton, E.A., and Armstrong, D.T., 1986, Increased mortality during early embryonic development after *in vitro* fertilization of rat oocytes, *J. Reprod. Fertil.* 77: 401-409.

Weick, R.F., Dierschke, D.J., Karsch, F.J., Butler, W.R., Hotchkiss, J., and Knobil, E., 1973, Periovulatory time courses of circulating gonadotropic and ovarian hormones in the rhesus monkey, *Endocrinol.* 93: 1140-1147.

Wiley, L.M., Yamami, S., and Van Muyden, D., 1986, Effect of potassium concentration, type of protein supplement, and embryo density on mouse preimplantation development *in vitro*, *Fertil. Steril.* 45: 111-119.

CHAPTER 14

ANALYSIS OF EMBRYOTOXIC EFFECTS IN PREIMPLANTATION EMBRYOS

HORST SPIELMANN

1. INTRODUCTION

In human pregnancy, it has been particularly difficult to obtain data on the action of drugs and environmental chemicals during the period before implantation of the embryo in the uterus. Scientific interest in the action of xenobiotics in early pregnancy is focusing on the earliest stage of gestation since recent epidemiological evaluations are not only indicating that 60% to 75% of all fertilized oocytes are dying during normal pregnancy but moreover that 50% of them are already dying before implantation (Biggers, 1981; Kline and Stein, 1985). This view is supported by the experience with human *in vitro* fertilization, which even in the most proficient hands has a low survival rate of embryos that are transferred back to the mother (Edwards and Steptoe, 1983).

One may speculate that the high level of fetal wastage in human pregnancy is normal and that it may be related to chromosomal aberrations, which were found in 40% to 65% of all first trimester spontaneous abortions (Boué *et al.*, 1975; Kline and Stein, 1985) and also in cleavage stage human embryos after *in vitro* fertilization (Angell *et al.*, 1983; van Blerkom *et al.*, 1984). However, it is so far unclear to what extent toxic chemicals, to which the mother is exposed when she is generally unaware of being pregnant, are additionally contributing to the early embryo mortality.

Although methods for culturing and transplanting mammalian embryos during the preimplantation period were developed almost 30 years ago (McLaren and Biggers, 1958), teratologists have hardly used them to analyse the mechanism of action of drugs on early embryos (Spielmann, 1976; Spielmann and Eibs, 1978). In teratology, the preimplantation period is of little

Horst Spielmann Max v. Pettenkofer-Institut, Bundesgesundheitsamt (BGA), P.O. Box 33 00 13, 1 Berlin 33, West Germany.

relevance since most teratologists hold that cleavage stage embryos have not undergone true differentiation (Wilson, 1977) and that the final outcome of embryotoxic agents on the preimplantation embryo depends on the number of cells killed. Above a certain proportion, the embryo dies, below that number, the remaining cells multiply to replace those lost and subsequent development is essentially normal (Austin, 1973); this is often referred to as the "all-or-nothing" law. Developmental biologists, on the other hand, have proven that during mouse embryogenesis cell commitment into two groups of cells, inner cell mass (ICM) and trophoblast cells, can already be demonstrated before the blastocyst stage (Johnson, 1977).

Using cyclophosphamide as the embryotoxic agent and applying methods recently developed for the study of early mammalian embryology *in vitro*, we have been able to demonstrate that the cells of the mouse blastocyst exhibit a differential sensitivity towards maternal drug treatment in a dose-related manner and not according to an "all-or-nothing" law. We have also found that the transfer of an embryotoxic drug into the embryo before implantation can already be detected shortly after treatment (Eibs and Spielmann, 1977; Spielmann *et al.*, 1977, 1981a,b; Spielmann and Eibs, 1978; Spielmann and Jacob-Müller, 1981). Our results have been confirmed by other investigators using treatment of preimplantation mouse embryos with embryotoxic agents *in vivo* (Fabro *et al.*, 1984; Giavani *et al.*, 1984; Kola and Folb, 1986a,b) or *in vitro* (Iannaccone, 1984; Iannaccone *et al.*, 1984; Katayama and Matsumoto, 1985). In addition, several investigators have provided evidence for the ability of the preimplantation embryo to metabolize xenobiotics, both in the mouse (Galloway *et al.*, 1980; Filler and Lew, 1981; Pedersen *et al.*, 1985) and in the rabbit (Balling *et al.*, 1985).

In the present report, the methods that are currently available for studying the action of toxic drugs during the preimplantation period in the mouse are described as they are carried out in our laboratory. Particular reference is made to more recently established cytogenetic methods (Krüger *et al.*, 1985; Vogel *et al.*, 1985). We mainly refer to the mouse since preimplantation development in this species seems to be more comparable to the human situation than (*e.g.*) the rabbit with its much larger blastocysts and since most toxicological studies on early embryos have been performed in the mouse. The results of toxicological studies on preimplantation embryos both *in vivo* and *in vitro* are discussed with respect to the methodological approach and an evaluation of the methods is attempted.

2. SENSITIVE TOXICOLOGICAL ENDPOINTS

2.1. *In Vitro* Culture of Preimplantation Embryos Before and During Implantation

In Vitro Culture During Preimplantation Development. Retrieval of cleavage stage embryos from the oviduct or uterus as well as handling during toxicological procedures is performed in simple phosphate-buffered media, such as PB-1 (Spielmann and Eibs, 1978). Culture conditions for supporting cleavage of preimplantation embryos are well established. In inbred strains of some species, *e.g.*, mouse and rabbit, a high percentage of embryos placed in culture at the 1-cell, 2-cell or 4-cell stages will reproducibly develop into blastocysts. The *in vitro* culture of preimplantation embryos is not only an established method in human and veterinary reproductive clinics but has also helped to elucidate biochemical and genetic characteristics of early mammalian development.

In toxicological studies on preimplantation embryos, a dose-related inhibition of development is very easily obtained after treatment with physical or chemical agents. Cleavage and development to the blastocyst stage are the critical endpoints in such investigations, but they are not as sensitive as subsequent development after transplantation *in vivo* or during implantation *in vitro* (Spielmann and Eibs, 1978) and also not as sensitive as cytogenetic or biochemical parameters. *In vitro* culture during the preimplantation period has not only been used to detect stage specific differences in biochemical and genetic maturation of preimplantation embryos (Epstein, 1975) but also to study stage specific sensitivities against toxic agents in cleavage stage embryos (Spielmann and Eibs, 1978). However, the observation of a dose-related inhibition of development in culture is not an adequate result in toxicological studies. Nevertheless, the *in vitro* culture of preimplantation embryos is a prerequisite for most of the more sophisticated toxicological methods, since determination of the highest non-toxic concentration of a certain treatment, often incorrectly referred to as "no-effect-level" (NOEL), is essential in every study on cultured embryos.

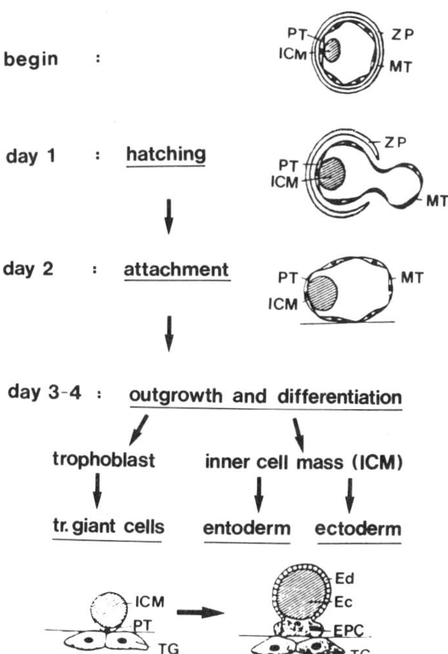

Figure 1. Diagram describing differentation of mouse blastocysts during implantation *in vitro* in medium NCTC-109 supplemented with 10% fetal calf serum. Culture period - 96 hr; Ec - ectoderm, Ed - entoderm (endoderm), EPC - ectoplacental cone, ICM - inner cell mass, MT - mural trophoblast, PT - parietal trophoblast, TG-trophoblast giant cells, ZP - zona pellucida.

In Vitro Culture During Implantation. In vitro culture of mouse embryos during the time of implantation has successfully been applied by several groups to study genetic, biochemical and toxicological problems (Spielmann and Eibs, 1978). In our laboratory, the following conditions are routinely used with mouse blastocysts of the strain NMRI: temperature - 37°C; culture medium - NCTC-109 (M.A. Bioproducts) supplemented with 10% fetal calf serum (FCS); gas phase - 5% CO_2 in a humidified air atmosphere; culture period - 96 hr. In NCTC-109 and similar media, development and differentiation of mouse blastocysts proceed through several characteristic steps (Fig. 1): hatching from the zona pellucida, attachment to the surface of the culture dish and outgrowth of 3 characteristic cell types: a trophoblast layer consisting of trophoblast giant cells and an inner cell mass (ICM) containing 2 germ layers (ectoderm and endoderm). In this culture system, the following endpoints are recorded: hatching, attachment and outgrowth of the trophoblast and development of the two layers of the ICM with two types of cells. Early blastocysts and morulae should initially be cultured for 24 hr in Whitten's medium (Whitten, 1971) supplemented with 10% FCS to reach the expanded blastocyst stage since optimal results can only be reached in NCTC-109 with late blastocysts. A lower success rate was obtained with rat blastocysts in NCTC (Spielmann *et al.*, 1980).

The culture of mouse blastocysts during implantation can be used to demonstrate a differential sensitivity of the two groups of cells in the blastocyst (trophoblast and ICM) towards previous treatment of the embryo either *in vivo* or *in vitro*. Figure 2 demonstrates on the one hand that after treatment with cyclophosphamide (CPA) *in vivo* at the 4-cell to 8-cell stage

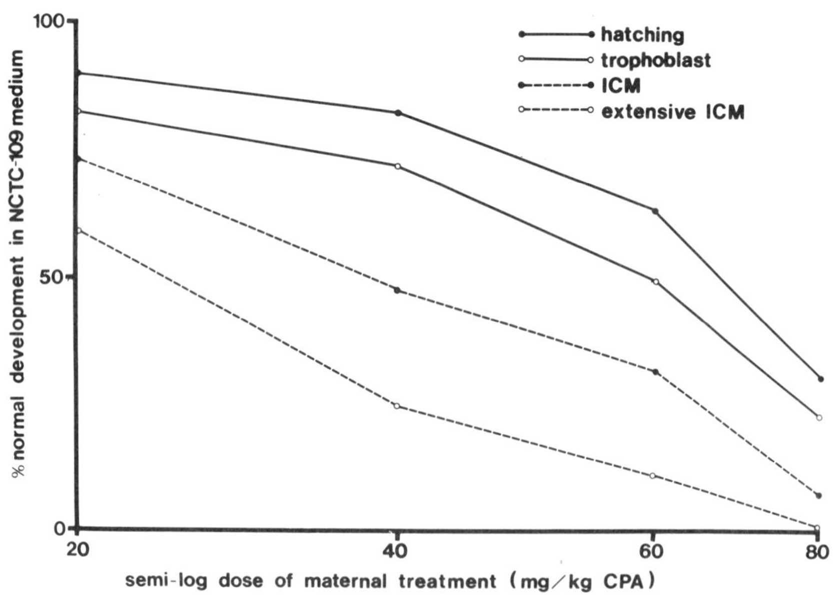

Figure 2. Dose response plot (semilogarithmic) for cyclophosphamide (CPA) treatment of mice on day 2 of pregnancy and differentiation during implantation in culture of blastocysts obtained from the uterus 24 hr after treatment of the mother animal (culture period 96 hr).

and transfer of blastocysts to NCTC, development of the ICM is the most sensitive parameter for detecting embryotoxic effects (Eibs and Spielmann, 1977). Figure 2 additionally proves that in a semilogarithmic dose response plot, differentiation of CPA-treated mouse embryos was inhibited in a dose-dependent manner for all parameters studied. This result illustrates that xenobiotics are acting on preimplantation embryos according to basic principles in toxicology and not according to an unclear "all-or-nothing" law.

For the evaluation of toxic effects on preimplantation embryos, the *in vitro* culture of treated embryos beyond implantation has several advantages in comparison to the embryo transfer technique. The culture system requires fewer embryos, and it is faster and more precise since maternal factors (litter size, *etc.*) are not involved. It finally allows calculation of clear dose-response relations and particularly toxicologically effective dose levels (ED_{50}) or concentrations (EC_{50}), which are difficult to obtain after transfer of treated embryos to foster-mothers (Spielmann and Eibs, 1978; Fabro *et al.*, 1984; Iannaccone, 1984; Katayama and Matsumoto, 1985).

2.2. Cytological and Cytogenetic Tests

Determination of the Cell Number in Morulae and Blastocysts. In studies on the action of drugs before implantation *in vivo*, the number of embryos in each uterine horn is routinely determined and also their developmental stage, which can easily be identified up to the 8-cell stage. Subsequently, however, at the morula and blastocyst stages morphology can be misleading, since the number of cells (blastomeres) per embryo can vary considerably. After CPA-treatment of pregnant rats and mice, we found in morphologically normal blastocysts a dose-related reduction of the cell number (Fig. 3; Spielmann and Eibs, 1978). In preimplantation embryos, the cell number is determined according to Tarkowski's (1966) method for chromosome preparation in early embryos. After incubation in a hypotonic medium, all embryonic cells are spread on a cover slip. The nuclei and also the number of cells in mitosis can be identified after staining with 5% Giemsa. A dose-related reduction in the cell number of blastocysts that appear normal under the light microscope has also been observed in rats treated with chlorambucil (Giavini *et al.*, 1984) and in mice treated with chlorpromazine (Kola and Folb, 1986a). These toxicological data confirm that blastulation, the first step of morphological differentiation, is not dependent on the presence of a particular number of cells and that blastomeres, having lost the capacity to divide at a normal rate, can still form the blastocyst cavity (Eibs and Spielmann, 1977). Unfortunately, in several investigations on the action of drugs in early pregnancy, the cell number of exposed blastocysts was not determined.

Determination of the Cell Number of the ICM. Figure 3 additionally shows a dose-related decrease in the cell number of the ICM in blastocysts of CPA-treated mice. The decrease was significantly greater than the reduction of the total cell number of the blastocysts (Spielmann *et al.*, 1981b). This result proves that *in vivo* the two groups of cells in blastocysts can, in a dose-related manner, exhibit a differential sensitivity against toxic agents. Immunosurgery, an elegant and simple technique (Solter and Knowles, 1975), was used to perform this complex toxicological experiment on blastocysts (Spielmann *et al.*, 1981b).

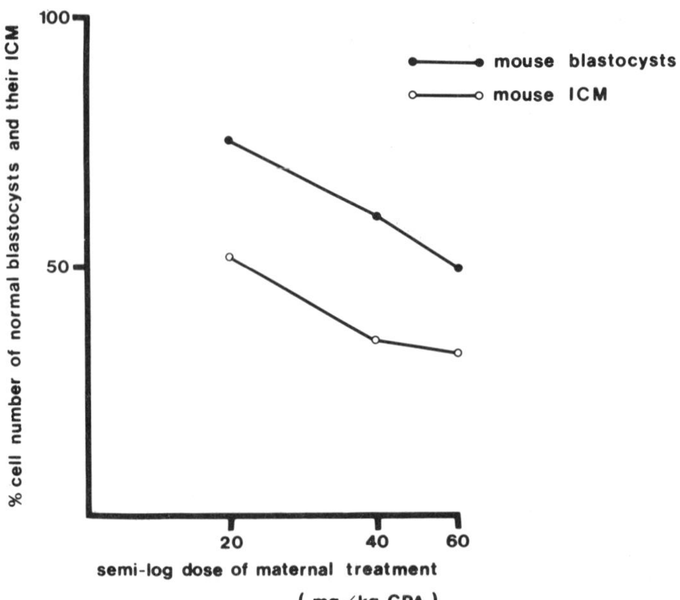

Figure 3. Dose response plot (semilogarithmic) for cyclophosphamide (CPA) treatment of mice on day 2 of pregnancy and decrease in cell numbers of blastocysts and of their inner cell mass (ICM) 42 hr after treatment on day 5 (isolation of ICM by immunosurgery).

Cytogenetic Evaluation. In our investigations on long term effects of CPA-treatment during the preimplantation period, we found a retarded embryolethal effect during organogenesis that resulted in a dose-related increase in the resorption rate at term (Spielmann *et al.*, 1977). In more recent studies on chlorambucil (Giavini *et al.*, 1984), chlorpromazine (Kola and Folb, 1986a) and methylnitrosourea (MNU) (Iannaccone, 1984; Takeuchi, 1984) it was confirmed that exposure of preimplantation embryos to xenobiotics can result in toxic effects long after the exposure, indeed after birth in the case of MNU. Since these particular drugs are also mutagenic, the long-term effects may result from chromosomal damage that was insufficiently repaired by preimplantation embryos, rather than from an unusually long halflife of the drug which then acts on the embryo after implantation (*e.g.*, Eibs *et al.*, 1982). Therefore, several groups have started to conduct cytogenetic studies on preimplantation embryos.

Structural Chromosomal Aberrations and Micronuclei. Structural and numerical chromosome aberrations can be analysed in carefully prepared chromosomal preparations according to Tarkowski (1966) (Fig. 4a). The same preparations can be used to determine micronuclei (Fig. 4b); these are DNA-positive particles within the cytoplasm of interphase cells and originate from chromatin which has been lagging in the anaphase (Schmid, 1975).

Sister Chromatid Exchange (SCE). Analysis of the rate of sister chromatid exchanges (SCE), which is a sensitive indirect indicator of persistent DNA-

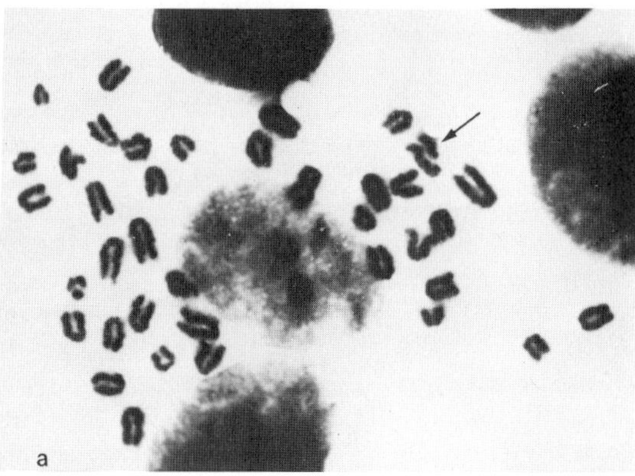

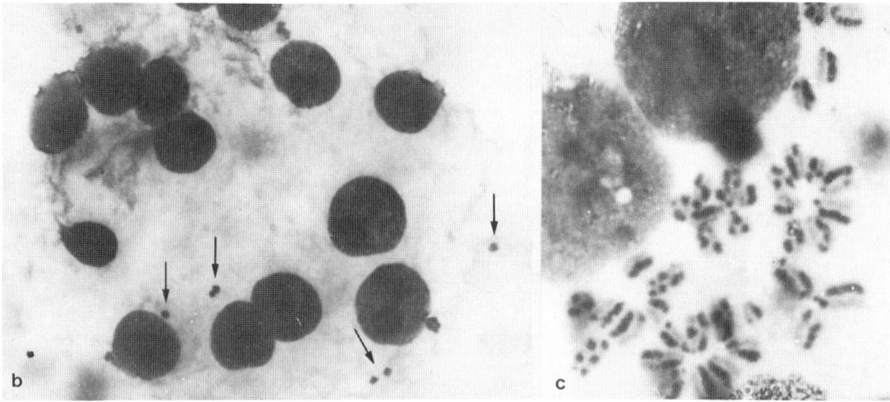

Figure 4. Chromosomal aberrations in mouse blastocysts. (a) Structural aberration - metaphase showing a chromatid break (arrow) after mitomycin C treatment *in vivo* (x 1500); (b) Micronuclei (arrows) in a mouse blastocyst (day 3) consisting of 15 cells 24 hr after cyclophosphamide (CPA) treatment *in vivo* (80 mg/kg on day 2) (x 500); (c) Sister chromatid exchange (SCE) in mouse blastocyst (x 1500) was induced *in vitro* by culture in the presence of sera containing active metabolites of CPA from rats which had been treated with CPA (see Table I; according to Vogel *et al.*, 1985). Magnifications given are original optical magnifications.

lesions, has been very difficult in preimplantation embryos. The technique is based on a differential staining of sister chromatids using incorporation of the thymidine analogue 5-bromo-2-desoxyuridine (BrdU) into DNA. For this purpose, labelling of the chromatids with BrdU during two cell cycles and subsequent staining with 33258 Hoechst plus Giemsa has been established as a standard technique for many cell types (Perry and Wolff, 1974). The SCE assay is particularly promising for early mammalian embryos, which contain limited numbers of cells. However, it is well known that in contrast to other differentiating cells, even at very low concentrations of BrdU, cleavage stage mouse embryos taking up the nucleoside are arrested in development (Garner, 1974; Golbus and Epstein, 1974). It has so far been impossible to determine the SCE-rate in preimplantation mouse embryos *in vivo*; it was finally R. Pedersen's group that established the determination of the SCE-rate in preimplantation mouse embryos *in vitro* in their studies on cytochrome P-450 activity (Galloway *et al.*, 1980; Bennett and Pedersen, 1984; Pedersen *et al.*, 1985). More recently, Katayama and Matsumoto (1985) used SCE as a parameter to detect embryotoxic effects in mouse blastocysts after exposure *in vitro* and we have measured SCE-rates in preimplantation mouse embryos exposed either *in vitro* (Fig. 4c; Krüger *et al.*, 1985; Vogel *et al.*, 1985) or after CPA-treatment *in vivo* (Spielmann *et al.*, 1985).

The following conditions have been used successfully in our laboratory to determine the frequency of SCE in preimplantation mouse embryos of the strain NMRI. To analyse SCE in 4-cell and 8-cell embryos and in morulae and blastocysts, the embryos have to be cultured for a total of 48 hr, approximately 2 cell cycles, in the dark. During the first cell cycle, the medium is supplemented with BrdU and during the second 24 hr with thymidine. BrdU concentrations are critical and may have to be tested for each cleavage stage, since the sensitivity of cleavage stage mouse embryos to BrdU changes during preimplantation development (Garner, 1974; Golbus and Epstein, 1974). Four-cell and 8-cell embryos are cultured in the dark for 24 hr in Whitten's medium (Whitten, 1971) supplemented with 0.3% BSA (= W-BSA) and with $1x10^{-6}$ M BrdU (Sigma Chemical Co.), then for another 24 hr in W-BSA supplemented with $1x10^{-6}$ M thymidine (Sigma). Morulae and blastocysts are cultured in the dark for 24 hr in W-BSA supplemented with $2x10^{-6}$ M BrdU, then for an additional 24 hr in NCTC-109 supplemented with 10% FCS and with $1x10^{-6}$ M thymidine. Three hr before termination of the culture, colcemid (Serva, Heidelberg) is added to give a final concentration of 0.1 μg/ml. Chromosome preparations are carried out according to Tarkowski (1966) with hypotonic treatment in 0.5% sodium citrate and fixation in ethanol/glacial acetic acid (3:1). For SCE-staining, the fluorescence-plus-Giemsa technique according to Perry and Wolff (1974) is performed in the following manner: Chromosome preparations are stained with fluorochrome H 33258 (Hoechst) ($5x10^{-2}$ μg/ml) for 30 min at room temperature. They are then exposed to UV-light (8 W, 254 nm) for 60 min and incubated at 60°C for an additional 60 min. The final staining with 5% Giemsa is carried out for 10 min. The preparation can also be used to determine micronuclei and chromosomal aberrations in preimplantation mouse embryos (Fig. 4c).

This procedure can be used to determine the influence of exposing preimplantation embryos to xenobiotics either *in vitro* or *in vivo* when the embryos are removed from the mother at appropriate time intervals (Fig. 5; Spielmann *et al.*, 1985). Since there are only a few publications on SCE-frequencies in preimplantation embryos, the method still requires further evaluation. Bennett and Pedersen (1984) found a considerable strain variation in SCE-frequency in zygotes of different strains of mice, ranging from one per cell for C57BL/6J to 12 per cell for BALB/c. The same group reported SCE-frequencies at the blastocyst stage between 4.5 per cell for AKR/J and 16 per cell for Dub/ICR (Pedersen *et al.*, 1985). In 4-cell and 8-cell embryos, as well as in morulae and blastocysts of the NMRI strain, we found an SCE-frequency of 25 per cell (Krüger *et al.*, 1985; Vogel *et al.*, 1985). Strain-specific differences in SCE-frequency of preimplantation mouse embryos exposed to toxic agents may indicate strain-specific differences in repair of DNA lesions by the early embryos (Bennett and Pedersen, 1984). Determination of SCE-frequencies in embryos exposed before implantation may, therefore, help to better understand both short term and long term effects of treatment with xenobiotics.

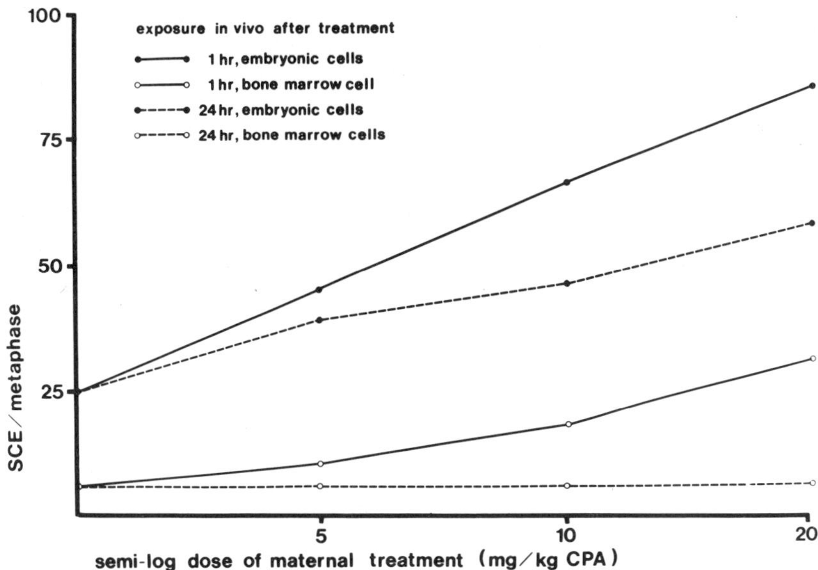

Figure 5. Dose response plot (semilogarithmic) for cyclophosphamide (CPA) treatment of mice on day 2 of pregnancy and sister chromatid exchange frequency (SCE) in blastocysts and in bone marrow of the mother animal 1 hr and 24 hr after treatment.

Cytological Methods to Evaluate Embryonic Viability. In addition to cytogenetic analysis, staining with fluorescein diacetate (FDA) has recently been used to evaluate the viability of mouse blastocysts after exposure *in vitro* to MNU (Iannaccone *et al.*, 1982; Iannaccone, 1984) and *in vivo* to chlorpromazine (Kola and Folb, 1986a). FDA tests both for the presence of esterase activity and for cell membrane integrity. In these three studies, no changes in viability could be detected by FDA in exposed blastocysts, even at concentrations that inhibited subsequent development either *in vivo* after transfer to foster mothers (MNU) or during implantation *in vitro* (chlorpromazine); however, development of a more sensitive test for viability holds promise for the selection of viable embryos in transplantation experiments, both in animals and humans.

2.3. Transplantation of Preimplantation Embryos

The embryo transfer technique is a very useful method in reproductive toxicology, since it allows analysis of whether embryotoxic effects observed at term after preimplantation treatment were induced by maternal effects or by direct effects of the agent on the embryo. Experience and skill are required to obtain statistically relevant data, including dose-response relations on the survival of embryos that were exposed to toxic treatment before transfer (Spielmann, 1976; Spielmann and Eibs, 1978). In many toxicological investigations, only a single group of blastocysts was transplanted after exposure to a concentration of a toxic agent that neither inhibited development nor reduced the viability of the embryos. Despite many efforts, term malformations could not be induced by transferring treated preimplantation embryos to foster mothers, either after exposure *in vivo* to drugs which have to be activated (*e.g.*, CPA: Spielmann *et al.*, 1977) or *in vitro* to direct acting agents such as MNU (Iannaccone *et al.*, 1982; Iannaccone, 1984) and 4-nitroquinoline-1-oxide (4-NQO) (Katayama and Matsumoto, 1985). As expected, in these studies both a dose-related increase in resorption rate and a decrease in birth rate were observed. In a very elaborate study one year after birth, offspring developing from blastocysts exposed to MNU *in vitro* were found to have a 3-fold higher mortality rate than offspring developing from blastocysts exposed to solvent (Iannaccone, 1984). No chromosomal abnormalities could be discovered in either surviving group of offspring. These data may illustrate that the embryo transfer technique, although neglected in recent years, still holds promise for the study of longterm effects of *in vitro* exposure.

The embryo transfer technique has been used quite successfully to study genetic effects modifying the embryotoxicity of drugs after implantation, as suggested by Marsk *et al.* (1971). These investigators used blastocyst transfer to study the influence of the genetic background of embryos or mothers in cortisone-induced cleft palate during organogenesis. The same approach has been used to show that the embryonic and not the maternal genome is responsible for the teratogenic sensitivity of NMRI embryos to 2,3,7,8-tetrachlorodibenzo-p-dioxin (TCDD) (D'Argy *et al.*, 1984). In addition, the embryo transfer technique has been used to analyze the site of action of drugs interfering with implantation such as clomiphene citrate (Birkenfeld *et al.*, 1985).

2.4. Biochemical Studies

To study the molecular basis of early mammalian development as well as its genetic control, very sophisticated biochemical studies have been carried out on preimplantation embryos (generally in the mouse), covering the whole area of molecular biology (*e.g.*, Magnuson and Epstein, 1981; Johnson, 1981). [See also Chapter 7 (Ed.).] These investigations proved that preimplantation embryos are sufficiently able to carry out synthesis and degradation of intermediates and macromolecules of metabolic pathways that are essential for rapidly growing tissues. Using early mammalian embryos, methods have, therefore, been devised for studying the action of toxic agents on specific metabolic pathways. Preimplantation embryos have also been exposed to specific metabolic inhibitors in culture to analyse developmental changes in different metabolic pathways. Presently, however, the measurement of a dose-related inhibition of DNA synthesis during exposure to another drug interfering with DNA replication is not necessarily expanding our knowledge about the action of xenobiotics on early embryos. Unfortunately, there are no systematic investigations in which the most sensitive *in vitro* methods for studying toxic effects in preimplantation embryos (*e.g.*, *in vitro* culture during implantation and cytogenetic analysis) are compared in sensitivity to biochemical methods (*e.g.*, DNA-synthesis).

Biochemical studies demonstrating a measurable activity of enzymes that are metabolizing xenobiotics in preimplantation embryos have become very important, particularly with respect to the irreversible long-term effects of teratogenesis, mutagenesis and carcinogenesis (Galloway *et al.*, 1980; Filler and Lew, 1981; Balling *et al.*, 1985). However, one should not forget that direct acting toxic agents can reach the embryo before implantation (Sieber and Fabro, 1971) and that metabolites of drugs which have to be activated, for example CPA (Spielmann *et al.*, 1981a), can reach the embryo after activation by the mother without any delay even at the 4-cell stage.

2.5. General Toxicological Procedure

As a basic rule in prenatal toxicology, the evaluation of embryotoxic agents is performed at term according to routine teratology, since the general public is interested in irreversible long-term effects of exposure during pregnancy, such as intrauterine death (or resorption rate), malformations and postnatal effects, rather than in embryotoxic effects of some unrealistic culture conditions. Although many investigations are providing information on dose-related toxic effects on preimplantation embryos *in vitro*, they do not allow a risk evaluation since dose-related embryolethal effects at term (ELD_{50}) of the same treatment are not determined. In addition to examination of embryos at term, it is sometimes necessary to search for retarded embryotoxic effects after implantation, *e.g.*, during organogenesis as shown for CPA (Spielman and Eibs, 1978).

If it seems appropriate, the effects of treatment of early embryos *in vivo* or *in vitro* can be evaluated before implantation using the toxicological methods discussed in the present chapter. Figure 6 summarizes morphological and cytogenetic endpoints that can routinely be analyzed in preimplantation mouse embryos and also the time intervals when evaluation should be performed. Treatment and evaluation of the toxic effects can, of course, be carried out at any time during preimplantation development. Depending on the results, additional embryo transfer experiments and biochemical studies may be appropriate.

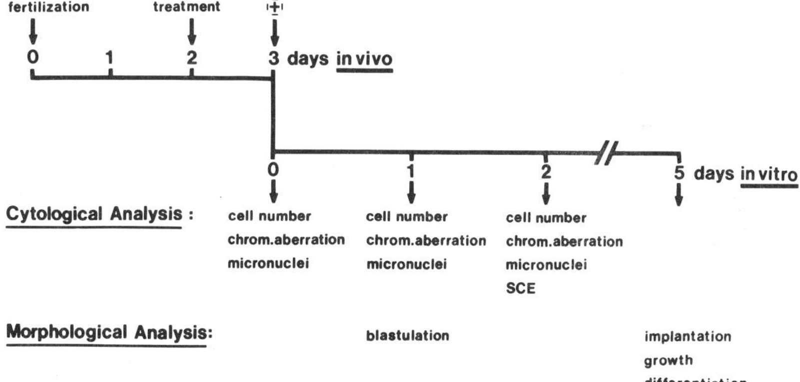

Figure 6. Diagram explaining cytogenetic and morphological endpoints for evaluating toxic effects in mouse blastocysts exposed to embryotoxic agents during the preimplantation period *in vivo*. Toxicological endpoints are identical for evaluation of exposure *in vitro*. The times of treatment and of analysis may be changed according to the requirements of the problem to be solved.

3. EXPOSURE OF EMBRYOS *IN VITRO*

3.1. Culture of Preimplantation Embryos in the Presence of Toxic Agents

Many toxicological studies have been carried out on the exposure of preimplantation embryos *in vitro* to drugs, environmental chemicals and X- and UV-irradiation and also on the embryotoxicity of culture media. A dose-related inhibition of development can usually be observed and also changes in biochemical parameters. Such data are relevant for comparison with results on adult or other embryonic tissues (*e.g.*, during organogenesis), particularly in the analysis of unique toxicological characteristics of early embryos, *e.g.*, DNA-repair, metabolism of xenobiotics or specific sensitivities of the embryonic cells.

In a study on the effects of fractions of blood sera on cultured 2-cell mouse embryos, SCE-frequency was a more sensitive parameter than embryonic growth, chromosomal aberrations and micronuclei formation (Saito *et al.*, 1984). In investigations on the toxicity of heterologous sera (Krüger *et al.*, 1985) and of sera containing CPA-metabolites (Vogel *et al.*, 1985) on preimplantation mouse embryos, we also found that SCE-rate is a more sensitive parameter than differentiation of the embryos in culture.

To study the effects of UV- and X-irradiation *in vitro* on early mouse embryos, both cytogenetics and differentiation in culture have been analysed. In an investigation on mouse egg sensitivity to X-irradiation *in vitro* between fertilization and first cleavage, differentiation of the ICM was the most sensitive developmental parameter (Jacquet *et al.*, 1983). Superadditive toxic effects were observed after combined treatment of preimplantation mouse embryos *in vitro* with X-rays and caffeine (Müller *et al.*, 1985). However, no superadditive effects were seen after combined exposure to X-rays and cadmium (Müller *et al.*, 1982) using development in culture, cell number and induction of micronuclei as toxicological endpoints.

Acute and long-term effects of exposure to heavy metals *in vitro* have been studied by several groups. Although treatment with cadmium chloride inhibited cleavage of early mouse embryos, morulae exposed to 10^{-5} M $CdCl_2$ developed into normal fetuses after transfer to foster mothers (Pedersen and Lin, 1978). Blastocysts developing from male and female mouse gametes exposed during fertilization to 0.4 μM cadmium chloride were transferred to surrogate dams; no toxic effects except a loss of embryos during implantation were found at term (Schmid *et al.*, 1983). Treatment of preimplantation mouse embryos with mercuric compounds revealed that embryonic sensitivity changes during development (Matsumoto and Spindle, 1982) and that methylmercuric chloride is 200 times more toxic than mercuric chloride with respect to cell proliferation and 50 times with respect to inhibition of protein synthesis (Katayama *et al.*, 1984). Mouse morulae and blastocysts were able to recover from 1 hr exposure *in vitro* to a toxic concentration of nickel chloride (400 μM) as shown by development *in vivo* after transfer to pseudopregnant dams (Storeng and Jonsen, 1984). Potassium dichromate (10^{-6} M) inhibited differentiation of the ICM of cultured mouse blastocysts and increased their SCE-rates (Ijima *et al.*, 1983).

In a comparison of the effects of 7 direct acting toxic chemicals (exposure time 24 hr) on differentiation of mouse blastocysts in culture (Katayama and Matsumoto, 1985), formation of the ICM was the most sensitive morphological endpoint; however, at even lower concentrations, the frequencies of chromosome aberrations and SCE were increased significantly. In this study, 4-nitroquinoline-1-oxide (4-NQO) was the most toxic chemical, followed by mitomycin C, bleomycin, methylmercury chloride, diethylstilbestrol, mercuric chloride and ochratoxin A. In blastocysts exposed to 4-NQO at concentrations that increased SCE-frequency, developmental retardations were seen *in vivo* after transfer to surrogate dams. Exposure of preimplantation mouse embryos to the direct acting mutagens methylmethanesulfonate, MMS (Fabro *et al.*, 1984) and MNU (Iannaccone, 1984) revealed an inhibition of DNA-synthesis at concentrations that induced long-term effects *in vivo* in the offspring after transfer to foster mothers but which did not inhibit development to the blastocyst stage *in vitro*. In both studies neither cytogenetic analysis nor embryo culture during implantation were performed.

Dose-related toxic effects have also been found in mouse and rabbit blastocysts treated in culture with drugs and hormones (nicotine, prostaglandins and inhibitors, and estradiol) at concentrations that can hardly be attained *in vivo* (Uehara *et al.*, 1984; Biggers *et al.*, 1978; Balling and Beier, 1985).

3.2. Culture of Preimplantation Embryos in the Presence of Activated Metabolites or Metabolizing Systems

In the environment of the female genital tract, preimplantation embryos are exposed to the same chemicals as the mother, including metabolites produced in the maternal organism, since these freely pass into the luminal fluid of the tract (Sieber and Fabro, 1971). *In vitro* studies on toxic chemicals which must be activated have to take this into account. Since metabolically active systems like the liver S-9-mix of the Ames test are embryotoxic, there are no reports on the exposure of preimplantation embryos to S-9-mix. There is only one study on implanting mouse embryos (day 7), which showed a significant increase in SCE in the presence of liver S-9-mix and benzo(a)pyren (BP) (Galloway *et al.*, 1980). To imitate the situation *in vivo*, mouse blastocysts were incubated for 1 hr with endometrial cell microsomes which oxidized the BP metabolite BP-trans-7,8-dihydrodiole to more toxic BP-7,8-dihydrodiol-9, 10-epoxides (Iannaccone *et al.*, 1984). After transfer of exposed blastocysts to surrogate mothers, a reduced implantation rate was seen; however, cytogenetics or differentiation *in vitro* were not analysed.

To establish an *in vitro* test system in which sera containing active metabolites of chemicals from exposed individuals can be tested for toxic effects on preimplantation embryos, we have cultured preimplantation mouse embryos in sera from CPA-treated animals containing its active metabolites (Vogel *et al.*, 1985). Among the endpoints tested (development in culture, cell number, chromosomal aberrations and SCE), SCE-frequency was the most sensitive parameter, indicating embryotoxic effects after only 1 hr of exposure to rat serum containing CPA metabolites (Table I).

3.3. Potential of Preimplantation Embryos to Metabolize Xenobiotics

The onset of mixed-function oxidase activity (microsomal cytochrome P-450) coincides with blastocyst formation in the mouse (Filler and Lew, 1981) as proven by active metabolites of aromatic hydrocarbons which were identi-

Table I
Effects of Rat Serum Containing Active Metabolites of Cyclophosphamide (CPA) on the Cytogenetics of Mouse Morulae and Blastocysts *in Vitro*

Exposure in Whitten's medium	Cytological endpoints after culture of exposed embryos in BrdU/thymidine for 48 hr			
	Cell number[a]	Mitotic index	Chromosomal aberrations[a,d]	SCE/metaphase[a,e]
24 hr 10% FCS[b]	49 ± 13 (23)	5.1	0 (50)	25.4 ± 1.9 (25)
24 hr 10% RS[c]	56 ± 9 (20)	4.7	0 (50)	38.6 ± 2.6** (15)
1 hr 10% RS-CPA	50 ± 12 (14)	4.7	0 (50)	66.0 ± 2.9** (10)
3 hr 10% RS-CPA	49 ± 13 (14)	4.7	0 (50)	75.3 ± 2.6** (10)
6 hr 10% RS-CPA	49 ± 14 (16)	5.0	5* (50)	120.5 ± 4.5** (10)
24 hr 10% RS-CPA	NS[f]	NS[f]	9 (1)	NS[f]

[a]No. of determinations in parentheses.
[b]FCS = Fetal calf serum.
[c]RS = Rat serum.
[d] = χ^2-test.
[e] = t-test.
[f]NS= non-scorable.
* = Significant ($p < 0.01$).
** = Significant ($p < 0.001$).

fied by high pressure liquid chromatography in the culture medium. In the same study, formation of glucuronic acid or sulfate ester conjugates could not be detected. Cytochrome P-450 activity was also measured in mouse blastocysts using the SCE assay as a sensitive indicator of DNA damage induced by reactive intermediates of BP (Pedersen *et al.*, 1985). Analysis of metabolites formed during incubation of rabbit blastocyst (day 5 to 6) with ^{3}H-diethylstilbestrol indicated activity of conjugative enzymes and of a specific embryonic mono-oxygenase which is not found in adult rabbits (Balling *et al.*, 1985). In toxicological evaluation of early pregnancy, it should also be noted that preimplantation embryos have enzymes for repairing UV- and X-irradiation-induced DNA damage (Pedersen and Cleaver, 1975; Spielmann and Eibs, 1978; Bennett and Pedersen, 1984) and also (in the pig, rabbit, rat and mouse) for metabolizing steroid hormones (Perry *et al.*, 1976; George and Wilson, 1978; Wu and Matsumoto, 1985).

4. EXPOSURE OF EMBRYOS *IN VIVO*

4.1. Toxicological Evaluation at the End of Gestation

Evaluation at term of pregnant animals exposed to toxic agents before implantation usually shows that the embryos are either dead or alive and not malformed (Wilson, 1977; Spielmann and Eibs, 1978). Claims of investigators that malformations can be induced by early treatment could either not be repeated by other groups or were due to a long halflife of the drug or its metabolites (*e.g.*, Eibs *et al.*, 1982). Storeng and Jonsen (1981, 1983) recently reported that treatment of pregnant mice before implantation with nickel chloride and also with ricin not only increases the resorption rate but also the number of fetuses with exencephaly. However, no toxic effects were induced by transferring preimplantation mouse embryos to foster mothers after exposure to $NiCl_2$ *in vitro* (Storeng and Jonsen, 1984). Takeuchi (1984) found that the application of MNU to pregnant mice on days 2, 3 or 4 of gestation at a dose of 10 mg/kg induced malformations, including exencephaly and cleft palates. Again, when blastocysts were exposed to toxic concentrations of MNU in culture, no malformations could be found after transfer to foster mothers but only an increased mortality of the offspring (Iannaccone, 1984).

In several studies on effects of exposure of embryos during the preimplantation period, a dose-related increase in resorption rate and slight developmental retardations were observed with (*e.g.*) the metals Al, Co, Mo, V, W (Wide, 1984;), lanthanum (Abramzuk, 1985), oil retort water (Gregg *et al.*, 1981) and 2,2-dichlorobiphenyl (Török, 1978). Recently, in a toxicological risk evaluation at term, we have not only taken into account the embryolethal effect of treatment before implantation (ELD_{50}) but also the maternal LD_{50} (MLD_{50}) and the teratogenic dose during organogenesis (TD_{50}) as shown in Table II (Spielmann *et al.*, 1985). A comparison of ELD_{50} with the MLD_{50} reveals a high risk only for caffeine and CPA and no risk for heavy metals. Furthermore, as expected, the ELD_{50} was always higher than the TD_{50} which is determined in routine toxicological testing.

Table II
Treatment During the Preimplantation Period (Day 2) and Risk Evaluation at Term (Day 17)

Substance used in treatment	MLD_{50}[a] (mg/kg)	ELD_{50}[b] (mg/kg)	$\frac{MLD_{50}}{ELD_{50}}$	TD_{org}[c] (mg/kg)	$\frac{ELD_{50}}{TD_{org}}$
Methylmercurychloride	9.5	9.5	1.0	2.0-4.0	3.2
Cadmium chloride	10.0	10.0	1.0	4.0	2.5
Lead nitrate	180.0	120.0	1.5	15.0	8.0
Ethanol	10,000.0	6000.0	1.66	6000.0	1.0
Caffeine (4 x 1/4 dose)	700.0	85.0	8.2	100.0	0.85
Cyclophosphamide	560.0	45.0	12.4	10.0	4.5
Vinblastine sulfate	6.0	3.0	2.0	0.4	7.5
Doxycycline HCl	3000.0	3000.0	1.0	2300.0	1.3
Phenobarbital-Na	300.0	300.0	1.0	175.0	1.7

[a]MLD_{50} = maternal LD_{50}.
[b]ELD_{50} = embryonic LD_{50}.
[c]TD_{org} = teratogenic dose during organogenesis.

Table III
Analysis of Embryotoxicity of Cyclophosphamide (CPA) Given on Day 2 of Gestation Using Cytological and Morphological Endpoints at Different Times After Treatment

		CPA (mg/kg)							
Development after treatment[a]	Endpoint	0	5	10	20	40	60	80	160
24 hr *in vivo*	Cell number	31	28	26	26	23	--	15	12
	% Structural chromosome aberrations	0	0	0	0	--	--	28	50
	% Micronuclei	0	0	1	1	2	--	12	18
24 hr *in vivo* + 48 hr *in vitro*	SCEs/metaphase	25	39	46	58	--	--	--	--
24 hr *in vivo* + 120 hr *in vitro*	ICM (% of control)	100	91	69	40	23	--	--	--
15 days *in vivo* (at term)[c]	% Retardation[b]	0	0	0	0	2	20	25	--
	% Embryolethality	4	7	9	18	37	79	95	--

[a]One cell cycle *in vivo* = 12 hr; one cell cycle *in vitro* = 24 hr.
[b]Percentage of surviving embryos.
[c]= Day 17 of gestation.

4.2. Toxicological Evaluation Before Implantation

Many of the more recent studies have incorporated into their testing procedure not only the examination of the embryos at term but also before implantation (Spielmann and Eibs, 1978). Using the techniques that are described above (Section 2) in a series of studies on the high resorption rate in mice treated with lead before implantation, Jacquet *et al.* (1977) and Wide (1978, 1983) could not find toxic effects on preimplantation embryos but there were indications for an interference of lead with the action of steroid hormones on the endometrium. Our data at term shown in Table II suggest a similar mechanism. In investigations on the effects of toxic chemicals on preimplantation rat embryos, Giavini and coworkers have always combined the evaluation of the embryos at term and before implantation. They found a reduced cell number in blastocysts from animals exposed to actinomycin and chloramphenicol, cadmium, lead and copper and chlorambucil but no changes after methylmercury exposure (Giavini *et al.*, 1979, 1980, 1984, 1985). Surprisingly, an increase in micronuclei could be detected in blastocysts after chlorambucil treatment (Giavini *et al.*, 1984). For toxic agents that indicated a risk at term from treatment before implantation (Table II), we analysed toxic effects on embryos before implantation, as shown for CPA in Table III and for mitomycin C (MMC) in Table IV. SCE-frequency, although determined 72 hr after treatment of the mother, was the most sensitive parameter for CPA (Spielmann *et al.*, 1985; see also Fig. 5), whereas development during implantation *in vitro* and chromosomal aberrations were more affected than SCE-frequency by treatment with MMC.

Table IV
Analysis of Embryotoxicity of Mitomycin C (MMC) Given on Day 2 of Gestation Using Cytological and Morphological Endpoints at Different Times After Treatment

Development after treatment[a]	Endpoint	MMC (mg/kg) 0	0.5	1	5	10	20
24 hr *in vivo*	Cell number	28	29	29	25	18	--
	% Structural aberrat- ions in metaphase	2	8	8	10	22	--
24 hr *in vivo* + 48 hr *in vitro*	SCEs/metaphase	25	36	46	66	72	--
24 hr *in vivo* + 120 hr *in vitro*	ICM (% of control)	100	80	50	0	--	--
15 days *in vivo* (at term)[c]	% Retardation[b]	0	0	0	0	12	--
	% Embryolethality	4	--	--	13	50	100

[a]One cell cycle *in vivo* = 12 hr; one cell cycle *in vitro* = 24 hr.
[b]Percentage of surviving embryos.
[c]= Day 17 of gestation

4.3. Pharmacokinetics of the Transfer of Drugs into Early Embryos

Pharmacokinetic studies on the passage of xenobiotics into preimplantation embryos have only been carried out on rabbit blastocysts, which are unusually large with more than 10,000 cells (Daniel, 1964) compared to only 100 cells in mouse, rat and human blastocysts (McLaren, 1985). According to the basic studies of Sieber and Fabro (1971), which were extended by Thithapandha (1980), a variety of xenobiotics and their metabolites can enter the rabbit blastocyst. After application to the maternal animal, most drugs with a molecular weight of less than 600 can rapidly traverse the blastocysts and reach concentrations similar to maternal plasma levels. However, the concentrations used in many studies on the exposure of preimplantation embryos *in vitro* are significantly higher. The experiments with rabbits furthermore revealed that the transfer of xenobiotics into blastocysts is influenced by the degree of ionization and lipid solubility.

In a study on the transfer of ^{14}C-CPA into mouse blastocysts, we found identical pharmacokinetics of the radioactive label in maternal blood and fallopian tubes; however, because of their small size, measurable amounts of ^{14}C could not be detected in groups of up to 100 preimplantation embryos (Spielmann *et al.*, 1981a). Using an indirect approach, *i.e.*, flushing embryos from the genital tract of the mother at different time points after treatment and culturing them during implantation, we were able to show that differentiation was inhibited in embryos taken from the mother 10 min after treatment (Fig. 7; see also SCE increase at 1 hr treatment in Fig. 5). In the only report on human pharmacokinetics, the antibiotic cefotan reached identical concentrations in serum and in the uterine tubes 1 hr after i.v. administration (Daschner *et al.*, 1982).

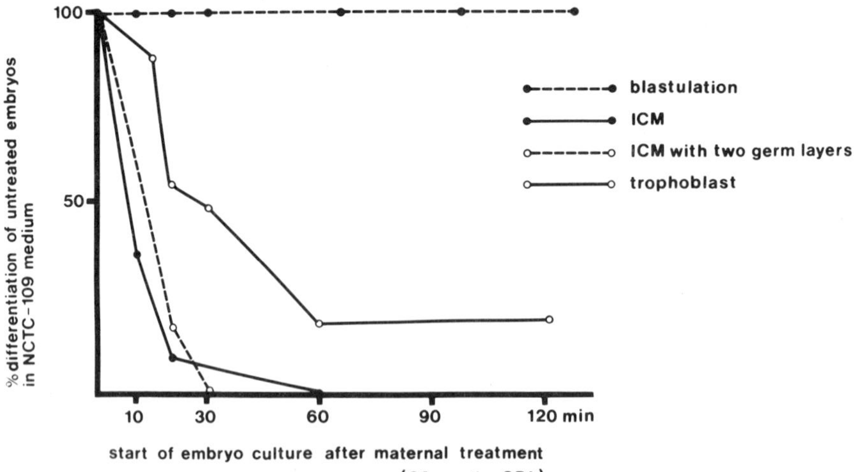

Figure 7. Indirect pharmacokinetic analysis of the transfer of cyclophosphamide (CPA; 60 mg/kg) into mouse embryos on day 2 of gestation (4-cell and 8-cell stage). The embryos were removed at various time intervals from oviducts and cultured in Whitten's medium for 48 hr and subsequently in NCTC-109 for 96 hr. Differentiation of the ICM was inhibited in embryos removed only 10 min after treatment.

5. CONCLUSIONS

The increasing knowledge of early human pregnancy should stimulate research on endogenous and exogenous factors that contribute to the high rate of embryonic loss before implantation. For toxicological studies, it is important to note that the action of chemical agents on cleavage stage embryos follows normal dose-response relations. Among the toxicological endpoints that can be analysed, morphological methods including *in vitro* culture and transplantation are sufficiently standardized. However, long-term effects at the end of pregnancy and in the offspring may have been underestimated so far and research should be intensified. The sensitivity of cytogenetic methods must be improved so that the effects of xenobiotics on repair mechanisms can be analysed. To allow a better extrapolation to early human pregnancy, the pharmacokinetics of drug transfer into embryos before implantation and also drug metabolism in cleavage stage embryos should not only be studied in the rabbit but also in the mouse or rat. Methods developed for molecular genetics of cleavage stage embryos, *e.g.*, transgenic mice, have so far not been introduced to toxicology and these may considerably expand our knowledge.

ACKNOWLEDGMENTS

Our investigations have been supported by the Deutsche Forschungsgemeinschaft (DFG) for many years and also by the Dept. of Research and Technology (BMFT) through the CMT (Carcinogenesis, Mutagenesis, Teratogenesis) program. Most of the cytogenetic studies were performed in collaboration with Dr. Richard Vogel, who contributed to the manuscript.

6. REFERENCES

Abramczuk, J.W., 1985, The effects of lanthanum chloride on pregnancy in mice and on preimplantation mouse embryos *in vitro*, *Toxicology* 34: 315-320.

Angell, R.R., Aitken, R.J., van Look, P.F.A., Lumbdsen, M.A., and Templeton, A.A., 1983, Chromosomal abnormalities in human embryos after *in vitro* fertilization, *Nature (London)* 303: 336-338.

Austin, C.R., 1973, Embryo transfer and sensitivity to teratogenesis, *Nature (London)* 244: 333-334.

Balling, R., and Beier, H.M., 1985, Direct effects of nicotine on rabbit preimplantation embryos, *Toxicology* 34: 309-313.

Balling, R., Haaf, H., Maydl, R., Metzler, M., and Beier, H.M., 1985, Oxidative and conjugative metabolism of diethylstilbestrol by rabbit preimplantation embryos, *Dev. Biol.* 109: 370-374.

Bennett, J., and Pedersen, R.A., 1984, Early mouse embryos exhibit strain variation in radiation-induced sister-chromatid exchange: relationship with DNA repair, *Mutation Res.* 126: 153-157.

Biggers, J.D., 1981, *In vitro* fertilization and embryo transfer in human beings, *N. Engl. J. Med.* 304: 336-342.

Biggers, J.D., Leonov, B.V., Baskar, J.F., and Fried, J., 1978, Inhibition of hatching of mouse blastocysts *in vitro* by prostaglandin antagonists, *Biol. Reprod.* 19: 519-533.

Birkenfeld, A., Mootz, U., and Beier, H.M., 1985, The effect of clomiphene citrate on blastocyst development and implantation in the rabbit, *Cell Tissue Res.* 241: 495-503.

Boué, J.G., Boué, A., and Lazar, P., 1975, Retrospective and prospective epidemiological studies of 1,500 karyotyped spontaneous human abortions, *Teratology* 12: 11-26.

Daniel, J.C., 1964, Early growth of rabbit trophoblast, *Amer. Natur.* 98: 85-98.

D'Argy, R., Hassoun, E., and Dencker, L., 1984, Teratogenicity of TCDD and the congener 3,3',4,4'-tetrachloroazoxybenzene in sensitive and nonsensitive mouse strains after reciprocal blastocyst transfer, *Tox. Lett.* 21: 197-202.

Daschner, F., Petersen, E., Langmaak, H., and Trennhäuser, M., 1982, Antibiotic prophylaxis in gynecology: cefoxitin concentrations in serum, myometrium, endometrium and salpinges, *Infection* 10: 341-342.

Edwards, R.G., and Steptoe, P.C., 1983, Current status of *in vitro* fertilization and implantation of human embryos, *Lancet* 2: 1265-1269.

Eibs, H.G., and Spielmann, H., 1977, Inhibition of postimplantation development of mouse blastocysts *in vitro* after cyclophosphamide treatment *in vivo*, *Nature (London)* 270: 54-56.

Eibs, H.G., Spielmann, H., and Hägele, M., 1982, Teratogenic effects of cyproterone acetate and medroxyprogesterone treatment during the pre- and postimplantation period of mouse embryos. I., *Teratology* 25: 27-36.

Epstein, C.J., 1975, Gene expression and macromolecular synthesis during preimplantation development, *Biol. Reprod.* 12: 82-105.

Fabro, S., McLachlan, J.A., and Dames, N.M., 1984, Chemical exposure of embryos during the preimplantation stages of pregnancy: mortality rate and intrauterine development, *Amer. J. Obstet. Gynecol.* 148: 929-937.

Filler, R., and Lew, K.J., 1981, Developmental onset of mixed-function oxidase activity in preimplantation mouse embryos, *Proc. Natl. Acad. Sci. USA* 78: 6991-6995.

Galloway, S.M., Perry, P.E., Meneses, J., Nebert, D.W., and Pedersen, R.A., 1980, Cultured mouse embryos metabolize benzo(a)pyrene during early gestation: genetic differences detectable by sister chromatid exchange, *Proc. Natl. Acad. Sci. USA* 77: 3524-3528.

Garner, W., 1974, The effect of 5-bromodeoxyuridine on early mouse embryos *in vitro*, *J. Embryol. Exp. Morphol.* 32: 849-855.

George, F.W., and Wilson, J.D., 1978, Estrogen formation in the early rabbit embryo, *Science* 199: 200-201.

Giavini, E., Prati, M., and Vismara, C., 1979, The effects of actinomycin D and chloramphenicol on the rat preimplantation embryos, *Experientia* 35: 1649-1650.

Giavini, E., Prati, M., and Vismara, C., 1980, Effects of cadmium, lead and copper on rat preimplantation embryos, *Bull. Environm. Contam. Toxicol.* 25: 702-705.

Giavini, E., Bonanomi, L., and Ornaghi, F., 1984, Developmental toxicology during the preimplantation period: embryotoxicity and clastogenic effects of chlorambucil in the rat, *Teratogen. Carcinogen. Mutagen.* 4: 341-348.

Giavini, E., Vismara. C., and Broccia, M.L., 1985, Effects of methylmercuric chloride administered to pregnant rats during the preimplantation period, *Ecotox. Environm. Safety* 9: 189-195.

Golbus, M.S., and Epstein, C.J., 1974, Effect of 5-bromodeoxyuridine on preimplantation mouse embryo development, *Differentiation* 2: 134-149.

Gregg, C.T., Tietjen, G., and Hutson, J.Y., 1981, Prenatal toxicology of shale oil retort water, *J. Toxicol. Environ. Health* 8: 795-804.

Iannaccone, P.M., 1984, Long-term effects of exposure to methylnitrosourea on blastocysts following transfer to surrogate female mice, *Cancer Res.* 44: 2785-2789.

Iannaccone, P.M., Tsao, T.Y., and Stols, L., 1982, Effects on mouse blastocysts of *in vitro* exposure to methylnitrosourea and 3-methylcholanthrene, *Cancer Res.* 42: 864-868.

Iannaccone, P.M., Fahl, W.E., and Stols, L., 1984, Reproductive toxicology associated with endometrial cell mediated metabolism of benzo(a)-pyrene: a combined *in vitro, in vivo* approach, *Carcinogenesis* 5: 1437-1442.

Iijima, S., Spindle, A., and Pedersen, R.A., 1983, Developmental and cytogenetic effects of potassium dichromate on mouse embryos *in vitro*, *Teratology* 27: 109-115.

Jacquet, P., Gerber, G.B., Leonard, A., and Meas, J., 1977, Plasma hormone levels in normal and lead-treated pregnant mice, *Experientia* 33: 1375-1377.

Jacquet, P., Kervyn, G., and DeClercq, G., 1983, Studies on mouse-egg radiosensitivity from fertilization up to the first cleavage, *Mutation Res.* 110: 351-365.

Johnson, M.H., (ed.), 1977, *Development in Mammals*, Vol. 1, North-Holland, Amsterdam.

Johnson, M.H., 1981, The molecular and cellular basis of preimplantation mouse development, *Biol. Rev.* 56: 463-498.

Katayama, S., and Matsumoto, N., 1985, Toxic effects of chemicals on mouse post-blastocyst development - a trial to establish a testing system for embryotoxicity, *Acta Obst. Gynaec. Jap.* 37: 421-430.

Katayama, S., Kubo, H., and Matsumoto, N., 1984, Acute effects of mercuric compounds on preimplantation mouse embryos *in vitro*, *Acta Obst. Gynaec. Jap.* 36: 1957-1962.

Kline, J., and Stein, Z., 1985, Very early pregnancy, in: *Reproductive Toxicology* (R.L. Dixon, ed.), Raven Press, New York, pp. 251-265.

Kola, I., and Folb, P.I., 1986a, Chlorpromazine inhibits the mitotic index, cell number, and formation of murine blastocysts, and delays implantation of CBA mouse embryos, *J. Reprod. Fertil.* 176: 527-536.

Kola, I., and Folb, P.I., 1986b, The effects of cyclophosphamide on alkaline phosphatase activity and on *in vitro* post-implantation murine blastocyst development, *Devel. Growth and Differentiation* (in press).

Krüger, C., Vogel, R., and Spielmann, H., 1985, Preimplantation mouse embryos cultured in mouse, rat and human sera: differentiation and sister chromatid exchange, *Experientia* 41: 1599-1601.

Magnuson, T., and Epstein, C.J., 1981, Genetic control of very early mammalian development, *Biol. Rev.* 56: 369-408.

Marsk, L., Theorell, M., and Larsson, K.S., 1971, Transfer of blastocysts as applied to experimental teratology, *Nature (London)* 234: 358-359.

Matsumoto, N., and Spindle, A., 1982, Sensitivity of early mouse embryos to methylmercury toxicity, *Toxicol. Appl. Pharmacol.* 64: 108-117.

McLaren, A., 1985, Early mammalian development, *Prog. Clin. Biol. Res.* 163A: 29-40.

McLaren, A., and Biggers, J.D., 1958, Successful development and birth of mice cultivated *in vitro* as early embryos, *Nature (London)* 182: 877-878.

Müller, W.U., Streffer, C., and Kaiser, U., 1982, The combined treatment of preimplantation mouse embryos *in vitro* with cadmium ($CdSO_4$, CdF_2) and X-rays, *Arch. Toxicol.* 51: 303-312.

Müller, W.U., Streffer, C., and Wurm, R., 1985, Supraadditive formation of micronuclei in preimplantation mouse embryos *in vitro* after combined treatment with X-rays and caffeine, *Teratogen. Carcinogen. Mutagen.* 5: 123-131.

Pedersen, R.A., and Cleaver, J.E., 1975, Repair of UV damage to DNA of implantation-stage mouse embryos *in vitro*, *Exp. Cell Res.* 95: 247-253.

Pedersen, R.A., and Lin, T.P., 1978, Cadmium toxicity in preimplantation mouse embryo development, in: *Developmental Toxicology of Energy-Related Pollutants* (D.D. Mahlum, M.R. Sikov, P.L. Hachett, and F.D. Andrew, eds.), U.S. Dept. of Energy, Washington, D.C., pp. 600-613.

Pedersen, R.A., Meneses, J., Spindle, A., Wu, K., and Galloway, S.M., 1985, Cytochrome P-450 metabolic activity in embryonic and extraembryonic tissue lineages of mouse embryos, *Proc. Natl. Acad. Sci. USA* 82: 3311-3315.

Perry, P., and Wolff, S., 1974, New Giemsa method for the differential staining of sister chromatids, *Nature (London)* 268: 156-158.

Perry, J.S., Heap, R.B., Burion, R.D., and Gadsby, J.E., 1976, Endocrinology of the blastocyst and its role in the establishment of pregnancy, *J. Reprod. Fertil.*, Suppl. 25: 85-104.

Saito, H., Berger, T., Mishell, D.R., and Marrs, R.P., 1984, The effect of serum fractions on embryo growth, *Fertil. Steril.* 41: 761-765.

Schmid, B.P., Hall, J.L., Goulding, E., Fabro, S., and Dixon, R., 1983, *In vitro* exposure of male and female mice gametes to cadmium chloride during the fertilization process, and its effects on pregnancy outcome, *Toxicol. Appl. Pharmacol.* 69: 326-332.

Schmid, W., 1975, The micronucleus test, *Mutation Res.* 31: 9-13.

Sieber, S.M., and Fabro, S., 1971, Identification of drugs in the preimplantation blastocyst and in the plasma, uterine secretion and urine of the pregnant rabbit, *J. Pharmacol. Exp. Ther.* 176: 65-75.

Solter, D., and Knowles, B.B., 1975, Immunosurgery of mouse blastocyst, *Proc. Natl. Acad. Sci. USA* 72: 5099-5102.

Spielmann, H., 1976, Embryo transfer technique and action of drugs on the preimplantation embryo, in: *Current Topics in Pathology*, Vol. 62 (E. Grundmann, and W.H. Kirsten, eds.), Springer Verlag, Berlin, pp. 87-103.

Spielmann, H., and Eibs, H.G., 1978, Recent progress in teratology: A survey of methods for the study of drug actions during the preimplantation period, *Arzneim. Forschg./Drug Res.* 28: 1733-1742.

Spielmann, H., and Jacob-Müller, U., 1981, Investigations on cyclophosphamide treatment during the preimplantation period. II. *In vitro* studies on the effects of cyclophosphamide and its metabolites 4-OH-cyclophosphamide, phosphoramide mustard, and acrolein on blastulation of four-cell and eight-cell mouse embryos and on their subsequent development during implantation, *Teratology* 23: 7-13.

Spielmann, H., Eibs, H.G., and Merker, H.J., 1977, Effects of cyclophosphamide treatment before implantation on the development of rat embryos after implantation, *J. Embryol. Exp. Morphol.* 41: 65-78.

Spielmann, H., Jacob-Müller, U., and Beckord, W., 1980, Immunosurgical studies on inner cell mass development in rat and mouse blastocysts before and during implantation *in vitro*, *J. Embryol. Exp. Morphol.* 60: 255-269.

Spielmann, H., Habenicht, U., Eibs, H.G., Jacob-Müller, U., and Schimmel, A., 1981a, Investigations on the mechanism of action and on the pharmacokinetics of cyclophosphamide treatment during the preimplantation period in the mouse, in: *Culture Techniques* (D. Neubert, and H.J. Merker, eds.), Walter de Gruyter & Co., Berlin, pp. 435-443.

Spielmann, H., Jacob-Müller, U., Eibs, H.G., and Beckord, W., 1981b, Investigations on cyclophosphamide treatment during the preimplantation period, I. Differential sensitivity of preimplantation mouse embryos to maternal cyclophosphamide treatment, *Teratology* 23: 1-5.

Spielmann, H., Krüger, C., and Vogel, R., 1985, Embryotoxicity testing during the preimplantation period, in: *Concepts in Toxicology*, Vol. 3 (F. Homburger, ed.), S. Karger AG, Basel, pp. 22-28.

Storeng, R., and Jonsen, J., 1981, Nickel toxicity in early embryogenesis in mice, *Toxicology* 20: 45-51.

Storeng, R., and Jonsen, J., 1983, Inhibitory effect of ricin on the development of preimplantation mouse embryos, *J. Toxicol. Environ. Health* 12: 193-202.

Storeng, R., and Jonsen, J., 1984, Recovery of mouse embryos after short-term *in vitro* exposure to toxic nickel chloride, *Tox. Lett.* 20: 85-91.

Takeuchi, I.K., 1984, Teratogenic effects of methylnitrosourea on pregnant mice before implantation, *Experientia* 40: 879-881.

Tarkowski, A.K., 1966, An air-drying method for chromosome preparation from mouse eggs, *Cytogenetics* 5: 394-400.

Thithapandha, A., 1980, Characteristics of drugs that penetrate the preimplantation blastocyst, *Biochem. Pharmacol.* 29: 1663-1668.

Török, P., 1978, Delayed implantation and early developmental defects in the mouse caused by PCB: 2,2'-dichlorobiphenyl, *Arch. Toxicol.* 40: 249-254.

Uehara, S., Villee, C.A., and Hoshiai, H., 1984, Effect of estradiol, prostaglandin E_2, and prostaglandin $F_{2\alpha}$ on incorporation of ^{3}H-uridine by preimplantation mouse embryos *in vitro*, *Tohoku J. Exp. Med.* 144: 305-313.

Van Blerkom, J., Henry, G., and Porreco, R., 1984, Preimplantation human embryonic development from polynuclear eggs after *in vitro* fertilization, *Fertil. Steril.* 41: 686-693.

Vogel, R., Krüger, C., Granata, I., and Spielmann, H., 1985, Development and sister-chromatid-exchange of mouse morulae and blastocysts cultured in rat serum containing active metabolites of cyclophosphamide, *Tox. Lett.* 28: 23-28.

Whitten, W.K., 1971, Nutrient requirements for the culture of preimplantation embryos *in vitro*, in: *Advances in the Biosciences*, Vol. 6 (G. Raspé, ed.), Pergamon Press, Oxford, pp. 129-141.

Wide, M., 1978, Effect of inorganic lead on the mouse blastocyst *in vitro*, *Teratology* 17: 165-170.

Wide, M., 1983, Retained developmental capacity of blastocysts transferred from lead-intoxicated mice, *Teratology* 29: 293-298.

Wide, M., 1984, Effect of short-term exposure to five industrial metals on the embryonic and fetal development in the mouse, *Environm. Res.* 33: 47-53.

Wilson, J.G., 1977, Current status in teratology, general principles and mechanism derived from animal studies, in: *Handbook of Teratology*, Vol. 1 (J.G. Wilson, and C.L. Fraser, eds.), Plenum Press, New York, pp. 47-74.

Wu, J.T., and Matsumoto, P.S., 1985, Changing 17-hydroxysteroid dehydrogenase activity in preimplantation rat and mouse embryos, *Biol. Reprod.* 32: 561-566.

Chapter 15

APPLICATIONS OF ANIMAL EMBRYO CULTURE RESEARCH TO HUMAN I V F AND EMBRYO TRANSFER PROGRAMS

SUSAN HEYNER

1. INTRODUCTION

The motivation for many reproductive biologists to study early developmental processes is the specter of overpopulation. The figures are grim; for example, India is estimated to have a billion inhabitants by the year 2000. Similarly, birth rates remain extremely high throughout Africa, a continent that is facing continual shortfalls in food production. However, the other side of the coin is the problem of infertility. In traditional societies, particularly those of Asia and Africa, a childless woman is the object of scorn and pity. Even in Western societies where adoption has been common, the combination of increased rates of infertility and the difficulty of finding appropriate adoptive babies has exacerbated the problems of the childless couple. Thus, many reproductive biologists have come to realize that although the global imperative of halting population growth is still of paramount importance, the individual right to reproduction must be recognized, and in cases of infertility, ameliorated in so far as is possible. Furthermore, the study of normal processes, albeit *in vitro*, may lead to important new insights regarding *in vivo* early reproductive processes, and enhance our ability to develop new and acceptable contraceptive techniques.

On 25th July 1978, the birth of a baby girl to a female without functional fallopian tubes generated world-wide interest. The reasons are many; foremost, this birth was the result of a new treatment for a formerly intractable condition, that of blocked fallopian tubes, in a woman who was otherwise fertile. Since this birth represented the alleviation of a state that had proven difficult, if not generally impossible to correct, it gave hope to thousands of infertile women. It also gave rise to ethical, philosophical and religious debate, for the embryo that developed into a human being had been derived

Susan Heyner Department of Obstetrics and Gynecology, Albert Einstein Medical Center, Northern Division, York and Tabor Roads, Philadelphia, Pennsylvania 19141, USA.

from an oocyte that was recovered by laparoscopy, fertilized in the laboratory, and grown in a simple culture system until the 8-cell stage was reached, and the embryo was then transferred back to the maternal reproductive tract.

2. STUDIES ON ANIMAL IVF, EMBRYO CULTURE AND TRANSFER

Although mammalian *in vitro* fertilization (IVF) has been a relatively recent advance, there is a long history of short-term culture of preimplantation embryos followed by transfer to a recipient female. Nearly a century ago, Walter Heape (Heape, 1890) made the first successful transfer of fertilized rabbit ova from the uterus of one strain, to the upper end of the fallopian tube of another rabbit from a different strain, which had already been mated. The recipient mother bore 6 offspring, 4 recognizably of the maternal strain, while the other two were "strangers" whose phenotypic characteristics marked them as the young resulting from the transfer procedure.

2.1. Embryo Culture

Culture of preimplantation mammalian embryos has progressed greatly in the past 30 years, and this technique is becoming a part of veterinary and medical practice, as well as having a significant impact upon the cattle industry. Human *in vitro* fertilization programs developed from the knowledge gained from numerous animal studies. In fact, it is fair to say that without the experience gained in the laboratory, particularly with hamsters and mice, human *in vitro* fertilization-embryo transfer (IVF-ET) programs would not be around today. One of my colleagues has an office that is notable for the absence of ornate diplomas and other reassurances of technical and educational achievements: rather, it is hung with illustrations of mice. He feels that this is totally appropriate, since he owes his livelihood to *Mus musculus*, and so displays her upon his office walls. A large debt is owed to innumerable studies with mice, hamsters, rabbits and other mammalian species in terms of knowledge gained that can be applied to the human situation.

It is of interest to focus on the aspects of animal studies that have contributed in particular to the human programs. In almost all species that have been studied to date, there appears to be a stage of preimplantation development that is difficult to grow *in vitro*. In the mouse, this is at the one-cell stage; in the bovine, around the 8-cell stage; and in the hamster, between the 2- and the 8-cell stage [see Chapters 1, 9, 11 and 12 (Ed.)]. Human and other primate embryos do not appear to have a block *in vitro* corresponding to those encountered in other species. The absence of a specific stage block has proven to be most fortunate for IVF-ET programs, and somewhat unexpected. The absence of the block in human and non-human primates may be due to the use of highly enriched media for culture of these embryos, which would render the observation of a block less likely (Bavister and Boatman, personal communication). Indeed, it became apparent from early studies that human embryos are relatively easy to grow *in vitro*. In this respect, they are similar to 2-cell mouse embryos, which tolerate a fairly wide range of osmolality and pH, although optimal development occurs between 250 and 280 mOsmols and a pH of 7.2 - 7.3 (Brinster, 1972).

Culture media used for human IVF-ET are quite variable between programs. The Bourn Hall Clinic (Bourn, U.K.) still uses Earle's medium containing pyruvate and 8% serum (Edwards, 1985) while many other programs, particularly in the U.S.A., use a modification of Ham's F10 medium. All the media have in common the components pyruvate or lactate, since glucose alone cannot support the growth of pronucleate eggs.

A major limitation of human IVF-ET programs is the lack of material for obtaining data on embryo culture requirements. Human eggs and embryos are precious and few, and therefore there is little leeway for changing conditions in order to optimize conditions for their culture. Controlled studies to evaluate different media can only be carried out in the largest programs, which have big samples to study, and in which there is a record of obtaining large numbers of oocytes. It is well known that most media used in IVF-ET programs are based on the ionic composition of serum (Bavister, 1981), while newer studies on media for the cultivation of mammalian embryos base their formulations on the environment of the reproductive tract. Only one IVF group, in Australia, has developed a medium based on the composition of human fallopian tube fluid, although this would appear to be the most logical approach (Quinn *et al.*, 1985a). Indeed, these authors claim to have achieved an enhanced pregnancy rate by using this new medium.

One way to avoid the problems inherent in the limitations of human programs is to incorporate routine testing of culture media using another mammalian embryo system. Thus, the ability of 2-cell mouse embryos to develop to the blastocyst stage in medium that is to be used for patients provides a widely-used quality control assay for culture procedures. Ideally, *in vitro* fertilization would be included in this assay, but except for certain strains of mice, it is difficult to achieve high rates of fertilization and cleavage in the laboratory, due to the 1-cell block in this species. A potentially more sensitive assay is the use of 1-cell mouse zygotes, which should develop into fully expanded blastocysts in 5 days. A recent report (Quinn *et al.*, 1985b) indicates that the use of this assay resulted in higher and more consistent patient pregnancy rates. The mouse assay is particularly useful for providing a check on established procedures, such as the preparation of the medium and the particular lot number of the plasticware; further, it provides a means for evaluating an individual woman's serum for the ability to support embryonic growth. A recent report has extended the usual application of the mouse assay, to evaluate the effect of peritoneal fluid from infertile patients with and without endometriosis on the *in vitro* cleavage of 2-cell mouse embryos. The investigators concluded that heat-inactivated peritoneal fluid from patients with endometriosis was more toxic to 2-cell mouse embryos than similar concentrations of peritoneal fluid from infertile patients not suffering from endometriosis (Morcos *et al.*, 1985). Thus, the mouse assay is valuable not only as a routine quality control measure, but is useful in addition for screening biological fluids that may affect embryo growth. However, it does not provide a means of improving media for use in human programs. Mouse strains differ in their capacity to develop in a particular medium, and media that are widely used for the growth of human embryos are not optimal for the mouse embryo (Dandekar *et al.*, 1985). Nevertheless, the use of the mouse assay remains an important, if not crucial, component of any IVF-ET program.

2.2. Oocyte Maturation

A significant problem in human IVF programs is that the use of hormonal stimulation regimens often results in the production of immature oocytes. Frequently, immature oocytes, as judged by the criteria of an unexpanded cumulus and corona, and presence of a germinal vesicle, are recovered in numbers equal to those of morphologically mature oocytes. There is a need for improved methods of maturing human oocytes *in vitro*. The problem of maturing oocytes *in vitro* is not restricted to the human programs; it is also encountered in studies on large farm animals that are aimed at improving reproductive performance and stock quality. Animal studies, whether in farm animals or non-human primates, are very likely to provide methods that lead to an improved success rate of maturation of human oocytes.

2.3. Cryopreservation

Cryopreservation is another area that is gaining increased attention, and which has developed from animal studies in a number of species. Cryopreservation is already used in a number of human IVF-ET programs to store embryos that were not transferred into the uterus on the first cycle, and there are reports of live offspring resulting from these procedures. The advantages of storing additional embryos are obvious; not only is the woman spared the ordeal of another retrieval procedure, but even more important, the uterus is likely to be in a more receptive phase in a natural cycle, rather than one in which there have been unusual levels of endocrine stimulation. As with the other procedures used in IVF programs, the first attempts to cryopreserve preimplantation embryos were made in animals. In 1953, Smith attempted to freeze rabbit embryos, and concluded that the technique was feasible (Smith, 1953). In 1972, two laboratories independently reported successful cryopreservation of mouse embryos (Whittingham *et al.*, 1972; Wilmut, 1972), and subsequently, cryopreservation procedures have been used successfully to store preimplantation embryos of a number of species. In some species, notably the cow and the sheep, embryos can only be frozen successfully at particular developmental stages; however, mouse embryos can be frozen at all stages from the unfertilized oocyte to the blastocyst (Whittingham, 1971, 1977). In the human, embryos at the 4-cell and 8-cell stages have survived cryopreservation, and there is no reason to suppose that other stages cannot also be stored successfully.

The major problem with conventional slow cooling methods for cryopreservation is the formation of ice crystals. Vitrification is a new method, in which the biological material is equilibrated with cryoprotectant, and then supercooled, to avoid ice crystal formation. This method has already been used successfully to preserve 2-cell mouse embryos (Rall and Fahy, 1985), and hamster oocytes (Baker and Heyner, unpublished observations). Vitrification is clearly a powerful new approach to the problem of post-thaw damage, and its adoption by IVF-ET programs can be anticipated in the immediate future. A more speculative use of cryopreservation would be the technique in which a 2-cell embryo is split in two, one half being allowed to grow *in vitro* prior to embryo transfer, while the other half is preserved for future use. This technique of embryo splitting was developed in the mouse (Tarkowski and

Wroblewska, 1961), but it has been applied subsequently to a number of different species. The technique allows the number of embryos to be amplified, an important consideration in a species that produces few eggs.

2.4. Fertilization *In Vitro*

The successful achievement of fertilization *in vitro* in the human owes a debt to pioneering animal studies; for a review of the technical aspects of these procedures see Gwatkin (1977) and Rogers (1978). Two studies are of particular relevance to human IVF. The report of Fraser and Drury (1975), which showed that there is a dose-response relationship between the incidence of polyspermy and sperm concentration in the insemination medium, has had an important impact on human programs. A major concern of IVF programs is that each oocyte should be penetrated by a single spermatozoon, and it is mandatory that fertilization be monitored for evidence of two and not more than two pronuclei. Thus, the report of Fraser and Drury (1975), on the critical importance for monospermic penetration of the concentration of spermatozoa in the insemination medium, has been translated into generally accepted norms for sperm concentrations for IVF, and has proved exceedingly useful. Thadani (1982) studied the effects of sperm concentration on *in vitro* fertilization in the mouse, and showed that live young could be produced at very low sperm:egg ratios. These data have been confirmed by successful fertilization of human oocytes *in vitro*, using low numbers of sperm from oligospermic men.

3. PRESENT STATUS AND PROSPECTS OF HUMAN IVF-ET

IVF programs began as a treatment for female infertility. However, as greater experience has been gained with the techniques, it has become clear that male factor infertility is an indication for IVF-ET when other conventional treatments have proven unsuccessful. Indeed, the application of these procedures has been shown to be as successful in terms of pregnancy as when IVF-ET has been applied to women with tubal infertility (Hewitt *et al.*, 1985). Recent reports indicate that the use of IVF procedures has successfully circumvented the problems of oligospermia, asthenospermia, autoimmunity and hostile cervical mucus (Cohen *et al.*, 1986; Hewitt *et al.*, 1985; Van Uem *et al.*, 1985).

Although fertilization rates *in vitro* are of the order of 85% in many programs, and thus are probably close to the *in vivo* situation, a distressing aspect of human IVF-ET programs is their apparently low success rate as judged by the ultimate criterion, that of a live birth. The most difficult issue facing personnel involved with IVF-ET is the likelihood of achieving a pregnancy following the replacement of one or more embryos. Although delivery rates appear to be depressingly low as seen from the vantage point of the prospective recipient, in fact, they are not too different from those recorded for *in vivo* fertilization. In a recent review (Kline and Stein, 1985), the authors have assembled data from studies carried out during the past 3 decades, and have concluded that a realistic figure for human reproductive failure between the period of fertilization and implantation is probably of the order of 75%, although this estimate may be on the low side. Therefore it is of particular interest to note that some of those clinics which have sufficiently large numbers of patients to enable statistically significantly conclusions to be drawn, report pregnancy rates that closely approximate the *in vivo* estimates. Given the high fertilization rates that have been recorded for human IVF, it may be possible, in the future, to maximize pregnancy rates through a better understanding of the uterine factors that predispose toward successful implantation. Experimental studies in a species such as the pig, which is known to have a high rate of peri-implantation mortality (Hunter, 1980), could have important consequences for embryo transfer procedures in humans.

A more speculative aspect of IVF-ET is the possibility of predicting the sex of the offspring before reimplantation of the embryos into the maternal environment. Although of obvious commercial value in terms of cattle, sheep or pigs, sex determination is of potentially enormous import in developing or third world societies, where the sex of a child may determine how many offspring are ultimately produced. The first steps towards this kind of sex determination have already been achieved by Singh and Jones (1982), who used a cDNA probe to identify the sex-determining region on the Y chromosome from a variety of animals. Subsequent studies have shown that this probe can be used to determine the sex of mouse blastocysts (Edwards, 1985), paving the way for similar studies on human embryos in culture. Unfortunately, the probe used in these studies does not bind to the human Y chromosome. Thus, methods for determining the sex of human preimplantation embryos must await the development of the appropriate probes. The use of probes such as that described above has significance for the detection of genetic diseases; homozygous affected, and heterozygous carriers, could be identified at very early stages of development, and only the homozygous normal embryos replaced. In this way, a mutant gene could be removed from the gene pool.

4. CONCLUSIONS

I would like to reiterate that enormous advances in the understanding of basic developmental processes have taken place during the past 25 years. These have been due largely to the development of techniques that have allowed the stages of early mammalian development to be directly observed and manipulated *in vitro*. The results of studies on animals such as the mouse have provided biologists and clinicians with tools that can be applied directly to human problems. *In vitro* fertilization is a recognized treatment for infertility; in addition, it has the potential to allow techniques such as gene therapy to be applied in the human.

5. REFERENCES

Bavister, B.D., 1981, Analysis of culture media for *in vitro* fertilization and criteria for success, in: *Fertilization and Embryonic Development in vitro* (L. Mastroianni, Jr., and J.D. Biggers, eds.), Plenum Press, New York, pp. 41-60.

Brinster, R.L., 1972, Cultivation of the mammalian egg, in: *Growth, Nutrition and Metabolism of Cells in Culture*, Volume II (G. Rothblat, and V. Cristofalo, eds.), Academic Press, New York, pp. 251-286.

Cohen, J., Edwards, R.G., Fehilly, C., Fishel, S., Hewitt, J., Purdy, J., Rowland, G., Steptoe, P., and Webster, J., 1985, *In vitro* fertilization: a treatment for male infertility, *Fertil. Steril.* 43: 422-432.

Dandekar, P.V., Spindle, A.I., Martin, M.C., and Glass, R.H., 1985, Development of mouse embryos in five different media, *Fertil. Steril.* 41: 61S (abs. no. 142).

Edwards, R.G., 1985, *In vitro* fertilization and embryo replacement: opening lecture, *Ann N.Y. Acad. Sci.* 442: 1-22.

Fraser, L.R., and Drury, L.M., 1975, The relationship between sperm concentration and fertilization *in vitro* of mouse eggs, *Biol. Reprod.* 13: 513-518.

Gwatkin, R.B.L., 1977, *Fertilization Mechanisms in Man and Mammals*, Plenum Press, New York, pp. 33-51.

Heape, W., 1890, Preliminary note on the transplantation and growth of mammalian ova within a uterine foster-mother, *Proc. Roy. Soc. (London)* 48: 457-458.

Hewitt, J., Cohen, J., Krishnaswamy, V., Fehilly, C.B., Steptoe, P.C., and Walters, D.E., 1985, Treatment of idiopathic infertility, cervical mucus hostility, and male infertility: artificial insemination with husband's semen or *in vitro* fertilization? *Fertil. Steril.* 44: 350-355.

Hunter, R.H.F., 1980, *Physiology and Technology of Reproduction in Female Domestic Animals*, Academic Press, London, pp. 214-218.

Kline, J., and Stein, Z., 1985, Very early pregnancy, in: *Reproductive Toxicology* (R.L. Dixon, ed.), Raven Press, New York, pp. 251-265.

Morcos, R.N., Gibbons, W.E., and Findley, W.E., 1985, Effect of peritoneal fluid on *in vitro* cleavage of 2-cell mouse embryos: possible role in infertility associated with endometriosis, *Fertil. Steril.* 44: 678-683.

Quinn, P., Kerin, J.F., and Warnes, G.M., 1985a, Improved pregnancy rate in human *in vitro* fertilization with the use of a medium based on the composition of human tubal fluid, *Fertil. Steril.* 44: 493-498.

Quinn, P., Warnes, G.M., Kerin, J.F., and Kirby, C., 1985b, Culture factors affecting the success rate of *in vitro* fertilization and embryo transfer, *Ann N.Y. Acad. Sci.* 44: 195-204.

Rall, W.F., and Fahy, G.M., 1985, Ice-free cryopreservation of mouse embryos at -196° C by vitrification, *Nature (London)* 313: 573-575.

Rogers, B.J., 1978, Mammalian sperm capacitation and fertilization *in vitro*: A critique of methodology, *Gamete Res.* 1: 165-223.

Singh, L., and Jones, K.W., 1982, Sex reversal in the mouse *(Mus musculus)* is caused by a recurrent reciprocal crossover involving the X and an aberrant Y chromosome, *Cell* 28: 205-216.

Smith, A.U., 1953, *In vitro* experiments with rabbit eggs, in: *Mammalian Germ Cells* (Ciba Foundation Symposium) (G.E.W. Wolstenholme, ed.), Churchill, London, pp. 217-232.

Tarkowski, A.K., and Wroblewska, J., 1967, Development of blastomeres of mouse eggs isolated at the 4- and 8-cell stage, *J. Embryol. Exp. Morphol.* 18: 155-180.

Thadani, V.M., 1982, Mice produced from eggs fertilized at a very low sperm:egg ratio, *J. Exp. Zool.* 219: 277-283.

Van Uem, J.F.H.M., Acosta, A.A., Mayer, J., Ackerman, S., Burkman, L.J., Veek, L., McDowell, J.S., Bernardus, R.E., and Jones, H.W., 1985, Male factor evaluation in *in vitro* fertilization: Norfolk experience, *Fertil. Steril.* 44: 375-383.

Whittingham, D.G., 1971, Survival of mouse embryos after freezing and thawing, *Nature (London)* 233: 125-126.

Whittingham, D.G., 1977, Fertilization *in vitro* and development to term of unfertilized mouse oocytes previously stored at -196° C, *J. Reprod. Fertil.* 49: 89-94.

Whittingham, D.G., Leibo, S.P., and Mazur, P., 1972, Survival of mouse embryos frozen to -196° C and -269° C, *Science* 178: 411-414.

Wilmut, I., 1972, Effect of cooling rate, warming rate, cryoprotective agent and stage of development on survival of mouse embryos during cooling and thawing, *Life Sci.* 11: 1071-1079.

APPENDIX I

BARRY D. BAVISTER

1. INTRODUCTION

As mentioned in several chapters in this book, our ability to sustain normal embryo growth *in vitro* is far from satisfactory, except in a very few specific instances. In this Appendix, the Editor outlines some of the technical problems relating to embryo culture, particularly those problems that may contribute to poor development of embryos *in vitro*, which was discussed in the Preface. Most of these problem areas have been cited by one or more of the authors in this book. Clearly, the major problem is the unsuitability of most presently-available culture media for supporting normal embryonic development. The vast literature on the culture of somatic cells may well contain information that will help us to design more appropriate media for embryos, for there must be some points of similarity between these cell types. However, we should tread warily along this comparative path, since preimplantation embryos have evolved in a highly specialized environment and undoubtedly possess some unusual metabolic and nutritional quirks, such as the well-known inability of early cleavage stage mouse embryos to utilize the glycolytic pathway (see Chapters 1 and 12). Moreover, mammalian embryos, especially the cleavage stages, are delicate creatures that are sensitive to insult. If we focus our attention as much as possible on the analysis of end-points for development, including biochemical, physiological, genetic and morphological criteria, we may obtain useful clues about the design of appropriate culture media for obtaining normal development. In describing how to design culture media for refractory cells, Ham (1982) said: "*A major key to progress is to forget what we think we know about cellular growth*

Barry D. Bavister Department of Veterinary Science, University of Wisconsin, Madison, Wisconsin 53706 USA.

requirements and instead to ask the cells we are studying to tell us at every step exactly what conditions they consider to be optimal. The cells know far better than we do what their actual requirements are, and to the extent that we try to impose our wills on them instead of listening to their needs, they will resist our efforts". Hopefully, the information contained in the preceding chapters, and perhaps some of the material discussed below, will help us to become better listeners.

2. CONDITIONS FOR EMBRYO CULTURE

A wide variety of culture media, many of which are available from commercial sources, have been designed for supporting growth and function *in vitro* of many different kinds of somatic cells. In contrast, relatively few media are available that will support normal-looking growth of preimplantation embryos from early cleavage stages to blastocyst, and most often, normal embryonic function is not maintained (see Chapter 1 and Chapters 8 through 13). In order to design better culture environments for mammalian early embryos, we urgently need information on the conditions prevailing *in vivo*, especially within the oviduct, and on the metabolism of embryos placed in an *in vitro* environment (see Chapters 1, 5, 6 and 11 through 13). As with culture of somatic cells, progress in understanding the needs of embryonic cells in culture will undoubtedly be aided by the development of serum-free media, as described in Chapter 8, so that effects of (*e.g.*) growth factors on embryogenesis (see Chapters 9 and 10) can be analysed in detail. A large amount of useful information about serum-free media is contained in a book edited by Mather (1985).

2.1. Culture Media and Components

Simple or Complex Media? Simple culture media are so-called "balanced" salt solutions, usually supplemented with energy substrates and protein (BSA). Complex media are mostly designed for somatic cell culture and usually contain amino acids, vitamins, trace metals and a variety of "nutrient" substances; these media are most often used in conjunction with blood serum. Standard formulations of both types are sub-optimal (or downright unsuitable) for embryo culture; the "ideal" recipe probably lies somewhere between these two approaches. A problem common to both types is that their ionic composition is based on that of blood serum, which differs substantially in some respects from the ions found in female reproductive tract fluids (Bavister, 1981). When culture media are formulated using the oviductal ionic composition as a model (Tervit *et al.*, 1972; Quinn *et al.*, 1985), significant improvements in embryo development or pregnancy rates may be achieved. Adjustment of ionic composition is not the total answer to problems with embryo culture, however; early stage embryos may require growth factors such as those described in Chapters 8, 9 and 10. Moreover, substantial changes in the ionic composition of the culture medium failed to overcome blocks to development in hamster embryos (Chapter 11). Some other components of the *in vitro* environment that most likely need to be investigated are outlined in the following sections. A very informative outline of the main features and benefits of culture media and supplements is given by Ham and McKeehan (1979).

There may be a relationship between refractoriness (partial or total) to culture conditions *in vitro* and the stage at which embryos normally enter the uterus. In general, "uterine" stages of embryo development survive and grow in culture more readily than "oviductal" stages of the same species. For example, mouse embryos are easily cultured from the 8-cell (uterine) stage, when they first become capable of utilizing glucose (Whitten, 1957; Brinster and Thomson, 1966), but earlier (cleavage) stages will not develop *in vitro* without pyruvate, lactate or similar substrates (Biggers *et al.*, 1967; and see Chapter 1). Hamster (and rat) embryos enter the uterus at the 8-cell stage; hamster embryos can only be grown *in vitro*, albeit with some difficulty, from this stage onward and also show quite wide tolerance to different culture conditions (Chapter 11). In addition, hamster embryos undergo a major change in their dependence on amino acids between the early and late 8-cell stages (Bavister *et al.*, 1983a). Cattle embryos, which enter the uterus at about the 8- to 16-cell stage, usually block at just this stage *in vitro* but develop relatively easily in culture after this stage (Chapters 9 and 12). Similarly, rhesus monkey embryos can develop normally up to the 8- to 16-cell stage in a simple culture medium containing a few amino acids, but growth beyond this point seems to require much more complex conditions (Bavister *et al.*, 1983b; Boatman *et al.*, 1987; and see Chapter 13). Rabbit embryos do not show a block to development in culture, but there is evidence of a transition in their nutritional requirements for growth *in vitro* at the blastocyst stage, which is when rabbit embryos enter the uterus (Chapter 10). Furthermore, rabbit embryos undergo substantial changes in their amino acid transport characteristics (Miller and Schultz, 1983) and glucose metabolism pathways (Fridhandler, 1961) between the morula and blastocyst stages.

It would appear that growth requirements for oviductal stage embryos are relatively simple but somewhat restricted, whereas additional factors (or substantially modified conditions) are needed (in most species) for development of uterine stages. The more advanced embryos (uterine stages) seem to show wider tolerance of culture conditions (*e.g.*, nutrients, energy substrates) and appear to be more like somatic cells in this respect. This generalization holds true in the rabbit, the only species other than the mouse in which culture requirements for embryos have been systematically examined (Chapters 1 and 10). In view of this, we should perhaps consider whether simple culture media (probably modeled on oviduct fluid composition) should be used preferentially to support growth of oviductal stage embryos *in vitro*, while complex media are more appropriate for uterine stages.

Water Quality. Water is the major component of culture media and is the most likely source of contaminants that can block or retard embryo growth *in vitro*. Most investigators have learned by rigorous training (or by bitter experience!) that one cannot be too careful about the treatment of water for preparing culture media. Several authors emphasized the importance of using ultrapure water (see Chapters 8, 10, 11 and 13), although Kane (Chapter 10) pointed out that the degree of water treatment required depends somewhat on the amount of protein included in the culture medium. Various methods for preparing pure water can be used, including multiple distillation in glass and the new cartridge-filter types of water purifiers such as "reverse osmosis" treatments. Treatment of water first by reverse osmosis, then by Milli-Q® filtration produces water with minimal contaminants (Giles *et al.*, 1985) that is suitable for critical culture work (Bavister and Andrews, 1987).

A crucial point is that the conductivity meter provided on most filter-type systems is only a very rough guide to water purity: if conductivity (actually, resistance) is 18 megOhms or more, then the system is working satisfactorily, but the treated water may well require further processing before it is suitable for use in culture media. Dissolved organic and other non-ionic substances will not register on the conductivity meter, and may be present in minute amounts, yet they can cause embryotoxic effects. Endotoxins are found in a wide variety of culture medium ingredients and are a common contaminant in filtration-type water purifiers (Anon., 1984).

The obvious question is: how do we know when the water is pure enough? The answer is: when it works! Since we usually don't want to waste valuable embryos for testing the quality of water and other ingredients in our culture media, some type of bioassay is desirable that uses expendable embryos or non-embryonic cells. Many human *in vitro* fertilization laboratories routinely use a mouse 2-cell embryo culture test for this purpose (for details, see Ackerman *et al.*, 1984). This bioassay method has been criticized for a number of reasons, the main one being that it is relatively insensitive because inbred mouse 2-cell embryos are very tolerant of culture conditions. In one study, mouse 2-cell embryos grew to the blastocyst stage equally well in Ham's F-10 prepared with distilled water or tap water (Silverman *et al.*, 1987). A more stringent test is the growth of 1-cell mouse embryos (Quinn *et al.*, 1985). In either case, the test takes several days to perform and can be quite expensive because of the cost of maintaining a mouse colony. In our laboratory, we routinely use a bioassay based on the sensitivity of hamster spermatozoa to contaminants in the culture medium (Bavister and Andrews, 1987). Essentially, hamster sperm are incubated in the culture solution being tested and their motility (% and quality) is estimated over a 6 hr period. A "sperm motility index" value is calculated for semi-quantitative comparison of different culture treatments. This bioassay is relatively rapid and inexpensive, and has been used to detect contaminants in water, culture dishes, disposable syringe filters and even in the syringes themselves. The ultimate test, of course, is the ability of a batch of culture medium to support growth of the particular type of embryo under investigation.

Oil Overlays. The use of a layer of oil to cover small drops of culture medium reduces evaporation and consequent increases in osmotic pressure. Additionally, by retarding gas exchange, the oil overlay helps to maintain the pH of the medium during brief periods when the culture dish is removed from the incubator. This practice is widespread; however, care must be exercised to minimize the risk of contamination of the culture medium with impurities from the oil. One solution is to use silicone oil (we use oil from Aldrich Chemical Co., cat. no. 14615-3), which is conveniently sterilized by vacuum aspiration through a Nalgene filter (Nalge Co., cat. no. 4500020, 0.2 μm). The oil is then equilibrated for several hours with 157 mM NaCl solution (prepared with culture-grade water). Alternatively, mineral oil, which is much less expensive, can be used but usually needs more thorough washing. This may be conveniently done using a separating funnel as described in Chapter 11. Contaminants in oil overlays are often manifested by changes in the profile of culture drops after prolonged incubation (see Chapter 13).

Energy Substrates. The availability of energy substrates to cultured cells deserves some attention as a potential source of problems. We know that 1- to

4-cell mouse embryos cannot use glucose, at least as their sole energy substrate, but require pyruvate, oxaloacetate or lactate (Biggers *et al.*, 1967; and see Chapter 1). Apart from studies with rabbit embryos (see Chapters 1 and 10), there have been few thorough investigations of energy substrate requirements in embryos of other species. It may be that early cleavage stage embryos of some animals are even more restrictive than mouse embryos in this respect. Lack of energy substrate could account for the very rapid cessation of development seen in cultured hamster 2-cell embryos (Chapter 11). It is possible that pyruvate and/or lactate are not provided in appropriate concentrations to be utilized adequately by early cleavage stage embryos of some species. Some imbalance in the culture environment might reduce the availability of energy substrates to embryos; an example of this is the discovery by Brinster (1965) that the pH of the culture medium has a pronounced effect on the development of mouse embryos *in vitro*, due to changes in the amount of available pyruvic acid (non-ionized form). Davis and Day (1978) reported that *in vitro* development of 4-cell pig embryos was actually inhibited by pyruvate.

An intriguing possibility is that early mammalian embryos are especially sensitive to the "Crabtree effect", in which glucose exerts a repressive effect on oxidative phosphorylation (Crabtree, 1929). Since early stage embryos (at least in the mouse) cannot metabolize glucose, this situation could deprive them of a usable energy source. This may be why other metabolic pathways, such as the pentose phosphate pathway in mouse embryos (see Chapter 12), are apparently so important. The Crabtree effect is amplified (or is actually caused) by low availability of inorganic phosphate (Koobs, 1972); this could be very significant in the case of the hamster, since in our laboratory hamster embryos have routinely been cultured in modified Tyrode's solution, which is very low in phosphate and high in glucose (Chapter 11).

Protein. Exogenous proteins *may* play several roles in supporting preimplantation embryo development *in vitro*. There may be *specific* proteins or peptides, such as growth factors (see Chapter 8) and factors of embryonic origin (Chapter 9), that are needed to regulate particular developmental events. Inclusion of semi-defined protein preparations, usually bovine serum albumin (BSA), in the culture medium may be helpful, but there is little, if any, evidence that the presence of BSA or other proteins in the culture medium is required for embryo growth, at least up to the morula stage. A requirement for BSA to support growth of rabbit embryos was found to be due to an unidentified, protein-bound growth factor of low mol. wt. (Kane, 1985). BSA also carries fatty acids, which can serve as energy substrates for growth of rabbit embryos (Kane, 1979) [see also Chapter 10]. Fatty acids can also stimulate or inhibit growth of hamster cells *in vitro* (Nilausen, 1978). The ability of BSA to support growth of mouse 1- and 2-cell embryos to the blastocyst stage was virtually abolished by extraction with trichloroacetic acid (Quinn and Whittingham, 1982). Wiley *et al.* (1986) found that using a mixture of BSA and bovine serum immunoglobulins enhanced *in vitro* development of mouse 2-cell embryos over that obtained with BSA alone. The cell number in cultured embryos was dependent on the protein source that was used. The potential value of albumin as a chelating agent has been pointed out (Chapter 10), but for this effect, amino acids can usually be substituted for protein. A synthetic polymer (PVA) can be used in place of BSA for growth of hamster embryos from the 8-cell to blastocyst stages, although viability of the cultured embryos was not assessed (Carney, unpublished data). In some strains of mice, exogenous protein is not required for growth of embryos from the 2-cell to blastocyst stages (Brinster, 1968; Cholewa and Whitten, 1970; Kuzan *et al.*, 1982), or for blastocysts to hatch from the zona pellucida (Spindle and Pedersen, 1973). In one study, development of fetuses after blastocyst transfer was not impaired by having grown the embryos in protein-free medium (Caro and Trounson, 1984). There may be negative effects of using protein preparations for embryo culture: some of these preparations can cause reduction of embryo development *in vitro* (Caro and Trounson, 1984). Alternatively, exogenous proteins may simply lack any growth-promoting properties. In a study involving large numbers of human IVF patients, Caro and Trounson (1986) found no difference in oocyte fertilization, early embryo development or pregnancy rates whether the culture medium (T6) contained maternal serum (10%) or no protein at all.

There appears to be little, if any, justification for using blood serum as a supplement in culture media for growth of embryos up to about the 8-cell stage; during this phase of development, the requirements for exogenous nutrients seem to be minimal. Provided that some potentially important amino acids are included in the medium, serum is unnecessary for normal growth of cleavage stages of rhesus embryos (Bavister *et al.*, 1983b, 1984). There is also little, if any, scientific rationale for the use of human fetal cord serum (HCS) for culture of *cleavage* stage embryos of any species: this material is highly variable in quality and frequently embryotoxic (Shirley *et al.*, 1985). Moreover, there is no evidence showing any beneficial effect of HCS on these early stages of development (however, Leung *et al.* [1984] reported a significantly higher pregnancy rate after transfer of human IVF embryos that were cultured in medium with HCS *vs.* maternal serum). The general rationale for using HCS stems from its stimulatory effects on *post-hatching* mouse embryos (Hsu, 1979), which are obviously not comparable to cleavage stage embryos in their nutritional needs and growth characteristics.

In summary, there is no conclusive case to be made for including or for omitting protein in the preparation of media for the culture of preimplantation embryos; rather, there may be some specific needs that depend on the species of embryos and on their stage of development, and perhaps on the purity of other ingredients of the medium.

Amino Acids and Vitamins. A requirement for specific amino acids in the development of rabbit and hamster preimplantation embryos (8-cell to blastocyst stages) has been demonstrated (Chapters 1, 10 and 11), and a beneficial effect of amino acids for growth of rhesus monkey embryos is inferred (Chapter 13). However, the presence of certain amino acids appears to *reduce* growth of hamster embryos (Chapter 11). More information is needed on: the specific amino acid requirements of embryos from different species, the stage(s) of embryo development at which exogenous amino acids become needed, their optimal concentrations and the relative beneficial/detrimental effects of different amino acids. In addition, the functional role(s), *e.g.*, metabolism, of the required amino acids needs to be elucidated. Several amino acids are present in eggs and embryos in much greater concentrations than would appear to be necessary for protein synthesis or for energy production (Petzoldt *et al.*, 1973; Schultz *et al.*, 1981).

A specific requirement for vitamins in growth of embryos up to the beginning of the blastocyst stage has not been shown (Chapters 10, 12). However, rabbit and hamster embryos need vitamins for blastocyst expansion and for hatching from the zona pellucida (Chapters 10, 11). Investigations of the role of vitamins in blastocyst development need to be performed, similar to those referred to above for amino acids.

A final point is that studies on the roles of amino acids and vitamins, or indeed on any other potential growth-stimulating factors, should include embryo viability assessment by embryo transfer, otherwise subtle effects on embryos may not be detected (see Sections 4 and 5).

Growth Factors. Studies with mouse, cattle and rabbit embryos have demonstrated directly or indirectly that growth factors can play an important role in embryo development *in vitro* (Chapters 8, 9 and 10). This avenue of research promises to be very rewarding, based on results obtained to date. Two different approaches are suggested. The first is the identification of growth-stimulating factors that are produced by the embryos themselves (auto-regulation?) and/or that are found in female tract secretions or other natural sites (*e.g.*, those associated with BSA: Chapter 10). The second approach is the examination of possible regulatory effects of "known" growth factors, such as EGF, PDGF, TGF, *etc.* (Chapter 8). Of course, these two approaches are mutually compatible and will probably overlap considerably.

Hormones. As pointed out by Kane (Chapter 10), it is remarkable that no stimulatory effect of any hormone on *in vitro* growth of preimplantation embryos has yet been reported, in spite of the abundance and variety of hormones that must be present in the female tract secretions during embryo development. Evidence obtained with embryos of the pig (Chapter 6) and other species shows a regulatory role for estradiol during differentiation from the morula to the blastocyst stage, but there is no evidence for a direct effect of *extra*-embryonic estradiol on development. However, estradiol can stimulate uptake of amino acids by mouse blastocysts (Smith and Smith, 1971). Since embryos of some species have been found to metabolize steroids (Wu and Matsumoto, 1984, 1985), perhaps endogenous estradiol acting as an auto-regulator of embryo growth/differentiation is of primary importance, as suggested by Sengupta *et al.* (1977). It is quite possible that effects of *extra*-embryonic steroids, and other hormones, may be masked by growth-limiting effects of other factors in the culture *milieu.*

2.2. Physico-Chemical Factors

Gas Phase. The composition of the gas mixture used for embryo culture is critically important since it plays a major regulatory role in metabolism and pH control (Chapter 12). There has long been a tendency to use the "standard" 5% CO_2 in air mixture for embryo culture because (a) it works (for some species) and (b) it is difficult to demonstrate the superiority of other gas mixtures (with a very few notable exceptions). However, there is ample evidence that the role of the gas phase composition needs re-evaluation.

a. carbon dioxide. In one case, the % CO_2 was shown to be a potent regulator of embryo growth *in vitro*; we suggested that this is due to changes in intracellular pH (Chapter 11) [see also section on *pH*]. Earlier studies have shown that CO_2 can have substantial stimulatory effects on metabolism, including fatty acid synthesis and ATP-ase activity (Hastings and Longmore, 1965; Fanestil *et al.*, 1963; McLimans, 1972). Several studies have indicated that embryo growth *in vitro* is improved by placing a number of embryos together in one culture drop (Wright *et al.*, 1978; Wiley *et al.*, 1986; and see Chapter 10), and it is possible that a localized increase in CO_2 might account in part for this effect. Similarly, part of the stimulatory effect of "feeder cell" monolayers on embryo growth might be due to CO_2 production [see also discussion of O_2]. Such effects would be enhanced when an oil overlay is used in the culture system. From the viewpoint of adverse effects on embryo culture, the possibility of contamination of bottled CO_2 by carbon monoxide should also not be overlooked (McLimans, 1972).

b. oxygen. Although a few studies have shown a beneficial effect of lowered pO_2 on embryo growth *in vitro* (see Chapter 12), reduction of pO_2 in most cases has not produced any significant increase in development. One reason for this lack of effect may be the insensitivity of commonly-used end-points for evaluating embryo growth (Chapters 10, 11; and see Section 5). Numerous publications attest to the detrimental effect of atmospheric oxygen on cells in culture (*e.g.*, Richter *et al.*, 1972; Packer and Fuehr, 1977; Bradley *et al.*, 1978; Held and Sönnichsen, 1984). In some studies, pO_2 has been shown to exert a regulatory effect on cell function in culture (*e.g.*, Beckman *et al.*, 1982; Knighton *et al.*, 1983). The use of atmospheric O_2 levels (about 160 mm Hg partial pressure) is untenable when attempting to mimic normal conditions, since the highest physiological pO_2 (in systemic arterial blood) is only about 95 mm Hg. Moreover, most body tissues are in equilibrium with a pO_2 of about 40 mm Hg or less, in some locations perhaps as low as 2 to 5 mm Hg, depending on the rate of cell metabolism (Grant and Smith, 1963; Balin *et al.*, 1984). In practical terms, a gas phase for culture containing about 5% O_2 or less is probably closer to the natural situation experienced by most cells *in vivo*. Even lower values may be more appropriate for preimplantation embryos *in vivo*, since they develop in an environment that is somewhat distant from blood capillaries, and they are probably adapted to relatively hypoxic conditions.

Judging from the substantial literature demonstrating a regulatory effect of pO_2 on growth and function of somatic cells *in vitro*, it would be surprising if (eventually) we are not able to find an important role for O_2 in controlling embryo growth. *In vitro* maturation of hamster oocytes is blocked by 21% O_2 (Gwatkin and Haidri, 1973). Mouse 1- and 2-cell embryos are particularly sensitive to atmospheric pO_2; about 5% O_2 is optimal for their development *in vitro* (Whitten, 1971; Quinn and Harlow, 1978). Exposure of mouse embryonic (fetal) cells to atmospheric O_2 caused an increased frequency of chromosomal abnormalities (Parshad and Sanford, 1971). Reference has already been made (see section on *Energy Substrates*) to the possible contribution of high pO_2 in the Crabtree effect, which may underlie some of the difficulties with embryo culture. Reduction of local pO_2 could be one role of somatic cell monolayer cultures in facilitating embryo growth *in vitro*; such an effect could help explain the lack of substrate cell specificity that has been observed (Chapter 10).

Extrapolation of data obtained with somatic cell cultures to the preimplantation embryo culture situation must be approached very cautiously because of differences in cell requirements and in culture parameters. For example, data on somatic cells are usually obtained with dense (or confluent) cultures, which will obviously affect their responses to changes in gas conditions, *etc.*, to a different extent than in the case of a few embryonic cells. It should be mentioned that there is a possibility of toxic hyperoxia when preimplantation embryos are incubated in plastic culture dishes. Many of the plastics used in the manufacture of these dishes avidly bind oxygen and will release it into the culture medium, possibly in toxic amounts (Chapman and Sturrock, 1969). This effect may be particularly important if the gas pO_2 is already on the high side for supporting embryo growth. An obvious way to counteract some of the detrimental effects of hyperoxia on cultured cells is to incorporate an anti-oxidant into the medium; perhaps this should become a standard ingredient of the medium formulation.

In summary, although a regulatory effect of O_2 has not been found for embryos of most species, as culture conditions are improved it would seem prudent to consider the possibility that pO_2 may become a rate-limiting factor in the growth and differentiation of preimplantation stages of development.

Temperature and Light. Embryos may be adversely affected during handling by physical trauma, such as lowered temperature and short wavelength light, which would frustrate subsequent attempts to grow them in culture. Embryos of the mouse, which are the most widely studied embryos, do not appear to be very susceptible to brief exposure to room temperature. This may not be true of embryos from other species, but little data are available on this topic. However, sheep oocytes are known to be permanently damaged by briefly lowering the temperature to around 26°C and there was a striking reduction of blastocyst formation *in vivo* following this exposure *in vitro* (Moor and Crosby, 1985). We might also take care that the microscope stage on which embryos are examined does not become overheated in case embryos are exposed to elevated temperatures for even brief periods of time, for this is likely to be lethal. The bulbs used in some stereo microscopes can produce stage temperatures in excess of 40°C.

Damaging effects of short wavelength light, *e.g.*, from fluorescent lights, on early embryo development have been documented. Hirao and Yanagimachi (1978) showed that brief exposure of hamster oocytes to light from fluorescent lamps was detrimental to fertilization; Daniel (1964) showed that development of rabbit embryos was adversely affected by short wavelength light. However, another study showed that exposure of 2-cell mouse embryos to fluorescent light for 30 min did not affect their ability to develop *in vitro* (Kruger and Stander, 1985). It is not widely known that brief exposure of *culture media* to short-wavelength light can not only cause deterioration of the medium but also induce formation of cytotoxic products. For example, peroxides are produced by photo-oxidation of tyrosine and tryptophan; such oxidation reactions can be reduced or eliminated by adopting a variety of precautionary procedures (MacMichael, 1986). HEPES buffer is usually the buffer of choice for embryo handling media because of its low toxicity; however, exposure of HEPES-buffered medium (RPMI 1640) to daylight caused it to become cytotoxic, the effect being directly proportional to the concentration of HEPES (Spierenburg *et al.*, 1984). In view of the widespread use of fluorescent lights to illuminate laboratories and culture rooms, it seems prudent to cover the lights with red filters (Hirao and Yanagimachi, 1978) or to use other types of lighting.

pH. Most studies on the effects of pH on embryo growth indicate that embryos are relatively insensitive to changes in pH of the culture medium, *i.e.*, roughly pH 6 to 8 (Chapters 10-12). However, this may reflect a relative impermeability of embryonic cell membranes to H^+; modification of the culture medium with permeant acids may give a quite different result (Chapter 11). Virtually no attention has been given to the possibility of changes in pH_i in cultured embryos as a possible contributing factor to poor growth and development. In some organisms, changes in pH_i mediated by the external *milieu* are major regulators of metabolism (Hand and Carpenter, 1986).

3. EQUIPMENT

A detailed description and critique of equipment used in embryo culture experiments would require a whole book in itself. Two major items of equipment that deserve brief mention, however, are incubators and microscopes.

3.1. Incubators

The incubator is critically important for providing constant temperature and gas environment. Typical laboratory incubators are designed for large-scale culture of cells; as a result, they can have some shortcomings for the culture of preimplantation embryos. The standard 5 to 6 cu. ft. (142 to 170 liter) incubators are cumbersome. They regulate CO_2, temperature, humidity and (sometimes) O_2. In terms of gas control, incubators usually operate in one of 3 modes: continuous gas flow, automatic (CO_2 sensor) and gas processor (cyclic gas injector). While good quality models of all 3 types will maintain the internal environment at reasonably constant levels if left undisturbed, the various types and brands may vary considerably in their ability to re-establish equilibrium after each door opening/closing sequence. This problem is considerably worsened if frequent opening of the incubator occurs for examination of embryo development. In some instances, it may take an hour or longer to re-establish normal pH in the culture medium. By this time, the viability of embryos may have been compromised. There are several ways to reduce or eliminate this problem. Automatic (CO_2 sensor) incubators usually have faster recovery times than other types and less tendency to overshoot the CO_2 set-point. Gas-processor incubators can be very prone to overshoot if the (manual) purge control is used at every (frequent) door open/close sequence. Using small gas-filled desiccators or similar chambers (such as the Flow Modular units) to isolate groups of dishes within the incubator will considerably reduce the detrimental effects of door-opening. It would be best if smaller capacity incubators than the standard large ones (*e.g.*, 1 cu. ft.) were available.

3.2. Microscopes

For rapid examination of the stage of development of embryos, a stereo ("dissecting") microscope is sufficient. However, examination with a compound microscope will give much more information about embryos, *e.g.*, the presence (and number) of nuclei in the blastomeres, the general cytoplasmic appearance and the presence of small irregularities or "blebs" (Chapter 13). An inverted microscope is advantageous here since considerable detail can be seen without removing embryos from the culture dish and embryo culture can be continued after the examination. Some models can be fitted with a "wrap-around" plastic chamber that can maintain a constant internal temperature, which helps prevent damage to embryos while under observation; with a little modification to this system (Bavister, 1987), the culture dish can be continuously gassed with CO_2 in air, allowing unlimited time for observation of the embryos (Chapter 11). Using only a stereo microscope for evaluation of embryos can be misleading because subtle indications of abnormal development may not be visible.

4. EXPERIMENTAL DESIGN

The importance of correct experimental design was emphasized by several authors (see Chapters 5, 10, 11 and 14). There is usually so much variability between embryos from different donors that results of embryo culture will be compromised unless all embryos from a single donor are distributed among all treatments (= 1 replicate). Part of this variability is due to embryos being at slightly different stages of development in different females, *q.v.* the differential developmental abilities of early *vs.* late 8-cell bovine and hamster embryos (Chapters 9 and 11). Failure to take these differences into account may well obscure small but meaningful differences in responses to culture treatments.

5. EVALUATION OF RESPONSES TO EMBRYO CULTURE

This topic was thoroughly discussed by Biggers *et al.* (1971). As mentioned in the Preface and in several chapters (8, 10, 11 and 13), cultured embryos tend to have fewer cells than corresponding-age embryos grown *in vivo*, longer cell division times, smaller (less inflated) blastocysts and reduced viability. Although the stage of development reached within a given period of time is a semi-quantitative end-point that can be rapidly and easily assessed, it is clear that this is not adequate by itself as a measure of the response to a treatment. Additional end-points need to be evaluated, primarily fetal development following embryo transfer (Chapters 1, 11, 13 and 14), counting of cell numbers (Chapters 10, 11, 14), measurement (preferably non-invasive) of metabolism (Chapters 5, 12) and cytogenetic parameters such as SCE-rates (Chapter 14). The need for more rigorous evaluation of responses is emphasized by studies (see Chapter 14) in which toxic agents produced no apparent detrimental effects in embryos cultured up to the blastocyst stage; however, symptoms appeared in these treated embryos after transfer.

6. BLOCKS TO DEVELOPMENT

The blocks to development *in vitro* shown by embryos of several species (cattle, pig, outbred mouse, rat, hamster: see Chapters 1, 9, 10, 11 and 12) remind us that we have much to learn about optimal culture conditions.

Elucidating the reason(s) for these blocks is one of the most challenging and potentially rewarding areas of research. While it is unlikely that a single cause will be found for the blocks in all species, there may be some common features that can help resolve the problem. For example, we may discover that hyperoxia or energy substrate availability are contributing factors that, once eliminated, pave the way for the true causes of the blocks to be identified. A renewed attack on these blocks, with a number of investigators using several species and a variety of approaches, may provide the long-overdue breakthroughs. Only after these blocks are eliminated will we be able to fully elucidate the metabolic pathways and the regulation of development in embryos from a substantial number of species, so that the similarities and differences in preimplantation embryogenesis can be more fully appreciated.

7. REFERENCES

Ackerman, S.G., Swanson, R.J., Stokes, G.K., and Veeck, L.L., 1984, Culture of mouse preimplantation embryos as a quality control assay for human *in vitro* fertilization, *Gamete Res.* 9: 145-152.

Anon., 1984, Endotoxin contamination: is it your water? in: *Art to Science in Tissue Culture*, Vol. 3, no. 2, Hyclone Laboratories, Logan, UT, pp. 4-6.

Balin, A.K., Fisher, A.J., and Carter, D.M., 1984, Oxygen modulates growth of human cells at physiologic partial pressures, *J. Exp. Med.* 160: 152-166.

Bavister, B.D., 1981, Analysis of optimal conditions for *in vitro* fertilization and criteria for success, in: *Fertilization and Embryonic Development In Vitro* (L. Mastroianni, and J.D. Biggers, eds.), Plenum Press, New York, pp. 41-60.

Bavister, B.D., 1987, A device for maintaining cultured cells in a constant CO_2 atmosphere on the stage of an inverted microscope, *In Vitro*, (submitted).

Bavister, B.D., and Andrews, J.C., 1987, A rapid sperm motility bioassay procedure for quality-control testing of water and culture media, *J. In Vitro Fertil. and Embryo Transfer* (submitted).

Bavister, B.D., Leibfried, M.L., and Leiberman, G., 1983a, Development of preimplantation embryos of the golden hamster in a defined culture medium, *Biol. Reprod.* 28: 235-247.

Bavister, B.D., Boatman, D.E., Leibfried, M.L., Loose, M., and Vernon, M.W., 1983b, Fertilization and cleavage of rhesus monkey oocytes *in vitro*, *Biol. Reprod.* 28: 983-999.

Bavister, B.D., Boatman, D.E., Collins, K., Dierschke, D.J., and Eisele, S.G., 1984, Birth of rhesus monkey infant following *in vitro* fertilization and non-surgical embryo transfer, *Proc. Natl. Acad. Sci. USA.* 81: 2218-2222.

Beckman, B., Belegu, R.D., Belegu, M., Katsuoka, Y., and Fisher, J.W., 1982, Hypoxic enhancement of murine erythroid colony formation, in: *Experimental Hematology Today* (S.J. Baum, G.D. Ledney, and S. Thierfelder, eds.), S. Karger, New York, NY, pp. 53-59.

Biggers, J.D., Whittingham, D.G., and Donahue, R.P., 1967, The pattern of energy metabolism in the mouse oocyte and zygote, *Proc. Natl. Acad. Sci. USA.* 58: 560-567.

Biggers, J.D., Whitten, W.K., and Whittingham, D.G., 1971, The culture of mouse embryos *in vitro*, in: *Methods in Mammalian Embryology* (J.C. Daniel, Jr., ed.) Freeman, San Francisco, pp. 86-116.

Boatman, D.E., Morgan, P.M., and Bavister, B.D., 1987, Culture of *in vitro* fertilized rhesus monkey oocytes up to peri-implantation stages of embryo development, *Biol. Reprod.* (submitted).

Bradley, T.R., Hodgson, G.S., and Rosendaal, M., 1978, The effect of oxygen tension on haemopoietic and fibroblast cell proliferation *in vitro*, *J. Cell Physiol.* 97: 517-522.

Brinster, R.L., 1965, Studies on the development of mouse embryos *in vitro*. II. The effect of energy source, *J. Exp. Zool.* 158: 59-68.

Brinster, R.L., 1968, Effect of glutathione on the development of two-cell mouse embryos *in vitro*, *J. Reprod. Fertil.* 17: 521-525.

Brinster, R.L., and Thomson, J.L., 1966, Development of eight-cell mouse embryos *in vitro*, *Exp. Cell Res.* 42: 308.

Caro, C.M., and Trounson, A., 1984, The effect of protein on preimplantation mouse embryo development *in vitro*, *J. In Vitro Fertil. and Embryo Transfer* 1: 183-187.

Caro, C.M., and Trounson, A., 1986, Successful fertilization, embryo development, and pregnancy in human *in vitro* fertilization (IVF) using a chemically defined culture medium containing no protein, *J. In Vitro Fertil. and Embryo Transfer* 3: 215-217.

Chapman, J.C., and Sturrock, J., 1969, The oxygen tension around mammalian cells growing on plastic Petri dishes and its effect on cell survival curves, *Brit. J. Radiol.* 42: 399.

Cholewa, J., and Whitten, W.K., 1970, Development of 2-cell mouse embryos in the absence of a fixed nitrogen source, *J. Reprod. Fertil.* 22: 553-555.

Crabtree, H.G., 1929, Observations on the carbohydrate metabolism of tumours, *Biochem. J.* 23: 536-545.

Daniel, J.C., 1964, Cleavage of mammalian ova inhibited by visible light, *Exp. Cell Res.* 29: 515.

Davis, D.L., and Day, B.N., 1978, Cleavage and blastocyst formation by pig eggs *in vitro*, *J. Anim. Sci.* 46: 1043-1053.

Fanestil, D.D., Hastings, A.B., and Mahowald, T.A., 1963, Environmental CO_2 stimulation of mitochondrial adenosine triphosphatase activity, *J. Biol. Chem.* 238: 836-842.

Fridhandler, L., 1961, Pathways of glucose metabolism in fertilized rabbit ova at various pre-implantation stages, *Exp. Cell Res.* 22: 303-316.

Giles, K., Gabler, R., and Wilkins, F., 1985, The effect of water purification methods on the viability and growth of cultured protoplasts, *Amer. Biotechnol. Laboratory* 3: 42-45.

Grant, J.L., and Smith, B., 1963, Bone marrow gas tensions, bone marrow blood flow, and erythropoiesis in man, *Ann. Intern. Med.* 58: 801-809.

Gwatkin, R.B.L., and Haidri, A.A., 1973, Requirements for the maturation of hamster oocytes *in vitro*, *Exp. Cell Res.* 76: 1-7.

Ham, R.G., 1982, Importance of the basal nutrient medium in the design of hormonally defined media, in: *Growth of Cells in Hormonally Defined Media*, Volume 9 (G.H. Sato, A.B. Pardee, and D.A. Sirbasku, eds.), Cold Spring Harbor Conferences on Cell Proliferation, Cold Spring Harbor Laboratory, pp. 39-60.

Ham, R.G., and McKeehan, W.L., 1979, Media and growth requirements, in: *Methods in Enzymology*, Volume LVIII (W.B. Jakoby, and I.H. Pastan, eds.), Academic Press, New York, pp. 44-93.

Hand, S.C., and Carpenter, J.F., 1986, pH-induced metabolic transitions in *Artemia* embryos mediated by a novel hysteretic trehalase, *Science* 232: 1535-1537.

Hastings, B.A., and Longmore, W.J., 1965, Carbon dioxide and pH as regulatory factors in metabolism, *Adv. Enzyme Reg.* 3: 147-159.

Held, K.R., and Sönnichsen, S., 1984, The effect of oxygen tension on colony formation and cell proliferation of amniotic fluid cells *in vitro*, *Prenatal Diagn.* 4: 171-179.

Hirao, Y. and Yanagimachi, R., 1978, Detrimental effect of visible light on meiosis of mammalian eggs *in vitro*, *J. Exp. Zool.* 206: 365-370.

Hsu, Y.C., 1979, *In vitro* development of individually cultured whole mouse embryos from blastocyst to early somite stage, *Dev. Biol.* 68: 453-461.

Kane, M.T., 1979, Fatty acids as energy sources for culture of one-cell rabbit ova to viable morulae, *Biol. Reprod.* 20: 323-332.

Kane, M.T., 1985, A low molecular weight extract of bovine serum albumin stimulates rabbit blastocyst cell division and expansion *in vitro*, *J. Reprod. Fertil.* 73: 147-150.

Knighton, D.R., Hunt, T.K., Scheuenstuhl, H., Halliday, B.J., Werb, Z., and Banda, M.J., 1983, Oxygen tension regulates the expression of angiogenesis factor by macrophages, *Science* 221: 1283-1285.

Koobs, D.H., 1972, Phosphate mediation of the Crabtree and Pasteur effects, *Science* 178: 127-133.

Kruger, T.F., and Stander, F.S.H., 1985, The effect of fluorescent light on the cleavage of two-cell mouse embryos, *S. Afr. Med. J.* 68: 744-745.

Kuzan, F.B., Pomeroy, K.O., and Seidel, G.E., 1982, Polyvinyl alcohol as a macro-molecular substitute for bovine serum albumin in mouse embryo culture medium, *Biol. Reprod.* 26: 65A.

Leung, P.C.S., Gronow, M.J., Kellow, G.N., Lopata, A., Speirs, A.L., McBain, J.C., duPlessis, Y.P., and Johnston, I., 1984, Serum supplement in human *in vitro* fertilization and embryo development, *Fertil. Steril.* 41: 36-39.

MacMichael, G., 1986, The adverse effects of UV and short-wavelength visible radiation on tissue culture, *Amer. Biotechnol. Laboratory* 4: 30-31.

Mather, J.P., (ed.), 1984, *Mammalian Cell Culture*, Plenum Press, New York.

McLimans, W.F., 1972, The gaseous environment of the mammalian cell in culture, in: *Growth, Nutrition and Metabolism of Cells in Culture*, Volume 1 (G.H. Rothblat, and V.J. Cristofalo, eds.), Academic Press, New York, pp. 137-170.

Miller, J.G.O., and Schultz, G.A., 1983, Properties of amino acid transport in preimplantation rabbit embryos, *J. Exp. Zool.* 228: 511-525.

Moor, R.M., and Crosby, I.M., 1985, Temperature-induced abnormalities in sheep oocytes during maturation, *J. Reprod. Fertil.* 75: 467-473.

Nilausen, K., 1978, Role of fatty acids in growth-promoting effect of serum albumin on hamster cells *in vitro*, *J. Cell Physiol.* 96: 1-14.

Packer, L., and Fuehr, K., 1977, Low oxygen concentration extends the lifespan of cultured human diploid cells, *Nature (London)* 267: 423-425.

Parshad, R., and Sanford, K.K., 1971, Oxygen supply and stability of chromosomes in mouse embryo cells *in vitro*, *J. Natl. Cancer Inst.* 47: 1033-1035.

Petzoldt, U., Briel, G., Gottschewski, G.H.M., and Neuhoff, V., 1973, Free amino acids in the early cleavage stages of the rabbit egg, *Dev. Biol.* 31: 38-46.

Quinn, P., and Harlow, G.M., 1978, The effect of oxygen on the development of preimplantation mouse embryos *in vitro*, *J. Exp. Zool.* 206: 73-80.

Quinn, P., and Whittingham, D.G., 1982, Effect of fatty acids on fertilization and development of mouse embryos *in vitro*, *J. Androl.* 3: 440-444.

Quinn, P., Warnes, G.M., Kerin, J.F., and Kirby, C., 1985, Culture factors affecting the success rate of *in vitro* fertilization and embryo transfer, in: *In Vitro Fertilization and Embryo Transfer*, *Ann. N.Y. Acad. Sci*, Volume 442 (M. Seppälä, and R.G. Edwards, eds.), N.Y. Acad. Sci., New York, pp. 195-204.

Richter, A., Sanford, K.K., and Evans, V.J., 1972, Influence of oxygen and culture media on plating efficiency of some mammalian tissue cells, *J. Natl. Cancer Inst.* 49: 1705-1712.

Schultz, G.A., Kaye, P.L., McKay, D.J., and Johnson, M.H., 1981, Endogenous amino acid pool sizes in mouse eggs and preimplantation embryos, *J. Reprod. Fertil.* 61: 387-393.

Sengupta, J., Dey, S.K., and Dickmann, Z., 1977, Evidence that "embryonic estrogen" is a factor which controls the development of the mouse preimplantation embryo, *Steroids* 29: 363-369.

Shirley, B., Wortham, J.W.E., Jr., Witmyer, J., Condon-Mahony, M., and Fort, G., 1985, Effects of human serum and plasma on development of mouse embryos in culture media, *Fertil. Steril.* 43: 129-134.

Silverman, I.H., Cook, C.L., Sanfilippo, J.S., Schultz, G.S., Yussman, M.A., and Hilton, F.K., 1987, Ham's F-10 constituted with tap water supports mouse conceptus development *in vitro*, *J. In Vitro Fertil. and Embryo Transfer* (in press).

Smith, D.M., and Smith, A.E.S., 1971, Uptake and incorporation of amino acids by cultured mouse embryos: estrogen stimulation, *Biol. Reprod.* 4: 66-73.

Spierenburg, G.T., Oerlemans, F.T.J.J., van Laarhoven, J.P.R.M., and de Bruyn, C.H.M.M., 1984, Phototoxicity of N-2 Hydroxyethylpiperazine-N'-2 ethanesulfonic acid buffered culture media for human leukemic cell lines, *Cancer Res.* 44: 2253-2254.

Spindle, A.I., and Pedersen, R.A., 1973, Hatching, attachment, and outgrowth of mouse blastocysts *in vitro*: fixed nitrogen requirements, *J. Exp. Zool.* 186: 305-318.

Tervit, H.R., Whittingham, D.G., and Rowson, L.E.A., 1972, Successful culture *in vitro* of sheep and cattle ova, *J. Reprod. Fertil.* 30: 493-497.

Whitten, W.K., 1971, Nutrient requirements for the culture of preimplantation embryos *in vitro*, in: *Schering Symposium on Intrinsic and Extrinsic Factors in Early Mammalian Development*, *Advances in the Biosciences*, Volume 6 (G. Raspe, ed.), Pergamon Press, Oxford, pp. 129-141.

Whitten, W.K., 1957, Culture of tubal ova, *Nature* (London) 179: 1081-1082.

Wiley, L.M., Yamami, S., and Van Muyden, D., 1986, Effect of potassium concentration, type of protein supplement, and embryo density on mouse preimplantation development in vitro, *Fertil. Steril.* 45: 111-119.

Wright, R.W., Watson, J.G., and Chaykin, S., 1978, Factors influencing the *in vitro* hatching of mouse blastocysts, *Anim. Reprod. Sci.* 1: 181-188.
Wu, J.T., and Matsumoto, P., 1984, Possible function of 17 beta-hydroxysteroid dehydrogenase (17 beta-HSD) in preimplantation hamster embryos, *Biol. Reprod.* 30 (Suppl 1): 45a.
Wu, J.T., and Matsumoto, P.S., 1985, Changing 17 beta-hydroxysteroid dehydrogenase activity in preimplantation rat and mouse embryos, *Biol. Reprod.* 32: 561-566.

APPENDIX II

1. SUPPLIES FOR EMBRYO CULTURE EXPERIMENTS

1.1. Culture Media

Prepared media:

Brinster's Medium (BMOC 3) (Chapter 1): *GIBCO*
CMRL-1066 (Chapter 13): *GIBCO*
Ham's F10 medium (Chapters 1, 5, 11, 15): *Flow Laboratories, Inc.*
Ham's F12 medium (Chapters 5, 8): *Flow Laboratories, Inc.*
Ménézo's B2 Medium (Chapter 9): *Api-System*
NCTC-109 (Chapters 8, 14): *Whittaker M.A. BioProducts, Inc.*

Media Components and Supplements:

Antibiotic-Antimycotic solution (Chapter 12): *GIBCO*
Bovine Serum Albumin (BSA) (Chapter 8): *Miles Laboratories* (now *Miles Diagnostics*);
Fraction V (Chapters 10, 11, 12, 13): *Sigma Chemical Co.*
Calf serum (Chapter 13): *Hyclone Laboratories*
Garamycin (Chapter 13): *Schering Corp.*
Horse serum (Chapters 11, 13): *Hyclone Laboratories*
100X MEM-vitamins (Chapter 8): *GIBCO*
Polyvinylalcohol (PVA) (Chapters 10, 11): *Sigma Chemical Co.* (cat. no. P8136)
Sodium lactate syrup: *Sigma Chemical Co.*
Vitamin supplement used in Minimal Essential Medium (Chapter 11): *GIBCO*

1.2. Culture Supplies

Culture dish, Linbro (Chapter 8): *Flow Laboratories, Inc.*
Culture dishes, plastic (Chapters 10, 11, 13): *Falcon*

Culture flasks, plastic (Chapter 11): *Falcon*
Disposable culture box, Multi-Vial, *Nunclon Delta* (Chapter 9): *Nunc*
Filters, *Millipore* Millex-GV and Sterivex-GS (Chapters 11, 13): *Millipore Corp.*
Filters, *Nalgene* (Chapters 8, 11, 13, Appendix I): *Nalge Co.*
Multiplates, Lux SAS (Chapter 10): *Miles Scientific*
Organ culture dishes, plastic (Chapter 11): *Falcon*
Paraffin oil (Chapter 10): *Merck Co., Inc.*; (Chapter 13, Appendix I): *Mallinkrodt, Inc.*
Petri dish, plastic (Chapter 11): *Falcon*
Silicone oil (Chapters 11, 13, Appendix I): *Aldrich Chemical Co.*
Tissue culture dish, plastic, *Costar* (Chapter 2): *Costar*

1.3. Probes and Stains

The following are from Sigma Chemical Co.:
α-Amanitin (Chapter 3)
Amiloride (Chapter 5)
5-Bromo-2'-desoxyuridine (BrdU) (Chapter 14)
Cycloheximide (Chapter 3)
Cytochalasin B (Chapter 3)
5,5-Dimethyl-2,4-oxazolidine dione (DMO) (Chapter 11)
Fluorescein diacetate (FDA) (Chapters 11, 14)
Hoechst fluorescent DNA-binding dye (Chapters 11, 14)
Horseradish Peroxidase (HRP) (Chapter 3)
Nafoxidine (Chapter 6)
Ouabain (Chapters 4, 5)
Phenazine ethosulfate (PES) (Chapter 12)
Phenylmethylsulfonyl fluoride (PMSF) (Chapter 5)

The following are from Molecular Probes, Inc.:
Carboxyfluorescein (Chapter 3)
Carboxyfluorescein Diacetate (CFDA) (Chapter 3)
Fluorescein (Chapter 3)
Indocarbocyanine dyes (Chapter 2)
Lucifer Yellow CH (Chapter 3)
Rhodamine-conjugated dextran (Chapter 3)

Radiolabelled tracers (Chapter 5): *New England Nuclear*
Steroids (cortisol, estradiol, progesterone) (Chapter 6): *Merck Co. (Darmstadt)*

1.4. Gonadotropin Preparations

FSH (Chapter 10): *Reheis Chemical Co.*
FSH-P (Follicle Stimulating Hormone) (Chapters 9, 10): *Burns-Biotec*
Human Chorionic Gonadotropin (hCG)
(Chapter 2): *Organon, Inc.; (Ekluton*, Chapter 6): *Vemie*
LH (Chapter 10): *Burns Biotec*
Pergonal (*hMG*, Chapter 13): *Serono Laboratories, Inc.*

Pregnant Mare's Serum Gonadotropin (PMSG):
(Chapter 2): *Calbiochem Biochemicals*
(*Seragon*, Chapter 6): *Ferring*
(*Gestyl*, *Organon*, Chapter 11): *Diosynth, Inc.*
Urofollitropin: (Chapter 13): *Serono Laboratories, Inc.*

1.5. Miscellaneous

Acrylamide (Chapter 5): *Boehringer Mannheim Biochemicals*
Borosilicate glass storage bottles (*Pyrex*, Chapters 11, 13): *Corning Glassworks*
Dimilume (Chapter 6): *Packard Instrument Co., Inc.*
HDL (Chapter 8): *Meloy Laboratories*
Hypnorm (Chapter 10): *Janssen Pharmaceutica*
Norit A, Serva No. 30890 (Chapter 6): *Serva Fine Biochemicals*
Nonidet P-40 (Chapter 5): *Sigma Chemical Co.*
Prosil-28 (Chapters 11, 13): *SCM Specialty Chemicals*
Protosol (Chapter 5): *New England Nuclear*
Radiolabelled tracers (Chapter 5): *New England Nuclear*
Sephadex (Chapters 9, 10): *Pharmacia*
Soluene 350 (Chapter 6): *Packard Instrument Co., Inc.*
Staphylococcus aureus cells (*IgGsorb*): *The Enzyme Center, Inc.*
Teflon-lined caps (Chapters 11, 13): *Corning Glassworks*
UM10 ultrafiltration (Chapter 9): *Amicon Corp.*
X-ray film (Chapter 5): *Eastman-Kodak*

2. EQUIPMENT

Fractionation apparatus (Chapter 5): *ISCO*
Gel filtration-electrophoresis (Chapter 9): *Pharmacia*
for Iontophoretic microinjection (Chapter 3):
chart recorder (Brush 2200): *Gould, Inc.*
electrometers (Model M-707 Microprobe System): *W-P Instruments, Inc.*
microelectrode capillary tubing: *W-P Instruments, Inc.* (cat. no. 1B1OOF)
microelectrode puller (vertical): *Narishige Scientific Laboratory*
micromanipulators (De Fonbrune, pneumatic, ref. 8187): *Ch. Beaudoin*
pulse generator (Anapulse Stimulator, Model 301-T): *W-P Instruments, Inc.*
for Ion transport measurements (Chapter 5):
micromanipulators: *Brinkmann (Brinkmann Instruments); Leitz (E. Leitz, Inc.); Narishige (Labtron Scientific Co.)*
micrometers, injection: *Gilson Medical Electronics; Stoelting Co.*
micropipette forge: *Curtin Matheson*
micropipette puller: *Kopf Instruments*
stereomicroscope: Wild (Leitz) model M5A: *E. Leitz, Inc.*; Zeiss Model SR: *Rainin Instrument Co.*
Microscope, inverted *Diaphot* (Chapters 9, 11, 13): *Nikon, Inc.*
Milli-Q water purifier (Chapters 8, 10, 13 and Appendix I): *Millipore Corp.*

Modular Unit (Chapter 10): *Flow Laboratories*
Nanopure II water purifier (Chapter 8): *Barnstead Co.*
for QO_2 measurement (Chapter 5):
  oxygen chambers: *Wilber Scientific*
  oxygen monitors and probes: *Microelectrodes, Inc.; YSI, Inc.*
  temperature probes (thermistors): *Omega Engineering, Inc.*

3. MANUFACTURERS AND DISTRIBUTORS OF SUPPLIES AND EQUIPMENT

Aldrich Chemical Company, P.O. Box 355, Milwaukee, Wisconsin 53201, USA
Amicon Corporation, Scientific Systems Division, Danvers, Massachusetts 01923, USA
Api-System, La Balme-les-Grottes, Montalieu-Vercieu, France

Barnstead Company, Division of Sybron Corporation, 225 Rivermoor Street, Boston, Massachusetts 02132, USA
Ch. Beaudoin, Paris, France
Boehringer Mannheim Biochemicals, 7941 Castleway Drive, P.O. Box 50816, Indianapolis, Indiana 46250, USA
Brinkmann Instruments, Inc., Cantiague Road, Westbury, New York 11590, USA
BDH Chemicals, Ltd., Poole, Dorset, BH124 NN, United Kingdom
Burns-Biotec (now **Schering Animal Health),** 10409 I Street, Omaha, Nebraska 68127, USA

Calbiochem Biochemicals, Behring Diagnostics, P.O. Box 12087, San Diego, California 92112, USA
Corning Glassworks, Corning, New York 14831, USA
Costar, 205 Broadway, Cambridge, Massachusetts 02139, USA
Curtin Matheson Scientific, Inc., 9999 Stuebner Airline, Houston, Texas 77038, USA

Diosynth, Inc., 3432 West Henderson Avenue, Chicago, Illinois 60618, USA

Eastman Kodak Company, 343 State Street, Rochester, New York 14650, USA
The Enzyme Center, Inc., 36 Franklin Street, Malden, Massachusetts 02148, USA

Falcon Plastics, (Division of Becton Dickinson & Co.), 1950 Williams Drive, Oxnard, California 93030, USA; Cockeysville, MD 21030, USA
Ferring, Kiel, Federal Republic of Germany
Flow Laboratories, Inc., 7655 Old Springhouse Road, McLean, Virginia, 22102, USA; P.O. Box 17, Second Avenue, Industrial Estate, Irvine, Ayrshire, Scotland, KA12 8NB, UK

GIBCO (Grand Island Biological Company), 3175 Staley Road, Grand Island, New York 14072, USA; **GIBCO Ltd.,** 3 Washington Road, Sandyford Industrial Estates, Paisley, Renfrewshire, Scotland PA3 4EP
Gilson Medical Electronics, Box 27, 3000 West Beltline, Middleton, Wisconsin 53562, USA

Gould, Inc., (Instruments Division), 3631 Perkins Avenue, Cleveland, Ohio 44114, USA

HyClone Laboratories, 1725 South State Highway 89-91, Logan, Utah 84321, USA

Instech Laboratories, Inc., 151 Gibraltar Road, Horsham, Pennsylvania 19044, USA

ISCO (Instrumentation Specialties Company), 4700 Superior Avenue, Lincoln, Nebraska 68504, USA

Janssen Pharmaceutica, Life Sciences Products Division, Turnhoutseweg 30, B-2340, Beerse, Belgium; 40 Kingsbridge Road, Piscataway, New Jersey 08854, USA

D. Kopf Instruments, 7324 Elmo Street, Tujenga, California 91042, USA

Labtron Scientific Company, 400 Rabro Drive, Hauppage, New York 11788, USA

E. Leitz, Inc., 24 Link Drive, Rockleigh, New Jersey 07647, USA

Mallinckrodt, Inc., P.O. Box M, Paris, Kentucky 40361, USA

Meloy Laboratories, 6715 Electronic Drive, Springfield, Virginia 22151, USA

Merck & Company, Inc., P.O. Box 2000, Rahway, New Jersey 07065, USA; Frankfurterstrasse 250 D-6100, Darmstadt, Federal Republic of Germany

Microelectrodes Inc., Oak Hill Park, Londonderry, New Hampshire 03053, USA

Miles Diagnostics, 195 West Birch Street, Kankakee, Illinois 60901, USA

Miles Scientific, 2000 North Aurora Road, Naperville, Illinois 60566, USA

Millipore Corporation, 80 Ashby Road, Bedford, Massachusetts 01730, USA; Millipore S.A., Zone Industrielle, 67120 Molsheim, France; Millipore GmbH, Siemensstrasse 20, 6078 Neu-Isenburg, West Germany; Millipore (U.K.) Ltd., 11-15 Peterborough Road, Harrow, Middlesex HA1 2YH, England

Molecular Probes, Inc., 24750 Lawrence Road, Junction City, Oregon 97448, USA

Nalge Company, (Division of Sybron Corporation), Rochester, New York 14602, USA

Narishige Laboratory, Tokyo, Japan

New England Nuclear, 549 Albany Street, Boston, Massachusetts 02118, USA; **Du Pont de Nemours (Deutschland) GmbH,** NEN Division, Postfach 401240, D-6072 Dreieich, West Germany

Nikon Inc., Instrument Group, 623 Stewart Avenue, Garden City, New York 11530, USA

Nunc, 4000 Roskilde, Denmark; US distributor **Vanguard International,** 1111-A Green Grove Road, Neptune, New Jersey 07753, USA

Omega Engineering, Inc., 1 Omega Drive, Box 4047, Stamford, Connecticut 06907, USA

Organon, Inc., 375 Mt. Pleasant Avenue, West Orange, New Jersey 07052, USA

Packard Instrument Co., Inc., 2200 Warrenville Road, Downers Grove, Illinois, USA

Pharmacia Biotechnology International AB, S-751 82 Uppsala, Sweden; **Pharmacia, Inc.,** 800 Centennial Avenue, Piscataway, New Jersey 08854, USA

Rainin Instrument Company, Mack Road, Woburn, Massachusetts 01801, USA

Reheis Chemical Company, 235 Snyder Avenue, Berkeley Heights, New Jersey 07922, USA

Schering Corporation, Kenilworth, New Jersey 07033, USA

SCM Specialty Chemicals, P.O. Box 1466, Gainesville, Florida 32602, USA

Serono Laboratories, Inc., Randolph, Massachusetts 02368, USA

Serva Fine Biochemicals, Inc., P.O. Box A, Garden City Park, Long Island, New York, 11040, USA

Sigma Chemical Company, P.O. Box 14508, St. Louis, Missouri 63178, USA; Sigma Chemical Co., Ltd., Fancy Road, Poole, Dorset BH17 7NH, England; Sigma Chemie GmbH, Grünwalder Weg 30, D-8024 Deisenhofen, West Germany

Stoelting Company, 1350 Kostner Avenue, Chicago, Illinois 60623, USA

Vemie, Kempen, Federal Republic of Germany

Whittaker M.A. Bioproducts, Inc., Building 100, Biggs Ford Road, Walkersville, Maryland 21793, USA

Wilber Scientific, 37 Leon Street, Boston, Massachusetts 02115, USA

Wild Heerbrugg Instruments Inc., 465 Smith Street, Farmingdale Long Island, New York 11735, USA; CH-9435 Heerbrugg, Switzerland

W-P Instruments, Inc., P.O. Box 3110, New Haven, Connecticut 06515, USA

YSI Inc., P.O. Box 279, Yellow Springs, Ohio 45387, USA

INDEX